LIABILITY FOR OIL POLLUTION AND COLLISIONS

OTHER TITLES IN THIS SERIES

Shipbuilding Contracts
SECOND EDITION
by Malcolm A. Clarke
1992

Time-Barred Actions
SECOND EDITION
Edited by Francesco Berlingieri
1993

Liability for Damage to the Marine Environment
Edited by Colin de la Rue
1993

The York-Antwerp Rules
SECOND EDITION
by N. Geoffrey Hudson
1996

The Modern Law of Marine Insurance
Edited by D. Rhidian Thomas
1996

The Bill of Lading: A Document of Title to Goods
by Michael Bools
1997

Shipbrokers and the Law
by Andrew Jamieson
1997

Ship Arrest Handbook
Edited by Paul Smith
1997

Douglas & Geen on the Law of Harbours, Coasts and Pilotage
by Richard Douglas, Peter Lane and Monica Peto
1997

LIABILITY FOR OIL POLLUTION AND COLLISIONS

BY

Z. OYA ÖZÇAYIR
Ph.D., LL.M.
Lawyer, Member of İzmir Law Bar, Türkiye

FOREWORD BY PAUL BARTLETT
Director, Sedgwick Energy and Marine Limited

|L|L|P|

LONDON HONG KONG
1998

LLP Reference Publishing
69–77 Paul Street
London EC2A 4LQ
Great Britain

SOUTH EAST ASIA
LLP Asia Limited
Sixth Floor, Hollywood Centre
233 Hollywood Road
Hong Kong

British Library Cataloguing in Publication Data
A catalogue record for this book
is available from the
British Library

ISBN 1 85978 158 6

Typeset in 10/12 Plantin
by Mendip Communications Ltd
Printed in Great Britain by
WBC Limited
Bridgend, Mid-Glamorgan

To my parents

FOREWORD

Recent years have seen an explosion of interest by the general public in matters affecting our environment. In a number of cases, particularly those involving major commercial concerns, emotions can run high, leading perhaps to a reaction disproportionate to the gravity of the incident.

We no longer live in a blameless society; accidents do indeed happen—but surely someone must be responsible. Pollution generally is high profile but no more so than in the shipping sector, where even a minor spill can adversely impact on the natural environment, local commercial activities, tourism, etc. and criticism of the vessel or cargo owners can rapidly reach chronic proportions, severely impairing their business, thus making an already very complicated matter even more complex.

Various governments, agencies and insurance markets have sought to construct some method of bringing calm and reasonable restitution out of the chaos which follows a maritime incident, but shipping is by its very nature an international business consistently battling with a variety of laws operating in numerous jurisdictions. Whilst a great deal of progress has been made to consolidate good law into international agreements and treaties, these have not been by any means readily accepted by all concerned, leaving the participants in an incident no choice other than to search through a great deal of documentation, wrestling with numerous and sometimes conflicting legal principles, at some considerable cost in both time and resources.

Liability for Oil Pollution and Collisions is, I believe, the first work of its kind in this area and provides considerable enlightenment on all aspects of liability and the law relating to this subject in one complete volume. It will be of great assistance to all practitioners involved in shipping, insurance and the associated services.

PAUL BARTLETT
Director, Sedgwick Energy and Marine Limited

PREFACE

Oil pollution may happen in a variety of ways: a tanker may sink and oil may leak, a tanker may be grounded and oil may spill, a tanker may catch fire and burnt oil may spill into the sea.

These are not the only ways that oil pollution happens. In many cases the cause of oil pollution is collision involving oil tankers. Spills of the size like those resulting from the *Torrey Canyon*, *Exxon Valdez* or *Amoco Cadiz* incidents are exceptional and it is not possible to predict where and when the next incident may occur or the particular problems it will create. As long as maritime transport continues collisions and oil pollution will take place. Such incidents may be followed by a variety of claims and resolution of them involves complicated procedures. In maritime transactions the parties involved are frequently of different nationalities and the places at which the contract is made or is to be performed are found in states where different laws are in existence.

In many cases the incident which gives rise to a variety of claims happens on the high seas or in the territorial waters of a state which does not have any connection with the contract of carriage.

The aim of this book is to give a general overview of the issues that one may face while dealing with oil pollution and collision incidents. It examines the international efforts made in collision and oil pollution in order to prevent these incidents from happening, and the compensation schemes applied when they cannot be prevented.

The book is divided into five parts. In case of an oil pollution or collision claim, a shipowner will inevitably seek to recover his losses from the owners of the other vessel involved or from her insurers. At this stage the identity of shipowners becomes quite important. The first part of the book is confined to the question of ownership. The second part discusses collision liability. Under the insurance aspects of collision liability it tries to demonstrate the relation between hull insurance and P&I cover with regard to collision liability risk.

In the oil pollution part, apart from international and regional marine pollution conventions, the unique US Oil Pollution Act 1990 has been examined. In order to examine the liability of shipowners with regard to oil pollution and collision one has also to consider the issue of limitation of liability. Limitation of liability is frequently a significant factor in determining the most appropriate jurisdiction.

Therefore, the limitation regimes of countries which apply different measures of limitation of liability under different international limitation conventions have to be compared. Hence, part four not only examines the 1976 Limitation Convention, but also discusses the 1957 Limitation Convention and the conflict of law problems that may arise.

Finally, the last part deals with jurisdiction. A maritime incident may take place in different maritime zones such as high seas or territorial waters of a state and in turn it will lead to problems of jurisdiction over these areas. This part examines the 1982 United Nations Convention on the Law of the Sea and definition of jurisdictional zones. Within Admiralty jurisdiction, beneficial ownership and restrictions on the exercise of jurisdiction are examined. The topics such as arbitration agreements, *forum non conveniens* and the Civil Jurisdiction and Judgments Act 1982 are discussed as they have to be considered when one is dealing with jurisdiction problems.

Both private and public law aspects of maritime law have important international components since shipping and other maritime activities are essentially international in scope. Because of its international character, maritime law is in need of international uniformity. It is not only international organisations that will unify the maritime law, governments should also take an active role.

I believe that certain essential duties are the responsibility of a maritime administration, such as passing the necessary laws, rules and regulations, setting adequate training standards, the establishment of competent authorities, and the punishment of violators. Despite the need for international uniformity, the US Oil Pollution Act 1990 is also in force which is more comprehensive and farther reaching than international conventions concerning oil pollution liability. Many observers believe that, despite the references to limitation of liability in most cases, following on oil spill shipowners face unlimited liability under the US Oil Pollution Act. Therefore since the entry into force of the legislation, the inadequacy of insurance cover remains a critical issue. The problem of unlimited liability, I believe, is not related only to US oil pollution legislation. The effects of termination of the voluntary oil pollution schemes, TOVALOP and CRISTAL, and their effects on liability of shipowners need to be considered. In countries which have not ratified or acceded to the Civil Liability Convention, TOVALOP was usually the principal mechanism for compensating victims of oil pollution from tankers under the applicable local law. There were even cases where TOVALOP and especially the TOVALOP Supplement could be involved in responding to a spill in a Civil Liability Convention country. The disappearance of TOVALOP and CRISTAL will have a significant effect on the handling of marine pollution compensation claims from tankers in non-convention states. In the case of pollution in the waters of a non-convention country, and in the absence of the voluntary agreements, the claim would be brought under the domestic legislation of that state against, probably, the only vessel of the company, and a flag of convenience company with no assets.

One may think that even if a state is not a party to the Civil Liability Convention, being a member of the 1976 Limitation of Liability Convention for Maritime Claims will solve the limitation of liability problems. But the 1976 Limitation Convention, Article 3(b), provides that the Convention does not apply to "claims for oil pollution damage within the meaning of the International Convention on Civil Liability for Oil Pollution Damage, dated 29 November 1969 or of any amendments or Protocol thereto which is in force". However, exclusion of claims for oil pollution damage under the 1976 Limitation Convention does not depend on inclusion of claims under the 1969 Civil Liability Convention. The Civil Liability Convention contains special rules relating to limitation of liabilities under the Convention and these rules are not of general application. As there are gaps in the international regime for limitation of oil pollution liabilities there is a need for supplementary legislation in states becoming parties to the 1976 Limitation Convention. In the UK this problem is addressed by paragraph 4(2) of Part II of Schedule 7 to the MSA 1995. The paragraph provides that "the claims excluded from the Convention by paragraph (b) of Article 3 are claims in respect of any liability incurred under section 1 of the Merchant Shipping (Oil Pollution) Act 1971."

November 1997 Z. OYA ÖZÇAYIR

ACKNOWLEDGEMENTS

I am grateful to Professor Robert Merkin for his guidance, support and encouragement during the writing of this book.

I would like to thank the Alexander Maxwell Law Scholarship Trust (c/o Sweet & Maxwell, 21 Alva Street, Edinburgh EH2 4PS) for their financial assistance in the form of a scholarship which enabled me to convert my PhD thesis into this book.

I also owe special thanks to the International Tanker Owners Pollution Federation staff for providing invaluable information, to librarian Ms Deborah Ansell for her kind help and to Ms Suzanne Lucas for her hospitality.

I should also thank Mr. Faramarz Amiri who provided invaluable information technology expertise.

OUTLINE TABLE OF CONTENTS

PART 5 – JURISDICTION OVER THE SEA AND SHIPS

APPENDICES

DETAILED TABLE OF CONTENTS

PART 4 LIMITATION OF LIABILITY

PART 5 JURISDICTION OVER THE SEA AND SHIPS

TABLE OF CASES

TABLE OF UK STATUTES

TABLE OF INTERNATIONAL LEGISLATION

TABLE OF STATUTORY INSTRUMENTS

TABLE OF CONVENTIONS

TABLE OF TREATIES

TABLE OF AGREEMENTS

TABLE OF ABBREVIATIONS

ATCMs	The Antarctic Treaty Consultative Party Meetings
CEP	The Committee for Environmental Protection
CERCLA	The Comprehensive Environmental Response, Compensation and Liability Act of 1980 (USA)
CLC	The International Convention on Civil Liability for Oil Pollution Damage 1969 and its Protocols
CMI	The Comité Maritime International
CMV	Contingent Valuation Methodology
COFR	Certificates of Financial Responsibility
COGSA	Carriage of Goods by Sea Act
CRISTAL	Contract Regarding a Supplement to Tanker Liability for Oil Pollution
DOI	The Department of the Interior (USA)
DOJ	The Department of Justice (USA)
EEZ	Exclusive Economic Zone
FC	The International Convention on the Establishment of an International Fund for Compensation for Oil Pollution Damage 1971 and its Protocols
FWPCA	The Federal Water Pollution Control Act (The Clean Water Act) (USA)
FY	Financial Year
GIS	German International Register
GRT	Gross Registered Tonnage
HNS	The International Convention on Liability and Compensation for Damage in Connection with the Carriage of Hazardous and Noxious Substances by Sea 1996
IAEA	The International Atomic Energy Agency
IFR	Interim Final Rule
ILC	International Law Commission
ILO	International Labour Organisation
IMCO	Inter-Governmental Maritime Consultative Organisation
IMDG	International Maritime Dangerous Goods Code
IMF	International Monetary Fund
IMO	International Maritime Organisation

IOPCF	International Oil Pollution Compensation Fund
ITF	International Transport Workers' Federation
LDC	The International Convention on the Dumping of Wastes at Sea 1972
LNG	Liquefied Natural Gas
LOT	Load on Top (System)
LPG	Liquefied Petroleum Gas
MARPOL 73/78	The International Convention for the Prevention of Pollution From Ships 1973 and its 1978 Protocol
MCA	Maritime Conventions Act 1911 (UK)
MEPC	The Marine Environment Protection Committee
MSA	Merchant Shipping Act (UK)
N/A	Not Applicable
NDR	National Driver Register
NIS	Norwegian International Ship Register
NOAA	The National Oceanic and Atmospheric Administration of the Department of Commerce
NPFC	National Pollution Funds Centre
NPRM	Notice of Proposed Rulemaking
NRDA	Natural Resource Damage Assessment
OCSLA	The Outer Continental Shelf Lands Act Amendments (USA)
OECD	The Organisation of Economic Co-operation and Development (Europe)
OILPOL	The International Convention for the Prevention of Pollution of the Sea by Oil 1954
OPA	Oil Pollution Act 1990 (USA)
OPRC	The International Convention on Oil Pollution Preparedness, Response and Co-operation 1990
OSLTF	Oil Spill Liability Trust Fund
P&I	Protection and Indemnity
QI	Qualified Individual
SBT	Segregated Ballast Tanks
SDR	Special Drawing Right
TAPAA	The Trans-Alaska Pipeline Authorisation Act (USA)
TOVALOP	Tanker Owners' Voluntary Agreement Concerning Liability for Oil Pollution
UNCLOS III	United Nations Conference on the Law of the Sea
UNCLOS 1982	United Nations Convention on the Law of the Sea 1982
UNCTAD	United Nations Conference on Trade and Development
UNEF	United Nations Emergency Force
UNEP	United Nations Environment Programme
UNESCO	United Nations Educational Scientific and Cultural Organisation
UNKRA	United Nations Korean Reconstruction Agency
UNSCO	United Nations Suez Canal Clearance Operation
VLCC	Very Large Crude Carrier

CHRONOLOGICAL LIST OF TREATIES

1910 The International Convention for the Unification of Certain Rules of Law with Respect to Collisions Between Vessels (1910 Brussels Convention)

1954 International Convention for the Prevention of Pollution of the Sea by Oil (OILPOL 1954)

1957 International Convention Relating to Limitation of the Liability of Owners of Seagoing Ships

1958 Geneva Conventions on the Territorial Sea and Contiguous Zone

1958 Geneva Convention on the Continental Shelf

1958 Geneva Convention on Fishing and Conservation of the Living Resources of the High Seas

1958 Geneva Convention on the High Seas

1969 International Convention Relating to Intervention on the High Seas in Cases of Oil Pollution (Intervention Convention 1969)

1969 International Convention on Civil Liability for Oil Pollution Damage

1969 Bonn Agreement for Co-operation in Dealing with Pollution of the North Sea by Oil

1970 International Convention on the Establishment of an International Fund for Compensation for Oil Pollution Damage

1971 Agreement Between Denmark, Finland, Norway and Sweden concerning Co-operation in Measures to Deal with Pollution of the Sea by Oil (1971 Copenhagen Agreement)

1972 Convention for the Prevention of Marine Pollution by Dumping from Ships and Aircraft (Oslo Dumping Convention)

1972 International Convention on the Dumping of Wastes at Sea (LDC 1972)

1973 International Convention for the Prevention of Pollution from Ships and its 1978 Protocol (MARPOL 73/78)

1974 Convention between Denmark, Finland, Norway and Sweden on the Protection of the Environment (Stockholm Convention)

1974 Helsinki Convention on the Protection of the Marine Environment of the Baltic Sea Area

1976 Convention for the Protection of the Mediterranean Sea Against Pollution (Barcelona Convention)

1976 Convention on Limitation of Liability for Maritime Claims

1977 Protocol to the Barcelona Convention for the Prevention of Pollution of the Mediterranean Sea by Dumping from Ships and Aircraft

1978 Kuwait Regional Convention for Co-operation on the Protection of the Marine Environment from Pollution (Kuwait Convention)

1979 Protocol to Kuwait Convention (1978) Concerning Regional Co-operation in Combating Pollution

1980 The Rome Convention on the Law Applicable to Contractual Obligations

1980 Convention for Co-operation in the Protection of the Marine and Coastal Environment of the West and Central African Region (Abidjan Convention)

1981 Abidjan Protocol Concerning Co-operation in Combating Pollution in Cases of Emergency

1981 Convention for the Protection of the Marine Environment and Coastal Areas of the South-East Pacific (Lima Convention)

1982 Convention for Protection and Development of the Marine Environment of the Wider Caribbean Region (Jamaica Convention)

1982 Regional Convention for the Conservation of the Marine Environment of the Red Sea and the Gulf of Aden (Jeddah Convention)

1982 Jeddah Protocol Concerning Regional Co-operation in Combating Pollution by Oil and Other Harmful Substances in Cases of Emergency

1983 Bonn Agreement for Co-operation in Dealing with Pollution of the North Sea by Oil and Other Harmful Substances

1983 Convention for the Protection and Development of the Marine Environment of the Wider Caribbean (Cartagena Convention)

1983 Cartagena Protocol Concerning Co-operation in Combating Oil Spills in the Wider Caribbean Area

1984 Protocols to 1969 Civil Liability and 1971 Fund Conventions

1985 Convention for the Protection, Management and Development of the Marine Environment of the Eastern African Region (Nairobi Convention)

1986 United Nations Convention on Conditions for Registration of Ships (1986 UN Convention)

1986 Convention for the Protection of the Natural Resources and Environment of the South Pacific Region (Noumea Convention)

1986 Noumea Protocol Concerning Co-operation in Combating Pollution Emergencies in the South Pacific Region

1990 International Convention on Oil Pollution Preparedness, Response and Co-operation (OPRC)

1991 Protocol on Environmental Protection to the Antarctic Treaty (Madrid Environmental Protocol)

1992 Protocols to Civil Liability and Fund Conventions

1992 Helsinki Convention on the Protection of the Marine Environment of the Baltic Sea Area (Helsinki Convention)

1992 Convention on the Protection of the Black Sea Against Pollution (Bucharest Convention)

1993 The North West Pacific Agreement

1994 The Asian Seas Agreement

1996 International Convention on Liability and Compensation for Damage in Connection with the Carriage of Hazardous and Noxious Substances by Sea (HNS)

PART 1

OWNERSHIP

CHAPTER 1

THE CONCEPT OF OWNERSHIP

IN GENERAL

Long before the introduction of the concept of nationality states always considered some ships as their own and treated others as alien. States wanted to be able to enact regulations applicable to "their" ships whenever and wherever it was possible.[1] In treaties and state practice and in legal literature the word "ship" is employed in three connotations. First, it serves as a generic name for a moveable property which is suited for navigation. Secondly, it is frequently employed as a collective noun for designating particular persons who have an interest in that moveable property. This constitutes the ownership. Thirdly, ship is the word often used when reference is made to the ship's flag state.[2] In this context the definition of a ship becomes quite important in order to have the clearest possible understanding of the objects to which international shipping law applies.

There has always been discussion on the definition of ownership of vessels. Ships may be owned individually, by bodies corporate or by governments of states. Their ownership will be established according to the municipal law of the state whose flag they fly. The responsibility of the state linked to the flag of the ship gives rise to the concept of national ownership.[3] According to the Advisory Opinion of the International Court of Justice of 1960[4] national ownership is based on where the ship is registered, which determines the flag flown, and not the individual owner's nationality. Obviously state responsibility will arise to protect the interests

1. H. Meyers, *The Nationality of Ships* (1967), p.1.
2. *Ibid.*, p.8.
3. Although nationality is generally determined by the flag of the vessel, there are different aspects of the law of merchant shipping. These aspects are governed by different factors. This view is stated in the observation made by Hackworth: "The term 'nationality' of a merchant vessel is used in various senses. The sense in which it is used in a given case depends upon the purpose for which national character is to be ascertained. In matters pertaining to privileges or rights under national laws, the usual practice is to look to the registry, enrolment, or other documentation of the vessel. In matters of protection, the usual practice, in addition to considering the documentation of the vessel, is to inquire into the nationality of the owner". See Hackworth, *Digest of International Law*, Vol. II, p.724. Cited in Nagendra Singh, *Maritime Flag And International Law* (1978), p.23, n.1.
4. ICJ, Advisory Opinion, 8 June 1960, p.150.

of nationals who own the ship and the flag. But there is also state responsibility for the enforcement of international conventions to which the flag state is a party.[5]

KINDS OF OWNERSHIP

The answer to the question who can own a ship cannot ignore the overall concept of national ownership. Under national ownership registering states, those that grant the flag, are internationally responsible for their national fleets. Therefore the legal regime of merchant shipping in the framework of public international law is based on a national ownership concept. Apart from national ownership the vessel may be owned by a natural person who is a national of a state, a body corporate registered according to the laws of the municipal state in which it is situated or a government department or organ of the state which registers the vessel and grants the flag. In such cases the ownership will be individual. Apart from national and individual ownership merchant ships are also known to be operated by international organisations such as the United Nations.[6]

5. In *The Colonel Lloyd Aspinal* an American merchant vessel carrying Government papers was seized by a Spanish vessel. The US Secretary of State protested to the Spanish authorities. He demanded that the Spanish Government should "offer to the United States a suitable apology for the indignity to the flag of the United States" and also requested "that proper compensation be made to the owners and to all other persons who have suffered by the seizure or detention". Therefore it is possible to divide the United States' complaint into two parts: (1) complaint based on the protection of the flag; (2) complaint based on compensation to owners for injury suffered. See N. Singh and R. Colinvaux, *British Shipping Laws. Shipowners* (1967), Vol. 13, p.1.

6. Nagendra Singh, pp.3–4.

CHAPTER 2

REGISTRATION

HISTORY OF SHIP REGISTRATION

The registration of ships has its origin in the laws of Imperial Rome. It was widespread in the city states of mediaeval Italy. In England registration started with a statute of Charles II in 1660.[1] Navigation Acts aimed to prevent foreign vessels taking advantage of the commercial privileges enjoyed by vessels flying the British flag. In other words the first registrations were not to regulate standards but to ensure that certain trades were reserved for the benefits of British citizens. Towards the end of the eighteenth century Navigation Acts also sought to restrict entitlement to the flag to ships built within the British dominions.[2]

THE LEGAL CONCEPT OF SHIP REGISTRATION

One of the fundamental principles of public international law is the freedom of the high seas. The high seas were defined in Article 86 of the United Nations Convention on the Law of the Sea 1982 (UNCLOS 1982) as all parts of the sea that are not included in the exclusive economic zone, in the territorial sea or in internal waters of a state, or in the archipelagic waters of an archipelagic state. Article 87 of the UNCLOS provides the freedom of the high seas. "The high seas are open to all States, whether coastal or land-locked." Apart from being open to all nations, the high seas could be used by both coastal and land-locked states for the purpose of navigation, overflight, laying submarine cables and pipelines, constructing artificial islands and other installations permitted under international law, fishing and scientific research.[3]

1. 12 Car.2, c.18.
2. See Abbott's *Law of Merchant Ships and Seamen* (14th ed., 1901), for the history of British ship registration. Cited in N.P. Ready, *Ship Registration* (2nd ed., 1994), p.3, n.12.
3. UNCLOS 1982, Art. 87:

"Freedom of the high seas

1. The high seas are open to all States, whether coastal or land-locked. Freedom of the high seas is exercised under the conditions laid down by this Convention and by other rules of international law. It comprises, *inter alia*, both for coastal and land-locked states:

   (a) freedom of navigation;
   (b) freedom of overflight;

These freedoms, and others which are recognised by the general principles of international law, shall be exercised by all states with reasonable regard to the interests of other states in their exercise of the freedom of the high seas. Under the freedom of the high seas vessels belonging to all nations have unrestricted access to all parts of the sea that are not included in the territorial sea or internal waters of a state. Freedom of the high seas applies not only to coastal states but also to states that are land-locked. Freedom of access should not lead to abuse, therefore international law provides a framework for the exercise of that freedom. But these rules are enforced by individual states through the jurisdiction exercised over their national vessels. In this respect there are two main rules[4]; first, jurisdiction over a vessel on the high seas resides solely with the state to which the vessel belongs; secondly, all vessels using the high seas must possess a national character.

If a ship does not have a nationality it does not have any protection in international law. How can a ship have a nationality? Registration is generally, but not always, the test of a vessel's nationality.[5] This is the public law function of registration. Under public law a ship has the sovereignty of the state whose flag she flies. Under private law a ship is a moveable property and one or more persons have rights in it, which have protection under the law. The distinction between public law and private law functions of registration was explained in *Liverpool Borough Bank* v. *Turner* with reference to the Merchant Shipping Act 1894:

"There are two points of public policy which may be suggested in these acts relating to shipping: the one a policy regarding the interests of the nation at large, relating to the question who shall be entitled to the privileges of British flag . . . ; the other policy being similar to that which gave rise to the acts for registration of titles to land—the object being to determine what should be a proper evidence of title in those who deal with the property in question."[6]

The effect of registration may differ in various legal systems. In general priority in registration determines which right should prevail when two rights are in conflict

---

(c) freedom to lay submarine cables and pipelines, subject to Part VI;
(d) freedom to construct artificial islands and other installations permitted under international law, subject to Part VI;
(e) freedom of fishing, subject to conditions laid down in section 2;
(f) freedom of scientific research, subject to Parts VI and XIII.

2. These freedoms shall be exercised by all States with due regard for the interests of other States in their exercise of the freedom of the high seas, and also with due regard for the rights under this Convention with respect to activities in the Area."

4. N.P. Ready, p.1.

5. "The concept of registration is closely related to nationality but it is not necessarily synonymous with it. The nationality of a ship refers to the State which has authority and responsibility over, and in respect of, the ship, with the flag symbolizing the State of nationality in each case. Registration, however, is the act by which this nationality is conferred on the ship. In international texts and treaties the terms 'registration' and 'nationality' are often apposed. Thus, in the United Nations Convention on the Law of the Sea the ships of a State are referred to alternately by the expressions 'vessel of its registry', 'vessel flying its flag' or 'vessels having the nationality'": United Nations. Economic and Social Commission for Asia and the Pacific, *The Guidelines for Maritime Legislation* (1991), Vol. I, p.29.

6. *Liverpool Borough Bank* v. *Turner* (1860) 29 L.J.Ch. 827.

with each other. For instance, if a ship is sold to two different purchasers in good faith, title in the ship is not acquired by the first purchaser, but by the purchaser who registers his title first, irrespective of whether he was the first or second in time to sign his contract. Registration does not create a right which does not exist or create a presumption in this respect. If a contract of sale is void, its registration does not render it valid.

CONDITIONS FOR REGISTRATION

Registration is the administrative act by which nationality and the collateral rights and duties are conferred on a ship. The term is generally used to describe the attribution of national character to a vessel. With registration a vessel enters the public records of a state. A ship may be registered if it meets the relevant national requirements. With registration in a state the ship comes within national jurisdiction of that state. The state assumes authority over the ship to exercise the power inherent in the jurisdiction of the flag state. The state undertakes the national and international responsibilities of a flag state in relation to that ship.

Until about the end of the eighteenth century, the nationality of ships was not fully developed and merchant vessels were referred to as belonging to the subjects of the state, not to the state itself. With the development of the institution of nationality of ships series of bilateral treaties of commerce and navigation were concluded. But not all those treaties satisfied the conditions of the equality of the contracting parties. After 1830 various states made bilateral treaties and recognised mutually that the nationality of a vessel was to be decided in accordance with the laws of the country under whose flag it was sailing. And other states were obliged to recognise that nationality. Around the mid-nineteenth century flag recognition clauses became more abstract. Article VII of the Friendship, Commerce and Navigation Treaty concluded on 4 February 1859 between the United States and Paraguay is a good example of these clauses. Under Article VII it was stated that:

> "All vessels which, according to the laws of the United States of America, are to be deemed vessels of the United States of America, and all vessels which, according to the laws of Paraguay, are to be deemed Paraguayan vessels, shall, for the purposes of this treaty, be deemed vessels of the United States of America and Paraguayan vessels, respectively".[7]

The wording of this clause shows that flag recognition is still limited by the phrase "for the purposes of this treaty". In some treaties there is no reference to the place where the vessel may be found: "Vessels under the flag of either party and carrying the papers required by its law in proof of nationality, shall be deemed to be vessels of that party".[8] Under this type of treaty a vessel is regarded as having the nationality of the state whose flag it lawfully flies, without any reference to the

7. B.A. Boczek, *Flags Of Convenience, An International Legal Study* (1962), p.97.

8. Treaty of FCN with the Federal Republic of Germany, 29 October 1954, Art. XIX(1); 7(2) U.S.T. 1839; T.I.A.S. 3593, cited in B.A. Boczek, p.98, n.24.

question of who owns it or by which country's nationals it is manned or where it was built. This principle was declared in the second sentence of Article 5, para. 1, of the Geneva Convention on the High Seas 1958: "Ships have the nationality of the State whose flag they are entitled to fly." The same statement was made in Article 91 of the UNCLOS 1982.

Under the UNCLOS 1982, Article 92 prohibits double nationality, and thus double registration must be avoided, in order to ensure that a ship does not fall under the jurisdiction of more than one state.[9]

KINDS OF REGISTRATION LAW

As stated above, by Article 5 of the 1958 Geneva Convention on the High Seas "Each State shall fix the conditions for the grant of its nationality to ships, for the registration of ships in its territory, and for the right to fly its flag . . ."[10] Before the 1986 United Nations Convention on Conditions for Registration of Ships this Convention was the only international treaty governing the issue of ship registration and nationality.

The matter of fixing the conditions of registration of ships lies within each state's exclusive competence. This principle, which was set out in Article 5, also found substantiation in the opinion of the Permanent Court of Arbitration at The Hague in the *Case of the Muscat Dhows: Great Britain* v. *France*.[11] This case was about the legality of granting the right to fly the French flag to dhows[12] which were owned by subjects of the Sultan of Muscat. By using the French flag the subjects of the Sultan could evade the regulations and inspection of Muscat. Great Britain contended that France's rights were restricted by a treaty of 1892. And the dhows, being owned by the subjects of Muscat, should fly the British flag. The court before

9. UNCLOS 1982, Art. 92 says:
"**Status of ships**
1. Ships shall sail under the flag of one State only and, save in exceptional cases expressly provided for in international treaties or in this Convention, shall be subject to its exclusive jurisdiction on the high seas. A ship may not change its flag during a voyage or while in a port of call, save in the case of a real transfer of ownership or change of registry.
2. A ship which sails under the flags of two or more States, using them according to convenience, may not claim any of the nationalities in question with respect to any other State, and may be assimilated to a ship without nationality."

10. Geneva Convention on the High Seas 1958, Art. 5 states:
"1. Each State shall fix the conditions for the grant of its nationality to ships, for the registration of ships in its territory, and for the right to fly its flag. Ships have the nationality of the State whose flag they are entitled to fly. There must exist a genuine link between the State and the ship; in particular, the State must effectively exercise its jurisdiction and control in administrative, technical and social matters over ships flying its flag.
2. Each State shall issue to ships to which it has granted the right to fly its flag documents to that effect."

11. [1916] *Hague Court Reports* 93.

12. The "dhows" (in French, "*boutres*") are vessels of Arab build, 200–500 tons, which ply the Indian Ocean and the Red Sea. They are owned and operated mainly by the subjects of the Sultan of Muscat, a British protectorate; Westlake, "The Muscat Dhows", *Law Quarterly Review* 23:83 (1907) cited in B.A. Boczek, p.100, n.32.

dealing with the question whether France's rights were limited by treaty provisions, proclaimed the right of a state to grant its nationality to such ships and stated as follows:

"Whereas, generally speaking, it belongs to every sovereign to decide to whom he will accord the right to fly his flag and to prescribe the rules governing such grants, and whereas therefore, the granting of the French flag to subjects of His Highness the Sultan of Muscat in itself constitutes no attack on the independence of the Sultan ... for these reasons [the Court] decides and pronounces as follows: 1. Before the second of January 1892, France was entitled to authorise vessels belonging to subjects of His Highness the Sultan of Muscat to fly the French flag, only bound by own legislation and administrative rules."

In the court's view the execution of the state's rights with regard to granting nationality to ships could be restricted only by way of a treaty. This case is accepted as a leading case which supports the principle of international law that each state decides what conditions it will impose for the registration of its vessels under its flag.

Since each state has the duty to register ships according to its own national law, the law of registration of ships which brings with it the definition of ownership differs from state to state. There are as many variations of the law as there are states registering ships. By examining different states laws it becomes clear that there are three distinct schools of thought with regard to prescribing conditions for registration of merchant vessels[13]:

(1) Rigid law of registration: this group believes in firmly tying the ship to the registering state. Under its requirements a ship must be built in national shipyards, manned by captain and crew of its nationality, and its ownership should be substantially, if not entirely, national. Examples are Japan, Portugal.

(2) The relaxed view: relaxed law does not impose any conditions for registration. It permits foreign nationals to register ships within national territories almost on the same basis as their own nationals. Examples are Liberia, Panama.

(3) The balanced group: this group insists on ownership being in national hands on a majority basis at least. But it does not impose any other requirements concerning crew or other matters. An example is the United Kingdom.

13. Nagendra Singh, pp.4–9.

CHAPTER 3

MARITIME FLAG

HISTORY

Various authorities use the law of the ship's flag as the sole and definitive indicator of the applicable maritime law. Until 1919 the privilege of having a maritime flag was restricted to maritime states only. Land-locked countries with their nationally owned tonnage could not register ships to fly the national flag.[1] Therefore non-maritime states' citizens had to register their ships in ports of maritime states and fly the flag of the state where the vessel was registered. On 20 April 1921 the nations that had assembled in Barcelona for the Conference of Communication and Transit signed a Declaration, which incorporated the principles of the Treaty of Versailles 1919 and recognised the right to a flag of the shipping of a state having no sea coast. In this Declaration the term "maritime state" was extended to states without a coast and recognition was granted to the flags of such land-locked states whose vessels were registered "at some one specified place situated in their territory".[2] And a land-locked state which owned a merchant fleet could also be described as a maritime state for the legal purpose of operating ships.

In the nineteenth century, it became acceptable to apply the law of the flag to the relations between the master, crew, vessel and third parties. In this respect one of the most important conflict of laws decisions was *Lloyd* v. *Guibert*.[3] This leading case emphasised the importance of the law of the flag with respect to contracts entered into by the master in foreign ports: "... the flag of the ship is notice to all the world that the master's authority is that conferred by the *law of that flag* ...". "... Reason and convenience are certainly in favour of holding that the authority of the master to bind his owners should be fixed and uniform according to the *law of his flag*, which is known to both, rather than that it should vary according to the law of the port in which the ship may happen for the time to be."[4]

After ships are permitted to fly the maritime flag of a state they constitute part of the national merchant fleet. The operation of such a fleet on the high seas and in

1. For instance, the Swiss Federal Council refused Swiss nationals permission to use the Swiss national flag at sea.
2. Cited in Nagendra Singh, p.30, n.16. U.K. Treaty Series No. 29, Command Paper 1924.
3. (1864) 33 L.J.R. 242 (Q.B.), *aff'd.* (1865) L.R. 1 Q.B. 115.
4. *Ibid.*, p.248.

the territorial waters of the maritime states of the world creates rights and obligations with regard to the flag. Basically, the flag state will enforce the rules and regulations not only of its municipal law but of international law as well. Therefore a state which gives its flag to a merchant fleet has to be a sovereign state. If a maritime flag is granted by a political entity which is not fully sovereign and not recognised by the maritime community of sovereign states the ship may be liable to seizure, because the flag flown has not been accepted.[5] This point was accepted by the Privy Council in *Naim Molvan* v. *Attorney-General for Palestine*.[6] In this case the vessel *Asya* which sailed to Palestine was flying the Turkish flag which she was not entitled to fly. She did not have any papers on board either. She was arrested by a British destroyer 100 miles off the Palestine coast. The Palestinian court ordered the forfeiture of the vessel although she had been seized on the high seas. On appeal, the Privy Council rejected the proposition that the principle of freedom of the high seas extended to a ship possessing no nationality. It was held that:

> "Having no usual ship's papers which would serve to identify her, flying the Turkish flag, to which there was no evidence she had a right, hauling it down on the arrival of a boarding party and later hoisting a flag which was not the flag of any state in being, the *Asya* could not claim the protection of any state nor could any state claim that any principle of international law was broken by her seizure."

Registration of a ship and granting the flag confers on the registering state the right to protect the ship in all circumstances. It is difficult for a non-sovereign state to undertake such a function. A ship on the high seas is governed by national laws, therefore a flag state should be able to make those laws. And also a flag state would be required to sign international shipping agreements for the regulation of its fleet to carry its international trade.

MARITIME FLAG AND STATE RESPONSIBILITY

(a) How state responsibility arises

Flying the national flag is visual evidence of a ship's nationality. In other words a ship without a flag will be deprived of many of the benefits and rights available under the legal regime on the high seas. The national flag constitutes the primary source of state responsibility in relation to a ship. On the high seas only the flag state may exercise rights over the ship in relation to jurisdiction. This principle was stated in the *Lotus* case.[7] In this case it was held that "vessels on the high seas are subject to no authority except that of the state whose flag they fly".

The most famous modern judicial statement concerning the law of the flag was

5. Oppenheim, *International Law* (ed. Lauterpacht) (8th ed., 1995) Vol. I, p.596: "Every ship sailing without authority and hence without the flag of a State is liable to be captured, prosecuted and punished, including its confiscation."

6. [1948] A.C. 351.

7. [1927] PCIJ, Series A, No. 10, p.25.

made by the United States Supreme Court in *Lauritzen* v. *Larsen*,[8] where the importance of the law of the flag was stressed in relation to nationality. In this case the court referred to several contacts to determine the ship's nationality. These were: the place of the wrongful act, the law of the flag, the allegiance or domicile of the injured person, the allegiance of the defendant shipowner, the place of the contract, the inaccessibility of the foreign forum, the law of the forum and the shipowner's base of operations. Obviously it is important to remember that "the flag" is only one contact of several referred to by the Supreme Court in *Lauritzen* v. *Larsen*. But within these contacts the flag was of cardinal importance.[9]

(b) The principles of state responsibility

(i) The sovereign equality of states

State responsibility is a fundamental principle of international law. It arises out of the nature of the international legal system and the doctrines of state sovereignty and equality of states. In the maritime field the sovereign equality of states determines the responsibility of states. This responsibility may be in relation to the open sea, in the treatment of ships of a foreign flag or in allocating cargoes to the national flag or foreign flags. The principle of equality was frequently affirmed in many cases. The *Lotus*[10] case was one of them.[11]

8. (1953) 345 U.S. 571, 1953 AMC 1210.

9. "Perhaps the most venerable and universal rule of maritime law relevant to our problem is that which gives cardinal importance to the law of the flag. Each state under international law may determine for itself the conditions on which it will grant its nationality to a merchant ship, thereby accepting responsibility for it and acquiring authority over it. Nationality is evidenced to the world by the ship's papers and its flag. The United States has firmly and successfully maintained that the regularity and validity of a registration can be questioned only by the registering state.

This Court has said that the law of the flag supersedes the territorial principle, even for the purposes of criminal jurisdiction of personnel of a merchant ship, because it 'is deemed to be part of the territory of that sovereignty [whose flag it flies], and not to lose that character when in navigable waters within the territorial limits of another sovereignty.' On this principle, we concede a territorial government involved only concurrent jurisdiction of offenses aboard our ships. Some authorities reject, as a rather mischievous fiction, the doctrine that a ship is constructively a floating part of the flag-state, but apply the law of the flag on the pragmatic basis that there must be some law on shipboard, that it cannot change at every change of waters, and no experience shows a better rule than that of the state that owns her.

It is significant to us here that the weight given to the ensign overbears most other connecting events in determining applicable law." See W. Tetley, "The Law Of The Flag, 'Flag Shopping', And Choice Of Law" (1993) 17 *Tulane Maritime Law Journal* 156–157, n.128. *Lauritzen*, 345 U.S. at 584–585, 1953 AMC at 1220–1221 (cited in *United States* v. *Flores*, 289 U.S. 137, 155–159, 1933 AMC 649, 658–661 (1993)).

10. See n.7 *supra*.

11. "In conformity with the principle of the equality of independent States, all nations have an equal right to the uninterrupted use of the unappropriated parts of the ocean for the navigation and no State is authorized to interfere with the navigation of other States on the high seas in time of peace except in the case of piracy by law of nations or in extraordinary cases of self defence"; (Le Louis (1817) 2 Dudson 210, 243–244). According to Singh, with this statement Judge Moore affirmed the principle of equality but at the same time deduced from it the important doctrine of the freedom of the high seas. (Nagendra Singh, p.50).

(ii) Doctrine of the open sea

The doctrine of the open sea has been established as a consequence of the application of the principle of the equality of states.

(iii) Sovereign immunity

As a consequence of the equality of states, states cannot exercise jurisdiction over each other. Under this rule it is not possible to bring an action against a sovereign state unless that state agrees to submit to the jurisdiction of the courts of the state concerned.

(c) Aspects of state responsibility

The flag state has a responsibility to ships as instruments of navigation and to the crews that man the ships. This responsibility was set out in Article 10 of the Geneva Convention on the High Seas 1958.[12]

Under customary and conventional international law, with registration in a state, that state has jurisdictional rights over a ship and acts occurring on board.[13] In other words the state which has granted the ship the right to sail under its flag has the exclusive right to exercise legislative and enforcement jurisdiction over it on the high seas.[14] Therefore registration establishes the jurisdiction of the registering flag state and the application of its national laws to the events which take place on board a ship. For instance the validity of marriage on the high seas is the traditional example of the application of the law of the ship's flag or registry.[15]

12. 1958 Geneva Convention on the High Seas, Art. 10 says:

"1. Every State shall take such measures for ships under its flag as are necessary to ensure safety at sea with regard *inter alia* to:

(a) the use of signals, the maintenance of communications and the prevention of collisions;
(b) the manning of ships and labour conditions for crews taking into account the applicable international labour instruments;
(c) the construction, equipment and seaworthiness of ships.

2. In taking such measures each State is required to conform to generally accepted international standards and to take any steps which may be necessary to ensure their observance."

13. See *The Lotus*, n.7 *supra*.

14. 1958 Geneva Convention on the High Seas, Art. 6 says:

"1. Ships shall sail under the flag of one State only and, save in exceptional cases expressly provided for in international treaties or in these Articles, shall be subject to its exclusive jurisdiction on the high seas. A ship may not change its flag during a voyage or while in a port of call, save in the case of a real transfer of ownership or change of registry.

2. A ship which sails under the flags of two or more States, using them according to convenience, may not claim any of the nationalities in question with respect to any other State, and may be assimilated to a ship without nationality."

See UNCLOS 1982, Art. 92.

15. "A marriage on board a merchant ship would apparently be governed by the law of the flag; and where the flag, such as that of Britain or the United States, covers a number of legal systems, reference would have to be made to the law of the port of the registration of the ship." R.H. Graveson, *Conflict of Laws* 278 (7th ed. 1974), cited in W. Tetley, p.160, n.144.

When a master enters into a contract in a distant port the contract is subject to the law of the flag. Obviously by modern communication methods masters have contact with their employers and employers can also contact suppliers in foreign ports. Therefore now the flag of the ship is not the only connecting factor with regard to contracts on behalf of a ship in a foreign port.[16]

Where ships are involved in collisions on the high seas states may assume the collision and those responsible for it to be within their jurisdiction. The existence of such concurrent jurisdiction was upheld by the Permanent Court of International Justice in the *Lotus* case.[17] In this case the decision was based on the fact that the collision had "taken place" on the Turkish ship and that the officer responsible could therefore be prosecuted by Turkey as well as by France. The rule enunciated in this decision was reversed by the 1952 Brussels Convention for the Unification of Certain Rules Relating to Penal Jurisdiction. The Brussels rule was adopted in the 1958 High Seas Convention and in the United Nations Convention on the Law of the Sea (UNCLOS). Under both conventions penal or disciplinary proceedings may be taken against the master or other person in service of the ship only by the authorities of either the flag state or (if that be different) the state of which the particular person is a national. It is also provided that, even for investigation purposes, no arrest or detention of the ship can be ordered by authorities other than those of the flag state (Geneva Convention on the High Seas 1958, Article 11—UNCLOS 1982, Article 97).[18]

The authority of the state of which the particular person is a national shows that a state retains jurisdiction over its nationals wherever they may be, whether on foreign ships or anywhere else. But in the case of concurrent jurisdiction the expectation is that the flag state has primary jurisdiction (UNCLOS 1982, Article 94).[19]

16. See *Lloyd* v. *Guibert*, n.3 *supra*.
17. See n.7 *supra*.
18. 1958 Geneva Convention on the High Seas, Art. 11 says:
"1. In the event of collision or of any other incident of navigation concerning a ship on the high seas, involving the penal or disciplinary responsibility of the master or of any other person in the service of the ship no penal or disciplinary proceedings may be instituted against such persons except before the judicial or administrative authorities either of the flag State or of the State of which such person is a national.
2. In disciplinary matters, the State which has issued a master's certificate or a certificate of competence or license shall alone be competent, after due legal process, to pronounce the withdrawal of such certificates, even if the holder is not a national of the State which issued them.
3. No arrest or detention of the ship, even as a measure of investigation, shall be ordered by any authorities other than those of the flag State."
The same statement was made in UNCLOS 1982, Art. 97.
19. UNCLOS 1982, Art. 94 states:
"**Duties of the flag state**
1. Every State shall effectively exercise its jurisdiction and control in administrative, technical and social matters over ships flying its flag.
2. In particular every State shall:
(a) maintain a register of ships containing the names and particulars of ships flying its flag, except those which are excluded from generally accepted international regulations on account of their small size; and

The exclusiveness of the flag state's jurisdiction is not absolute. The principle of flag state's jurisdiction is subject to exceptions regarding other vessels. Under these exceptions other states share legislative or enforcement jurisdiction, or both, with the flag state.

(d) Exceptions to the exclusivity of flag-state jurisdiction

(i) Right of visit

In customary international law it is recognised that warships have a right of approach to ascertain the nationality of ships. But this right does not include the right to board or visit ships. There are, however, exceptions to this rule. Article 110 of the UNCLOS recognises that a warship encountering on the high seas a foreign ship, other than a public vessel enjoying immunity, is entitled to board it if there is reasonable ground for suspecting that the ship is engaged in piracy, in the slave trade, in unauthorised broadcasting with the flag state of the warship having jurisdiction under Article 109, or if the ship is without nationality or, though flying a foreign flag or refusing to show its flag, is of the same nationality as the warship. In such cases the warship may proceed to verify the ship's right to fly its flag. However, if the suspicions prove to be unfounded, and provided that the ship boarded has not committed any act justifying them, it shall be compensated for any loss or damage that it may have sustained.[20]

(ii) Slave trade

Under the Geneva Convention on the High Seas and the UNCLOS 1982[21] parties are required to prevent and punish the transport of slaves in ships authorised to fly their flags and to prevent the unlawful use of their flags for that purpose. The slave trade has also been referred to in the context of the right of visit. Therefore if a ship is reasonably suspected of being engaged in the slave trade, it may be visited and boarded. But in international law the slave trade is not analogous to piracy. Therefore if the suspicion of slave trading is well founded only the flag state may actually proceed to seize the ship or to arrest those on board. Other states may only report their findings to the flag state. Any slave taking refuge on board any ship, whatever its flag, shall *ipso facto* be free. But it is not clear what action may be taken if evidence of slave trading is discovered.

(b) assume jurisdiction under its internal law over each ship flying its flag and its master, officers and crew in respect of administrative, technical and social matters concerning the ship."

20. UNCLOS 1982, Art. 110.
21. Geneva Convention on the High Seas, Art. 13 and UNCLOS 1982, Art. 99.

(iii) Piracy

Parties to both the Geneva Convention on the High Seas and the UNCLOS 1982 are obliged to co-operate to the fullest possible extent in the repression of piracy on the high seas or in any other place outside the jurisdiction of any state. Piracy is strictly defined in international law and consists of any of the acts set out in Article 15 of the Geneva Convention on the High Seas 1958 and Article 101 of the UNCLOS 1982.[22] Every state may seize a pirate ship or aircraft and arrest the persons and seize the property on board even on the high seas. The fact that every state may arrest and try persons accused of piracy makes the crime quite exceptional in international law as so much emphasis is placed upon the sovereignty and jurisdiction of each particular state within its own territory.

(iv) Unauthorised broadcasting

Under Article 109 of the UNCLOS 1982 all states are to co-operate in the suppression of unauthorised broadcasting from the high seas. Unauthorised broadcasting means transmission of sound or TV from a ship or installation on the high seas intended for reception by the general public. Any person engaged in such broadcasting may be prosecuted by the flag state of the ship, the state of registry of the installation, the state of which a person is a national, any state where the transmission is received or any state where authorised radio communications are suffering interference.[23]

In the case of piracy and unauthorised broadcasting, states other than the flag state are given legislative jurisdiction and enforcement jurisdiction over ships on the high seas under the UNCLOS 1982.

(v) Hot pursuit

The right of hot pursuit is designed to ensure that a vessel which has infringed the rules of a coastal state cannot escape punishment by fleeing to the high seas. In practice this right means that a coastal state may extend its jurisdiction into the high seas in order to pursue and seize a ship which is suspected of infringing its laws. This right is comprehensively set out in Article 23 of the Geneva Convention on the High Seas 1958. Under Article 23 hot pursuit must be conducted in accordance with the following rules: there must be a good reason to believe that the ship has violated the laws and regulations of the coastal state in its internal waters or territorial sea. Hot pursuit must normally be commenced when the foreign ship

22. Art. 103 of the UNCLOS 1982 defines the term "pirate ship or aircraft" as follows: "A ship or aircraft is considered a pirate ship or aircraft if it is intended by the persons in dominant control to be used for the purpose of committing one of the acts referred to in article 101. The same applies if the ship or aircraft has been used to commit any such act, so long as it remains under the control of the persons guilty of that act."

23. M.N. Shaw, *International Law* (2nd ed., 1986), pp.321–322.

or one of its boats or other craft working as a team and using the ship pursued as a mother ship is within internal waters or the territorial sea.[24]

Until the adoption of the UNCLOS 1982 there was no right of hot pursuit from the continental shelf or the exclusive economic zone. Article 111 of the UN Convention is largely based on the language of Article 23 of the Geneva Convention. Under Article 111 hot pursuit of a foreign ship may be undertaken when the competent authorities of the coastal state have good reason to believe that the ship has violated the laws and regulations of that state. Article 111(1) adds archipelagic waters to the zones in which the pursued ship must be situated before hot pursuit can commence. Now, the right of hot pursuit applies *mutatis mutandis* to violations in the exclusive economic zone or on the continental shelf, including safety zones around continental shelf installations. The right also applies to the relevant rules and regulations of the coastal State applicable to such areas.[25]

The right of hot pursuit ceases as soon as the ship pursued enters the territorial sea of its own state or of another state.[26] Where a ship has been stopped or arrested outside the territorial sea in circumstances which do not justify the exercise of the right of hot pursuit, it shall be compensated for any loss or damage that may thereby have been sustained.[27]

(vi) Pollution

One of the exceptions to the exclusivity of flag state jurisdiction exists in favour of states whose coastline is threatened with serious pollution from a foreign shipping casualty on the high seas. Under Article 24 of the Geneva Convention on the High Seas 1958, every state is required to draw up regulations to prevent pollution of the seas by the discharge of oil from pipelines or resulting from the exploitation and exploration of the seabed and its subsoil. While doing so, states are required to take account of existing treaty provisions on the subject. Again in 1958 under Article 1 of the Convention on the Fishing and Conservation of the Living Resources of the High Seas it was declared that all states have the duty to adopt, or co-operate with other states in adopting, such measures for their respective nationals as may be necessary for the conservation of the living resources of the high seas. With these provisions states were encouraged to create agreements to protect the seas from pollution.

Under the International Convention Relating to Intervention on the High Seas

24. According to Brown Article 23 is badly drafted in at least two respects: "Firstly, Article 23(1) says that pursuit must be commenced when *the foreign ship or one of its boats* is within its internal waters or territorial sea. However, Article 23(3) provides that, 'Hot pursuit is not deemed to have begun unless the pursuing ship has satisfied itself that the ship pursued or one of its boats *or other craft working as a team and using the ship pursued as a mother ship* is within the limits'. Secondly, the clarity of Article 23 was not enhanced by its failure to specify the meaning of working as a team and using the ship pursued as a mother ship." See E.D. Brown, *The International Law of the Sea* (1994) Vol. I, p.296.

25. UNCLOS 1982, Art. 111(2).

26. UNCLOS 1982, Art. 111(3).

27. UNCLOS 1982, Art. 111(8).

in Cases of Oil Pollution 1969 the exception to the exclusiveness of the flag state jurisdiction in relation to pollution gained rapid recognition. This convention provided for action to be taken to end threats to the coasts of states.

Also under the Convention on Civil Liability for Oil Pollution Damage signed in 1969, the owners of ships causing oil pollution damage were to be liable to pay compensation. The 1969 Civil Liability Convention was supplemented in 1971 by the Convention on the Establishment of an International Fund for Compensation for Oil Pollution Damage. Both conventions are now known as the 1992 Civil Liability Convention and the 1992 Fund Convention. These agreements are only a small part of the treaties covering the preservation of the sea environment. Apart from international conventions a variety of regional agreements have also been signed.

Nearly 50 articles of the UNCLOS 1982 deal with the protection of the marine environment. Obviously flag states still have the competence to legislate for their ships but certain minimum standards are imposed on states.[28] The convention also provides that states are responsible for the fulfilment of their international obligations concerning the protection and preservation of the marine environment. And states are liable in accordance with international law. Under Article 235 they must also ensure that recourse is available in accordance with their legal systems for prompt and adequate compensation or other relief regarding damage caused by pollution of the marine environment by persons under their jurisdiction.[29] States are also under a basic obligation to protect and preserve the marine environment.[30]

28. See Arts. 192–238 of the UNCLOS 1982. These articles cover, *inter alia*, global and regional co-operation, technical assistance, monitory and environmental assessment, and the development of the enforcement of international and domestic law preventing pollution.

29. UNCLOS 1982, Art. 235 states:

"**Responsibility and Liability**

1. States are responsible for the fulfilment of their international obligations concerning the protection and preservation of the marine environment. They shall be liable in accordance with international law.

2. States shall ensure that recourse is available in accordance with their legal systems for prompt and adequate compensation or other relief in respect of damage caused by pollution of the marine environment by natural or juridical persons under their jurisdiction.

3. With the objective of assuring prompt and adequate compensation in respect of all damage caused by pollution of the marine environment, States shall co-operate in the implementation of existing international law and the further development of international law relating to responsibility and liability for the assessment of and compensation for damage and the settlement of related disputes, as well as, where appropriate, development of criteria and procedures for payment of adequate compensation, such as compulsory insurance or compensation funds."

30. UNCLOS 1982, Art. 192 states:

"**General Obligation**

States have the obligation to protect and preserve the marine environment."

(vii) Exceptional measures

In some cases states interfere with foreign ships on the high seas on the grounds of self-defence and necessity. In such cases the normally exclusive jurisdiction of the flag state is overridden by the right of the warship to act in self-defence.

In the *Virginius* case,[31] the United States registered vessel *Virginius* was seized by a Spanish warship on the high seas because the ship was carrying American and British and many weapons which they intended to use in the Cuban insurrection against Spain. The Spanish government argued that, as the vessel was Cuban-owned, she must have obtained her American papers by fraud and was not entitled to the protection of the American flag. The American government's argument was that the question whether or not the ship was entitled to fly the American flag was a matter purely for the United States. Great Britain, but not America, accepted that the arrest of the ship was justified on the grounds of self-defence. In *Duizar*,[32] the French Ministry of Defence argued that France's right to visit and search on the high seas ships suspected of carrying arms to Algeria during the emergency of 1956–62 was justified as a right of self-defence. But French action was opposed by many of the states whose ships were affected. The most notable claim to exceptional jurisdiction over foreign ships on the high seas can be seen in *US* v. *Gonzalez*.[33] In this case a United States court held that the "protective" principle of jurisdiction in international law justified the exercise of jurisdiction over foreign ships on the high seas in order to enforce narcotics legislation. In other words under this case jurisdiction over a foreign ship which attempted to smuggle drugs into the United States was based on the protective principle of jurisdiction in international law.

(viii) Rights under treaty

Apart from the exceptions stated above in principle the exclusiveness of flag state jurisdiction on the high seas is intact. But this principle can be varied by specific agreements. Exceptional rights of interference with foreign ships may be conferred by treaty. For instance the 1884 Convention for the Protection of Submarine Cables gave the warships of contracting states the right to stop and ascertain the nationality of merchant ships that were suspected of infringing the terms of the convention. Like the 1884 Convention the 1967 Convention on the Conduct of Fishing Operations in the North Atlantic gave a right to approach suspected miscreants in order to determine their nationality. Under conventions like these two, only enforcement of jurisdiction is extended. The capacity to prescribe rules and to punish violations of these rules rests with the flag state. Both enforcement and legislative jurisdiction are extended in only a few cases. In 1981 Haiti and the United States entered into an agreement to stop the clandestine immigration of

31. Moore, *Digest of International Law*, Vol. II, pp.895 *et seq*. Cited in I.A. Shearer, *Starke's International Law* (11th ed., 1994), p.221.
32. *Ignazio Messina & Cie.* v. *l'Etat* (1963–66) 70 RGDIP 1056 (1966).
33. (1985) 776 F. 2d 931.

residents of Haiti to the United States. The United States agreed to the presence of a Haitian liaison officer on board vessels implementing a co-operative programme of interdiction on the high seas. Each side undertook to prosecute traffickers and confiscate vessels of its own nationality engaged in such traffic.[34]

(e) Ships of uncertain nationality

Ships without nationality are in a curious position. A vessel without documents may be refused permission to enter foreign ports and engage in commercial activities. In the case of *The Merritt*, it is possible to see the disadvantages of an undocumented vessel. In *The Merritt*, an American owned, foreign built vessel was used in importing coal and iron from Canada to America. The vessel was in violation of the provisions of the Act of 1 March 1817. Under this Act the ship had to be either a United States vessel or a vessel of the nationality of the state whose goods were imported. In this case it was stated that:

"The case does not show that *The Merritt* has any of the evidences of being a British ship. She produces no register, or certificate, or document of any kind to entitle her to make that claim. The fact that she is foreign built does not prove it. Proof even that she was built in Great Britain would not establish it. Pirates and rovers may issue from the most peaceful and most friendly ports. The documents a vessel carries furnish the only evidence of her nationality. Of these *The Merritt* is entirely destitute, so far as the case shows ..."[35]

The statelessness of a ship will not entitle other states to assert jurisdiction over it. Such ships enjoy the protection of no state. It was lawful to seize a stateless ship on the high seas. This point was made in *Naim Molvan* v. *Attorney General for Palestine*.[36] In this case the Privy Council said that "the freedom of the open sea, whatever those words may connote, is a freedom of ships which fly the flag of a state which is within the comity of nations ... No question of comity nor any breach of international law can arise if there is no state under whose flag the vessel sails".

Vessels which fly two flags can also be accepted as stateless ships, because when a vessel flies two flags it has been treated as outside the protection of the law of either nationality it is claiming. In order to prevent ships from sailing under more than one flag the International Law Commission proposed that such ships should not be permitted to claim any of the nationalities in question with respect to any other state and "may be assimilated to a ship without nationality".[37] This rule was

34. Agreement to Stop Clandestine Migration of Residents of Haiti to the United States, Exchange of Letters, 23 September 1981 (1981) 20 I.L.M. 1198. See E.D. Brown, Vol. I. p.313 and p.332, n.228.

35. 17 Wall. (81 U.S.) 582, 586–587 (1873). See also *Naim Molvan* v. *Attorney General for Palestine*, *supra*, n.6 and *The Virginius* n.31 *supra*.

36. In this case Akehurst argues that it is possible that arbitrary confiscation or destruction of a stateless ship would entitle the national state of the shipowners to make an international claim. See M. Akehurst, *A Modern Introduction to International Law* (6th ed., 1987), pp.183–184. n.6 *supra*.

37. E.D. Brown, Vol. I, p.291.

adopted in Article 6(2) of the Geneva Convention on the High Seas and is repeated in Article 92(2) of the UNCLOS 1982.[38]

In *United States* v. *Passos-Paternina*[39] the ship had narcotics on board and had two flags. When it was boarded by the US Coast Guard the master claimed both registries. It was held that the vessel was "assimilated to a vessel without nationality" under Article 6(2) of the Geneva Convention on the High Seas 1958.

38. UNCLOS 1982, Art. 92 states:
"**Status of Ships**
1. Ships shall sail under the flag of one State only and, save in exceptional cases expressly provided for in international treaties or this Convention, shall be subject to its exclusive jurisdiction on the high seas. A ship may not change its flag during a voyage or while in a port of call, save in the case of a real transfer of ownership or change of registry.
2. A ship which sails under the flags of two or more States, using them according to convenience, may not claim any of the nationalities in question with respect to any other State, and may be assimilated to a ship without nationality."
39. 918 F. 2d 979, 1991 A.M.C. 719, 722 (1st Cir. 1990).

CHAPTER 4

RELEVANT INTERNATIONAL LAW

INTRODUCTION

In international law there is general agreement that a merchant ship on the high seas has to possess a nationality to be able to prove its existence. Each state determines the conditions of acquisition of nationality according to its exclusive national policies. The basic principles of nationality and registration of ships can be found in the Geneva Convention on the High Seas 1958 and are elaborated in the United Nations Conventions on the Law of the Sea 1982. More specifically, an international agreement on Conditions for Registration of Ships was signed on February 1986.

GRANTING OF NATIONALITY

According to Article 5 of the Geneva Convention on the High Seas 1958 every state shall set the conditions for granting its nationality to ships, for the registration of ships in its territory, and for the right to fly its flag. The same Article provided that "there must exist a genuine link between the state and the ship; in particular, the state must effectively exercise its jurisdiction and control in administrative, technical and social matters over ships flying its flag". The provisions of Article 5 have been variously interpreted especially with respect to the concept of *genuine link* between vessel and the flag of registry. A similar statement was made in Article 91 of the UNCLOS 1982.[1] However, in Article 91 the "in particular" clause was omitted and the Article simply provided that "there must exist a genuine link between the State and the ship".

The controversy over the meaning and practical interpretation of the genuine link concept remained the central issue in the debate on open registries until the United Nations Convention on Conditions for Registration of Ships was adopted.

1. UNCLOS 1982, Art. 91 states:
"**Nationality of Ships**
Every State shall fix the conditions for the grant of its nationality to ships, for the registration of ships in its territory, and for the right to fly its flag. Ships have the nationality of the State whose flag they are entitled to fly. There must exist a genuine link between the State and the ship.
Every State shall issue to ships to which it has granted the right to fly its flag documents to that effect."

(a) The concept of genuine link

If states accepted the need for a genuine link as described in Article 5 of the Geneva Convention on the High Seas 1958 and in Article 91 of the UNCLOS 1982 there would be no conflict between the law of the flag and the law of the ownership of the vessel. But the practice is different from the theory. For instance, if a ship is registered in state A and is owned by a body corporate in state B, both sovereign states A and B would like to exercise the right of protection over the ship and this gives rise to complicated issues. State B, promoting the interests of its nationals, may wish to protest against the use of the ship in a danger zone to assist state C, a state unfriendly to B but friendly to state A which has the right to control the movement of the ship, being the flag state registering and owning the fleet of which the vessel is a part.

To avoid these types of complications the concept of genuine link was first introduced in 1956 into the deliberations of the International Law Commission (ILC) set up by the United Nations. In the draft 1958 Geneva Convention on the High Seas the requirement was included that a genuine link must exist between ship and registry to ensure effective control and jurisdiction over such vessel by its state of registry. But there was no definition of the precise meaning of genuine link and also of the basis for testing whether this link between ship and registry did or did not exist. The conference had several possibilities: the retention of the genuine link clause as it had been drafted by the ILC, the definition of the genuine link concept's essentials, or the rejection of the concept altogether. Another possibility was to refer the whole issue to a specialised body, like the Inter-Governmental Maritime Consultative Organisation (IMCO).

On the legal issues the states' delegates reflected directly or indirectly their interests on the problem. The traditional maritime countries of Europe wanted to check the practice of flags of convenience. Therefore they were in favour of the introduction of the genuine-link clause.[2] The clause, Article 29, drafted by ILC in 1956, states:

"Each State shall fix the conditions for the grant of its nationality to ships, for the registration of ships in its territory, and for the right to fly its flag. Ships have the nationality of the State whose flag they are entitled to fly. Nevertheless, for purposes of recognition of the national character of the ship by other States, there must exist a genuine link between the State and the ship."

The flag-of-convenience states, especially Liberia and Panama, opposed the principle of genuine link as it had been formulated by the ILC. The vagueness of

2. The hopes of these countries were best expressed by the government of Denmark. In its comment on Art. 29 it was stated:

"In this connection, the Danish authorities emphasize that the implementation of the proposed rules, which assumes the existence of certain guarantees or evidence of actual relationship of the ship with the State concerned, may serve to support the various endeavors of international shipping circles (including the Danish shipping trade) to prevent the nationality and registration of a ship from being established on the basis of such mere formalities as to come within the concept of 'flags of convenience'": UN.CLS.OR (Preparatory documents) 1:83.

the term "genuine link" was one of the arguments used against it by the flag-of-convenience countries.[3] Also a small group of states recommended further study of the issue and for various reasons they were against the retention of the genuine-link clause.[4] On the other hand the sponsors of the genuine link, Norway and the Netherlands, contended that the term was not too vague or too general. Their argument was that if equally vague phrases like "with reasonable regard to" had been found acceptable, then the genuine link clause should be treated in the same way and accepted as it was.[5]

The majority of delegates agreed that the formulation of the criteria for determining the existence of the genuine link between the state and the ship had great difficulties because there is a great diversity of domestic laws. In Germany's criticism it was pointed out that the draft of the ILC did not make nationality of the ships obligatory and there was no obligation for states to register all ships when a genuine link existed. In some situations ships could not be registered in states offering convenient conditions because of the lack of a genuine link. On the other hand ships could not qualify for registration in states with which they were connected by a genuine link, as conditions for registration were too stringent. At a later stage of the debate the German delegate proposed an amendment to Article 29. According to this proposal the genuine link would have to exist not only between the state and the ship but also between both of these and the shipowner.

The general agreement was that effective exercise of jurisdiction and control should be the responsibility of the state of the ship's nationality. The difference of opinion was that the opponents of flags of convenience connected this exercise to the concept of genuine link. The flag-of-convenience group of states recognised this principle but did not want to connect it with the vague concept of genuine link.

In the establishment of the genuine link requirement the draftsmen of the convention were strongly influenced by the judgment of the International Court of Justice in the *Nottebohm* case.[6] This case did not concern the nationality of a ship, but the nationality of an individual. In this case that of Friedrich Nottebohm, formerly a German national who had acquired Liechtenstein nationality, and property expropriated by Guatemala during the Second World War. The dispute in this case was whether Liechtenstein was entitled to exercise protection over Nottebohm against another state as a result of its unilateral grant of naturalisation or whether the nationality of an individual presupposed the existence of a substantive connection between the individual and the state whose nationality he claimed. The Court found that no such connection existed. The grant of nationality to an individual need not be respected by other states if there is no genuine connection between the state and the individual. Where a state is claimed

3. Liberia, U.N.CLS.OR 4:22.

4. Greece, the USA and India were among these countries. For instance, the Indian delegate believed that Art. 29 should not be included in a convention for the time being because the principle of genuine link was not stated precisely enough.

5. B.A. Boczek, pp.251–252.

6. [1955] I.C.J. Rep. 4.

to exercise diplomatic protection in respect of its nationals, in circumstances like those in this case, nationality should be the legal reflection of a factual link, "a genuine link", between the individual and the state. If the decision of the Court in this case applies to ships, a state will have the right to refuse recognition of a registration because a genuine link does not exist between the ship and the state of registration.[7]

In this case the question was whether "the naturalization conferred on Nottebohm can be successfully invoked against Guatemala, whether it can be relied upon as against that state, so that Liechtenstein is thereby entitled to exercise its protection in favor of Nottebohm against Guatemala".[8] The Court held that for the grant of nationality by state A to an individual to be effective in state B and for it to be recognised by state B that state A has the right to protect that individual, the nationality must be real and effective. It should be reinforced by the existence of a genuine connection between the individual and state A. The nationality granted should be based upon a genuine connection. Otherwise it will not constitute "a translation into juridical terms of the individual's connection with the state which made him its national".[9] And it would not be recognised by other states, and cannot be invoked against those other states as an effective and valid act within the ambit of international law.

As the *Nottebohm* judgment set out a general rule of international law in the case of individuals the question was whether its extension applied to cases dealing with the recognition of the nationality of ships or not. In the ILC's sessions there was no explicit reference to this case. But the reference could be found in the proposal made by the Netherlands government on the genuine link. The Netherlands government referred to the *Nottebohm* judgment in its comment on the provisional articles.[10] The judgment was also referred to in the "Reference Guide to the Law of the Sea adopted by the ILC", prepared by the United Nations Secretariat.[11] At the conference three references were made to the *Nottebohm* case. The first was made by New Zealand, which stated that an analogy could be drawn from that decision and the principle of genuine link embodied a parallel rule in regard to the nationality of ships.[12] Norway made the second reference and quoted the "genuine connection" of the *Nottebohm* judgment to support the necessary use of vague terms.[13] Guatemala, which was the respondent in the *Nottebohm* case, made the

7. See B.A. Boczek, pp.116–124 for a comprehensive examination of the *Nottebohm* case.

8. N.6 *supra*, p.21.

9. *Ibid.*, p.23.

10. ILC Yearbook 1956, II, 62.

11. The reference was as follows: "In connection with the question of the nationality of ships, it may be noted that, in the *Nottebohm* case, the International Court of Justice enunciated certain principles concerning the duty of one State to recognize the nationality legislation of another State. Although this case concerned the nationality of individuals ... the Court considered the matter of nationality in general terms. It is possible, therefore, that some of the principles laid down by the Court may be relevant to the question of the nationality of ships, in particular, to the question of the circumstances under which one State may or may not be obliged to recognize the national character of ships of other States."

12. UN.CLS.OR 4:65.

13. UN.CLS.OR 4:62.

third reference and stated that the "principles laid down in the *Nottebohm* case did not apply to ships".[14]

As stated by Ready the "over-extrapolation"[15] of the *Nottebohm* decision to the question of nationality of merchant vessels has led to much confusion. Considering the facts of the *Nottebohm* case it becomes clear that the decision in this case applies to individuals only; it does not make any reference to corporations or ships. Even the scope of application with regard to individuals is limited.[16] It is also quite difficult to extend considerations relevant to the determination of an individual's nationality to the question of state jurisdiction over a ship. In the case of a ship the international character of sea transportation should be taken into account, which is not the case for determination of an individual's nationality. Considering the fact that the requirement of certainty is much more important with respect to ships than with respect to individuals makes the *Nottebohm* theory inapplicable to the nationality of ships.

As a result it was not possible to reach an agreement on what elements constituted a genuine link and, as stated above, Article 5 was adopted. Under Article 5 it is for the state to maintain the link of proper legislation for its enforcement, and the non-recognition of nationality in the case of absence of genuine link was dropped in the final text of the convention. But although the solution adopted by the Geneva Conference does not follow the *Nottebohm* case it does not describe what is meant by the "genuine link". It does not describe the preconditions for granting nationality. And it is also not clear how nationality is granted in the absence of "genuine link". What happens if there is no genuine link between the ship and the flag state? Is the nationality valid or invalid? Neither Article 5 of the Geneva Convention nor Article 91 of the United Nations Convention provides any answer to these questions.[17] Since its codification into international maritime law in 1958, the genuine link concept has been widely interpreted and also caused disagreements between legal commentators.

The requirement of a genuine link was further discussed by the International Court of Justice in 1960 in the *Constitution of the Maritime Safety Committee of IMCO case*.[18] The Court was requested to deliver an advisory opinion in relation to the constitution of the Maritime Safety Committee of the Inter-governmental Maritime Consultative Organisation (now the IMO). The problem in relation to IMCO arose in 1948. At that time the organisation was still in a formative stage. The controversy was caused by the position of Panama. During the occurrence of the problem Panama was the only major flag-of-convenience nation. The

14. UN.CLS.OR 4:4.

15. See McDougal and Burke, *The Public Order of the Oceans, A Contemporary International Law of the Sea* (1962), pp.1029 *et seq.*, cited in N.P. Ready, p.11.

16. "In the first place, what is involved is not recognition for all purposes but merely for the purposes of the admissibility of the Application, and secondly, that what is involved is not recognition by all States but only by Guatemala": n.6 *supra*, p.17.

17. The provisions of Art. 5 are repeated, with slight differences, in Arts. 91 and 94(1) of the UNCLOS 1982.

18. [1960] I.C.J. Rep. 150.

Panamanian delegate was aware of the quasi-maritime status of his country. He supported the view for the establishment of IMCO but made a reservation as well: "his government was opposed in principle to any measure which might limit its registered tonnage or interfere with freedom of maritime trade".[19] Panama wanted to be a member of the "Main Working Group" of the conference and to become an original member of the IMCO Council. The 1948 Conference agreed to divide the 16 members of the Council into groups of six, six and four. It was provided that six members should be governments of the nations "with the largest interest in providing international shipping services", six "with the largest interest in international seaborne trade", and the remaining four should be governments "having a substantial interest in international seaborne trade". Despite the large tonnage registered under her flag Panama was not accepted as an important maritime nation and was also not proposed by the Main Working Party as a member of the Council, and consequently withdrew from the conference. The problem was whether countries which had a very large amount of tonnage registered under their flag would be considered eligible for privileges reserved for the largest maritime nation, and if they would be eligible to have seats in the various organs of IMCO.

The convention establishing IMCO did not come into force until 1958. During that period Liberia became the greatest flag-of-convenience nation and the third largest nation in terms of tonnage. The Panamanian fleet also grew but not as fast as Liberia. This situation caused concern to the traditional European maritime countries. In England this fear was expressed as follows: "by virtue of this technical ownership they [Panama and Liberia] may be able to participate in international discussions and to affect by their vote both maritime law and maritime safety regulations through for instance the proposed Intergovernmental Maritime Consultative Organization".[20]

During the first Assembly of the IMCO the European Maritime Nations decided to block the election of Panama and Liberia to organs of the United Nations agency. In other words they decided not to recognise Panama and Liberia as "true" maritime nations. On the other hand Panama and Liberia had just ratified the IMCO Convention and were determined to play major roles as maritime nations in the deliberations of the Assembly. The dispute was about Article 28(a) which was on the election of the Maritime Safety Committee, a major organ of IMCO. This Article required that at that time the eight "largest ship-owning nations" be elected to the Maritime Safety Committee. The Article provided that:

"(a) The Maritime Safety Committee shall consist of fourteen Members elected by the Assembly from the Members, governments of those nations *having an important interest in maritime safety, of which not less than eight shall be the largest shipowning nations*, and the remainder shall be elected so as to ensure adequate representation of Members,

19. Doc. E/CONF.4/SR. 4 Rev. (1948), p.33.

20. United Kingdom Chamber of Shipping, *Annual Report* (1955), *The New York Times*, 20 February 1956, p.46.

governments of other nations with an important interest in maritime safety, such as nations interested in the supply of large numbers of crews or in the carriage of large numbers of berthed and unberthed passengers, and of major geographical areas."

The dispute was over the interpretation of the phrase "large shipowning nations". Panama and Liberia, with the support of the USA and a few other non-European nations, argued that the phrase should be interpreted as the eight nations having the largest tonnage registered under their respective national flags, as they were listed in *Lloyd's Register of Shipping*. If this view was adopted both Liberia and Panama would become members of the Maritime Safety Committee because in *Lloyd's Register of Shipping* Liberia was placed third and Panama eighth among the merchant fleets of the world.[21] The opponents of this view were mainly all the European members of the IMCO. They supported the view that the phrase did not necessarily mean the eight nations which had the largest tonnage registered under their respective flags, but rather the nations to whose nationals the vessels really and actually belonged. Under the application of this interpretation the opponents of the flag-of-convenience countries were going to block Liberia and Panama from membership of the Maritime Safety Committee.

The nations opposing the flags of convenience argued that Liberia and Panama did not meet any of the criteria laid down in Article 28. Their argument was as follows: first, to determine "governments of those nations having an important interest in the maritime safety" one had to consider how far governments were interested in maritime questions and to what extent they were able to contribute to various fields in relation to maritime safety. Obviously given such a consideration Liberia and Panama would not fulfil the requirements, because their governments did not have the necessary technical knowledge and did not make any important contribution to maritime safety. Secondly, the delegations of traditional maritime countries based their interpretation on the notion of "genuine link" accepted at the Geneva Conference on the High Seas 1958. They agreed that under Article 28(a) a criterion was brought in but at the same time there was no guidance for its interpretation. Therefore the Assembly was under a duty to interpret it. The majority of the Assembly contended that if the phrase were to be interpreted in the light of the genuine link concept the largest shipowning nations would not be the same as those having the largest registered tonnage. A country's registered tonnage might not reflect its actual importance as a shipowning nation. For instance the United Kingdom delegate expressed the view that "vessels had really to belong to the countries in question, which was obviously not the case with Panama and Liberia". Large fleets flying flags of convenience would be adequately represented on the Maritime Safety Committee if those countries whose nationals had the owning and controlling interests in those ships were elected to the committee. This would be the only way to interpret the phrase in accordance with the letter and the spirit of Article 28.

21. IMCO/A.I/Working Paper 5 (1959) (Merchant Fleets of the IMCO Members according to the statistics in *Lloyd's Register of Shipping*, 1958) cited in B.A. Boczek, p.131, n.22.

Liberia, Panama, the United States and other countries supporting flags of convenience made the following comments on the interpretation of Article 28(a): in the view of the Liberian and United States delegates the two criteria of eligibility set out in Article 28 were not independent of each other. Therefore the largest shipowning nations were *ipso facto* those having an important interest in maritime safety.[22] The requirement of governments on maritime safety, "having an important interest in maritime safety", could be understood only with reference to the governments which had given proof of the widest expert knowledge of maritime safety matters. The United States delegate stated that an important interest in maritime safety was clearly established by active participation in international regulation of the problem of maritime safety or by adoption of codes based on internationally approved safety regulations.[23] In the view of Liberia and Panama even if the criterion of important interest in maritime safety were taken independently, Liberia and Panama were capable of contributing to maritime safety and had many highly qualified people in the service who were able to make a valuable contribution to the Maritime Safety Committee's work.[24]

According to the flag-of-convenience nations and the United States the eight largest shipowning nations, to be members of the Maritime Safety Committee, were the first eight nations shown in *Lloyd's Register of Shipping* because registered tonnage was the sole criterion for judging the size of a fleet. Even if the criterion of the nationality of the shipowners were acceptable Liberia and Panama would still have to be elected because practically all Panamanian and the majority of Liberian vessels were owned by national corporations, which still placed Liberia fifth on the list of *Lloyd's Register* and Panama eighth. Therefore "largest shipowning nations" could not mean that the vessels had to be owned by the state. Gross registered tonnage was the only evidence by reference to which the size of a shipowning nation could be determined.

The dispute was referred to the Court.[25] It was asked the meaning of the phrase "the largest shipowning nations" in Article 28(a) of the IMCO Convention. Did this phrase simply cover the states with the largest gross registered tonnage of shipping registered under their flags or did it cover the tonnage which had to be beneficially owned by nationals of the flag states? With regard to the requirement of an important interest in maritime safety under Article 28, the Court held that "in the context of the provision, possession of this interest is implied in relation to the eight largest shipowning nations as a consequence of the language employed. The particular condition of being one of the eight such nations describes the nature of the required interest in maritime safety and constitutes that interest".[26] For the interpretation of the words "the largest shipowning nations" the Court decided

22. The statement of the Liberian delegate; IMCO/A.I/SR.8 (1959), p. 4 and the statement by the United States delegate; *idem*, p.2.

23. The statement of the United States delegate; IMCO/A.I/SR.7 (1959), p.9.

24. The statements of the delegates of Liberia and Panama; IMCO/A.I/S.R.8 (1959), p.4; IMCO/A.I/S.R (1959), p.6.

25. See B.A. Boczek, pp.125–142 for detailed discussion of the case.

26. [1960] I.C.J. Rep. 161.

that tonnage possessed by a nation was the only practical means of measuring the size of shipowning nations. The next question was in what sense Article 28(a) contemplated tonnage as owned by or belonging to the largest shipowning nations. First, the court rejected the possibility that the phrase "shipowning nations" meant ownership by the state itself. According to this view the criterion of ownership by the state would displace the maritime nations whose fleets were primarily not owned by governments or their agencies.[27] Therefore the Court considered the two meanings of the phrase: (1) it might refer to the tonnage beneficially owned by the nationals of a state; or (2) it might refer to the registered tonnage of a state regardless of its private or state ownership. The Court reached the conclusion that the test of registered tonnage was the only one by which the size of a shipowning nation should be determined. It stated that "the absence of any discussion on their meaning as the draft article developed strongly suggests that there was no doubt as to their meaning, that they referred to registered ship tonnage".[28] It also concluded that "it is unlikely that when Article 28(a) was drafted and incorporated into the convention it was contemplated that any criterion other than the registered tonnage should determine which were the largest shipowning nations".[29] The test of the nationality of the beneficial owners of ships was rejected on the basis that such a test was not practical or certain and also difficult to ascertain.[30]

There were different views on the requirements of Article 28(a). The Court held that the concept of the genuine link was irrelevant for determining the meaning of this phrase and that those nations with the largest registered tonnage fell within the terms of the phrase, whether or not they were flags of convenience. The Court declined this opportunity of giving its support to the requirement of a genuine link.

Since the mid-1970s UNCTAD has concentrated on finding an acceptable interpretation of the constituent elements of the genuine link doctrine. UNCTAD believed that the purpose of the genuine link clause in Article 5 was to make sure that states could ensure the exercise of effective jurisdiction and control over vessels in their registry. And the interpretation of a "genuine link" should be made within the context of the material requirements for the registration of ships to reflect economic ties between vessel and flag state and therefore administrative control by the state. It was thought that this type of interpretation should include the following elements of genuine link[31]:

(i) the vessel or the company owning the vessel should be beneficially owned as to a substantial part by nationals of the flag state,

(ii) the principal place of business and effective management of the legal entity should be located in the flag state;

27. See B.A. Boczek, pp.149–150.
28. [1960] I.C.J. Rep. 170.
29. [1960] I.C.J. Rep. 169.
30. Such a test "finds no basis in international practice, the language of international jurisprudence, in maritime terminology, in international conventions dealing with safety at sea or in the practice followed by the Organization itself in carrying out the Convention": [1969] I.C.J. Rep. 169.
31. S.R. Tolofari, *Open Registry Shipping, A Comparative Study of Costs and Freight Rates* (1989), pp.30–31.

(iii) the principal officers of the shipping company should be nationals of the flag state;
(iv) the flag state should exercise financial control by subjecting the profits of the shipping company to taxation;
(v) the state of registry should exercise full and regular control over the standards of the vessel and qualifications and conditions of employment of the crew.

With regard to the genuine link concept it is possible to reach a conclusion that the sole and exclusive criterion of ownership of a ship will not always determine its nationality. What is needed to establish a genuine link between the state and the ship, and hence the nationality of the ship for the purposes of international law, is the creation of jurisdiction over a ship and control in administrative, technical and social matters.

Since the Geneva Convention on the High Seas 1958 came into force the "genuine link" requirement appears to have had little influence on state practice. Although the requirement is not linked to the effective exercise of jurisdiction by the flag state, as it is in Article 5, it is repeated in the UNCLOS 1982, Article 92.

Apart from comments on genuine link one of the points about the convention was that it had not solved the problem of flags of convenience. Like the 1958 Geneva Convention the 1982 United Nations Convention, Article 92, did not provide much guidance for the interpretation of the "genuine link". The expectation is that the international requirement of the "genuine link" will sooner or later affect states' practice as it affected UK merchant shipping legislation.

UNCTAD made effective studies of flags of convenience and the United Nations Convention on Conditions for Registration of Ships 1986 was adopted. Before examining this convention it would be useful to explain the flag of convenience and ownership relation.

(b) Flags of convenience

(i) In general

Flags of convenience have otherwise been called "flags of necessity", "free flags", "flags of opportunity", "practical flags", "facilitating flags", "shadow flags", "cheap flags", "open registry" and so on. Each of these terms signifies something which in the view of the different national groups is thought to be the dominant characteristic of the institution. The origin of the term "flag of convenience" is not clear but it is believed that its general usage started in the 1950s. This term was used in 1954 in a report of the Organisation of European Economic Co-operation (OEEC, now OECD) in relation to the flags of Panama, Honduras and Liberia.

In 1958 the Maritime Transport Committee of the OEEC defined flags of convenience as[32]:

32. S.R. Tolofari, p.14.

"The flags of such countries as Panama, Liberia, Honduras, and Costa Rica whose laws allow—and indeed make it easy for—ships owned by foreign nationals or companies to fly these flags. This is in contrast to the practice in the maritime countries (and in many others) where the right to fly the national flag is subject to stringent conditions and involves far-reaching obligations" (OEEC, 1958).

Sturmey defined an open registry as the flag of any country which permits persons or companies, other than those with a genuine link with the country, to register their ships in its ports.[33] Sturmey remarked that what constituted a genuine link was a matter of definition because there was a lot of casuistry employed in determining the genuineness of existing links in individual cases.

Under Metaxas's definition,[34] "flags of convenience are the national flags of those States with whom shipping firms register their vessels with a view to maximizing their private benefits and minimizing their private costs by avoiding:

(a) the economic and other regulations; and
(b) the conditions and terms of employment of factors of production that would have been applicable if their vessels were registered in countries of their national origin".

According to Boczek,[35] in the consideration of flags of convenience reference should be made to foreign control: "functionally, a 'flag of convenience' can be defined as the flag of any country allowing the registration of foreign-owned and foreign-controlled vessels under conditions which, for whatever the reasons, are convenient and opportune for the persons who are registering the vessels".

Obviously foreign control is a common feature of flags of convenience. But in some countries which are not commonly considered as flag-of-convenience countries the element of foreign control may still be present. For instance in the United Kingdom, before the amendments brought about by the Merchant Shipping Act 1988, a British ship could in effect be wholly controlled by non-United Kingdom interests. The control was through the mechanism of incorporating a shipowning company which would be "established under and subject to the laws of some part of Her Majesty's dominions, and [have its] principal place of business in those dominions". By this practice the ownership requirement was satisfied, with no regard to the nationality of the shareholders.

As it is difficult to provide a simple definition of flags of convenience, the United Kingdom government prepared a report, called the Rochdale Committee Report, under which six features were identified as being common to such flags:

(i) the country of registry allows ownership and/or control of its merchant vessels by non-citizens;
(ii) access to the registry is easy. A ship may usually be registered at a consul's office abroad. Equally important, transfer from the registry at the owner's option is not restricted;

33. *Ibid.*
34. B.N. Metaxas, *Flags of Convenience, A Study of Internationalization* (1985), p.14.
35. B.A. Boczek, p.2.

(iii) taxes on the income from such ships are not levied locally or are low. A registry fee and an annual fee, based on tonnage, are normally the only charges made. A guarantee or acceptable understanding regarding future freedom from taxation may also be given;

(iv) the country of registry is a small power with no national requirement under any foreseeable circumstances for all the shipping registered, but receipts from very small charges on a large tonnage may have a substantial effect on its national income and balance of payments;

(v) manning of ships by non-nationals is freely permitted; and

(vi) the country of registry has neither the power nor the administrative machinery effectively to impose any government or international regulations; nor has the country the wish or the power to control the companies themselves.

Given the fact that since the 1970s some policy changes were made by some flag-of-convenience countries the above features are not present in all of them. For instance now most of the flag-of-convenience countries impose age limits on vessels entering their registries for the first time.[36] Some flag-of-convenience countries also make efforts to ensure that ships under their registries comply with international regulations, particularly in safety matters.

Today flags of convenience are defined by reference to the existence or otherwise of a genuine economic link between a vessel and its country of registry. The Intergovernmental Working Group of UNCTAD prepared a report on the Economic Consequences of the Existence or Lack of a Genuine Link Between the Flag of Registry and stated[37]:

"The following elements are normally relevant when establishing whether a genuine link exists between a vessel and its country of registry:

(i) the merchant fleet contributes to the national economy of the country;

(ii) revenues and expenditure of shipping, as well as purchases and sales of vessels, are treated in the national balance-of-payments accounts;

(iii) the employment of nationals on vessels;

(iv) the beneficial ownership of the vessel.

(ii) Development of flags of convenience

When maritime history is examined it is possible to reach the conclusion that the shifts of maritime activity from one flag to another are as old as modern national states. The history of flags of convenience dates back to the Roman Empire, when Roman shipowners registered their ships under the Greek flag. In the sixteenth and

36. In 1971 the Liberian Government established a ship inspection programme under the Liberian Bureau of Maritime Affairs in New York. Under this programme all vessels on the Liberian register undergo inspections and matters relating to safety and crew qualification are given special attention. Also since 1975 vessels over 20 years old cannot be allowed a first-time registration. In Panama since 1977, when a Department of Maritime Safety was established to inspect Panamanian vessels, the registry also started to tighten its control over its vessels. The Bahamas started an inspection service in 1982.

37. TD/B/C.4/177–TD/B/C.4/AC.1/3 annex.

seventeenth centuries English shipowners transferred their vessels to the French and Spanish flags. The aim of the transfer was to avoid Spanish monopoly restrictions on trade with the West Indies. Again in the seventeenth century English fishermen off Newfoundland used the French flag in order to avoid fishing restrictions imposed by Great Britain. In the middle of the nineteenth century many British shipowners sailed their vessels under the Norwegian flag. In the nineteenth century there were also cases where shipowners changed flags to avoid the disadvantages of government decrees and protectionist policies and also to gain protection against piracy and confiscation. In 1919 a small cargo vessel, the *Belen Quezada*, was transferred from the Canadian to the Panamanian flag and thereafter engaged in rum running in an effort to avoid American Prohibition laws. The history of the origin and evolution of the present-day flags of convenience can be divided into five parts[38]:

(i) the 1922 Prohibition laws in the USA;
(ii) rising American labour costs for US flag ships and their consequent high operating costs;
(iii) the imminence of World War II in the inter-war years;
(iv) the US Neutrality laws of 1939; and
(v) competition in the world shipping markets after World War II.

Under the 1922 Prohibition laws in the USA the sale or carriage of alcoholic beverages aboard American-owned vessels was illegal. As a result of this Act, an American shipping company, United American Line, transferred two of its US-flag passenger ships, *SS Resolute* and *SS Reliance*, to the Panamanian registry. The aim of the transfer was to avoid Prohibition regulations preventing the sale of liquor on board American vessels. In the 1920s this prohibition caused a considerable number of transfers to the Panama flag. In America, apart from *SS Resolute* and *SS Reliance*, seven other ships were transferred in 1920 to Panama by a US company under the auspices of the Shipping Board.

Under US maritime law a large majority of the crew complement (75 per cent) of a US flag ship had to be American. In order to avoid being caught by this condition and to escape from rising operating costs due to union pressure, in America the flags of Panama and Honduras became flags of necessity. The reasons for moving towards a recourse to flags of convenience were different in America and Europe. In America rising operating costs of US flag ships was the primary motive. In Europe political instability was the main motivation towards flags of convenience. The fear of requisitioning of their vessels and heavy tax burdens led European shipowners to apply flags of convenience. In 1935 the 25 vessels forming the Esso Baltic fleet were transferred from the flag of the Free State of Danzig to the Panamanian registry.

In 1939 the US Neutrality Act sought to prevent American-owned and crewed vessels from trading in the European war theatre. The Act aimed to keep US flag ships from entangling the USA in war. Despite the neutrality laws the American

38. S.R. Tolofari, p.24.

government wanted to supply essential war materials. As a solution a large number of US owned vessels, mainly tankers, were transferred to the Panamanian flag under the auspices of the US government.

After the war years the desire and need to be competitive in the world shipping markets gave rise to the spectacular growth of flags of convenience. Obviously the reasons for registering a ship in a flag-of-convenience country vary from one owner to another. Today the primary reason for flags of convenience is to obtain cost economies and stay competitive in the industry.

The economic aspect of flags of convenience was set out in Grunsey's definition[39]:

"A flag of convenience is the flag of a State whose government sees registration not as a procedure necessary in order to impose sovereignty and hence control over its shipping but as a service which can be sold to foreign shipowners wishing to escape the fiscal or other consequences of registration under their own flag."

To include the fact that open registration is a service sold by open registry states to earn income is a fundamental truth. This idea underlines the fundamental logic behind a declared policy of open registration by a state.

In nearly all the definitions of flags of convenience reference is made to fiscal obligations of the shipowner's own state. Metaxas and Doganis bring another definition of flags of convenience, in other words open registry, to bring a balance between the above-stated definitions:

"An open registry state is a state which has a declared policy of selling its nationality to shipowners for the registration of ships under the state as a means of earning national income in the first instance, and not necessarily as a means for effective control and jurisdiction, by offering shipowners conditions of ship registration which they find commercially and economically attractive in comparison with the conditions under their own flags for the purpose of obtaining cost and fiscal advantages which make them internationally competitive."[40]

In this definition the primary motivation of a shipowner for having his ships registered in a flag-of-convenience country/open registry is the need to be competitive and not the desire to "evade" or "avoid" tax obligations under his national flag. Fiscal advantages are important elements of the attractions of open registry.

It is believed that taking part in competition in the market has great importance for a shipowner when considering open registry. But in a maritime company all decisions are taken in order to achieve the common goal of minimising private costs and maximising private revenue. Therefore choice of flag relates to investment and to cost/revenue considerations. In this regard it is not possible for a shipowner to choose a flag without considering the fiscal advantages. Maritime companies also consider the political and commercial aspect of the problem. In

39. K. Grunsey, *Flags of Convenience in 1978,* Transport Studies Group, Discussion Paper No.8 (1978).

40. S.R. Tolofari, pp.20–21.

this respect a multinational oil firm or a passenger ferry would prefer not to operate all its fleet under flags of convenience.

Once flags of convenience shipping was dominated by the flags of Panama, Liberia, and Honduras.[41] Although Honduras is no longer a major factor, Panama and Liberia are both the largest flags of convenience ship registries and the largest registries in the world. (See the table on the following page.) According to Lloyd's Register Panama continues to have the world's largest cargo carrying fleet, 4,473 vessels totalling 66.85 Mgt. In second place Liberia has 1,568 vessels totalling 58.27 Mgt. Cyprus lies in third place and overtakes Greece which was in third place in 1994. In 1995 Malta the register has shown the greatest growth with an increase of 14.6 per cent.[42] In December 1994 flags of convenience countries comprised 43 per cent of the world gross registered tonnage. For 1994 the International Transport Workers' Federation (ITF) Secretary stated that flags of convenience ships were in the majority. The leading countries in which the beneficial owners who utilise flags of convenience shipping are headquartered are Greece, the United States and Japan.[43]

(iii) Problems arising from flags of convenience

Since the juridical order of the maritime zones is based on the institution of the nationality of ships, it is necessary that this nationality be easily identifiable. In disputing claims or other situations where the identification of the ship is necessary the ship's nationality should be established with certainty and clarity. There has been universal agreement on the principle of states' competence to fix the conditions of the granting of nationality to vessels. But the scope of this competence and the implications of the principle caused some disagreements between the authorities of international law, and as a result the flag-of-convenience institution enlarged its application.

The complexity of ownership with regard to the ship's flag brings complexity to the problem. For instance, a maritime company might include a European single owner or a majority shareholder of one Panamanian company owning or controlling a vessel under a European flag. Or a traditional shipowning family company might own in part or in whole several Panamanian and Liberian companies with vessels registered under traditional maritime flag and/or flags of convenience. A ship may be beneficially owned in one country, directly owned by a company resident in another country, registered in the flag of a third country, managed by a company in a fourth country and even sub-chartered to the interests of yet another country. Beneficial ownership abroad exists when the national origin of the maritime firm is different from the country of registration of the ship.[44]

41. See ISM, "Guide to International Ship Registers and Ship Management Services" (1996), pp.3–55 for open registry countries.

42. *Ibid.*, p.3.

43. Herbert R. Northrup and Peter B. Scrase, "The International Transport Workers' Federation Flag of Convenience Shipping Campaign: 1983–1995" (1996) 23 *Transportation Law Journal* (University of Denver, College of Law) 371–372.

44. See B.N. Metaxas, p.11.

Table 1. Fleet Registration and Gross Tonnage in 1996 (Twenty Largest Fleet Registers)[45]

	Ships	*Gross Tonnage*	*% G.T.*		*Ships*	*Gross Tonnage*	*% G.T.*
Panama	6,105	82,130,668	13.5	United States	5,289	12,024,644	2.9
Liberia	1,684	59,988,708	12.1	Singapore	1,480	16,448,536	2.5
Greece	1,743	27,507,109	6.3	Philippines	1,617	9,033,849	2.0
Cyprus	1,632	23,798,904	4.9	Hong Kong	398	7,862,964	1.6
Bahamas	1,186	24,408,787	4.8	South Korea	2,327	7,557,931	1.5
Japan	9,399	19,200,927	4.6	Italy	1,348	6,594,302	1.4
Norway (NIS)	685	18,948,844	4.2	India	920	7,127,246	1.4
Russia	4,866	13,755,374	3.5	Taiwan	681	6,174,535	1.2
China	5,121	16,992,863	3.3	Germany	1,101	5,425,682	1.1
Malta	1,247	19,479,431	3.2	Turkey	1,114	6,425,682	1.1

In *Pacific Charger*[46] a formal investigation was held in New Zealand in 1981 to consider the stranding of the vessel. During the investigation it became clear that there were some 12 companies (excluding the nominal shipowner in Liberia) involved in the operation and financing of the ship concerned. Many of the crew did not understand who actually owned the ship or who they actually worked for.[47]

In considering a ship's flag a maritime company is not the only decisive body. There are other institutions which take part in the flag state choice of the maritime

45. *Ibid.*, p.373. Source: Lloyd's Register, *World Fleet Statistics* (1996).

46. *M. Pacific Charger—Report of Court* (January 1982), and Annex thereto, of the formal investigation into the shipping casualty of 21 May 1981, held in Wellington, New Zealand, 7 July–9 October 1981. See A.G. Corbet, "Who is the Shipowner?" *Seaways*, The Journal of the Nautical Institute, December 1993, p.16.

47. The companies involved in this case were as follows: the *Pacific Charger* was owned by Ocean Chargers Company Limited; the company and the ship were both registered in Liberia; the company was a wholly owned subsidiary of Kansai Steamship Company Limited which was incorporated in Japan (where the ship was built), the ship was time-chartered to the Kansai Steamship Company, which had an operation contract for the ship with Crusader Swire Container Service Limited, a company incorporated in England; the Ocean Chargers Company also had a management contract with Harmony Maritime Company Incorporated of Taiwan; the general agents in Japan for the Harmony Maritime Company were Seiwa Navigation Corporation; the agents in Taiwan appointed by Harmony Maritime Limited for appointing the Chinese crew officers to the ship were the Union Maritime Company Incorporated; the agents in Hong Kong appointed by the Harmony Maritime Company for appointing the Burmese crew ratings to the ship were Ocean Services Corporation Limited, Far East Division; the general agents in Japan for Crusader Swire Container Service Limited were John Swire & Sons (Japan) Limited; the agents in New Zealand for Crusader Swire Container Service Limited were P&O, New Zealand Division; the shipowner's (Ocean Chargers Company Limited) agents in New Zealand, appointed after stranding, were Russell & Somers (Wellington) Limited; the ship was subject to two mortgages, the first to Sumisho Lease Company Limited and the second to Kurushima Dockyard Company Limited; the second mortgage was assigned to Sasebo Heavy Industries Company Limited, the ship's builder. *Ibid.*

company. During bareboat and long-term charters the chartering company becomes the decision-making body regarding the flag of the ship during the charter. Metaxas refers to another group which also has a great effect in the flag choice.[48] This group is the charterer in the main maritime markets. This group consists of companies which are usually users of maritime services. The aim of these firms is to minimise their sea transport costs and maximise profits in the markets where their final products are sold. As a result of this objective if tonnage which is beneficially owned abroad can provide sea transport services at a lower price than the minimum required by nationally registered tonnage the preference would be to charter tonnage belonging to firms that use the device of tonnage beneficially owned abroad. In cases where maritime companies have heavy financial commitments financial institutions like banks have a decisive influence on the flag-state choice.

(iv) Opposition to flags of convenience

Safety

For years vessels registered under flags of convenience were involved in maritime disasters. Most of these ships do not meet the international legal requirements for the technical seaworthiness of ships. The casualty records of open registry fleets show that the casualty rate for the flag-of-convenience fleet is substantially higher than that of the regulated fleet (see Appendix I).

In the study made by Metaxas and Doganis[49] it was concluded from the casualty records for 1958–1972 that "the analysis of casualty statistics ... has shown that the casualty rates of the flags of convenience fleet, taken as a whole, are significantly higher than those of regulated fleets .... There is clearly a distinct relationship between casualty rates and flag of registry. The convenience flags all have poor safety records in comparison to the world average and, more particularly, in comparison to the safety records of major regulated fleets." During the above-mentioned period and the following years the smaller flags of convenience fleets, like the Lebanese, Somali, Singaporean and Cypriot fleets, had a worse safety record than that of the larger flags of convenience fleets such as Liberia and Panama. In the above-mentioned study this difference was based on the larger number of owners in the first group than the second who wish to avoid regulatory standards. For instance, for many years Panama had the reputation for permitting the operation of sub-standard ships. As a result of this practice Panamanian ships had a very high accident rate compared with vessels which flew the traditional maritime countries' flags. After increasing pressure in 1977 the Panamanian government enacted laws and provided for stricter inspection of vessels flying its flag. Casualties involving flags of convenience fleets, such as the *Torrey Canyon* in 1968, the *Amoco Cadiz* in 1978, the *Exxon Valdez* in 1989 and the *Scandinavian Star* in 1990, continued into the 1990s.

48. See B.N. Metaxas, p.12.
49. See S.R. Doganis and B.N. Metaxas, *The Impact of Flags of Convenience* (1976), pp.79, 103.

In the international field the aim is to put pressure on the flag-of-convenience states to enforce effective jurisdiction and control over their ships and also to observe high standards of safety. In 1981 the UNCTAD Secretariat prepared a report and identified 10 reasons why under open-registry flags safety standards were not observed as much as under the flags of states which have genuine economic links with vessels[50]:

(1) Real owners are not readily identifiable (partly because of difficulties in identifying, partly because of lack of incentive to identify) and are therefore in a good position to take risks by comparison with owners in normal registries who are living under the eyes of maritime administration;

(2) Real owners can change their identities by manipulating brass-plate companies and consequently avoid being identified as *repeated* sub-standard operators or risk takers;

(3) Since the master and other key shipboard personnel are not nationals of the flag State, they have no need or incentive to visit the flag State and can avoid legal action;

(4) Owners who reside outside the jurisdiction of the flag State can defy the flag State by refusing to testify at an inquiry by the flag State and avoid prosecution;

(5) Since open-registry owners do not have the same interest in preserving good relations with the flag State, they do not feel the need to co-operate with inspectors of the flag State;

(6) Open-registry shipping lacks the union structure which is so essential to the application of safety and social standards in countries of normal registry; namely, a national trade union of the flag State representing basically the interests of national seamen on board vessels owned by owners who have economic links with the flag State;

(7) Open-registry owners are in a better position to put pressure on masters and officers to take risks, since there is no really appropriate government to which shipboard personnel can complain;

(8) Port State control is weaker because the port State can only report sub-standard vessels and practice to a flag State which has no real control over the owner;

(9) Owners can suppress any signs of militancy among crews by virtue of their freedom to change nationalities of crews at whim;

(10) Enforcement of standards is basically inconsistent with the operation of a registry with the sole aim of making a profit.

The best-known open-registry vessel casualty is the *Amoco Cadiz*. The *Amoco Cadiz* accident happened on 16 March 1978 off the coast of Brittany. In this case the vessel flew the Liberian flag. After the accident about 400 kilometres of the coast were directly affected. The amount of oil spilled into the sea from the *Amoco*

50. *Action on the Question of Open Registries*, TD/B/C.4/220, cited in N.P. Ready, p.26, n.17.

Cadiz was about twice that spilled from the supertanker *Torrey Canyon* in 1967 and about six times that spilled from the tanker *Argo Merchant* in 1976. As with the *Amoco Cadiz*, the *Torrey Canyon*, the *Argo Merchant* and the *Sea Empress* all flew the Liberian flag as well. Very recently,[51] the *Diamond Grace* flew the Panamanian flag (see Appendix I, Total Losses by Registration; and Appendix II, Total Losses by Registration and Shiptype Category).

With reference to the early stages of development of the flags of convenience some might argue that the flag-of-convenience fleet has brought social benefits as well as additional social costs. Obviously the lower freight rates helped the development of dry-cargo and tanker markets. In the countries of registry there were more employment opportunities. Since the beginning of the flag-of-convenience institution there have also been major changes in the attitudes of the major flag-of-convenience countries. In Liberia vessels seeking registration must be no more than 20 years old. Vessels exceeding that age limit may be accepted for re-registration under certain conditions. In Panama vessels over 20 years of age are subject to a special inspection before the Permanent Certificate of Registry can be issued. The Bahamas generally apply a 12 year age limit.[52] The general view on flags of convenience and safety is that comprehensive international safety standards to regulate the operation of ships of any flag should be formulated. There should be stricter international enforcement of these standards.

Labour

After the transfer of American ships to the Panamanian and Honduran flags organised labour opposition to flags of convenience began in the 1930s. In 1957 it also became clear to the business leaders of the advanced maritime states that if no steps were taken to control the situation the flag-of-convenience institution would cause serious problems. In 1948 the International Transport Workers' Federation (ITF), which unites more than 400 trade unions and represents over five million transport workers, adopted a resolution, in which it threatened to boycott ships transferred to the Panamanian flag. In 1958 the ITF Congress decided to start a worldwide boycott of open registry ships. The main objectives of the campaign were as follows[53]:

(a) to establish by international governmental agreement a genuine link between the flag that a ship flies and the nationality or residence of its owners, managers and seafarers, and so to eliminate the flag-of-convenience system entirely;

(b) to ensure that seafarers who serve on flag-of-convenience ships, whatever their nationality, are protected from exploitation by shipowners.

The sub-standard conditions of ships with regard to labour qualifications caused an accident in 1974. In the early months of 1974 a Liberian registered cargo ship,

51. 2 July 1997.
52. See N.P. Ready, p.23.
53. *Ibid.*, p.25.

Seagull, was lost with all hands off Sicily. As a result of investigations it became clear that only the ship's master had bridge watch-keeping qualifications. The chief officer was a promoted radio officer. There was no second officer. As a result of disqualification of the crew only the master had enough knowledge and experience to deal with bad-weather emergency situations.

ITF's campaign began to have effect in 1976. The *Globtik Venus* had a Filipino crew and a Bahamas flag at Le Havre. Under the nationality of the crew and flag the vessel became the symbol for the ITF to challenge shipowners flying convenience flags to provide better pay and conditions for the seamen. The crew refused to sail her from Le Havre. In the end a compromise was reached, but this case showed the importance of the threat to blacklist flags of convenience.[54]

During the ITF campaign ITF prepared an agreement called the ITF Collective Agreement, which contained terms and conditions for the employment of seafarers. It also covered minimum wages, holidays, hours and special conditions and also contributions by the owner to the ITF's Seafarers' International Assistance, Welfare and Protection Fund. Shipowners who operate vessels under flags of convenience employ their crews under the ITF Collective Agreement. When a shipowner signs this agreement he will have a "Blue Certificate". If a shipowner fails to provide such a certificate to an ITF inspector industrial action will be taken against the vessel. In other words ITF "blacks" the ships which cannot produce a blue certificate to one of its inspectors. Under this practice port workers, like dockers and tug crews, refuse to let the vessel leave until the ITF conditions are met.

A Blue Certificate provides that:

(a) The shipowner signs the ITF Collective Agreement or an ITF approved national agreement. Under the ITF Collective Agreement the average of European wage levels is taken as a base for the rates of pay. Under this determination inflation is taken into account but the nationality of the crew is irrelevant;

(b) Each crew member receives an individual contract of employment. This contract incorporates the current ITF wage scale. And if necessary the authorities of the labour-supplying country also receive such contracts;

(c) If crew members are not members of a union they affiliate to the ITF and enrol in the ITF's Special Seafarers' Department;

(d) For each seafarer the shipowner makes an annual contribution to the Federation's Seafarers' International Assistance, Welfare and Protection Fund;

(e) The Federation is informed of future crew or contract changes. It has access to all records;

(f) If there is a difference between the ITF rate and the previous rate it is possible to claim for the difference between the ITF rate and the previous rate to the date of commencement of the ITF agreement.

54. See Christopher Hill, *Maritime Law* (4th ed., 1995), p.21.

The ITF agreement requires wages to be paid at the level unilaterally established by it as equal to wages on the European average standard, which, since, 1994, has been US$856 per month for an able-bodied seaman. The ITF is now making efforts to standardise total crew contracts so that one document will apply to all nationalities. The concept of "total crew costs" provides a minimum total cost for an able-bodied seaman now set at US$1,100 per month. To obtain a blue certificate when the total crew costs is utilised by the ship, the ITF also requires a collective bargaining agreement from the crew's home country where the affiliated union has recommended the issue of the certificate subject to the approval of the ITF national headquarters and the incorporation of the regulations specified by the ITF.[55] As delay to a ship costs large amounts of money most of the time the shipowners agree to pay because legal action is also time consuming and the owner does not want to waste his money.

Between 1976 and 1980 the Court of Appeal dealt with five cases about union action against flags of convenience. In the first case, *Camellia Tanker Ltd.* v. *International Transport Workers' Federation*,[56] the dispute was between the union and the shipowners about the wages being paid on the *Camellia*. The tanker was blacked and unable to sail as a result of disagreement. The shipowners' application for an injunction was refused, because the union's action revealed a dispute between the employers (the owners) and workers (the union) and was therefore connected with the terms and conditions of employment of the crew of the *Camellia*.

In the second case, *Star Sea Transport Corporation of Monrovia* v. *Slater*,[57] a bulk carrier, the *Camilla M*, which flew a flag of convenience, was blacked under the instructions of the ITF. After blacking the Indian crew of the carrier refused to sign ITF articles and was discharged. The Greek crew which was engaged later also refused to sign ITF articles. The union insisted that the crew be employed on ITF conditions, so the blacking continued. The Court of Appeal held that the statutory immunity did not apply. In this case the court, as a result of the impossible and unreasonable demand of the union, indicated that the action was taken not for a genuine trade reason but as a part of a political campaign against flags of convenience.

In *N. W.L. Ltd.* v. *Woods*,[58] and *N. W.L. Ltd.* v. *Nelson*[59] the ITF sent instructions to black the *Nawala*, a bulk carrier which flew a flag of convenience. The case mainly concerned a threatened boycott at an English port of a Hong Kong flag vessel with Chinese crew and suspected beneficial ownership elsewhere than in Hong Kong. The crew was entirely satisfied with its terms and conditions and actually opposed the intervention of the ITF. It was argued that the motive of the ITF was to promote its policy of abolishing flags of convenience and was not to pursue any legitimate trade object. In other words, the argument was whether in

55. Herbert R. Northrup and Peter B. Scrase, pp.378–379.
56. [1976] I.C.R. 274.
57. [1978] I.R.L.R. 507.
58. [1979] I.C.R. 744.
59. [1979] I.C.R. 755.

such a situation the unilateral action of the ITF justified the plea of trade union immunity. In both cases the Court of Appeal held that the immunity applied, and it suggested that trade unions could lawfully take action against flags of convenience. In the House of Lords this view was also supported. In the House of Lords it was accepted that in both the *Nawala* cases the dispute was between employers and workers and was connected with the terms and conditions of employment of the crew of the *Nawala*. Lord Diplock made it clear that the law was widely framed, and that any demand about terms and conditions of employment, however reasonable, would attract immunity. According to the ITF the crew should be paid ITF rates. It was held to be irrelevant that the crew was content with the existing terms and conditions. In this respect the blacking was in furtherance of the dispute.

In the House of Lords the limitations on statutory immunity were considered in relation to *Star Sea Transport Corporation of Monrovia* v. *Slater*. As stated above the decision in this case made it clear that demands in trade disputes should be reasonable or statutory immunity would not apply. But such a limitation was rejected in the House of Lords.[60] In *N.W.L. Ltd.* v. *Woods* the Lords dealt with motive limitation and seemed to suggest that motive is now an irrelevant consideration.[61] The House of Lords' decisions on union action against flags of convenience shows that trade union action against flags of convenience is protected by British law. The ITF would have been entitled to the protection given to trade unions acting in a trade dispute in the United Kingdom. Parliament clearly understood that statutory immunity would extend to disputes about flags of convenience, as shown by the relevant clause in the Trade Union and Labour Relations (Amendment) Bill which extends immunity to inducing breach of commercial contracts.[62] In the Employment Act 1980 an individual's legal

60. It was held by Lord Diplock in the House of Lords that: "if a demand on an employer by the union is about terms and conditions of employment the fact that it appears to the court to be unreasonable because compliance with it is so difficult as to be commercially impracticable or will bankrupt the employer or drive him out of business, does not prevent its being a dispute connected with terms and conditions of employment. Immunity under section 13 is not forfeited by being stubborn or pig-headed.' K. Ewing, "Union Action Against Flags of Convenience—The Legal Position in Great Britain" (1980) 11 *Journal of Maritime Law and Commerce* 506–507.

61. The traditional view of the English courts has been that where a dispute was pursued, not for a genuine trade reason but for some purpose such as political or personal reasons, or in the interest of union prestige, then the immunity will not apply. This motive limitation is essentially a judicial development which is not provided for in the language of the statute. In the House of Lords Lord Scarman said: "All that the subsection requires is that the dispute connected with one or more of the matters it mentions. If it be connected, it is a trade dispute, and it is immaterial whether the dispute also relates to other matters or has an extraneous, e.g. political or personal motive." He continued as follows: "A dispute may be political or personal in character and yet be connected with, for example, the terms and conditions of employment of workers." Lord Diplock took a similar view and stated that: "Even if the predominant object were to bring down the fabric of the present economic system by raising economic wages to unrealistic levels, or to drive Asian seamen from the seas except when they serve in ships beneficially owned by nationals of their own countries, this would not, in my view make it any less a dispute connected with terms and conditions of employment and thus a trade dispute, if the actual demand that is resisted by the employer is as to the terms and conditions on which his workers are to be employed". *Ibid.*, p.507.

62. Mr John Prescott M.P. said in the House of Commons that this provision was necessary in order to protect the rights of workers in "ships not belonging to traditional maritime countries but flying flags

immunity from suit in relation to "secondary action", where pressure is put on a person who is not directly a party to the trade dispute, was removed. This Act had in 1983 applied to the blacklisting of a ship. In the *Hoegh Apapa* case[63] a Liberian registered vessel docked in Liverpool. The vessel had a Filipino crew. The ITF persuaded the tug operators to "black" the ship and prevent her sailing. The shipowners had chartered the ship under a time-charter contract. Under the time charterparty the shipowners were responsible for time lost due to ITF's action. Therefore the shipowners claimed damages from ITF for the tort of interfering with the performance of a contract. The House of Lords held that the ITF had *prima facie* committed the tort. The actions of ITF would have been protected under the Trade Union and Labour Relations Act 1974. But under the Employment Act 1980 ITF's actions were secondary because the dispute was with the shipowners. The charterers who had contracted for the tugs were not party to that dispute.

The definition of trade dispute was further narrowed under the Employment Act 1982. In section 15 the immunity of the trade unions and employers' associations was removed. Under this section the union has only the same limited immunities as an individual. By section 16 the union is given a special limit of liability for actions in tort. The limit does not apply to personal injury claims or those arising out of its ownership or use of land. Under section 16 for a union with fewer than 5,000 members the limit is £10,000 and for a union with 100,000 or more members the limit is £250,000.

As the ITF boycott campaign grew in strength the interests of two different groups started to clash. On the one hand, there is the right of unions and workers to secure satisfactory working conditions. On the other hand, there is the right of shipowners to trade their vessels internationally without being detained by extra-legal action in countries which have no connection with the owner, the crew or the union. As is well known, in order to engineer the boycotts the ITF requires the co-operation of local or national maritime, tugboat, or longshore and other dock owners' unions, national laws which permit boycotts of this nature and a crew of inspectors.

In most cases the shipowners found themselves asking their lawyers whether an injunction could be obtained to prevent an ITF boycott. The answer depended on the local law. If the local law appeared to be very heavily weighted in favour of unions and employees against the employers then there was no redress to prevent a boycott because such boycott was permitted by the local law. Consequently, the question was whether there was an alternative remedy of damages and/or

---

of convenience. Liberia, Panama and Cyprus are among those countries which have no traditional maritime interest and which offer tax avoidance, cheap labour and a lack of safety legislation." He also stated that the immunity, as amended, would enable trade unions to take lawful action to stop the "exploitation" of workers on such ships: *ibid.*, p.508.

63. *Merkur Island Shipping Corporation* v. *Laughton, Shaw and Lewis* [1983] 2 Lloyd's Rep. 1. See also *The Uniform Star* [1985] 1 Lloyd's Rep. 173.

restitution of what had been paid, after the boycott had been lifted and the vessel had sailed. While a shipowner seeks legal advice there may be long delays which will be very expensive for him, so in some cases shipowners find that it is less expensive to agree to the ITF terms than to seek court action.

The first successful case against the ITF was the *Universe Sentinel* case, under which the vessel was detained by ITF boycott at a Welsh port, with the ITF demanding full ITF worldwide conditions. The vessel flew the Liberian flag but was beneficially owned in the United States and had a mixed crew, including Indonesians. After signing up on ITF terms and making the necessary payments the shipowners claimed the money back. The issue was whether the demand was connected with terms and conditions of employment. The court held that the Welfare Fund[64] payment was not connected with the terms and conditions of the employment and therefore refundable.[65]

The *Saudi Independence* was concerned with the conflict between domestic law pertaining to the legality of strikes and the chosen foreign law in a contract of employment. The vessel sailed under the Saudi Arabian flag and employed a Filipino crew under Filipino employment contracts. The contracts were subject to the law of the Philippines. While the vessel was at a Dutch port the crew sought the assistance of the ITF and went on strike with its advice. The owners commenced proceedings against both the ITF and the crew and sought an injunction restraining the strike. The court held that under Filipino law, the law of the employment contracts, the strike was unlawful and therefore the injunction was granted. Both the Dutch Court of Appeal and the Dutch Supreme Court upheld the decision. According, to the ITF the legality of the action taken by it should be decided under Dutch law where the act of inciting the strike occurred. However, the Dutch Supreme Court confirmed the appeal court decision by stating that even if Dutch law had applied to the ITF's conduct such conduct would have been unlawful under Dutch law as it was inducing a strike which was unlawful under the applicable foreign law, that of the Philippines.[66]

Following the Employment Act 1980 which contained legislation to limit secondary industrial action, the shipowners believed that they were more likely to be successful in litigation in the United Kingdom. This belief has turned out to be wrong in the *Evia Luck* case. In this case the plaintiff's vessel, *Evia Luck*,[67] was manned by 10 Greek nationals under contracts of employment governed by Greek law and by 20 Filipino nationals under contracts of employment governed by

64. The Welfare Fund was established in 1965 with the purpose of providing assistance and welfare disbursements to seamen. It is a distinct financial entity. A small percentage of its revenue accrues from subscription payments made by members of the ITF's Seafarers' Department and by those seamen who are covered by ITF agreements but who do not belong to any union or to a union affiliated with the ITF.

65. Herbert R. Northrup, and Peter B. Scrase, p.414; *Universe Tankships Inc., of Monrovia* v. *Int'l Transport Workers Federation, and others* [1983] A.C. 366 (HL).

66. Hoge Raad, 16 December 1983 [1985] *Nederlandse Jurisprudentie* 311; [1984] *Schip en Schade* 25, cited in *ibid.*, n.116 (Netherlands).

67. *Dimskal Shipping Co., SA* v. *Int'l Transport Workers Federation* [1992] A.C. 152 (HL).

Filipino law. The vessel flew the Panamanian flag. On 28 February 1983 when the vessel berthed in Sweden an agent of the ITF boarded her pursuant to the flags of convenience campaign and informed the master, officers and crew that the vessel would be blacked unless ITF employment contracts were entered into by the plaintiffs with the crew and the plaintiffs signed certain documents and paid over certain sums to ITF. On 7 March, the plaintiffs complied with ITF's demands and signed the necessary documents which were stated to be subject to English law. The vessel sailed from Swedish waters and the plaintiffs sought a declaration that the contracts were void on the grounds of duress. They claimed restitution of money paid under the agreement and claimed damages for the torts of intimidation and interference with contractual rights. The owners sued the ITF in London rather than in Sweden where the boycott had occurred. The ITF sought a stay on grounds of *forum non conveniens*. In February 1986 the court in London ruled that the ITF's application failed and that the case could continue in London. The restitution claim was pursued to the House of Lords and both parties agreed that it was subject to the English law of restitution. According to ITF the legitimacy of the duress applied should be tested in accordance with the domestic law of the place where the boycott happened, that is Sweden. On the other hand, the owners argued that, as the parties had made the agreements entered into under threat of boycott subject to English law, it should apply to all aspects of the claim. Both the Court of Appeal and the House of Lords considered that the ITF, having chosen English law, could not complain at English law being applied to the whole situation. The owners abandoned their claim for damages as under the English double actionability rule[68] they needed to show not only that the act in question was tortious under English law, but that the claim could also be actionable in the foreign country where the act was committed. And under Swedish domestic law the boycott was lawful. This case clearly established that when foreign law is considered under English law only the foreign domestic law is applied. Foreign law's private international law rules do not have any application. Therefore, in this case the English court did not consider Swedish private international law rules under which the owner's claim for damages would be successful.

In the *Nervion*[69] as no system of law was chosen the question of what should happen where the ITF insisted under threat of boycott on choosing a system of law favourable only to the ITF but not to the shipowner was left open. The Swedish Commercial Court, in the absence of any express stipulation, ruled in favour of the law of the flag, which was Panama. In accordance with Panamanian law, the contracts entered into as a result of industrial action were void by reason of duress. Under Swedish law this was not the case. Consequently, shipowners won their claim in restitution. Following the *Evia Luck* and *Nervion* cases the ITF inserted a specific Swedish choice of law clause into agreements entered into under boycott in Sweden. However, in the *Phillips Arkansas* the Swedish court affirmed that the

68. Double actionability rule in *Chaplin* v. *Boys* [1971] A.C. 356 (HL), cited in *ibid.*, p.416, n.120.
69. (HD 1987: 152) Swedish Supreme Court, 1987 No. 152; [1987] NJA [Sweden] 885, cited in *ibid.*, p.415, n.118.

agreements were effectively avoided under Liberian law, being the law of the flag. Therefore, the ITF's attempt was not successful.[70]

Economic aspect

As stated above under flags of convenience taxes are lower, and labour laws and safety requirements are less strict. When a vessel is registered under a flag of convenience it is owned by a corporation, and this corporation is specially established for that purpose and it does not have any asset other than the vessel. In cases where the shipowner has more than one ship generally a separate owning company will be established in respect of each vessel. Under flag-of-convenience practice beneficial ownership is concealed. This enables the shipowner to escape tax liabilities in the country of his establishment. Another attractive point for the shipowner is the decrease in operating costs. If a vessel is registered in one of the traditional maritime countries the owner will be restricted to employing crew members who are nationals of the country concerned. Obviously employing such crew will bring with it negotiations with local trade unions on rates of pay and manning levels. On the other hand if a ship is registered under a flag of convenience, apart from restrictions imposed by ITF, the shipowner has complete freedom to determine the nationality of his crew and their rate of pay.

In 1981 a report was prepared by the UNCTAD Secretariat on open registries. In this report the UNCTAD Secretariat blamed the freedom from fiscal obligations in flag-of-convenience countries for the liberalisation of tax concessions and increases in Government assistance in the traditional maritime nations.[71] Greece followed this policy in repatriating the substantial number of Greek-owned ships operated under flags of convenience. Under Law 2687/53 for the Protection of Foreign Capital and Ministerial decisions Greek owners were enabled to operate ships under the Greek flag in a virtually tax free environment. As with Greece some other European countries established "international" or "second" registers in order to stem the loss of employment and share of the shipping market.[72] There are also registers that fall between the flags of convenience and the second registers. In these "ship by ship" basis registers, the ITF recognises that a majority of vessels on the register are national but there are a

70. See *ibid.*, p.418. *Phillips Arkansas* (AD 10/92) Swedish Labour Court, No. 10, 1992.

71. In the UNCTAD Secretariat Report it was provided that: "... most of the traditional maritime countries allow shipowners various concessions ... to defer or eliminate tax liability, which reduce the effective tax level below the statutory rate level, and in many cases are reported to result in an effective rate of zero ... in Western Europe the concessions are believed to be particularly liberal, and shipowners who not only operate ships, but also buy and sell on a large scale do not appear to have any difficulty in minimizing their taxes to a low level or even avoiding taxes altogether": *Action on the Question of Open Registries*, TD/B/C.4/220. para. 26.

"... there is no doubt that the existence of open registries is the major cause of the distortions that governments have been forced to make to their fiscal regimes ... Shipowners in the traditional maritime countries have come to look to their governments automatically for subsidies, whatever the causes of their problems—whether due to low-cost competition, inefficient operations, failure to rationalize, or irresponsible overbuilding": para. 38.

72. For detailed information on second registers see ISM, pp.56–67.

substantial number of flags of convenience ships also on these registers. Each ship is examined individually.[73] Second registers employ seamen at reduced rates, and often employ third world personnel as ordinary seamen and national personnel as officers. In the Danish International Register there are no restrictions on the nationality of the ship's crew except that the master must be a Danish citizen. However, the Danish Maritime Authority will in limited circumstances be able to grant a dispensation on the nationality of the master. The most successful second register of Norway was created in 1987. In this registry it is not necessary for foreign-owned ships to establish a Norwegian shipowning company. The owner must appoint an authorised representative in Norway and entrust a significant part of the onshore management functions to a Norwegian based shipowning or management company. The Special Tax Act for seafarers resident abroad was amended as at 1 January 1990 and provides an exemption from tax for non-Nordic residents.

The ITF leadership and some national unions are opposed to second registers and national legislation that permits their operation. In 1993 the ITF designated the Spanish Second Register (Canary Islands) as flags of convenience following a request from the ITF's Spanish affiliates. In Germany the German unions requested that the ITF designate the German International Register (GIS) as a flag of convenience. Recently, the ITF designated the GIS as a flag of convenience. This decision came after pressure from German seafarers who in early 1996 lost a court case attempting to have the register ruled illegal under domestic legislation. The decision was effective from April 1995. Consequently, now GIS flagged ships will face boycott action unless they are covered by the ITF approved collective agreement. Unlike Germany the Norwegian and Danish unions have opposed pressure from the ITF and some of its affiliates to designate their second registers as flags of convenience. ITF has adopted policies which demand the right for unions in the second register country to bargain for non-domiciled seamen wages and conditions on these registers. The ITF has given special notice to the Norwegian Second Register that unless the authorities negotiate acceptable agreements with the Norwegian ITF affiliates the register will be added to the flags of convenience list.[74]

73. Any nationally owned vessels on these registers are automatically classified as flags of convenience by the ITF.

74. Under the ITF official list the following countries are stated as flags of convenience: Antigua and Barbuda, Bahamas, Belize, Bermuda, Burma, Canary Islands (Spain), Cayman Islands, Cook Islands, Cyprus, Gibraltar, Honduras, Lebanon, Liberia, Malta, Marshall Islands, Mauritius, Netherlands Antilles, Panama, St Vincent, Sri Lanka, Tuvalu, Vanuatu.

"Ship by ship" registers are as follows: Hong Kong, Philippines (foreign owned ships bareboat chartered to Philippines), Singapore (foreign owned ships without approved agreements). The "second register" status depends on ownership being from the flag country and on agreements acceptable to that country's unions.

"Second register" countries are Denmark (DIS), Germany (GIS), Isle of Man (UK), Kerguelen (France), Luxembourg (Belgian ships only), Madeira (Portugal), Norway (NIS). http://www.pacifier.com/~rboggs/FOC.HTML

(v) Campaigning against flags of convenience

The ITF is currently taking at least one action a week against flags of convenience vessels.[75] In November 1995 the union introduced a blacklist of shipowners, ship managers and crewing agencies. Two companies, Adriatic Tankers and the Indonesian Crewing Agency, PT Beraya Bintang Samudera, were the first names on the list. The vessels of these companies have been tracked and targeted for action by ITF inspectors worldwide and no ITF affiliate organisation has been allowed to deal with them.[76] Adriatic Tankers employs about 2,500 seafarers aboard its fleet of chemical tankers, VLCCs and bulk carriers. The crews include Russian, Sri Lankan, Greek, Vietnamese, Polish, British, Korean, Burmese, Chinese, Yugoslav and Filipino seafarers. The union claims that Adriatic Tankers owe some US $5 million in back wages. According to an ITF report one of Adriatic Tankers' ships, the Panamanian-registered tanker *Ionian Jade*, was auctioned in January 1996 and the ship's Russian and Maldivian crew received payments totalling some US $342,000 to cover back wages, legal costs and food provision. In a recent case in February 1996, the US Coast Guard found 21 mechanical and safety problems on a Romanian flag vessel that limped into Los Angeles port. According to the ITF inspector the crew were owed back wages and several individuals had injuries that had not been treated.[77]

ITF also carried on its activities against flags of convenience in 1997. Between 12 and 16 May 1997 transport trade unions organised boycotts, strikes and demonstrations against substandard and flag-of-convenience shipping throughout Europe. This action was co-ordinated by the ITF. In 1996 action by ITF and its 95 inspectors in 36 countries resulted in the payment of more than US $35 million in back pay to crews of flag-of-convenience shipping.[78] The first day of the week of action resulted in a ship boycott in Norway. The vessel was the *Blue L*, a 42,000 tonnes deadweight Cypriot-registered bulk carrier loading fertiliser in Porsgrunn, Norway. The ship had a crew of 25 Ukrainians and was managed from Greece.[79] In England, in Sheerness another Cypriot flag vessel, the *Pacific Koru*, was under arrest with the ITF initiating a US $80,000 claim for back pay on behalf of the 25 Polish crew.[80]

During the action week inspectors from the ITF visited 570 ships and signed about 90 ITF wage agreements, including a number covering fleets of ships operated by individual owners. The Panamanian-registered *Sharpnes* was involved in the first action of the campaign in Portugal organised by the Portuguese transport unions. Meanwhile, the German-owned *Christine Schulte*, a 6,300 tonnes

75. Ken Cottrill, "Campaigning Against Flagging Out", *The Baltic*, V.5, Issue 3, May 1996, p.62.
76. *Ibid.*
77. *Ibid.*
78. ITF (9597 release) "European Week of Action", http://www.itf.org.uk/english/797ewal.html (30 May 1997).
79. ITF (12 May 1997), "Successful start to action week", http://www.itf.org.uk/english/woa1.html (30 May 1997).
80. *Ibid.*

vessel, was held in the Croatian port of Rijeka on behalf of a mixed nationality crew. The Cypriot-flag *Moniuszko* was boycotted in Gothenburg, Sweden, to press for an ITF agreement for the Polish crew.[81] The *Pacific Rose II*, a member of the same fleet as the *Pacific Koru*, was arrested on behalf of the Polish crew in Palermo, Italy, with a back wage claim of US $46,000.[82]

A rise in numbers of seafarers abandoned in foreign ports has led the ITF to call for new legally binding regulations for their protection. This move was part of action week. In the nine months from July 1996 to March 1997 the ITF became aware of 54 separate cases of abandonment involving hundreds of seafarers. Of the 119 vessels abandoned since July 1995, 62 were registered in Panama. Hence, the figures show that there is a close correlation between the worst flags of convenience and ship losses. Flags of convenience are still at the top of the list.[83]

81. ITF (13 May 1997), "European actions proliferate as week heats up", http://www.itf.org.uk/english/woa2.html (30 May 1997).

82. *Ibid.*

83. ITF (14 May 1997), "ITF calls for action over crew abandonment. Right to boycott in Germany legally upheld", http://www.itf.org.uk/english/woa4.html (30 May 1997).

CHAPTER 5

THE UNITED NATIONS CONVENTION ON CONDITIONS FOR REGISTRATION OF SHIPS 1986 (1986 UN CONVENTION)

INTRODUCTION

In 1974 UNCTAD's Committee on Shipping unanimously adopted a resolution which stated, *inter alia*, that the economic consequences for international shipping of the existence or lack of a genuine link between vessel and flag of registry was a matter suitable and ripe for harmonisation. After this resolution the open registry issue was considered in UNCTAD at the meetings of the Committee on Shipping and also in special working committees and groups.

The developing countries proposed to eliminate open registries by the establishment of an obligatory economic link between the ship and the flag state. This view was supported by the socialist countries. Until early 1985 the OECD countries strongly opposed this proposition. Despite the opposition of the developed countries a conference of plenipotentiaries met for the first time in July 1984. After the three preparatory meetings, in the fourth session a compromise was reached on the adoption of the convention in February 1986.

The convention is the outcome of a long debate which started in UNCTAD as a move to eliminate flags of convenience. There was also a need to assist developing countries in their efforts to develop their commercial fleets.

The developed and the socialist countries had their own organisations such as OECD and UNCTAD. But the developing countries did not have such organisations, so they asked the UNCTAD Secretariat itself to fulfil this role. There may be arguments on the effect of the convention with regard to flags of convenience. Obviously one of the weak points of the convention is that until it comes into force it has no formal effect. When it enters into force it is going to be binding only on those countries which are parties to the convention. It is a possibility, in fact more than a possibility, that countries which benefit from the flag-of-convenience institution will not become contracting parties. There is also another possibility which has to be taken into account: even a shipowner in a contracting party's country will not be completely prevented from registering his ships in another country which is not a party to the convention.

Another difficulty is about the enforcement of compliance with the convention. Enforcement of such a convention can be secured through port state control or users' boycotts. Effective port state control can only be achieved if the major trading nations agree to operate it. Obviously a users' boycott is not practically possible. No one can expect that charterers will boycott the ships of countries which do not respect the convention. If countries which do not respect the convention offer lower charter rates which are below the market rates for ships flying the flags of contracting parties the charterers will ignore non-application of the convention.

THE CONVENTION

The convention spells out the conditions a contracting state shall require to be fulfilled before it accepts a vessel on its registers. It will enter into force one year after it has been ratified by at least 40 states with a combined tonnage of not less than 25 per cent of the world's total. The convention aims to strengthen the genuine link between the ship and the flag state. It aims to ensure that states effectively exercise jurisdiction and control over their ships, not only in relation to administrative, technical, economic and social matters, but also with regard to identification and accountability of shipowners and operators.[1]

The original proposals which would have had the effect of eliminating flag-of-convenience shipping were rejected and they are not in the convention. Also the port states have no role with regard to the convention's enforcement. The convention refers to a genuine link between the flag state and the ship like the 1958 Geneva Convention on the High Seas and the 1982 United Nations Convention on the Law of the Sea. But the wider concept of a genuine economic link was not approved.

The convention has important articles such as those stating that the flag state shall have an efficient and competent maritime administration. It covers long-needed measures for ready identification and accountability of those responsible for the operation and management of vessels. In general the convention covers articles on the genuine link, although the term is never used, civil and third party responsibility, transparency, labour exploitation and safety standards. The convention consists of mandatory, if the term is used, and recommendatory, if the term is used, clauses.

THE PROVISIONS OF THE CONVENTION

The convention comprises 22 articles and three appendices.

In the preamble to the convention there is a reference to the genuine link

1. In UNCTAD's *Report on Action on the Question of Open Registries* the importance of identifiability of owners was stressed in relation to the establishment of a genuine link: "... it can be concluded that the lack of knowledge on the part of open registry countries of the identities of the real owners of ships on their registers, and their practice of tolerating the nomination of brass-plate companies as owners, stems directly from their lack of economic linkage with the vessels concerned": UNCTAD 1981, TD/B/C.4/220, p.3.

requirements of the Geneva Convention on the High Seas 1958 and the United Nations Convention on the Law of the Sea 1982.[2] According to the seventh paragraph it is for each state to decide on its own conditions for accepting a ship on its register. Although the convention lists sets of conditions required by all states, detailed national conditions must be decided by the individual state. Obviously the national conditions of a contracting state must not contravene the convention's mandatory conditions.[3]

In Article 1, a mandatory article, the objectives of the convention are stated:

> "For the purpose of ensuring or, as the case may be, strengthening the genuine link between a State and ships flying its flag, and in order to exercise effectively its jurisdiction and control over such ships with regard to identification and accountability of shipowners and operators as well as with regard to administrative, technical, economic and social matters, a flag State shall apply the provisions contained in this Convention."

Under Article 1, the purpose of the convention is to ensure that where a genuine link between a ship and its flag does not exist such a link is established on the basis of the provisions of the convention. And in cases where such links are weak or not very clear they should be strengthened. A genuine link between a ship and its flag owners should be identifiable. If the owner is not identifiable control of the conduct of the ship will not be effective and therefore the link between ship and registry will not be genuinely established. In the earlier drafts there was a suggestion that a flag state should apply as a minimum the provisions contained in the convention. But "as a minimum" was deleted. The requirement under this Article is that a flag state must apply the mandatory provisions of the convention.

Ships trading in domestic waters are excluded from the definition of "ship" in Article 2. The definition requires that the ship is engaged in transport, so that mobile offshore units like drilling rigs and flotels are excluded.

The scope of the convention is explained in mandatory Article 3. This Article states that the convention shall apply to all ships as defined in Article 2. This statement means that the convention will also apply to ships which are flying the flags of non-contracting states. But such an application cannot be warranted. Therefore the convention can only be intended to apply to ships registered in states which are parties to the convention.

Article 4, a mandatory article, is largely based on the United Nations

2. The United Nations Convention on Conditions for Registration of Ships 1986 says:
"*Recalling also* that according to the 1958 Geneva Convention on the High Seas and the 1982 United Nations Convention on the Law of the Sea there must exist a genuine link between a ship and a flag State and conscious of the duties of the flag State to exercise effectively its jurisdiction and control over ships flying its flag in accordance with the principle of the genuine link."

3. The United Nations Convention on Conditions for Registration of Ships 1986 says:
"*Reaffirming*, without prejudice to this Convention, that each State shall fix the conditions for the grant of its nationality to ships, for the registration of ships in its territory and for the right to fly its flag."

Convention on the Law of the Sea 1982.[4] Paragraph 4 of Article 4 prohibits registration in more than one state. The prohibition is subject to the relevant mandatory and recommendatory paragraphs for bareboat chartered vessels in Articles 11 and 12.

Under Article 5, a mandatory article, the flag state must have a competent and adequate national maritime administration. Although it is one of the strongest provisions of the convention the terms "competent" and "adequate" are not defined. There is no requirement that the maritime administration be located within the flag state. This was proposed but rejected. The requirement that ships must comply with "applicable" international safety and anti-pollution rules and regulations and that the flag state must implement these met with some opposition from the developing countries. The industrialised countries accepted the term "applicable", but they would have preferred "generally accepted". The paragraph as finally agreed encourages flag states to adopt the IMO and ILO conventions and recommendations. Paragraph 3(b) requires periodic surveys of the ships. And relevant documents must be carried on board. At the preparatory stage there was a proposal on co-operation between flag and port states. Under this practice the port state would be able to check that a visiting ship complied with the provisions of the convention. In the end this proposal was not accepted and the port states had no role regarding the convention's enforcement. In practice this is a big gap for the security of the enforcement of the convention.

When the above-mentioned conditions[5] are considered together it is possible to think that they explain the meaning of "competent and adequate". But the wording is not very clear. For example, what is the difference between compliance with the laws and regulations concerning registration and compliance with the principles of registration? And also who would judge the adequacy and competency of the administration? Or who would determine what is adequate or competent? Even if "competent and adequate administration" was defined under Article 5 how would an administration enforce the rules of this convention? On the other hand, if Article 5 is examined with Article 2,[6] it is possible to reach the conclusion that the ownership, manning and management provisions cannot be effectively enforced if the maritime administration is incompetent and inadequate.

4. See Art. 92 of the UNCLOS 1982.

5. Under Art. 5, para. 3, a flag state has to ensure that ships flying its flag satisfy the following conditions: compliance with the laws and regulations of the State concerning registration of ships; compliance with applicable international rules and standards as concerns the safety of ships and persons, the prevention of pollution and frequency and standards of survey; provision on board of documents evidencing the right to fly the flag, and other valid relevant documents, and compliance by the owners of the ships with the principles of registration in accordance with the laws and regulations of the State and the provisions of this Convention.

6. The United Nations Convention on Conditions for Registration of Ships, 1986, Art. 2 states:

> "'National maritime administration' means any State authority or agency which is established by the State of registration in accordance with its legislation and which, pursuant to that legislation, is responsible, *inter alia*, for the implementation of international agreements concerning maritime transport and for the application of rules and standards concerning ships under its jurisdiction and control."

Article 6 on identification and accountability is partly mandatory and partly recommendatory. This Article is of major importance, as under it it should be possible to identify the person or persons who can be held financially accountable for the management and operation of the ship.

The proposal on this Article which stated that states should not accept on their registers any ships owned by companies which issue bearer shares was not accepted. Obviously bearer shares have many advantages for the shipowner. Because of their transferability they make the identification of the shareholders effectively impossible, so in many cases the share capital of the owning company will be represented by bearer shares. If a country does not permit the capital of the companies to be represented by bearer shares companies can reach the same result in a different way. They will register the shares in the name of an offshore company, whose capital is itself represented by bearer shares. As a result for beneficial interests the company will have the advantage of being impenetrable. For instance, in Liberian corporations there is no register of directors or officers. Thus, if a third party wants to find out the identity of those persons it is not possible, because the only way to ascertain the identity of those persons would be by inspection of the corporate books of the company, and a third party does not have access to these books.

(a) The genuine link

An attempt was made to insist on the genuine link through strong provisions covering both ownership and management. These provisions were going to constitute the frame of a genuine link. This has been avoided. On ownership Article 8 contains both mandatory and recommendatory paragraphs. Under Article 8 the convention provides that a flag state, in its laws and regulations, "shall include appropriate provisions for participation by that State or its nationals as owners of ships flying its flag or in the ownership of such ships and for the level of such participation". These laws and regulations need to be sufficient to permit the flag state effectively to exercise its jurisdiction and control over the ships flying its flag. Therefore this provision deals with both the genuine link and the responsibility aspects. Paragraph two includes the term "or in the ownership of such ships". When this term is considered it is possible to reach the conclusion that the convention covers countries like Greece where most Greek flag ships belong to companies which are incorporated abroad, but owned by Greek nationals.

The ownership condition in Article 7, a mandatory article, is optional. The conditions for national participation in vessel ownership and manning cover the most contentious issues during the whole process. After disagreements on this Article the developing countries made a suggestion. Under the suggestion the management article, Article 10, was to be obligatory. And either the article on ownership, Article 8, or parts of that on manning, Article 9, would apply. Therefore the conditions for national participation in vessel ownership and manning are an optional requirement. It is for the state of registry to choose whether to make national equity participation compulsory for vessel acceptance or to accept vessels on national manning. If the state desires it can impose two requirements at the same time. The economic link established between a ship and the flag state has to be transparent because the convention also has an article, Article 6, for the identification and accountability of the persons involved.

Article 7 leaves the decision on what is to be regarded as an appropriate level of participation entirely to the flag state to determine. In other words this Article confers on the state the right to decide which requirement suits its national interests and circumstances best. This is subject to an effective jurisdiction and control clause. After the establishment of certain minimum levels of participation by its nationals (either individuals or corporations) and permanent residents the flag state should be able to exercise effective jurisdiction and control over its shipping.

On manning of ships Article 9 has mandatory and recommendatory paragraphs. If the flag state has chosen not to make the relevant parts of Article 8 mandatory, then, according to Article 7, parts of Article 9 become mandatory. On manning, the convention provides that the state of registration "shall observe the principle that a satisfactory part of the complement consisting of officers and crew of ships flying its flag be nationals or persons domiciled or lawfully in permanent residence in that State". Examination of the terms used in paragraph 1 shows that it is not possible to describe this paragraph as restrictive. The term "observe the principle" is quite vague. The term "satisfactory" is not defined either. Therefore it is for the flag state to decide for itself what is a "satisfactory part of the complement". There is no minimum level for this satisfaction. Obviously this entails the conclusion that there is no possibility of legal challenge. The application of paragraph 1 is weakened by paragraph 2. The application of paragraph 1 is subject to certain considerations; the availability of qualified seafarers refers to low or no national participation among officers and crew, existing and legally enforceable multilateral or bilateral agreements like the EC and Scandinavian internal free labour market, and the sound and economically viable operation of the ships which refers to high crew costs. If the stated qualifications were misused the meaning of genuine link would be destroyed.

Under paragraph 3 the provisions can be implemented "on a ship, company or fleet basis". It means that the flag state has complete discretion whether it, in its national rules and regulations, will implement paragraph 1 for its whole fleet or on a ship-by-ship or company-by-company basis. If a state decides that a certain percentage of the crew shall be nationals or residents of, or domiciled in, that nation, the flag state also has full discretion whether this percentage is to apply for each single ship, each company or the whole merchant fleet under that flag. It is also possible to use this paragraph to destroy the whole intent of the provision. In Sturmey's view[7] these flexible provisions might be accepted as an incentive for responsible owners of open registry tonnage like the multi-national oil and mineral

7. S.G. Sturmey, "The United Nations Convention on Conditions for Registration of Ships" [1987] L.M.C.L.Q. 102.

companies. Such owners might try to satisfy the condition by promoting the education and training of enough nationals, in co-operation with the state of registration. But on the other hand the conditions might be satisfied in other ways which would not comply with the aim of the paragraph. Again in Sturmey's article[8] flexibility in Article 9 is assessed with regard to developing countries. In developing countries there is a shortage of qualified technical personnel for the high technology ships coming into service. In such circumstances national manning with qualified personnel could only be achieved by attracting technicians from land-based industries. Under such a practice, until enough technicians can be supplied, expatriate labour may be used.

Paragraph 4 provides that "the State of registration, in accordance with its laws and regulations, may allow persons of other nationalities to serve on board ships flying its flag in accordance with the relevant provisions of this Convention". This paragraph possibly means that those crew members who do not have to be nationals may be non-nationals. With regard to this paragraph one may conclude that the Convention makes it possible to employ non-nationals on board ships which are parties to the Convention.

If an open registry state decides to establish a genuine link through the manning provisions of Article 9 there would probably not be much change in that state, because the present open registries have the requirement under their laws that ships registered in them should employ a certain proportion of nationals subject to availability.

Paragraphs 4, 5 and 6 apply in all contracting states. Application will be in their respective mandatory or recommendatory way even if the ownership option is chosen. If a state decides to obey Article 8, some open registry countries include ownership requirements as well as manning requirements. For instance the Liberian Requirements for Vessel Registration and Recording of Instruments require that:

"with the exception of bareboat charter registration vessels may be registered (or re-registered) only in the ownership of a Liberian citizen or national. Corporations are nationals if formed in Liberia. The Deputy Commissioner may, however, waive the Liberian ownership requirement where (i) the vessel meets all other requirements for registration; (ii) it has been satisfactorily demonstrated that there is an absolute and genuine need for such waiver, and (iii) the owner of the vessel qualifies for, secures and maintains registration in the Republic of Liberia as a Foreign Maritime Trust or Corporation".

When the language of this legislation is considered it may be easier for a state like Liberia to choose to observe Article 8. And it will also be more convenient for the shipowner if this option is chosen.

Similarly under the Vanuatu Maritime Act the ownership requirement is not very different.[9]

8. *Ibid.*
9. The Vanuatu Maritime Act, Cap. 131 states:
"**Ownership Requirements**
A Vanuatu ship, in order to qualify as such, must be owned by a citizen or national of Vanuatu

The wording of these requirements may give the impression that it is not very easy to fulfil them. But in practice many shipowners set up two companies in order to fulfil these types of open registry ownership requirements. The incorporated company A in an open registry country owns the ship. Then another incorporated company, company B, in the same country owns all the shares in company A. And without considering your nationality you own all the shares in company A. For instance in Liberia the establishment of a Liberian corporation is accomplished through the offices of International Registries Inc., Reston, Virginia, or International Registries SA, Zurich, Switzerland. In Vanuatu the ownership requirement is fulfilled by establishing a Vanuatu international company such as a company designed for operations exclusively outside Vanuatu. On examination of the requirements under Articles 8 and 9 one might reach the conclusion that the provisions of these Articles are not new. These requirements already exist in the laws of some open registry countries.

(b) Civil and third party responsibility

Article 10 covers the role of the flag state with respect to the management of shipowning companies and ships. All contracting states have to comply with the mandatory paragraphs of this Article. It also has some recommendatory paragraphs. Under Article 10, paragraphs 1 and 2 of the Convention there is a requirement for one of three possible types of management representation in the flag state. These include: (a) the establishment of a shipowning (or subsidiary) company in the flag state—if a company is registered in the flag state and if there is no other requirement under the laws and regulations of that state then the company will be established there; (b) the establishment of a principal place of business in the flag state; or (c) the presence of a representative who is able to act legally on behalf of the shipowner. This possibility covers situations where the company is neither established nor has its principal place of business in the flag state. The representative referred to may be a juridical person. And such natural or juridical person "should be available for any legal process and to meet the shipowner's responsibilities in accordance with the laws and regulations of the State of registration". Again it is for the flag state to pass laws and regulations with regard to the Convention. When a state becomes a party to the Convention there will be a need for some laws and regulations on related topics. But the problem

---

which includes any corporation incorporated in Vanuatu, partnership or association of individuals. Strictly, therefore, all non-Vanuatu owners wishing to register a vessel under the Vanuatu flag must incorporate a Vanuatu company for that purpose. The nationality of the beneficial owners of the company is immaterial and if the company is an exempted company pursuant to the Companies Act of Vanuatu its records are not subject to public search. It is possible to overcome the Vanuatu ownership requirement by making application to the Commissioner or Deputy Commissioner for a waiver. The Act provides that, in exceptional circumstances, a waiver may be granted if it can be shown that the vessel meets all other requirements for registration and it is established that there is an absolute and genuine need for a waiver. The practice to date has been for a liberal interpretation of these provisions and waivers of ownership have been granted on practical or fiscal grounds."

after the enactment of these regulations is a test of their adequacy. Until they are tested in a court the flag state will be the only authority to judge their adequacy.

Another point about this Article is that the requirement of the location of management is optional and some open registry states provide that an agent or a locally incorporated legal entity is sufficient in meeting that requirement. This, as Tolofari points out, gives rise to the question how genuine is the representation and does the flag state really have the desire to control such a representative be it a person or a company?[10] If the location of effective management lies outside the jurisdiction and control of the flag state a link established on the basis of Article 10 is not genuine because the control which the state is expected to exercise under Article 10 is not effective. In other words, a nationally incorporated company or a local agent will not have ultimate control over the ships it represents. It would be established just to fulfil the national requirement under maritime law.

In paragraph 3 there is a recommendation about the security of seafarers' wages if the employers go bankrupt. Again it is for the flag state alone to decide how it will ensure that the financial obligations can be met. If a state's company laws, like those of Liberia, were to state that no company registered in the state need record its financial activities with any official state department except those required for initial registration what would be the effect of this requirement?

(c) Transparency

Article 11 provides that the register will include a great deal of information on the ship, its ownership and mortgages. Paragraph 1 also states that the establishment of a ship register must be in accordance with both national rules and this convention. If bareboat charter is permitted under national laws and regulations then it may be entered in the register. Therefore if the flag state does not have much interest in the financial aspects of the ship the legislation would reflect this view, and registration will not include any useful information concerning mortgages.

The aim of transparency is to find out the owner of the ship. Obviously even with developed legal and administrative systems it is difficult to establish ownership. Shipowners prefer to guard their privacy because of beneficial interests. Another problem which has to be considered under Article 11 is how to ensure that the flag state has been told the truth.

Paragraph 4 says that a state, before entering a ship in its register, "should assure itself that the previous registration, if any, is deleted". In other words a ship shall sail under the flag of one state only. Therefore the state into which a ship is sold or bareboat chartered is recommended to check that the right to fly the flag of the previous flag state has been revoked or suspended. But such a request will also bring some problems. As stated by Sturmey[11] when a state requests information on previous registration from the country concerned it may not be able to receive accurate information within a reasonable time. In this case the state has to go to the

10. See S.R. Tolofari, pp.334–336.
11. S.G. Sturmey, pp.107–108.

shipowner for information. And how is it possible to ascertain that the information given by the shipowner is accurate?

Under Article 6 it is not very clear whether the Article can easily achieve the transparency of ownership or not. If an open register deems that this provision is likely to destroy its financial position and attraction then it will not ratify the convention. Therefore the extent of control on the identification and accountability of the ownership will vary in different states in relation to their national interests.

When Articles 6 and 11 are examined together it becomes clear that the application of these Articles mainly depends on how effective or real the ownership genuine link requirement of Article 8 is in practice. Most of the provisions of the convention are subject to the escape clause, "in accordance with the laws and regulations of the flag state".[12] Under such a clause countries like Liberia and Panama can easily argue that the requirements under their maritime law on ownership are sufficient and therefore they do not need any further legislation. As states determine the extent to which they can truly observe the provisions on genuine link, under the lack of force in the provisions it is very difficult to enforce compliance with the requirements under the convention.

(d) Protection of the interests of labour-supplying countries

Article 14 aims to protect the interests of seafarers against exploitation. However, it does not place any obligation on shipowners. It gives the duty to the labour-supplying country[13] and the state to make sure that contracts and agreements entered into by shipowners/operators and seamen's unions or labour-supplying agencies are in the best interests of seafarers. In this context it is appropriate to refer back to Article 9(2)(b) concerning multilateral and bilateral agreements. The manning rules under Article 9 will not have much effect on labour-supplying countries as they cover quite vague terms. Article 14 should be examined together with the provision in Article 10 on an appropriate mechanism to cover wages and related moneys owed to seafarers. Paragraph 3 of Article 10 recommends that labour-supplying countries should ensure that the terms and conditions of employment do not lead to exploitation. And the flag states should secure seafarers' wages in some way. Obviously when labour has moved outside the jurisdiction the flag state will not have any information about the terms of the agreement and the labour-supplying state has no power of enforcement. The other possible problem is what would be the effect of this Article in a labour-supplying country which is not party to the Convention? Or what happens if a contracting party argues that the contract of employment is in the national interest?

12. See, the United Nations Convention on Conditions for Registration of Ships 1986, Art. 6, para. 3.

13. The practice of employing foreign labour is very old. During the 16th century Filipino crew were employed in the Spanish galleons and on other Spanish ships. In Britain the employment of foreign crews started in 1660 with permission of the law which stated that only three-quarters of a ship's complement had to be subjects of the King.

In Resolution 1 annexed to the Convention labour-supplying countries are encouraged to regulate crewing agencies in order to prevent abuses and contribute to the welfare of their nationals working on foreign vessels. In paragraph 2 co-operation among labour-supplying countries is recommended. In paragraph 3 various UN and other international bodies are called upon to help labour-supplying countries. This assistance also includes measures to attract foreign investment. In a sense it is possible to see such a study as a way of making these registries more "open". In practice some countries, especially the traditional maritime registers which have recognised that any attempt to abolish flags of convenience would not be successful, responded to the demands of shipowners to register their ships in lower-cost jurisdictions. For instance,[14] in France since 1986, French-owned solid and liquid bulk carriers have been registrable in Kerguelen Island, part of French Antarctic territory. Under this concession only 25 per cent of the crew has to be French nationals. As explained above, in 1987 the Norwegian Government established the Norwegian International Ship Register (NIS) based at Bergen. Denmark and Germany established their own international registers. Similarly Portugal established on offshore register on the island of Madeira (MAR). In 1990 Luxembourg established the Luxembourg International Ship Register.

Resolution 1 has led to four collective agreements:

(a) The "Nordic" agreement in respect of seamen recruited in the Scandinavian countries;
(b) The "Filipino and Indian" agreement concluded with the relevant unions in India and the Philippines, which facilitates the employment of nationals of those countries;
(c) The "Pakistani and Indonesian" agreement with the Pakistan Seamen's Union and the Indonesian Seafarers' Union;
(d) The "Polish" agreement with the Polish Seamen's Union.

(e) Measures to minimise adverse economic effects

The proposal on Article 15 was made by Panama. This Article was originally proposed as a part of Article 14, concerning labour-supplying countries. Article 15 urges the implementation of measures spelt out in Resolution 2. Resolution 2 contains appeals to various UN and other international bodies to provide assistance to minimise adverse economic effects resulting from the Convention. Resolution 2 is similar to Resolution 1, which covers problems of the labour-supplying countries. When the Resolutions are compared Resolution 2 is general and originally intended to help open registry countries. Both Resolutions are recommendatory and do not place any obligations on the contracting parties.

If an open registry country implements this Convention obviously it has to cover certain costs like the establishment of an adequate maritime administration. These

14. See N.P. Ready, pp.33–36.

types of economic effects are covered by Article 15. The need for this should be considered with the object of the Convention. As is common knowledge, the Convention is the outcome of a long debate which started in UNCTAD as a move to eliminate flag-of-convenience shipping. But the Convention will not have this effect. Nothing would stop the creation of new open registries if the demand existed. Therefore the main function of the Convention can be seen as granting international recognition to all registries which fulfil its conditions. As a result there is little need for Article 15 and Resolution 2 and also in practice this Article can be effective only if there is a fund that can provide such economic assistance.

BAREBOAT CHARTER REGISTRATION

In a bareboat charter a vessel registered in state A is chartered for a fixed period to nationals of state B. During the charter period the charterers operate the vessel under the flag of state B. During the period of charter the primary registration in state A is cancelled or suspended. At the end of the charter period the registration in state A becomes fully effective. In such a case state A is called the "flagging-out" state and state B is called the "flagging-in" state. The reason behind flagging out is the transfer of a ship from a national flag to a relatively low-cost flag. Flagging-out is mainly caused by the attractions of minimising costs in a relatively lower-cost country.

In relation to the changing economic costs governments and trade unions which in the past argued strongly against the practice of flagging-out, especially to open registries, changed their attitudes.[15] The change in strategies was expressed in ITF's Bulletin in 1988. ITF stated that "amongst its strategies for action ITF will be looking at ways of relaxing rules on existing registers or creating offshore or regional registers under conditions acceptable to ITF unions".[16] In the same year the ITF Maritime Policy Committee agreed to treat all former open registry vessels with temporary Philippine registration as flag-of-convenience ships.[17]

States approach bareboat charter registration in different ways. In some states the primary registration remains effective during bareboat charter registration in their countries. In others the primary registration has to be cancelled during the bareboat charter registration period. The important point is that bareboat charter registration can apply only if the laws of flagging-out and flagging-in countries do not conflict. The Bahamas and Panama are the two examples of open registry countries which permit bareboat charter registration with some differences. Under the Bahamas Merchant Shipping Act, section 4(4), if a foreign registered ship is bareboat chartered to any citizen of the Bahamas or body corporate established

15. For instance, in 1986 ITF made a proposal to raise its minimum wage of $821 per month for a crewman by 3.4% on more than 2,000 open registry vessels. But due to the difficult situation of the shipping industry ITF backed down on its proposal: *Lloyd's List*, 25 November 1986.

16. International Transport Workers Federation (1988), *Seafarers' Bulletin*, 3, p.17.

17. ITF *Seafarers' Bulletin* No. 3 1988.

under the laws of the Bahamas it can be registered as a Bahamian ship for the period of the bareboat charter and the registrar shall notify the proper officer of that foreign country of such registration as a Bahamian ship. Under section 25A, if a Bahamian ship is bareboat chartered to any citizen of a foreign country or to any body corporate established under the laws of a foreign country, the ship may be registered under the law of that foreign country. For the period of such registration the Bahamian registrar shall suspend the registry of that ship.[18] Similarly under Panamanian law foreign-flag vessels chartered for not less than two years may be registered in Panama for renewable two-year periods. During those periods primary registration is not abandoned. The difference under Panama law is that during registration in Panama primary registration remains effective.

Certain aspects of bareboat registration are explained in Articles 11 and 12 of the 1986 UN Convention.[19] Under Article 12, paragraph 1, a bareboat chartered-in vessel may fly the flag of the nation into which it has been bareboat chartered. But that is subject to suspension of the right to fly the previous flag as set out in Article 11, paragraph 5.[20] As a ship may sail under the flag of one state only, the state into which a ship is bareboat chartered is recommended to check that the right to fly the flag of the previous flag state has been rescinded or suspended.

By Article 12, paragraph 2, the Convention applies to a vessel bareboat chartered-in even if it was previously registered in a state not a party to the Convention. Paragraph 3 considers the bareboat charterer as owner. In other words the bareboat charterer assumes the same duties as an owner for the other rules of the Convention. With regard to jurisdiction paragraph 4 recommends that vessels bareboat chartered-in become subject to the new flag's jurisdiction and control. In other words a state should ensure that a ship bareboat chartered-in and flying its flag will be subject to its full jurisdiction and control.

Paragraph 5 states that "the State where the bareboat chartered-in ship is registered shall ensure that the former flag State is notified of the deletion of the

18. S. 4(4) of the Bahamas Merchant Shipping Act provides that a ship registered under the law of a foreign country shall, on application, if she is bareboat chartered to any citizen of the Bahamas or to any body corporate established under the laws of the Bahamas, in addition to being registered under the law of that foreign county, be registered for the period of the bareboat charter as a Bahamian ship and the registrar shall notify the proper officer of that foreign country of such registration as a Bahamian ship. S. 25A of the Act provides that a Bahamian ship bareboat chartered to any citizen of a foreign country, or to any body corporate established under the laws of a foreign country, may, on application to the proper office of that country, be registered under the law of that foreign country. Upon notification from the proper officer of the foreign country, the Bahamian registrar shall for the period of such registration suspend the certificate of registry of that ship and shall notify the proper officer of that suspension and of any mortgage instrument which is recorded in the register in respect of that ship. See N.P. Ready, pp.27–30.

19. A proposal that states should refrain from bareboat chartering-in from and bareboat chartering-out to open registry countries was not adopted.

20. The United Nations Convention on Conditions for Registration of Ships 1986, Art. 11, para. 5 states:

> "In the case of a ship bareboat chartered-in a State should assure itself that the right to fly the flag of the former State is suspended. Such registration shall be effected on production of evidence, indicating suspension of previous registration as regards the nationality of the ship under the former flag State and indicating particulars of any registered encumbrances."

registration of the bareboat chartered ship". This paragraph aims to ensure that no vessel may be considered as a stateless ship at the end of a bareboat charterparty.

If a ship is bareboat chartered from a state which is a party to the Convention into a non-contracting state, the Convention will not apply. On the other hand if a ship from a non-contracting state is bareboat chartered into a contracting state, the Convention will apply to that ship.

The examined provisions on bareboat charter registration demonstrate that there is no requirement for notification to the state of registration of the rate of charter hire. Article 12, para. 6 states that "all terms and conditions, other than those specified in this article relating to the relationship of the parties to a bareboat charter are left to the contractual disposal of those parties". As stated by Sturmey,[21] it is possible to establish a ship operating company in a country with national participation in the capital. This company could operate as the owner of the one or more bareboat chartered ships. The charter rate would be fixed at a level which provided no profit for actual operation. Therefore the genuine link requirement would be met with a national company involvement. But the only beneficiary would be the real owner of the ships.

21. S.G. Sturmey, p.108.

CHAPTER 6

NATIONALITY

Traditionally, jurisdiction over a ship has been connected with its nationality. Under international law, the concept of nationality comprises the rights and duties of a state *vis-à-vis* its ships. The nationality of a ship refers to the state which has authority over and responsibility for it. For instance, if a ship is classified as a British ship the British Government is able to exercise a large number of powers in relation to it and the application of the 1995 Merchant Shipping Act will subject such vessels to British safety and supervisory requirements. The exercise of these powers is not limited to British waters; they are exercisable in respect of any vessel flying the British flag, bearing in mind that there is increasing recognition of an element of port state control. The legislation of the state which governs the port in which a vessel is operating may retain some regulatory or supervisory powers regardless of the flag of the vessel.

The conditions governing nationality are usually formulated as requirements for registration. The concept of registration, though closely related to nationality, is not necessarily synonymous with it. The nationality of a ship refers to the state which has authority and responsibility over, and in respect of, it, with the flag symbolising the state of nationality in each case. Registration is the act by which this nationality is conferred on the ship. By registration nationality and collateral rights and duties are conferred on it. National shipping policy decides which conditions a ship has to meet in order to be granted nationality.

As a rule only the owner may apply for registration and the ship will be registered in his name. When registration by the bareboat charterer is allowed the ship is registered in the name of the bareboat charterer. When a ship is registered, the registering country assumes authority over the ship and undertakes the national and international responsibilities of a flag state in relation to that ship. Registration of a ship should be cancelled if the ship no longer meets the requirements for registration.

To protect the holders of registered rights against double registration two provisions have to be made: first, the registration of rights over a foreign ship or of rights which are already registered abroad should be prohibited; secondly, registration abroad of rights on a national ship should be invalid as regards the holders of rights in the national state.[1]

REGISTRATION OF SHIPS IN THE UNITED KINGDOM

Until 1988, the qualifications for ownership of a British registered ship were set out in Part I of the Merchant Shipping Act 1894. That Act, prior to its amendment by the Merchant Shipping Act 1988, stated that every ship owned by a British national had to be registered as a British ship. Therefore if a ship or part of it was owned by an alien it could not be registered as a British ship. The 1988 Act abolished the system of compulsory registration under the British flag for vessels wholly owned by British subjects. Under the 1988 Act a ship qualified for British registry may be registered outside the United Kingdom and the Overseas Territories. Under this legislation ships are not obliged to register; they are entitled to register.

In most cases ships are owned by companies. In a company the individual nationalities of relevant persons may be very diverse. A company may be incorporated in England, have English, Italian and German directors, employees from the Philippines, Liberia and France and shareholders from some other countries. In such a case the Merchant Shipping Act 1988 does not take account of the nationalities of individuals. If a company is registered in the United Kingdom and has its place of business here it is regarded as a United Kingdom company. With regard to the nature of shipping business it is difficult to identify the principal place of business of a company. In the *Polzeath*[2] a ship was owned by a company, the Lynn and Hamburg Steamship Co. Ltd., and incorporated under the English Companies Act. But it was controlled from Hamburg by the chairman of directors who was a naturalised British subject of German origin, the majority shareholder and resided there. Apart from him there were two British and one German directors. The chairman had the casting vote. Proceedings were instituted for forfeiture of the ship. The Court of Appeal held that the principal place of business of the Lynn and Hamburg Steamship Co. was not in England. Therefore the ship was forfeit to the crown. If a foreign person has a share in a British shipping company this does not disqualify the company because the law recognises the company independently of its individual shareholders. But the result is different if that person becomes the legal or beneficial owner of a British ship of his own free will. In such case he forfeits his interest.

The importance of the nationality of a ship in litigation became clear in *Oteri* v. *R*. In this case two naturalised Australian citizens were charged with stealing crayfish pots and tackle on the vessel *Providence*. The vessel was on the high seas

1. United Nations Guidelines for Maritime Legislation, Economic and Social Commission for Asia and the Pacific, Vol. I, 1991, p.34.
2. [1916] P. 241 85; L.J.P. 241; 116 L.T. 370; 32 T.L.R. 647; 13 Asp. M.L.C. 595 (CA).

beyond Australia's territorial waters. She was a fishing vessel and was not registered in any Australian or United Kingdom port. The Australian district court claimed that it had no jurisdiction to hear the charges. Although it was not registered under the Merchant Shipping Act the Privy Council accepted the *Providence* as a British ship for the purposes of jurisdiction regarding offences committed on board because under the law of the sea a ship owned by British subjects was a British ship and section 1 of the British Nationality Act 1948 designated naturalised Australians as British subjects. Similarly in *R.* v. *Anderson* an American was convicted of the manslaughter of a fellow seaman aboard a British ship. The ship was 300 yards from the shore, in the river Garonne in France. Anderson appealed on the basis that the English court had no jurisdiction over him. The court held that the jurisdiction of British law extends to the protection of British vessels, though in ports belonging to another country.[3] In *Chung Chi Cheung* v. *The King* a British subject cabin boy killed the captain, who was a British subject in the service of the Republic of China. The vessel was an armed Chinese public vessel and it was in Hong Kong territorial waters. The importance of this decision was the rejection of the "floating island" theory by the Privy Council. Under this theory a public ship of a state was accepted as a part of the territory of that nation.[4]

Under the 1988 Act, a ship more than 24 metres in length whose owners are not resident in the United Kingdom can only be registered in the British register if a representative person is appointed. A representative person may be a person resident here or a company which is incorporated and having its place of business here. The representative person receives legal documents relating to the enforcement of the public shipping law under the Merchant Shipping Acts in respect of the registered ship. A fishing vessel was eligible to be registered only if the vessel was British owned, was managed and its operations directed and controlled from within the UK and any charterer, manager or operator of the vessel was a qualified person or company and domiciled in the UK or a company incorporated in the UK and having its principal place of business here. These provisions gave rise to the *Factortame*[5] litigation which meant that the general entitlement to registration in

3. (1868) 11 Cox CC 198, 205 (C.C.R.). The court held that "there is no doubt that the place where the offence was committed within the territory of France, and that the prisoner was therefore subject to the laws of France, which the local authorities of that realm might have enforced if so minded; but at the same time, in point of law, the offence was also committed within British territory, for the prisoner was a seaman on board a merchant vessel under the British flag ...". This principle was expressed in Art. 27(1) of the UN Convention on the Law of the Sea 1982: "the criminal jurisdiction of the coastal State should not be exercised on board a foreign ship passing through the territorial sea to arrest any person or to conduct any investigation in connection with any crime committed on board the ship during its passage".

4. [1939] A.C. 160 (PC). The court held that "on the question of jurisdiction two theories have found favour ... One is that a public ship of a nation for all purposes either is, or is to be treated by other nations as, part of the territory of the nation to which she belongs ... The other theory is that a public ship in foreign waters is not, and is not treated as, territory of her own nation ... Their Lordships entertain no doubt that the latter is the correct conclusion."

5. *R.* v. *Secretary of State for Transport, ex parte Factortame Limited and Others (No 3)* [1991] 2 Lloyd's Rep. 648 (E.C.J.). See Christopher Hill, p.7.

the British register had to be reformulated and gave rise to the 1993 Merchant Shipping (Registration, Etc) Act.

By 1993 there existed three shipping registers. Under the 1894 MSA, Part I, the register was administered by HM Customs and Excise at individual ports all around the country. The Merchant Shipping Act 1983 provided for the registration of small ships. The reasons behind this registration was to remove many pleasure craft from the full register. The Small Ships Register was administered by the Royal Yachting Association under the Merchant Shipping (Small Ships Register) Regulations 1983. Under the amendments introduced by the Merchant Shipping Act 1988, Part II, the old fishing boat register under Part IV of the 1894 Act moved to a new centralised register of British fishing vessels. By section 1 of the 1993 Act a single central register of British ships is maintained in Cardiff by a Registrar General of Shipping and Seamen.

The 1993 Act preserved the voluntary registration regime introduced by the Merchant Shipping Act 1988. With the enactment of the Merchant Shipping Act 1995 all these provisions now form Part II of this Act.[6]

The Register of British Ships divided into four parts and corresponds to that under earlier legislation.[7] Part I covers vessels other than fishing vessels, small ships and bareboat charter ships which must be registered in the United Kingdom. Part II addresses fishing vessels and Part III small ships. Part IV covers ships registered in a country other than the United Kingdom and demise chartered to British charterers.[8]

The requirements under the 1995 Act are designed to ensure that a ship has a British connection. Therefore a ship is entitled to be registered only if it is owned by persons qualified to own British ships. The vessel and her owners must also meet such other conditions as may be imposed by the Registration Regulations issued under section 10. These Regulations ensure that qualified owners have a sufficient connection with Britain.

The term "sufficient connection" in this section is related to the principle "genuine link" stated in Article 4 of the Geneva Convention on the High Seas 1958 and Article 92 of the United Nations Convention on the Law of the Sea. As stated above, the need for a genuine link between the ship and the state of registry has been accepted in international law. Sections 3–5 and 14 of the Merchant Shipping Act 1988 set out requirements which related to the identity of the shipowner and the nature of its presence in the UK. Section 3 of the 1993 Merchant Shipping Act reaffirmed the 1988 provisions. Similarly the 1995 Merchant Shipping Act restated the need for sufficient connection. In this respect starting with the 1988 Merchant Shipping Act the UK gave more importance to the "genuine link"

6. The key change under the 1993 Act, apart from centralisation, was the transfer of all detailed provisions relating to merchant ship registration from the primary statute to regulations. Regulations made under this Act continue in force under the 1995 Merchant Shipping Act.

7. This Part reflects Part I of the 1894 Merchant Shipping Act.

8. This new category of registrable vessels was introduced by the section 7 of the 1993 Act.

requirement as discussed in international law for a very long time and believed that such provisions satisfied the international requirement of "genuine link".

Section 17 of the Act relates to ships on bareboat charter. This section re-enacts section 7 of the 1993 Act and sets out three requirements: the vessel should be chartered on bareboat charter terms to a charterer who is qualified to own British ships; the vessel should already be registered abroad; it must have been chartered so as to fulfil the requisite conditions for entitlement to registration under section 9 of the Act and under the Registration Regulations. For British public law purposes British regulations will apply to the bareboat chartered vessel. The ship will be considered a British ship and will be entitled to fly the British flag. The whole regulatory force of the Merchant Shipping Acts applying to British ships or British registered ships will be applied in the same manner to the bareboat chartered ships. The only limits would be British private law provisions and certain provisions which may be excluded from application or which may become applicable by orders in council that may be issued in due course. Private law considerations in respect of the vessel continue to be determined according to the law of the place in which the vessel was initially registered.

The 1995 Act separates the public law and private law provisions but despite this separation the private law role of registration has not been clarified, since under the 1995 Act there is no obligation to register. A ship which fails to register will not be entitled to the benefits of British ships and any of the statutory private law powers relating to British ships. There will not be any consequence other than this.[9]

CONDITIONS FOR SPECIAL CATEGORIES OF SHIPS

Only ships are registrable. Under section 742 of the Merchant Shipping Act 1894 "every description of vessel used in navigation not propelled by oars" is included. Therefore a vessel which is not used in navigation is not accepted as a ship and not registrable.

The rules for granting nationality cannot be uniformly applied to all ships. Certain ships may be exempted or may be subjected to different rules. In accordance with customary international law, state-owned ships have the nationality of their state. Under international law inland vessels are not required to fly a flag. There may be requirements for inland vessels under regional or bilateral agreements. But there is a need for regulations on the nationality of inland vessels. In most of the states there are no separate nationality requirements for pleasure craft. Under customary international law it is assumed that if a pleasure craft meets the nationality requirements for sea-going ships it automatically comes under the state's jurisdiction. If such an assumption is not supported by documents issued by the flag state another state may refuse to accept the nationality of the ship.

9. D.C. Jackson, *Enforcement of Maritime Claims* (2nd ed., 1996), p.479.

(a) Government-owned tonnage

Government-owned vessels are distinct from the naval vessels which constitute part of the armed forces of the state. Obviously a government-owned merchant fleet may be operated commercially. But if this fleet were to claim the privilege of sovereign immunity against the merchant fleets of other countries it would cause discrimination and raise complicated issues. Article 9 of the Geneva Convention on the High Seas 1958 provides that "ships owned or operated by a State and used only on government non-commercial service shall, on the high seas, have complete immunity from the jurisdiction of any State other than the flag State". This principle, which was re-stated in Article 96 of the 1982 United Nations Convention on the Law of the Sea, has an economic aspect too. If immunity from taxation were granted to state-owned vessels even when they were commercially operated this practice would cause a serious economic threat to the operation of privately owned vessels.

(b) International organisations

If a ship is owned by an international organisation which has a maritime flag of its own which is flown by the tonnage registered by it, state responsibility for the operation of such tonnage falls on the international organisation. In such a case national ownership ceases to exist. If the international organisation cannot grant its flag and uses the flags of the member states, the concept of national ownership is not distributed.

The development of international law on this issue is related to international case law. In 1954 the United Nations Korean Reconstruction Agency (UNKRA) had 10 wooden fishing trawlers. The trawlers were constructed in Hong Kong as a contribution to the reconstruction of the fishing industry of the Republic of Korea. In 1955 the vessels were to be delivered to Pusan, in Korea. The problem was the flag and registration. By reason of the vessels' ownership British registration was not available. As the vessels were still owned by UNKRA Korean registry was not available either. The transfer of ownership to Korean owners was not practically possible. Therefore it was decided that the United Nations should itself undertake the function of registration and should navigate the vessels to Pusan under the United Nations flag. Similarly in 1956 the vessels of the United Nations Emergency Force displayed the United Nations flag while in Egyptian ports. Vessels on charter to the UN Emergency Force (UNEF) have either displayed the United Nations flag alone or together with their respective flags.[10] The United

10. In 1957 an agreement on the status of UNEF was signed between the UN and the Government of Egypt, para. 20 of which states: "The Egyptian Government recognises the right of the Force to display within Egyptian territory the United Nations flag on its headquarters, camps, posts or other premises, vehicles, vessels and otherwise decided by the Commander. Other flags or pennants may be displayed only in exceptional cases in accordance with conditions prescribed by the Commander. Sympathetic consideration will be given to observations or requests of the Egyptian authorities concerning this last-mentioned matter": 8 February 1957 (A/3526).

Nations flag indicates that ships on assignment to UNEF are entitled to the privileges and immunities provided for in the relevant international instruments. In the United Nations Suez Canal Clearance Operation (UNSCO) the nationalities of the vessels involved were different. As in the case of UNEF the use of the United Nations flag was assembled by UNESCO.[11] Under the agreement between the United Nations and the Government of Egypt UNESCO vessels were entitled to the protection of the United Nations Organisation.

The right of an international organisation to fly its flag on a ship assigned for its purposes was recognised in Article 7 of the Geneva Convention on the High Seas 1958. The Article provided that the Convention's earlier articles on nationality "do not prejudice the question of ships employed on the official service of an inter-governmental organisation flying the flag of the organisation". In similar terms the United Nations Convention 1982, Article 93, provides that "the preceding articles do not prejudice the question of ships employed on the official service of the United Nations, its specialised agencies or the International Atomic Energy Agency, flying the flag of the organisation".

THE MECHANICS OF REGISTRATION

The purpose of the nationality document is to prove the nationality of the ship and consequently its right to fly a flag. As a principle the document is issued to registered ships. On registration a Certificate of Registration is issued to the ship. This certificate goes with the vessel and it must not be separated from it. The certificate is used in navigation and must be produced by the ship's master on demand. It is also an instrument to establish the identity of the shipowner.

The Certificate of Registry is *prima facie* evidence of title to ownership. However, if it were subsequently proved that the owner was not qualified to own a British vessel then, despite the certificate, the ship would not be legally British. *R.* v. *Bjornsen* is a good example of this practice. In this case a murder took place on a ship on the high seas. The ship flew the British flag. The port of the ship's registration was London and London was also the place of business. Therefore it was possible to assume that the ship was British. But overriding proof was produced that the owner was an alien, non-British, that the ship was foreign built and the officers and crew, including the accused man, were aliens. It was held that although the registration was British, as shown in the registration certificate, this was *prima facie* evidence only of the ship's nationality and of the apparent British nationality of the owner. This presumption could be displaced by admissible evidence showing that the owner was in fact a foreigner.[12]

11. The agreement of 8 January 1957 (A/3492) between the United Nations and the Government of Egypt stated that "the undertaking (UNSCO) would be under obligation to discharge their functions and regulate their conduct solely in the interest of the United Nations. In keeping with the United Nations responsibilities, the vessels would fly the flag of the United Nations in place of their national flags . . .".

12. (1865) 12 L.T. 473 (C.C.R.).

On examination of registration and flag it becomes evident that the flag a vessel flies can be recorded only as *prima facie* evidence, on the high seas, that the nationality of the vessel flying it corresponds to that of the flag. But on the other hand it is possible to reach the same conclusion for registration as well. For instance in *Rainbow Line Inc.* v. *M/V Tequlia* the Second Circuit said that "the only British contacts are the registration and nominal ownership of the vessel at the time the charterparty was breached. These British interests, however, are overshadowed by the fact that the real owners of the vessel, namely Windjammer and Burke, were American, and it is well settled that the courts will look through the façade of foreign registration and incorporation to the American ownership behind it".[13]

In *Chartered Mercantile Bank of India* v. *Netherlands Navigation Co. Ltd.*[14] an English company incorporated a Dutch company. It owned a vessel registered in the Netherlands under the Dutch flag. As the English company had control over the vessel it was held to be the effective owner. The ship herself was considered to be English and therefore subject to English law. The court held it was "absurd to suppose that the mere fact of carrying the Dutch flag makes her a Dutch ship. Pirates carried the flag of every nation but they were hanged by every nation notwithstanding".[15] The court then went on to say:

"Therefore the question whether a ship is bound to be registered according to English law depends upon whether she is an English ship. If she belongs absolutely and entirely to English owners, she is an English ship before she is registered, . . . whether she is registered or not . . . But it has always been held, and it seems to me to be the law, that she cannot evade liability if she is a British ship by the mere fact of her owner omitting to register her as a British ship . . . It seems to me that the nationality of a ship depends solely upon her ownership, and as to her liability it does not matter about her being registered . . . [T]he defendants cannot escape liability by saying their ship was not registered as a British ship, but she was registered as a Dutch ship. She was nevertheless an English ship, and the defendants are liable according to English law."[16]

On the contrary, in *Pereira* v. *Utah Transport* the court held that "the law of the flag governs all disputes".[17] Under the use of flags of convenience and bareboat charters the law of the flag becomes only one indicator or contact among many. Although it is possible to prove the contrary the real proof of a ship's nationality lies in its registration in the flag state. With regard to the international character of shipping business even the place of business of the owners of a vessel has to be considered to determine the nationality of the ship.

Section 13 of the Merchant Shipping Act 1995 says that "the certificate of registration of a British ship shall be used only for the lawful navigation of the ship, and shall not be subject to detention to secure any private right or claim". Any

13. 480 F. 2d 1024; 1973 A.M.C. 1431 (2d Cir. 1973).
14. (1883) 10 Q.B.D. 521 (CA).
15. *Ibid.*, Brett L.J., p.535.
16. *Ibid.*, p.536.
17. 764 F.2d 686, 1985 A.M.C. 2743 (9th Cir. 1985).

changes in the details given in the certificate must be promptly notified to the Registrar. Section 14(2) makes it an offence to fail to inform the Registrar of any change which may affect the British character of the vessel. Depending upon the circumstances of each individual case notification of any change shall take place as soon as practicable after the change occurs. "Any person who intentionally alters, suppresses, conceals or destroys a document which contains information relating to the British connection of a ship and which he has been required to produce to the registrar in pursuance of registration shall be guilty of an offence."[18]

18. Merchant Shipping Act 1995, s.14(3).

CHAPTER 7

DEFINITION OF OWNERSHIP UNDER CONTRACTS OF CARRIAGE BY SEA

IN GENERAL

When a shipowner directly or indirectly undertakes to carry goods by sea or to provide a vessel for the carriage of goods by sea the arrangement is known as a contract of affreightment. These contracts may be in different forms. The contract may be for the use of a whole vessel or the carrier may enter into a number of separate contracts with various cargo owners.

A charterparty is a contract for the use of an entire vessel, either for a particular voyage or a specified period of time. A charterparty is used either by liner operators or by shippers of large quantities of cargo. The charterparty itself constitutes the contract. Sometimes a shipowner employs his vessel in the liner trade and offers a carrying service to anyone who wishes to ship cargo. In such cases, where an entire vessel is not required, the contract will be on bill of lading terms. The bill of lading evidences the contract of carriage but itself does not constitute the contract.

Charterparties do not have any statutory form. They are just subject to supply and demand laws. In practice standard forms are widely used. Depending on the agreement between the parties existing clauses are amended and extra clauses are added. Many trades like grain, coal, timber and stone have their own forms of charterparty. These forms are negotiated between the organisation representing the merchants in that trade and others on behalf of shipowners. Much of charterparty law is a matter of interpreting the language of the particular charterparty involved. Considering the widespread use of standard form clauses it is of overriding importance that the meaning or legal effect of such clauses be certain and well understood.

In general charterparties for the carriage of goods may be divided into three categories: demise, voyage and time charterparties. It is also possible to classify them as charters by demise and charters not by way of demise. Under a demise charterparty the charterer hires the vessel itself from the shipowner but the shipowner does not provide any other services. In other words, demise charters are for the hire of the ship alone and the services of the crew are not included. This type of charterparty is relatively uncommon and fundamentally different from other types of charterparties. If the charterparty is not by way of demise the owner provides the ship to carry cargo but also provides master and crew. If the duration of the charter is on a voyage basis it is called a voyage charter. If it is based on a period of time it is called a time charter.

TYPES OF CHARTERPARTIES

(a) Voyage charterparty

The voyage charterparty is one of the oldest forms of contract for the carriage of goods. The shipowner undertakes to transport a specified cargo between designated ports. The remuneration, called freight, is calculated upon the quantity of cargo, not according to the time taken. Under a voyage charter all risks and delays at sea fall on the shipowner and some of those in port may also do so.

The parties to a voyage charterparty are called the carrier and the owners. The carrier undertakes the transportation. The owners do not necessarily own the vessel. They may have chartered the vessel or control her as managers and are therefore called chartered owners or disponent owners. The parties on the cargo side are called charterers. The charterer may not necessarily own the cargo. For example the charterers may be a computer company in London and may have brought the cargo from suppliers in Japan for delivery to Turkey. Each of these companies involved on the cargo side is often established as a separate company and regarded as a different legal entity. Servants and independent contractors who assist the parties in their performance of the contract are not mentioned in the charterparty. The only exception to this rule is the ship's agents and the brokers.

The charterer and the carrier perform the contract differently. The charterer may assign his rights to others. The carrier must accept performance from the new charterer. But the first charterer remains responsible. On the other hand the carrier's performance is deemed to be personal. Without a special arrangement the charterer does not have to accept performance by others.

Under a voyage charterparty apart from freight nothing is payable for the voyage. The shipowner provides the fuel at his own expense. The basis on which the freight is calculated is irrespective of the time the voyage actually takes. The risk of delay is on the shipowner. Therefore the charterer or the cargo owner determines the speed of loading and discharge. If the voyage takes longer than expected the shipowner cannot claim any extra freight from the charterer to compensate for the delay. He also cannot make use of the vessel for that period to earn freight elsewhere. If the voyage takes less time than expected the shipowner gains advantages because the freight is not reduced and it is possible to employ the vessel elsewhere. This is the fundamental distinction between voyage and time charterparties. In time charterparties it is the charterer rather than the shipowner who bears the risk of delay. Under a voyage charterparty the charterer controls the loading and discharging process and how much cargo is loaded. Therefore all voyage charterparties have clauses stipulating how much cargo the charterer must load and how quickly it must be loaded and discharged. All voyage charterparties have laytime and demurrage clauses. The aim of these clauses is to encourage cargo handling to be performed reasonably quickly.

In order to safeguard the freight earned by the owner the charterer will be under an obligation to load a full and complete cargo because the amount of freight is usually dependent on the amount of cargo loaded and dead freight is payable if the charterer does not load sufficient cargo. Demurrage and freight provisions are also contained in bills of lading contracts. When a voyage charterer loads his own cargo and bills of lading are issued to him the purchaser of the cargo will usually become liable for freight and demurrage. For instance under a cesser clause the shipowner is required to proceed against the receiver of the cargo, rather than the charterer, for demurrage incurred at discharge.

(b) Time charterparty

In a time charter the shipowner places his vessel at the charterer's disposal for a certain period of time. And in return he is paid hire at regular intervals. The shipowner is under an obligation to provide a seaworthy vessel as described in the charterparty, and engage and pay for the crew. The charterer pays for the hire and for bunker oil. In exchange for hire payments the charterer has the right to use the vessel wherever and for whatever cargoes he chooses. However, this freedom may be limited by the shipowner. Obviously under a time charter the charterer has more flexibility than when taking vessels on a single voyage charter but he also assumes many of the risks of delay. Thus, the charterer will assume the risks of bad weather at sea and in port and also be responsible for delays caused during loading and discharge, provided these are not caused by the vessel.

Under a time charterparty hire is calculated not on the amount of cargo carried but on the period for which the vessel is used, irrespective of how many voyages are completed and how much cargo is carried. Hence, the charterer bears the risk of delay. As the hire is calculated on a time basis laytime and demurrage clauses are not needed.

Under most time charters a clause entitles the charterer to have full use of the vessel within the limits stipulated in the charter. And the charterer undertakes that the master will comply with the charterer's orders and instructions.

In a time charterparty the ship is run by the shipowner. The navigation, management and maintenance are under the shipowner's control. But commercially the ship is operated under the charterer's orders. Under a time charterparty the shipowner may set geographical trading limits and specify the cargoes to be loaded or exclude certain cargoes not to be loaded or limit the ports to and from which the ship shall sail. For instance under consecutive voyage time charter the specific voyages to be undertaken by the ship are set out in the contract and no other voyages may be undertaken.

As the commercial exploitation of the vessel is for the charterer the charterer gives directions to the master and crew concerning its trading. The charterer will make arrangements with other shippers to place goods aboard the chartered vessel. Obviously the shipper will need bills of lading. Under a time charterparty the charterer is given the right to issue bills of lading which are to be signed by the master. The shipowner may become liable to the cargo owner on the terms of the bill of lading, which differ radically from the terms of the charterparty. The charterer is usually required to indemnify the shipowner against liabilities arising from signing bills of lading. In other words the charterer undertakes to indemnify the shipowner against any additional liability incurred by him with regard to bills of lading. The typical example of such a provision is found in clause 9 of the Baltime form[1]:

"The master to prosecute all voyages with the utmost dispatch and to render customary assistance with the vessel's crew. The master to be under the orders of the charterers as regards employment, agency, or other arrangements. The charterers to indemnify the owners against all consequences or liabilities arising from the Master, Officers or Agents signing Bills of Lading or other documents or otherwise complying with such orders, as well as from any irregularity in the vessel's papers or for overcarrying goods . . . ."

As a result the shipper will have a contract with the charterer against the shipowner. Under the indemnity clause the owner needs to have recourse to the charterer if he incurs any additional liability. If there is physical damage to the vessel as a result of its dispatch to an unsafe port or the charterer loading dangerous cargo the shipowner will be indemnified under the indemnity clause.

There are some limitations on the shipowner's right to recover. First, under a time charter the shipowner has responsibility for all matters relating to navigation and ship management. Therefore if there is a loss as a result of negligent navigation or unseaworthiness, even performed under the charterer's instructions, the shipowner will not be able to claim reimbursement for such losses because these issues fall within the responsibility of the shipowner.[2] Secondly, the master should not comply with orders which are clearly beyond the authority of the charterer.[3] Thus, if the master accepts instructions to proceed to a port which is obviously unsafe or signs bills of lading which are inconsistent with the terms of the charterparty the shipowner will not be able to rely on the indemnity clause. He will

1. The inclusion of such a clause was justified in the statement of Devlin J. in *Royal Greek Govt.* v. *Ministry of Transport* (1950) 83 Ll.L.R. 228 at p.234: "If the shipowner is to surrender his freedom of choice and put his master under the orders of the charterer, there is nothing unreasonable in his stipulating for a complete indemnity in return."

2. It was stated by Lloyd J. in *The Aquacharm* that "It is not every loss arising in the course of the voyage that can be recovered. For example, the owners cannot recover heavy weather damage merely because, had the charterers ordered the vessel on a different voyage, the heavy weather would not have been encountered. The connection is too remote. Similarly, the owners cannot recover the expenses incurred in the course of ordinary navigation, for example, the cost of ballasting, even though in one sense the cost of ballasting is incurred as a consequence of complying with the charterer's orders . . .": [1980] 2 Lloyd's Rep. 237 at pp.244–245.

3. In relation to this principle it was stated by Devlin J. in *Grace* v. *General Steam Nav. Co.* that "I cannot think that a clause in a time charterparty which puts the master under the orders of the charterers as regards employment is to be construed as compelling him to obey orders which the charterers have no power to give": (1950) 83 Ll.L.R. 297, at p.307.

not be able to claim reimbursement for such losses. Thirdly there is a need for proof of causation. If the chain of causation is broken by a negligent act of the crew the indemnity will not operate.[4]

(c) Demise charterparty

Demise or bareboat charterparties are leases of ships. Under such a charterparty a ship is hired empty. It has to be manned, provisioned and operated by the charterers for their account. The charterer takes possession and control of the ship and is responsible for its navigation, management and commercial exploitation. In *Baumwoll Manufactur von Carl Scheibler* v. *Furness*[5] a charterparty provided for the hire of the ship for four months. The charterer found the master and crew. Only the chief engineer was appointed by the owners. All wages were paid by the charterer. Bills of lading were signed by the captain and some of the charterer's agents, without any actual authority from the owner. They contained no reference to the charterparty. Under the bills of lading the ship was ordered to New Orleans where she loaded a cargo of cotton. On the voyage from New Orleans to Bremen the ship foundered, allegedly due to its unseaworthiness. The shippers sued the owner. In the report of the case it is not clear whether the charterers were also sued or not. The House of Lords held that the charterparty amounted to a demise because the possession and control of the ship vested in the charterer. Thus, the shipowner was not liable for the loss of the goods and the responsibility for the cargo lay with the charterer.

The charterer becomes the owner of the ship during the currency of the charter. The master and crew become his servants for all intents and purposes. In *Sandeman* v. *Scurr*[6] a ship was chartered to proceed to Oporto and there load a cargo. The charterparty gave the master authority to sign bills of lading at any rate of freight without prejudice to the charter. Goods were shipped to Oporto by persons ignorant of the charterparty, under bills of lading signed by the master. The Court of Queen's Bench held that the charterparty did not amount to a demise. Therefore the master's signature on the bill of lading bound the shipowner.

The only right of the registered owner is to receive the charter hire and take back the ship when the charter expires. Thus, under a demise charterparty the demise charterer is usually called a "disponent owner" or "owner *pro hac vice*". The crucial factor which will define a demise charterparty is the employment of the crew. The commonly used demise charterparty standard form is BARECON A. Clause 8 states that "the Vessel shall during the Charter period be in the full possession

4. It was stated by Donaldson J. in *The White Rose* that "a loss may well arise in the course of compliance with the charterer's orders, but this fact does not, without more, establish that it was caused by, and is in law a consequence of, such compliance, and in the absence of proof of such causation there is no right of indemnity": [1969] 3 All E.R. 374 at p.382.

5. [1893] A.C. 8; 62 L.J. Q.B. 201; 68 L.T. 1; 9 T.L.R. 71; 7 Asp.M.L.C. 263 (HL).

6. (1866) L.R. 2 Q.B. 86; 8 B. & S. 50; 36 L.J Q.B. 58; 15 L.T. 608; 15 W.R. 227; 2 Mar. L.C. 446; 41 Digest (Repl.) 216.

and at the absolute disposal for all purposes of the Charterers and under their complete control in every respect".

For all practical purposes the charterer acts as owner during the demise charter and he is responsible for all expenses incurred during the operation of the vessel and he is also responsible for insuring her. For the purposes of Hague and Hague-Visby rules the charterer is regarded as the carrier. Damage to the cargo will be the charterer's responsibility. He will be liable under bills of lading signed by the master. The terms of a demise charterparty would be inappropriate for incorporation into a bill of lading. As the master acts under the directions of the charterer rather than the shipowner the charterer is party to the carriage contract.

THE IDENTITY OF THE CARRIER UNDER BILLS OF LADING

Where a cargo is shipped aboard a chartered vessel the main question which has to be answered is whether the contract of carriage is made with the shipowner or the charterer. It is often difficult to identify whether it is the owner or the charterer who has undertaken the role of carrier under that contract. Does the master sign bills of lading on behalf of the owner or the charterer? The cargo owners will wish to know whom to sue if the cargo is lost or damaged. In practice where bills of lading are issued for goods on a chartered vessel very few bills are signed by the contractual carrier personally. The holder of the bill of lading has to trace the contractual carrier under the bill and then make sure that the carrier is bound by the signature of the agent who has signed it. In cases where the Hague or Hague-Visby Rules apply to the bill of lading contract, the carrier is entitled to the benefit of a one-year time limit after delivery of the cargo, by virtue of Article III(6). Therefore cargo owners have to ascertain quickly who the carrier is.

Where the vessel is demise chartered the carrier is likely to be the demise charterer. If the master signs the bill in his own name or the charterer signs "on master's behalf" it is likely that the bill will constitute a contract with the registered owner or demise charterer. If the charterer signs the bill of lading in his own name without indicating that he has done so as agent for the master or the owner he will be regarded as the carrier. The charterparty provisions that reveal the identity of the carrier will be relevant if they are expressly incorporated into the bill.

In some cases the express terms of the charterparty, such as identity of the carrier or demise clause, provide the identity of the carrier. In *The Berkshire*[7] the bill of lading contained a demise clause which specifically defined the registered owner as the carrier. The clause in question had the opposite effect in *The Venezuela*[8] where the charterer was to be the carrier. If the bill of lading does not contain any express provision which makes the position clear then the signature on the bill of lading and the terms of the charterparty need to be examined.

7. [1974] 1 Lloyd's Rep. 185.
8. [1980] 1 Lloyd's Rep. 393.

In time charterparties it is usual for the master to sign bills of lading as presented by the charterer. The charterer actually controls the terms of the bill of lading contract. Although it is less common, the same practice also occurs under a voyage charterparty. A master is generally deemed to have the implied authority of his employer to enter into contracts of carriage on behalf of his employer. Therefore in the absence of evidence to the contrary a master signs bills of lading as agent for and on behalf of his employer. If the vessel is demise chartered the master's employer will be the demise charterer; in other cases the employer is the registered owner. In *Sandeman* v. *Scurr*[9] the cargo interests managed to sue the shipowners for the part of the cargo which had leaked due to improper stowage. The bill of lading was signed by the master. In *Wagstaff* v. *Anderson*[10] the cargo owners sought to recover from the charterers in respect of cargo damage. It was stated that "prima facie the Master, in signing the bill of lading would be acting on behalf of the persons who were shipowners, and the shipowners would be the persons responsible for the carriage of goods". Consequently, the claim against the charterers failed.

If the bill of lading is signed by the charterer there will be considerable uncertainty as to the capacity in which he does so and much may depend upon the form of his signature. In general a charterer may have no authority to sign a bill on behalf of the owner unless the owner expressly confers such a power on him or actual or ostensible authority can be inferred on the facts. In *Tillmanns & Co.* v. *SS Knutsford Limited*[11] the master was under the charterers' orders. Furthermore the charterers were to indemnify the owners against all consequences of the master signing bills by order of the charterers. Following the charterers' signature of the bill of lading "for the Captain and the Owners" the owners argued that only the master could bind them by signing bills of lading. The court held that as the charterers did not cross out the words "for the Captain and the Owners" and then signed the bill they did not intend to make their own contract. Therefore the owners were bound by the signature.

There have been some restrictions on the master's ostensible authority. In *Grant* v. *Norway*[12] it was held that the master had no authority to sign a bill for non-existent goods. The owner was not bound if the bill signed by the master related to goods which had never been placed on board. This decision has been much criticised. If the master's authority to sign were limited to goods which were actually shipped, it would be very difficult for the indorsee to know whether this was the case or not. The rule in this case contains a well-known problem for cargo claimants.[13] Section 3 of the 1855 Bills of Lading Act attempted to deal with this problem, but it made statements in a bill of lading conclusive evidence only against the master or person signing the bill of lading, not against the carrier. In some cases

9. (1866) L.R. 2 Q.B. 86.
10. (1880) 5 C.P.D. 171.
11. [1907] 1 K.B. affirmed [1908] 2 K.B. 385; affirmed [1908–1910] All E.R.Rep. 549.
12. (1851) 10 C.B. 665; 138 E.R. 263.
13. See Paul Todd, *Contracts for the Carriage of Goods By Sea* (1988), pp.203–215 for a detailed discussion of the rule in *Grant* v. *Norway*.

Article III(4) of the Hague-Visby Rules will provide some assistance, but if no goods were shipped at all there was arguably no contract to which the Rules could attach.

Section 4 of the 1992 Carriage of Goods by Sea Act abolishes the effect of *Grant* v. *Norway*. The Act applies in favour of a holder of a bill of lading in good faith. Section 4 provides that a bill of lading which represents goods to have been shipped or received for shipment shall be conclusive evidence against the carrier of shipment or receipt, provided that it was signed by the master or by someone with authority to sign for the carrier.

In relation to the identity of the carrier most charterparties contain clauses which will assist in identifying the intentions of the owners and the charterers. In some cases the bill of lading expressly incorporates the relevant terms of the charterparty. These will be regarded as part of the bill of lading contract and will be helpful to determine the identity of the carrier. But the position will not be very clear if the charterparty contains relevant terms but these are not expressly incorporated into the bill of lading. In *Baumwoll Manufactur von Carl Scheibler* v. *Furness*[14] the charterparty provided that the captain and crew should be servants of the charterer and paid and appointed by him. Goods were shipped under bills of lading signed by the captain and some of the charterers' agents, without any actual authority from the owner. The bills of lading contained no reference to the charterparty and the shippers had no notice of its terms. Following the loss of cargo the shippers sued the owner. The House of Lords held that the owner was not liable, as the bills of lading had not been signed on his behalf. The decision was based on the fact that the charterparty was by demise, but surprisingly there was no use of the word "demise" within the charterparty. The main reason for treating this as a demise charter was based on the provisions on the employment of the captain and crew. The decision was not fair on the shippers, as they had no notice of the terms of the charter. In *Manchester Trust* v. *Furness, Withy & Co. Ltd.* the bills of lading referred to the charterparty but did not incorporate its terms. The owners and charterers agreed in the charterparty that in signing bills of lading the master should act as an agent of the charterers alone. The cargo owners sued the owners under the bills of lading for non-delivery and it was held that despite the charterparty provisions they were entitled to succeed. In the absence of the knowledge of the charterparty the master had authority to bind the owners.[15]

In general the Hague or Hague-Visby rules do not apply to a bill of lading in the hands of a charterer because under Article 1(b) such bill does not regulate the relations between a carrier and the holder. The relationship between the shipowner and charterer is governed exclusively by the terms of the charterparty. In cases where the charterparty does not include any express clause the courts will imply an indemnity. The implied indemnity covers the issue of onerous bills of lading. It also covers loss caused to the shipowners because of any order given by the charterers and obeyed by them.

14. [1893] A.C. 8.
15. [1895] 2 Q.B. 539.

In *Kruger & Co. Ltd.* v. *Moel Tryvan Ship Co. Ltd.*[16] the charterparty exempted the shipowners from liability for stranding and other accidents of navigation even when caused by the negligence of the master. The bills of lading signed by the master did not give the owners that exemption. The owners were liable to the indorsees of the bill of lading. They sued the charterers for breach of contract. It would have been possible to reach the same result under implied indemnity reasoning.[17]

In general the shipowner is the carrier, but sometimes the parties intend to give this role to the charterer. Under Article I(a) of the Hague Rules it is possible for the carrier to be either the shipowner or the charterer. The charterparty itself may provide that the master is authorised to sign the bills as agent on behalf of the charterer. The charterer will be bound by signed bills of lading. Even in cases where the charterer is apparently party to the bill of lading he may seek to transfer contractual liability to the shipowner. This object may be achieved by a demise clause in the bill of lading. Often a charterparty will also contain a demise clause but in principle a demise clause in the charterparty should not bind a shipper; it should bind its consignees or indorsees. In *Manchester Trust* v. *Furness, Withy & Co. Ltd.*[18] the *Boston City* was chartered under a time charterparty which contained a demise clause.[19] The charterers shipped coal aboard the vessel and the master issued bills of lading to them as shippers. Under the bills of lading Rio de Janeiro was declared as the port of destination. But the charterers induced the master to discharge at Buenos Aires without production of the bills of lading. The plaintiffs sued the shipowners for damages for non-delivery at Rio. The shipowners denied liability on the basis of the demise clause in the charterparty. The cargo owners won in spite of the demise clause.

In time and voyage charterparties there is a presumption that the master acts as agent of the shipowner, not the charterer. If a demise clause becomes effective under a charterparty the shipowner will not be liable and it binds the holder of the bill of lading who is unaware of the terms of the charterparty. Therefore the demise clause in a charterparty is ineffective against the plaintiff to rebut the presumption. The clause is effective only between shipowner and charterers. The shipowner can protect himself only by an explicit term in the bill of lading itself.

As stated above, in *Manchester Trust* v. *Furness, Withy & Co. Ltd.*, a demise clause was ineffective to bind an indorsee of a bill of lading. But a demise clause in a bill of lading can be effective. Clause 3 of Visconbill, a liner bill approved by BIMCO, can be given as an example. The "Identity of Carrier" clause provides that:

16. [1907] A.C. 272.
17. Paul Todd, *Modern Bills of Lading* (2nd ed., 1990), p.98.
18. See n.15, *supra*.
19. The demise clause was as follows: "The captain and crew, although paid by the owners, shall be agents and servants of the charterers for all purposes, whether of navigation or otherwise, under the charter. In signing bills of lading it is especially agreed that the captain shall only do so as agent for the charterers; and the charterers hereby agree to indemnify the owners from all consequences or liabilities (if any) that may arise from the captain signing bills of lading, or otherwise complying with the same."

"The contract evidenced hereby is between the Merchant [i.e. the shipper] and the Owner or demise Charterer of the vessel designated to carry the goods. No other person or legal entity shall be liable under this contract, and the protection of Article IV*bis* of the Hague-Visby Rules and any other statutory exemption from or limitation of liability shall inure also to the benefit of Stevedores and other Servants or Agents of the Carrier. For the purpose of this clause all such persons and legal entities are deemed to be parties to this contract, made on their behalf by the Carrier."

Under this clause it becomes clear that the contract is with the owner or demise charterer. It is not with the voyage or time charterer. The clause also attempts to protect stevedores and other people who are not parties to the contract.

THE CARRIAGE OF GOODS BY SEA ACT 1992

(a) The need for reform

The Carriage of Goods by Sea Act 1992 (COGSA 1992) replaced the Bills of Lading Act 1855. It applies only to documents subject to English law, which are issued after 16 September 1992. It has no effect on any bill of lading or other document issued before that date. The Act makes a number of important changes in the law regarding who can sue carriers under negotiable bills, seaway bills and delivery orders.

The problems relating to title to sue in respect of goods carried by sea are well known. In 1989, the Law Commission published Working Paper No. 112, and the Scottish Law Commission published Discussion Paper No. 83.[20] Both sought views on possible reforms to the law relating to the rights of those who buy goods which form part of a larger bulk. In these reports two problems were identified. The first problem was based on section 16 of the Sale of Goods Act 1979, which provides that, where there is a contract for the sale of unascertained goods, no property in the goods is transferred unless and until the goods are ascertained. This means that, if the seller becomes insolvent before property has passed, a buyer who has paid for the goods, and even one who has received a document of title, will merely have the rights of a general creditor. The second problem concerns section 1 of the Bills of Lading Act 1855. In normal domestic transactions the purchaser pays the price and gets immediate possession of the goods. Payment and transfer of title and risk are therefore immediate and contemporaneous. But in international transactions, a purchaser buys a cargo in a foreign country and may well have to pay the price on or before shipment. Despite the payment the purchaser will not actually obtain delivery of the cargo until the ship conveying the cargo has arrived at the discharge port. In some cases, after having paid for the cargo, the purchaser may wish to on-sell the cargo on before the ship arrives. The problems start when he sells the cargo on but then finds that it had actually been damaged or lost before

20. See J. Beatson, and J.J. Cooper, "Rights of suit in respect of carriage of goods by sea" [1991] L.M.C.L.Q. 196–208.

he sold it. In such circumstances who has the right to sue the carrier? Does the right to sue the carrier remain with the purchaser or is it transferred to the person to whom he sold it? Under the basic principle of English law only a person who was a party to a contract could sue on it. How could the parties, who were not originally parties to the bill of lading contract but who acquired title to the goods shipped under it by virtue of the transfer of the bill to them, sue the carrier under the bill of lading for loss of or damage to the goods in the course of carriage? Section 1 of the Bills of Lading Act 1855 was passed to remedy the problem that, in respect of a contract of carriage made between a shipper and a carrier, a third party such as an indorsee of a bill of lading was unable to sue for loss or damage caused by the carrier because he was not privy to the contract. The Act provided that "every consignee of goods named in a bill of lading to whom property in the goods therein mentioned shall pass, upon or by reason of such consignment or endorsement, shall have transferred to and vested in him all rights of suit, and be subject to the same liabilities in respect of such goods as if the contract contained in the bill of lading had been made with himself".

Over the years the difficulty has been caused by the use of the words "*upon or by reason of*[21] such consignment or endorsement". As stated in the joint report of the Law Commission and the Scottish Law Commission[22] there have been two views on the construction of these words. According to the narrow view the phrase "upon or by reason of" meant that property in the goods had to pass at the same time as the consignment or indorsement.[23] On the wide view it was sufficient if the property passed from the shipper to the indorsee under a contract in pursuance of which a bill of lading was indorsed in his favour.[24] On this view the rights to sue would be transferred even where property passed before or after the consignment or indorsement, providing that property passed at some stage. On the contrary, under the narrow view if the property did not pass "upon or by reason of" the consignment or endorsement itself, the right to sue did not pass to the consignee or endorsee under the Act. Consequently, if the endorsement itself did not pass title then the purchaser might not be able to sue the carrier even if the purchaser acquired the bill of lading. The narrow view brought the Bills of Lading Act into conflict with the Sale of Goods Act 1979. As stated above, under section 16 of the latter Act, if the goods are unascertained, that is to say unidentified, property cannot pass until those goods are ascertained. The problems first occurred in the case of *The Gosforth*.[25] This case was decided by the High Court in Rotterdam, applying its view of English law. There, a bulk consignment was sold by a merchant

21. Emphasis added.

22. Law Commission Report No. 196; Scottish Law Commission Report No. 130 (1991), "Rights of suit in respect of carriage of goods by sea", p.6.

23. The narrow view favoured by Scrutton is "If the property in the goods passes otherwise than upon or by reason of the consignment or indorsement the rights of suit do not pass to the receiver". Sir Alan Mocatta, Sir Michael Mustill, and others, *Scrutton on Charterparties and Bills of Lading* (19th ed., 1984), p.27.

24. Thomas Gilbert Carver, *Carver's Carriage by Sea* (Ed. Gilbert Colinvaux, 13th ed., 1982), p.98.

25. Sen. S. 1985 No. 981. See also, Paul Todd, pp.49–50 for detailed information on this case.

in South America on deferred payment terms to a merchant in Rotterdam. The original buyer obtained a bill of lading on terms that payment was to be made 15 days after transfer of documents. The seller was not paid. But the original buyer had obtained both the shipping documents and the property in the cargo. Then the bill of lading was handed to an independent third party in exchange for 13 delivery orders. The cargo which was the subject of this bill of lading was sold on to a number of different buyers to whom delivery orders were issued against payment. When the goods arrived the unpaid seller applied to attach the sub-buyers. Both the purchaser and sub-purchasers objected. But the Dutch court decided against them on the basis that, as they had no contract with the carrier as such, they were not entitled to bring a claim. The question in this case was whether the purchaser and sub-purchasers had ever obtained the property in the goods or not. The court's decision was based on section 1 of the Bills of Lading Act 1855 and section 16 of the Sale of Goods Act 1979. The property did not pass to the independent third party because he was acting as the original buyer's agent for the sub-sale. It did not pass to the sub-buyers because their goods as part of an undivided bulk were unascertained until breaking of bulk. As provided for in section 16 of the 1979 Act property in part of a bulk cargo could not pass until the relevant parts of the goods had been appropriated to the particular buyer. Since this apportionment had not taken place the rights of suit in the various parcels had not yet passed to the European buyers so as to entitle them to assert their rights. Therefore the seller still had the property in the goods and was clearly subject to the attachment.

Until the *Delfini* case[26] there was some judicial support for a wide view. But in this case the Court of Appeal rejected the "wide view". In *Delfini* a bill of lading for an unspecified portion of a cargo of bulk crude oil was endorsed to the final buyers after the vessel had arrived and discharged its cargo. The cargo was short delivered and the buyers sued the shipowners for breach of the bill of lading contract under section 1 of the Bills of Lading Act 1855. The judge at first instance held that once the ship had unloaded and sailed away the contract of carriage was discharged by performance despite the incomplete delivery. On the basis of this conclusion the bill of lading had ceased to be effective as a transferable document of title and therefore no contractual rights were acquired by the transferee under the 1855 Act. In the Court of Appeal the plaintiffs claimed that since delivery was to be made against bills of lading the contract of carriage was not discharged by delivery against letters of indemnity. The Court of Appeal upheld the judgment on different grounds. Basically the court held that the plaintiffs had no rights of suit under the Bills of Lading Act 1855 as the Act transferred the right of suit only where property in the goods passed "upon or by reason of the endorsement". As the property had passed before the endorsement there was no causal link between the endorsement and the parcel of property.

The Delfini gave unsatisfactory results which followed from the linking of the passing of property and the acquisition of contractual rights.[27] The decision in this

26. [1990] 1 Lloyd's Rep. 252.
27. See *The Report*, n.22 *supra*, p.7.

case seriously weakened the bill of lading as a commercially useful document. Particularly in the case of oil trade problems where there were long chains of buyers and comparatively short ocean voyages the title to sue became more problematic. Where a shipper sells a cargo of crude oil to five different receivers the cargo will be shipped in bulk. The shipper may issue five different separate bills of lading corresponding to each quantity sold and then pass those bills to the receivers in accordance with CIF contract principles. As the receiver's individual parcel is not ascertained or identified in the vessel's tanks the property cannot pass until the parcels are distributed to each receiver on outturn.[28]

Under the 1855 Act there were some financial problems as the Act transferred the right of suit only to a person who acquired the property. Consequently, the person who had a pledge could not sue. In the case of banks this approach caused serious problems, as banks normally finance by means of pledges and therefore cannot acquire the right of suit. However, in some cases the court has been ready to find that a party who does not have the right to sue under the Bills of Lading Act may nevertheless be able to sue the carrier as a result of the fact that an implied contract is concluded between them when the claimant offers to take delivery and the carrier agrees to give delivery of the cargo. In *Brandt* v. *Liverpool Brazil and River Plate*,[29] by the acts of presenting the bill of lading, payment of the freight and delivery of the cargo a contract was implied between the indorsees and the shipowners on the terms of the bill of lading. In this case the Court of Appeal held that an implied contract was created between Brandt and the shipowners with the result that Brandt could sue the shipowners for breach of a bill of lading. But the effectiveness of this approach has been diluted over the years.[30] In *The Aramis*[31] the Court of Appeal decided that the mere presentation of the bill of lading on discharge could not create an implied contract.

(b) Changes under the Carriage of Goods by Sea Act 1992

The Act applies to three kinds of documents. These are bills of lading; seaway bills, which are defined as documents which constitute a receipt for goods which evidences a contract for the carriage of goods by sea and which identifies the person to whom delivery is to be made; and ship's delivery orders, which are defined as documents given for the purposes of a contract for the carriage of goods by sea which constitute an undertaking by the carrier to a person identified in the document to deliver the goods to that person.

Section 2(1) of the Act provides that a bill of lading holder, or the person to whom delivery is to be made under a seawaybill or a ship's delivery order has transferred to him all the rights of suit under the contract of carriage as if he had

28. Richard W. Williams, "The New English Carriage of Goods by Sea Act 1992 (C.O.G.S.A. 1992)", September 1992, Issue 127, *Gard News*, p.8.
29. [1924] 1 K.B. 575.
30. See *The Aliakmon* [1983] 1 Lloyd's Rep. 203 and *The Elli 2* [1985] 1 Lloyd's Rep. 107.
31. [1989] 1 Lloyd's Rep. 213 (CA).

been a party to that contract. Hence, the lawful holder of a bill of lading has the right to claim from the shipowner for loss of or damage to the goods as if he were the original party to the bill of lading contract. The older method of transferring the contract to a consignee or endorsee who obtains the property in the goods by consignment or endorsement is abolished.

Even if the holder of the bill of lading does not himself suffer any financial loss he can sue the shipowner on behalf of the person who did suffer the loss. Thus, a bank holding the bill of lading endorsed in its favour can sue the shipowner on the terms of that bill of lading.

In some cases more than one person could sue for the same loss, but section 2(5) provides that a previous holder of a bill of lading may not sue if somebody else has become the lawful holder. In the case of a bill of lading which has been transferred away, the original contracting party and intermediate holder can no longer sue, but the original contracting party is still liable on the contract. Therefore once the shipper has passed the bill of lading to someone else he can no longer sue the shipowner. The Act does not affect the shipper's liabilities under the bill of lading. Therefore the shipper remains liable to the shipowner for his original obligations. If the shipper is a charterer he retains his rights under the charterparty. Where the relevant document passes down a chain the intermediate holder loses his contractual rights once the next party becomes the holder.

The Act defines a lawful holder of a bill of lading as someone who has become a holder in good faith. This person must be either the consignee named in the bill or an endorsee of the bill, a transferee of a bearer bill or a person who would have been such a consignee or endorsee or transferee had the transaction not taken place after the bill had ceased to be a transferable document of title. Where a person takes or demands delivery or makes a claim against the carrier, and he is a person who has title to sue under the Act, he becomes subject to any liabilities under the contract. Similarly, if the contractual rights pass to him after delivery, he will become liable when those rights are transferred to him.

PART 2

COLLISION

CHAPTER 8

INTRODUCTION TO THE LAW OF COLLISION

HISTORICAL BACKGROUND

The origins of maritime courts and law go back to ancient times. Ancient Greece had a developed system of maritime courts and law which originated on the Island of Rhodes in about 300 B.C. The maritime law of the Roman Republic was the same as that of Rhodes. But in Ancient Greece and Rome there was no law of collision. Until the nineteenth century there was no published discussion of collisions as an incident of commerce.

The first known written Roman law on collision was the *Lex Aquilia*.[1] Some collision cases mentioned in the Justinian Digest (533 A.D.) were governed by the ordinary rules of liability based on fault. Therefore as in other tort cases, contributory fault was a bar to recovery for negligent marine collision. The oldest known statement of marine collision law can be found in "Rhodian Sea Law", the maritime code of the later Roman Empire[2]:

> "If a ship in sail runs against another ship which is lying at anchor or has slackened sail, and it is day, all the collision and the damage regards the captain and those who are on board. Moreover, let the cargo too come into contribution. If this happens at night, let the man who slackened sail light a fire. If he has no fire, let him shout. If he neglects to do this and a disaster takes place, he has himself to thank for it, if the evidence goes to this. If the sailsman was negligent and the watchman dozed off, the man who was sailing perished as if he ran on shallows and let him keep harmless him whom he strikes."

This statement envisaged three possibilities: (a) ship A might hit ship B at anchor in daytime; ship B would win and ship A's cargo would contribute; (b) ship A might hit ship B at anchor at night where ship B had not lighted up or made a noise; ship B would lose; (c) ship A might hit ship B as the result of a negligent look-out; ship B would win.[3]

The Rhodian Sea Law did not specifically deal with cases of collision. It codified

1. N.J. Healy and J.C. Sweeney, "Basic Principles of the Law of Collision" (1991) 22 *Journal of Maritime Law and Commerce* 360.

2. D.R. Owen, "The Origins and Development of Marine Collision Law" (1977) 51 *Tulane Law Review* 760–761. For detailed information on the historical background of collision law see *ibid.*, pp.759–809.

3. *Ibid.*, p.761.

the negligence liability of the cargo owner at fault in shipwrecks. In marine collision law fault was first mentioned in the Digest of Justinian and later codes, where the determining question was whether there was a fault on the part of mariners or not. If there was they were liable. Where there was an inevitable accident there was no recovery.

In the Mediterranean city states there were many maritime codes, but only three of them dealt with collision. *The Constitutum Usus of Pisa* (c. 1160) mainly involved rules for collisions in harbour. *The Statutes of Ancona* (c. 1350) dealt with two cases of collision in harbour and, in the case of the navigator's personal fault, the navigator personally had to pay 10 per cent of the damage. *The Consolato del Mare* of Barcelona was the most important Mediterranean code which recognised the concept of fault and set out novel principles of division of damages.

In the Atlantic and North Sea the Rolls of Oleron (c. 1150) was the most important code in the development of all maritime law and soon became part of English law. The principles of this code are as follows:

"Art. XIV. If a vessel, being moored, lying at anchor, be struck or grappled with another vessel under sail, that is not very well steered, whereby the vessel at anchor is prejudiced, as also wines, or other merchandise in each of the said ships damnified. In this case the whole damage shall be in common, and be equally divided and appraised half by half; and the master and mariners of the vessel that struck or grappled with the other shall be bound to swear on the Holy Evangelists, that they did it not willingly or wilfully. The reason why this judgment was first given, being, that an old decayed vessel might not purposely be put in the way of a better, which will the rather be prevented when they know that the damage must be divided."

The Rolls of Oleron dealt only with harbour collisions. Under the Rolls equal division of damages is specified for the first time. The rule of equal division of damages in collision applied in the maritime world for more than 800 years. It was believed that it avoided expense and delay. Contribution by innocent cargo is considered under the Rolls but the basis for contribution was not clear. Under the Mediterranean codes the blame was determined easily but under the Rolls, for the first time, it was provided that it was impossible to apportion liability.

The Laws of Wisby showed similarities to the Rolls of Oleron. For the first time three articles under the Wisby rules were not limited to collisions in harbour.[4] *The Ordonnance de la Marine* of Louis XIV of France (1681) had a significant influence on the development of maritime law. *The Ordonnance* provided for equal division of damages where there was a collision between two ships under way, regardless of fault.

Obviously the above-explained maritime codes should be examined in relation to navigational needs. A complete system of collision law was not needed as ships were small and navigation and commerce were not developed. The most important codes were limited to anchored vessels. There were no rules for the prevention of collisions and for the division of loss.

4. *Ibid.*, pp.763–764.

DEVELOPMENT OF COLLISION LAW IN THE ENGLISH ADMIRALTY

(a) 1530–1789

Although 1360 is accepted as the founding date of the High Court of Admiralty it is possible to obtain continuous records of the court only from 1530. Before that date there was no record of any collision case or statement of collision law. After 1530 the history of collision law in the High Court consisted of a few cases which covered only the division of loss. During that period there was no development for the prevention of collisions in writing. During the period 1530–1789 the High Court of Admiralty did not develop collision law as there were not many collisions and there were limitations on the court's jurisdiction over maritime matters. Thus, during the first 11 years of the High Court of Admiralty out of 257 actions filed only five involved collision. Until 1538 there was no recorded judgment in a collision case. During the sixteenth century there were only eight recorded collision decisions. This number reached 46 during the seventeenth century, but in the eighteenth century it dropped again.

The most important case during this period was decided in 1789. In *The Petersfield and The Judith Randolph*[5] the equal division of damages rule was established where there was mutual but unequal fault. This rule remained as a principle in England until 1911 and in the United States until 1975.[6]

(b) 1789–1838

During this period two important events occurred. In 1798 regular reporting of the decisions of the High Court of Admiralty began and Sir William Scott, later Lord Stowell, came to preside over the High Court of Admiralty. Although he decided only five reported collision cases, *The Woodrop-Sims*, the most frequently cited collision case, was decided by him. The case was a sole fault case and in dicta Lord Stowell enunciated four basic principles of collision law in this case. These principles were as follows: (1) if there is an inevitable accident each vessel bears its own loss and there is no recovery; (2) when both vessels are to blame there is an equal division and damages are divided equally between vessels; (3) if the damaged vessel is solely to blame there is no right of recovery; (4) where the other vessel is solely to blame the damaged vessel is entitled to full recovery.[7]

5. 167 E.R. 596, Burrell 332 (Adm. 1789), cited in D.R. Owen, p.766, n.41.
6. *Ibid.*, pp.765–766.
7. The four principles were listed in Lord Stowell's judgment as follows:

"This is one of those unfortunate cases in which the entire loss of a ship and cargo has been occasioned by two vessels running foul of each other. There are four possibilities under which an accident of this sort may occur. In the first place, it may happen without blame being imputable to either party; as where the loss is occasioned by a storm, or any other vis major: In that case, the misfortune must be borne by the party on whom it happens to light: the other not being responsible to him in any degree. Secondly, a misfortune of this kind might arise where both parties are to blame; where there has been a want of due diligence or of skill on both sides: In such a case, the rule

The missing point in Lord Stowell's statement was the possibility of unascertainable fault, where the nature and extent of the fault is not clear.[8] Apart from the second principle the remaining principles are in force today. The principle of equal division of damages originated in the Rolls of Oleron and was abolished in England in 1911. Instead of equal division of damages Parliament adopted the comparative fault rule of the 1910 Brussels Convention.[9] In the United States the Supreme Court adopted the same rule in its decision in *United States* v. *Reliable Transfer Co.*[10]

(c) 1838–1911

Around 1840 modern collision law developed in relation to coincident events. Commerce expanded with the development in free trade and marine technology and also in relation to the Industrial Revolution. In 1837 Victoria ascended the throne and Britannia ruled the waves. Trinity House, a society incorporated by Henry VIII in 1514 for the promotion of commerce and navigation and which later was made responsible for pilotage and lighthouses, promulgated the first Collision Regulations in 1840. The jurisdiction of the High Court of Admiralty was expanded in 1840 and in 1861 to include collisions in local tidal waters. Between 1838 and 1867 Dr Stephen Lushington presided over the High Court of Admiralty. During Dr Lushington's presidency, for the first time in history, collisions became a matter of public concern and a subject of statistics. He created the rule of division of loss and also the Anglo-American collision law.

In 1840 the Steering Rules were formalised for the first time by Trinity House. The Steering Rules were the first official regulations, reciting three recognised rules for sailing vessels and promulgating two new rules for steamships.[11] Under the Rules each vessel was required to port her helm. This rule was the most important development in collision prevention up to that time.[12] The Steering Rules had no statutory authority but the Admiralty Court enforced them in civil actions arising out of collisions.

The first case under the Steering Rules was *The Duke of Sussex*. In this case Lushington made a clear statement that the steering rules would become binding

---

of law is, that the loss must be apportioned between them as having been occasioned by the fault of both of them. Thirdly, it may happen by the misconduct of the suffering party only; and then the rule is, that the sufferer must bear his own burden, it may have been the fault of the ship which ran the other down; and in this case the injured party would be entitled to an entire compensation from the other": (1815) 2 Dods. 83, 85; (1815) 165 E.R. 1422.

8. *Ibid.*, pp.767–768.

9. International Convention for the Unification of Certain Rules of Law with Respect to Collisions Between Vessels, 23 September 1910.

10. US Sup. Ct. 1975 A.M.C. 541; [1975] 2 Lloyd's Rep. 286.

11. One of these Rules required a steam vessel passing another vessel in a narrow channel to leave the other on her own port hand. The other regulation relating to steam ships required steam vessels on different courses, crossing so as to involve risk of collision, to alter course to starboard so as to pass on the port side of each other.

12. *Ibid.*, pp.768–769.

authority upon the owners of steam vessels.[13] The next case was *The Friends*.[14] In this case Lushington gave a broad application to an ostensibly narrow rule which was literally inapplicable to the case. This decision was inconsistent with Lushington's previous decisions. In *Hay* v. *Le Neve*[15] Hay's vessel was two-thirds at fault. The point of dispute was whether Hay's vessel should pay two-thirds or one-half of Le Neve's damages. The House of Lords resolved the issue without any discussion of cargo's position and awarded only half damages to the cargo on Le Neve's vessel as well as to the vessel. Lushington represented Hay and argued for half damages on Hay's behalf.

In *The Milan*,[16] Lushington's most questionable decision, he held that in a both-to-blame collision cargo could recover only half damages from the non-carrying ship. It was admitted by Lushington that this was contrary to the common law principle of joint and several liability to innocent persons. But he felt constrained to follow the rule in *Hay* v. *Le Neve* and other cases of mutual fault wherein cargo was awarded only half damages. As a result the court declined to adopt a rule of solidary liability in favour of cargo. This principle was preserved in the Judicature Act of 1873, in the Brussels Collision Convention of 1910 and in the Maritime Conventions Act of 1911.[17] As Lushington had argued for half damages in *Hay* v. *Le Neve* he could have been expected to declare that the rule was unjust in *The Milan*. But he said in *The Milan* that the rule was founded in equity and applied. After Lushington's retirement from the High Court of Admiralty in 1867 his contributions to the law of collision were documented in the text on collision law in the same year.[18]

(d) The Collision Regulations

As stated above the first official collision regulations were published in 1840 by Trinity House, London. In 1847 a statute came into force and gave effect to steam navigation rules. Under this statute a starboard side rule for ships navigating in narrow channels was introduced.[19] In 1848 Lights Regulations were made. In 1851 the statutory rules for steamships were repealed. They were re-enacted and

13. "The rule in question emanates from Trinity House; and although it cannot be said to constitute a *law per se*, it is nevertheless a rule to be observed, and it is important that it should be distinctly understood that in all future cases of this kind the Court will consider this rule of binding authority upon the owners of steam vessels; and if the masters of such vessels shall think fit not to comply with it, in so doing they will be guilty of unseamanlike conduct, and their owners will be responsible for the consequences that may result from their disobedience of it. The rule itself is drawn up with great precision, and may be understood without difficulty": *The Duke of Sussex*, 166 E.R. 575, 1 W. Rob. 274 (Adm. 1841).

14. 166 E.R. 651, 1 W. Rob. 478 (Adm. 1842).

15. II Shaw's Rep. 395 (HL 1824) (Scot.).

16. 167 E.R. 167, Lush. 388 (Adm. 1861).

17. It is argued in Owen's article that this situation largely caused the rejection of the Brussels Collision Convention of 1910 by the USA as a result of the opposition of cargo interests: D.R. Owen, p.771.

18. R. Lowndes, *The Admiralty Law of Collisions At Sea* (1867).

19. Steam Navigation Act 1846 (repealed).

applied to both steam and sail.[20] In 1854 the 1851 rules were replaced by new statutory rules. Under the new rules the port-to-port rule was applied to all ships and the starboard-side rule in narrow channels was retained for steamships only. In 1852 regulations on lights for vessels under way were made.

In 1862 all existing regulations were repealed.[21] A mandatory obligation to stand by after collision was introduced. In 1863 a completely new set of rules came into operation. Under these rules when steam vessels were crossing so as to involve risk of collision the vessel with the other on her own starboard side was required to keep out of the way. Steam vessels meeting end-on or nearly end-on were required to alter course to starboard. Every vessel overtaking any other had to keep out of the way of the vessel being overtaken. In 1880 some changes to the 1863 rules were brought into force. In 1884 a new set of regulations came into force, but these rules were not very different from the previous ones.

In 1889 the first International Maritime Conference was held to consider regulations for preventing collision at sea.[22] In 1910 a further maritime conference was held in Brussels and an international agreement was reached. The Regulations were revised at an International Conference on Safety of Life at Sea in 1948.[23] The revised rules came into force in 1954. In 1960 an International Conference on Safety of Life at Sea was held in London and at the Conference it was agreed that a new paragraph should be added to the rules as there was a considerable increase in the number of ships fitted with radar and collisions involving such vessels. The changes clarified the mariner's responsibility regarding the use of radar in reduced visibility or fog. These rules came into force in 1965. In 1972 the 1965 Regulations were made applicable to hovercraft, and an international conference was convened in London by IMCO. In the same year a compulsory traffic separation scheme in the Dover Strait and elsewhere was established. In 1977 the Collision Regulations and Distress Signals Order came into force and revoked the 1965 regulations and the 1972 order. Following the 1977 Order the regulations applied to British ships "on the high seas and all waters connected therewith navigable by sea-going vessel".

The 1977 Order was revoked and replaced by the Merchant Shipping (Distress Signals and Prevention of Collisions) Regulations 1983. The 1983 Regulations incorporated the International Regulations for Preventing Collisions at Sea

20. Steam Navigation Act 1851 (repealed).

21. See R.H.B. Sturt, *The Collision Regulations; the application and enforcement of the Merchant Shipping (Distress Signals and Prevention of Collisions) Regulations 1989* (1991), pp.1–10; A.N. Cockroft and J.N.F. Lameijer, *A Guide to the Collision Avoidance Rules; international regulations for preventing collisions at sea* (1990), pp. 14–16.

22. Some of the new provisions agreed at the Conference were as follows: a stand-on vessel should keep her speed as well as her course, a giving-way vessel should avoid crossing ahead of the other vessel, steamships should be permitted to carry a second white masthead light.

23. In this International Conference slight changes were made: the second masthead light was made compulsory for power-driven vessels of 150 feet or upwards in length, a fixed stern light was made compulsory for almost all vessels under way, the wake-up signal of at least five short and rapid blasts was introduced as an optional signal for use by a stand-on vessel.

1972.[24] The Merchant Shipping Regulations 1983 were largely revoked and replaced by the Merchant Shipping (Distress Signals and Prevention of Collisions) Regulations 1989. The 1989 Regulations and the following Merchant Shipping (Distress Signals and Prevention of Collisions) (Amendment) Regulations 1991, the Merchant Shipping (Signals of Distress) Rules 1992, the Collision Regulations (Seaplanes) Order 1989, the Collision Regulations (Seaplanes) (Amendment) Order 1990, and the Collision Regulations (Seaplanes) (Amendment) Order 1991 are revoked and replaced by the Merchant Shipping (Distress Signals and Prevention of Collisions) Regulations 1996[25] which came into force on 1 May 1996. The Regulations apply to United Kingdom vessels wherever they may be and other ships while within the United Kingdom or the territorial waters thereof; seaplanes registered in the United Kingdom and on the surface of water anywhere; and other seaplanes on the surface of water in the United Kingdom or the territorial waters thereof. In the 1996 Regulations "ships" include hovercraft.[26]

The Collision Regulations were intended to provide a comprehensive code to regulate the relation of ships with each other. The main object of the 1972 Regulations was to prevent collisions by prescribing what action should be taken by ships in close proximity so that each ship might anticipate the manoeuvres of the other ship with reasonable certainty and take the appropriate action necessary to avert collision. Therefore each rule was made for the benefit of the ship that had to obey it and for the other ship whose safety was endangered by her presence. Accordingly obedience to the Regulations is the primary requirement of good seamanship.

The 1972 Regulations do not interfere with the operation of special rules made by an appropriate authority for roadsteads, harbours, rivers, lakes or inland waterways connected with the high seas and navigable by seagoing vessels. But the special rules must conform as closely as possible to the Regulations. Provision is also made for areas governed by IMO Traffic Separation Schemes as well as for vessels of special construction which cannot comply with the Regulations (Rule 1).

Where any of the Collision Regulations are contravened[27] the owner of the vessel, the master and any person for the time being responsible for the conduct of the vessel shall each be guilty of an offence. Any person charged under these Regulations may show as a defence that he took all reasonable precautions to avoid the commission of the offence.[28]

To give a detailed commentary on the entire current Collision Regulations would be quite beyond the scope of this work, so the Regulations will be discussed within the referred cases.

24. The 1972 Rules were amended by IMO Resolutions A.464(XII) of 19 November 1981, A.626(15) of 19 November 1987, and A.678(16) of 19 October 1989 and A.736(18).

25. S.I. 1996 No. 75.

26. *Ibid.*, Art. 2.

27. Regulations include the 1996 Merchant Shipping Regulations and the International Regulations for Preventing Collisions at Sea 1972 as amended.

28. Merchant Shipping Regulations 1996, Art. 6.

CHAPTER 9

BASIS OF COLLISION LIABILITY

DEFINITION OF COLLISION

In collisions at sea every variety of vessel may be involved. They may be small motorboats or oceangoing supertankers. The damage may be minor or incredibly large. The damage resulting from a collision may not only be damage to ships but also death and personal injury, cargo loss and oil pollution.

The word "collision" simply defines a crash or accident between vessels. In common usage it may mean the running down by one vessel of another vessel, a buoy, a permanent or floating oil rig, or a quay or dock. In the maritime sphere it means the coming together of moving vessels. It has also been defined as an impact or sudden contact of a moving body with an obstruction in its line of motion, whether both bodies are in motion or one stationary and the other, no matter which, in motion.[1]

One may come across the term "allision"[2] while considering the definition of "collision". In *Wright* v. *Brown*[3] it was stated that: "In its strict sense, collision means the impact of two vessels both moving, and is distinguished from *allision*, which designates the striking of a moving vessel against one that is stationary. But in a broad sense, collision is used, to include allision, and perhaps other species of encounters between vessels, and between a vessel and other floating, though non-navigable, object. In some courts the term 'allision' used in a broader sense to include the contacts of moving vessels not only with stationary vessels or other floating structures, but also with piers, wharves, bridges and other shoreside installations."[4]

Collisions can happen with many and various objects. These objects may be wrecks, fixed objects ashore such as piers, dock walls and anchors of other vessels. Therefore the courts have to consider what constitutes a collision. When one looks at the collision cases it can be seen that the contact and/or impact between vessels

1. *Black's Law Dictionary* (5th ed., 1979), p.241.
2. Defined in *Black's Law Dictionary*, p.69, as "The running of one vessel into or against another, as distinguished from a collision, i.e., the running of two vessels against each other. But this distinction is not very carefully observed."
3. 4 Ind. 95, 97 (1953), cited in N.J. Healy and J.C. Sweeney, p.359, n.1.
4. See, e.g., *Georgia Ports Authority* v. *The Atlantic Towing Co.*, 1985 A.M.C. 332 (S.D. Ga. 1983). Cited in *ibid.*, n.2.

need not necessarily be between their respective hulls. A vessel may cause damage by its negligence to another vessel without actual contact. For instance a ship, by proceeding at an excessive speed, can cause another ship to sink.[5] Contact with an anchor or mooring chain of another vessel has been held to be contact within the meaning of collision, but damage so caused to the nets of a fishing vessel is not a collision. Damage caused to one vessel by the excessive wash from another vessel is not a collision because there is no actual contact. Contact between a vessel and a fixed and floating object is also not a collision claim. These types of questions arise when there are cases on the collision liability clause in the standard Hull Policy of Insurance, as the term "collision" is not defined in the clause. For the purposes of marine insurance a collision claim should be related to contact between one vessel and another. It does not cover the situation where the insured ship makes contact with any object other than another vessel. Therefore when a collision claim arises marine insurers first consider whether the given instance constitutes a collision or not. This consideration covers the examination of the term "vessel" and what constitutes a collision with another ship or vessel has been the subject of several judicial decisions.

THE CONCEPT OF LIABILITY FOR COLLISION

(a) Development of specific features in collision liability[6]

Strict liability (no fault): according to a commonly accepted medieval strict liability proposition man had absolute liability for damage caused by an act "... even though damage is the result of pure accident". Although this strict liability doctrine was adopted by the law of marine carriage and personal injury it was excluded for collision damage.

Peril of the sea (no fault): the peril of the sea doctrine allowed the loss to "lie where it falls". This doctrine applied where there was an "inevitable accident" case.

Equal division of damages (regardless of fault): this rule originally appeared in the Rolls of Oleron (c. 1150) and in the Laws of Wisby. The equal division rule developed before the development of the concept of fault. At that time mutual fault was not even conceived of.

General average: in a collision case ship and cargo each contribute to damages according to its value for division of damages on a general average basis. This contribution does not take fault into account.

Intentional fault: under the Codes of Oleron and Wisby an oath of good intentions was sufficient to exculpate the moving ship. The test of liability on the part of a moving ship striking an anchored ship was whether the act was wilful or

5. See *The Royal Eagle* (1950) 84 Ll.L.R. 543; *The Royal Sovereign* (1950) 84 Ll.L.R. 549.

6. For more information about the development of specific features of substantive collision law, see D.R. Owen, pp.778–782.

not. But, given the impiety of seamen this test became an unrealistic measure of liability.

Negligent fault: in the Roman Law of Justinian fault was established as a determinant of liability and mentioned in the collision law of Rhodes. In a later version of the Rolls of Oleron (c. 1300) "default of the mariners" was the determinant of liability where one ship at anchor broke loose and collided with another. Under the common law the concept of negligence was not recognised as a separate tort until the early nineteenth century. In the civil law the Roman idea of fault and negligence had been accepted three centuries earlier. Although negligence, "oversyghte", carelessness and fault were mentioned in the early collision judgments of the High Court of Admiralty in the sixteenth century the development of the concept of negligence remained insignificant. The High Court of Admiralty had difficulties in deciding questions of negligence. Despite these difficulties negligence became the test of collision liability long before it was recognised by the common law as a separate tort.

(b) Establishing fault in collision cases

Civil liability for collision is based on the existence of "fault" which contributes to or causes the collision. The mere fact that the collision occurred does not impose liability. If someone wants to establish liability depending on fault he must show that he has suffered loss or damage as a result of fault and the causation of fault for the given damage should also be proved. The definition of fault is not given under the Collision Regulations as fault is basic to the law of tort and a quite significant issue and the states that implement the Regulations would not like to amend their concept of this issue.

In *The Llanover* Viscount Maugham stressed the ordinary degree of skill and care while considering the fault. In his words, "a failure to exercise that degree of the skill and care which are ordinarily to be found in a competent seaman ... It is negligence not to take all responsible steps to avoid danger in navigation, and the nature of those steps must of course depend on the surrounding circumstances, and they may call for the utmost possible precautions".[7] In a given case the point that has to be examined is whether the collision was prevented by the exercise of ordinary care, caution and maritime skill or not. In *The Boleslaw Chrobry* the standard of skill and care was also examined. According to Brandon J.,[8] "the standard of skill and care to be applied by the Court is that of the ordinary mariner and not of the extraordinary one, and seamen under criticism should be judged by reference to the situation as it reasonably appeared to them at the time, and not by hindsight". Also in *The Martin Fierro*[9] it was stated that when a seaman is faced with a situation of danger created by the negligence of another the law requires that he exercise the skill and care of the ordinary prudent mariner.[10] If he fails to do so

7. (1945) 78 Ll.L.Rep. 461.
8. [1974] 2 Lloyd's Rep. 308, at p.316.
9. [1974] 2 Lloyd's Rep. 203.
10. S.T. Harley, "Collision Cases, 1974" [1975] L.M.C.L.Q. 160–161.

and his failure to exercise such skill and care contributes to a collision a finding of blame will be made against him. The people who should show the ordinary degree of skill and diligence are those who discharge their duty.

If a statutory duty is imposed and that duty is disregarded, criminal liability for the statutory penalty arises. There is no need for an inquiry into fault. But in cases of civil liability a breach of statutory duty may not necessarily be the cause of a collision. Until 1911 the Merchant Shipping Acts provided that if a breach of the regulations was shown to have occurred in the context of a collision, the vessel shown to have infringed regulations was deemed to be at fault. Under this presumption the vessel would be held liable even when the breach did not cause the collision or damage. The statutory presumption of fault rule was abolished by the Maritime Conventions Act 1911 as breach of a statutory duty may not necessarily be the cause of a collision. Since 1911, it has been necessary also to show that the breach of the regulations contributed to the collision, damage or loss. In this respect violation of a traffic separation scheme can be given as an example. In *The Estrella*,[11] the Cypriot vessel the *Setubal* collided with the Finnish vessel the *Estrella*. The *Setubal* was in breach of the recommendations on the traffic separation zone. But the court found that this breach was not causative of the eventual collision.[12]

The liability of the owner of a ship for damage caused to another as a result of a collision does not depend on ownership. The basis for this liability is that those navigating the vessel at the time of the collision were his employees and that the owner was obtaining the benefits of that employment. The size of the ship, type of cargo and nature of the employment is not important while considering the standard of care. Only the amount of care required will be different and will depend on the circumstances.

(i) Classification of fault[13]

Faults separated

If it is possible to draw a line between the initial fault of one vessel and the subsequent fault of the other, faults are separated. If one vessel is negligent for a considerable time and that negligence is apparent to the other, which could have avoided the collision by adopting proper measures, the latter may be the only one to blame.

In *The Boleslaw Chrobry*[14] the stern of the *Boleslaw Chrobry* came into contact with the starboard side of the *Melide*. Serious damage was sustained by both vessels. The *Boleslaw Chrobry* was not at fault in failing to see the *Melide* when she left the jetty. The *Melide* was about 1½ miles away, visibility was reduced by rain and there were other vessels in the vicinity of the *Melide*. But when the *Boleslaw*

11. [1977] 1 Lloyd's Rep. 525.
12. See also a similar case, *The Genimar* [1977] 2 Lloyd's Rep. 17.
13. See S. Mankabady, *Collision At Sea, A Guide to the Legal Consequences* (1978), pp.12–14.
14. See n.8 *supra*.

Chrobry had embarked on the turning manoeuvre it was her duty to keep a good look-out for any vessel going or coming up the river.[15] When the *Melide* altered her course to port it was reasonable to expect that the *Boleslaw Chrobry* would be aware of the *Melide*'s approach early enough to avoid collision by her action alone. The *Boleslaw Chrobry* was solely to blame for the collision.

Faults mixed

Faults are mixed when there is a considerable overlap between the faults of the two vessels involved in a collision.

In *The Sabine*[16] the *Ore Prince* went aground in the river Schelde on her way to Antwerp. Due to the failure of her steering gear the *Sabine* anchored up river. Late in the evening the *Ore Prince* was being assisted to refloat. The *Sabine* was also being assisted to turn in the river to proceed upstream and into Antwerp port. The *Sabine* was swinging round until after the *Ore Prince* had been refloated and was making headway up the river. The river had a width there of only around 900 feet. The *Sabine* herself was around 650 feet in length. The *Ore Prince*, instead of hanging back until the *Sabine* had completed her swing, tried to edge through the limited space between the *Sabine*'s stern and the river's edge. She failed and the two ships collided. The dominant cause of the collision was the failure of those on board both vessels to keep a proper look-out. Both ships were equally to blame, so their fault cannot be regarded a separate fault.

In *The Hagen*[17] the fault in going hard to starboard at the last moment was a consequence of a bad look-out and therefore it cannot be called a separate fault.

In *The Sea Star*[18] the *Sea Star*, fully loaded with crude oil, collided with the *Horta Barbosa* in ballast in the Gulf of Oman. The *Sea Star* altered course to starboard and turned a safe situation into one of acute danger. This improper manoeuvre was the result of a defective look-out. The *Horta Barbosa* was at fault as well. Her second officer was not at his post just before the collision. Therefore the *Horta Barbosa* could not properly react to the position of danger created by the *Sea Star*. The faults of both ships were the consequence of other faults, thus they were not separated.

In *The Elazig*[19] the court held that failure to keep a proper look-out was not a separate fault. Faults in keeping a speed of nine knots until far too late and in not going full astern at once were all interconnected.

15. It is a general rule that where a ship is turning in a river or channel she should not manoeuvre so as to interfere with the navigation of any ship on passage in the river or channel. To prevent such interference the keeping of a vigilant look-out is paramount. But this elementary precaution is too often forgotten.

16. [1974] 1 Lloyd's Rep. 465.

17. [1973] 1 Lloyd's Rep. 257, at p.265.

18. [1976] 1 Lloyd's Rep. 115. The *Sea Star's* appeal was dismissed: [1976] 2 Lloyd's Rep. 477.

19. [1972] 1 Lloyd's Rep. 355.

Faults interlocked

This situation was described in *The Volute*.[20] In certain situations the respective negligence of two vessels becomes so close that "the second act of negligence is so much mixed up with the state of things brought about by the first act, that the party secondly negligent . . . might invoke the prior negligence as being part of the cause of collision so as to make it a case of contribution".

In *The Sestrière*[21] both vessels had pilots on board and were manoeuvring to drop the pilots. The *Alonso* turned to a northerly heading and stopped her engines. The *Sestrière* was approaching the pontoon and the *Alonso* saw the possibility of a collision. The *Alonso* put her engines full ahead and went hard a-starboard. The *Alonso*'s owners blamed the *Sestrière* completely as she did not keep a proper look-out. And she failed to keep out of the way of the *Alonso* when it was on her starboard side. She was also blamed for not complying with the Collision Regulations (rules 19, 23, 27 and 29). The owners of the *Sestrière* counterclaimed that the *Alonso* was to blame in improperly going full ahead and hard a-starboard instead of either stopping her engines or going full astern. The basis of argument for greater blame of the *Sestrière* was that it was her initial fault which created the risk of collision and the *Alonso*'s fault was in her negligent reaction to the risk created by the *Sestrière*. On the other hand for the defendants the *Alonso*'s fault was very important and the fault of the *Sestrière* was more excusable as the people on board her could not reasonably have foreseen that the *Alonso* would take such action. It was obvious that the initial fault of the *Sestrière* brought about the situation of danger but the *Alonso*'s action was completely wrong and made the collision inevitable. Therefore both vessels were guilty of unseamanlike action and of faults which caused the collision. Blame was divided equally.[22]

(ii) Types of collision

Inevitable accident

"Inevitable accident" has been defined as a collision which a party could not

20. [1922] 1 A.C. 129, at p.144.
21. [1976] 1 Lloyd's Rep. 125.
22. In Brandon J.'s words: "It was argued for the plaintiffs that the greater blame should attach to the *Sestrière* on the ground that it was her initial fault which created the risk of collision, and that the fault of the *Alonso* lay only in her negligent reaction to the situation of risk so created. It was argued for the defendants, on the other hand, that the fault of the *Alonso* was a grave one, in that to go full ahead in the face of danger was so plainly wrong as to amount to rashness, and that the fault of the *Sestrière* was more excusable in that those on board could not reasonably have foreseen that the *Alonso* would take such action.

I think that there is force in both these arguments. It was certainly the initial fault of the *Sestrière* which brought about the situation of danger. But on the other hand, the action taken by the *Alonso* to deal with the situation was plainly wrong and made inevitable the collision which it was intended to avoid.

In the result, it seems to me that both ships were guilty of serious errors of appreciation due, I think, to an inadequate look-out which could, and should, have been avoided.

Having weighed both the culpability and the causative potency of the faults on either side, I have come to the conclusion that both ships were equally to blame": *ibid.*, pp.131–132.

possibly prevent by the exercise of ordinary care, caution and maritime skill.[23] In other words it was not intended and could not have been foreseen and avoided by the exercise of reasonable care and skill. The law does not demand an extraordinary degree of skill. What is demanded from those responsible for the navigation of vessels is the exercise of reasonable skill. However, it should also be noted that what is ordinary diligence is determined by the circumstances of the case. If a ship causes a threat of collision that cannot be avoided without "extraordinary" skill on the part of the other ship and this threat causes a collision it will be held that the collision was solely caused by the first ship.

Well known examples of inevitable accidents are where the collision is caused by the force of the wind in parting the lines of a moored or anchored vessel. Obviously it is not easy to establish an inevitable accident in this case as some conditions must be fulfilled: the force of the wind could not have been anticipated, the vessel has been moored and anchored properly, there should be no negligence on the part of those in charge of the vessel.[24] Again a collision resulting from sudden and unavoidable machinery breakdown can be accepted as resulting from inevitable accident in certain circumstances. It must be proved that the defect is latent and could not be discovered by any visual inspection which could reasonably be required according to the standards current at the time, could not be discovered by reasonable skill, and that the collision was the result of the defect or the breakdown and could not have been avoided by proper navigation after the problem started.

In inevitable accident cases it is impossible for one vessel to impose liability on the other. The loss lies where it falls. Therefore where a collision results from inevitable accident no liability arises on either side. The burden of proof has to be discharged by the vessel alleging that the accident was inevitable. Under Admiralty Court practice if it is obvious that it was an inevitable accident which caused the collision the court will dismiss the action with costs and each party pays its own costs.

Where the cause of a collision could not be determined, the plea of inevitable accident could not be sustained. In *The Merchant Prince*[25] the moving ship's steering gear jammed and the ship collided with an anchored vessel. Under Admiralty Court Practice the onus lies with the moving ship to rebut the charge of negligent navigation which meant that the accident was inevitable. But the Court of Appeal reversed the decision on the ground that the probable cause of the jamming was that the chain connecting the steering-gear with the rudder had been allowed to get too loose, causing kinks in the links. The Court of Appeal held that the defendants were obliged to "either show what was the cause of the accident, and show that the result of that cause was inevitable, or they must show all possible causes, one or other of which produced the effect, and must further show with regard to every one of those possible causes that the result could not have been avoided".

23. V. Dover, *A Handbook to Marine Insurance* (8th ed., 1975), p.521.
24. S. Mankabady, p.15.
25. [1892] Prob.Div. 179.

In *The Alletta and The England*[26] the *Alletta* had been moored alongside the jetty with her head downstream. She left the jetty loaded with spare parts for motor cars and proceeded right across the main channel in the River Thames to anchor on the south side. In doing so she crossed the course being taken by the *England*, which was proceeding downstream. The stem of the *England* struck the starboard side of the *Alletta*. The *Alletta* collided with a number of moored dumb barges and another moored vessel, the *Mare Liberum*, before she sank. Actions were brought by barge owners against the *England* and the *Alletta* and by the *England* against the *Alletta*. One of the issues was whether the collision with the barges was an inevitable consequence of the first collision or not. It was held that the *Alletta*'s manoeuvres after the collision were proper and seamanlike. As there was no new course between the first and the second collisions the second collision was completely foreseeable as a reasonable consequence of the first. Therefore in this case the plea of inevitable accident failed.

In *The James River Transport Inc.* v. *Nasenbulk*[27] two ships were lying at anchor in anchorage positions designated by the harbour authority in Sasebo, Japan. During a 56-knot typhoon the ships collided after dragging their anchors. It was held that neither ship was at fault, and each should bear its own damages.

Agony of the moment

In this class of accident one person chooses one danger which he believes to be smaller rather than continue to be exposed to one which is apparently imminent.[28] For the agony of the moment defence to succeed the danger must have been imminent and the time should be too short to form a deliberate alternative plan of action to avoid the threatening situation. If ship A puts ship B into grave danger with its negligent manner and the master of ship B does not have enough time to form an alternative plan of action to avoid the danger, the decision taken by the master is described as in the agony of the moment as it may be wrong and is the proximate cause of the action. In these circumstances ship A must bear the whole loss; the negligence of the master of ship B is not to be taken into account. In such a case vessel B will not be held to be at fault as the acts of its master were done in the "agony of the moment" arising out of imminent danger of a collision resulting from the negligent act or acts of vessel A.

In *The Bywell Castle*[29] the *Bywell Castle* was navigating down the River Thames at night and the *Princess Alice* was proceeding up river. The ships were very close to each other. The *Princess Alice* suddenly and negligently hard-a-ported her wheel. Also the *Bywell Castle* quickly but wrongly put her wheel hard-a-starboard. She struck the *Princess Alice* on her starboard bow and sank. Although the *Bywell Castle*'s master's action was wrong no blame attached to her and the action was found excusable.

26. [1965] 2 Lloyd's Rep. 479.
27. [1974] A.M.C. 575.
28. Robert Samuel-Theodore Chorley, *Chorley and Giles' Shipping Law* (8th ed., 1988), p.383.
29. (1879) 4 Prob. Div. 219 (CA).

One vessel is to blame

Where damage results wholly from the negligence of one ship she will be liable for all liabilities arising from the collision.[30]

Both vessels are to blame and degree of fault can be apportioned

Where each vessel involved in the collision is to some extent to blame, the liabilities arising will be apportioned between the colliding vessels according to the degree of their fault.[31]

Damage to one vessel by negligence of two other vessels

Where two or more vessels which are at fault cause damage to another vessel the loss of the innocent vessel or the loss by the owners of property on her is covered in full either from one vessel alone or from both vessels jointly.

Both vessels are to blame but the degree of fault cannot be determined

This is called inscrutable fault. Where the vessels are equally to blame or where they are to blame in differing degrees, but the varying degrees of blame cannot be fixed the old rule of maritime law remains unaffected and the aggregate damages must be equally divided. If it is not possible to apportion blame in differing degrees, liability shall be apportioned equally.

APPORTIONMENT OF LIABILITY IN COLLISION CASES

(a) The Brussels Convention of 1910

The International Convention for the Unification of Certain Rules of Law with Respect to Collisions Between Vessels (the 1910 Brussels Convention) was signed at Brussels on 23 September 1910. It covers collisions between ships and also other casualties in which one ship by fault in her navigation causes damage to another ship or to persons or property on board either, even though no collision occurs. But the question of limitation of liability was reserved for further consideration. Under common law it was the general rule of civil liability that there could be no recovery where a casualty was caused by mutual fault. Under Admiralty law, this rule developed, and where two ships were to blame for a collision damages were to be apportioned equally regardless of fault. In other words, the original rule in Admiralty on the apportionment of damages in collisions was equal division of

30. See *The Boleslaw Chrobry, The Aldebaran/The Esso Brussels* [1972] 1 Lloyd's Rep. 286; [1973] 2 Lloyd's Rep. 73 (CA). On appeal the *Esso Brussels* was found solely to blame.

31. See *The Maritime Harmony* [1982] 2 Lloyd's Rep. 400; *The Roseline* [1981] 2 Lloyd's Rep. 410; *The Martin Fierro* [1974] 2 Lloyd's Rep. 203; [1975] 2 Lloyd's Rep. 130 (CA); *The Statue of Liberty* [1970] 2 Lloyd's Rep. 151; [1971] 2 Lloyd's Rep. 277 (HL).

total damages. With the 1910 Brussels Convention this rule was abolished and the proportionate fault rule gained recognition.

Before the 1910 Brussels Convention there were three systems for allocating liability in collision cases. In the UK and the USA the rule of equal division of damages was used. In Belgium, Greece, Scandinavia and France loss was apportioned according to the degree or gravity of each party's fault. Other countries, such as Argentina, Chile, Italy, Germany, Mexico and Spain, permit neither of the two colliding vessels to recover damages. The proportionate fault rule was first suggested and approved at the 1899 London Conference of the CMI. The draft convention prepared by the CMI was then submitted to the Diplomatic Conference in 1905. The wording of Article 4, which provides that if two or more vessels are at fault the liability of each vessel is in proportion to its share of fault, was adopted by the Conference.[32] With this rule the degree of fault becomes the proper basis for apportionment of liability.[33]

(b) The British Maritime Conventions Act of 1911

The Maritime Conventions Act of 1911 (1911 Act, MCA 1911) was passed to carry into effect the 1910 Brussels Convention. Under section 9 the Convention applies to all British dominions or protected territories other than Canada, Australia, New Zealand, South Africa and Newfoundland. The Act does not apply in cases of damage between a ship and a craft or other structure which is not a vessel for the purposes of the Merchant Shipping Acts. Cases not covered by this Act are subject to the rules of common law. An actual collision between ships is not necessary for the application of the Act; damage to a ship caused by the wash of another ship will be covered under the Act.[34]

The most important changes in English collision law effected by the MCA 1911 were the repeal of the statutory presumptions of fault formerly contained in sections 419(4) and 422(2) of the Merchant Shipping Act 1894 and the replacement of the traditional rule in Admiralty of equal division of damages in

32. Art. 4 of the 1910 Brussels Convention reads as follows: "If two or more vessels are in fault the liability of each vessel is in proportion to the degree of the faults respectively committed. Provided that if, having regard to the circumstances, it is not possible to establish the degree of the respective faults, or if it appears that the faults are equal, the liability is apportioned equally.

The damages caused, either to the vessels or to their cargoes, or to the effects or other property of the crews, passengers, or other persons on board, are borne by the vessels in fault in the above proportion, and even to third parties a vessel is not liable for more than such proportion of such damage.

In respect of damages caused by death or personal injuries, the vessels in fault are jointly as well as severally liable to third parties, without prejudice however to the right of the vessel which has paid a larger part than that which, in accordance with the provisions of the first paragraph of this Article, she ought ultimately to bear, to obtain contribution from the other vessel or vessels in fault.

It is left to the law of each country to determine, as regards such right to obtain contribution, the meaning and effect of any contract or provision of law which limits the liability of the owners of a vessel towards persons on board."

33. For the development of the proportionate fault rule see D.R. Owen, pp.790–798 and F. Berlingieri, "Jurisdiction and An Overview of the Concept of Fault and Its Apportionment" (1977) 51 *Tulane Law Review*, 873–879.

34. *The Batavier III* (1925) 23 Ll.L.R. 21, 16 Asp. M.L.C. 563.

cases of mutual fault with rules for proportional liability under the MCA 1911. Prior to the passage of the MCA 1911 in certain circumstances proof of negligence was unnecessary in cases in which it was shown that a vessel infringed a collision regulation and the accident followed. At present an infringement of a collision regulation is no more than *prima facie* evidence of negligence on the part of the offending ship. In order to succeed in an action the plaintiff must prove that the act of negligence was the effective cause of the collision or at least that it contributed to the casualty. If the defendant can show that the infringement of a collision regulation had no effect on the collision no fault will be imputed to his vessel on the basis of infringement alone.

Section 1 of the Act covers the rules on division of loss and deals only with vessels at fault. This section does not apply to claims for loss of life or personal injury. Section 2 of the Act specifies that where there are two or more vessels to blame, liability for loss of life or personal injury shall be "joint and several". Section 3 deals with contribution between ships at fault. Section 4 repeals the provisions of the Merchant Shipping Act 1894 relating to statutory presumption of fault. Under section 5 claimants for loss of life or personal injury have the right to sue either *in rem* or *in personam* in any court having Admiralty jurisdiction.

In scope the Act seems to be wider than the Convention in two respects:

(1) The Act covers cases of fault not only in the navigation of ships but also in the management of them in other respects than navigation. For instance the Act covers a case in which a ship moored in harbour negligently discharges water or other liquids onto a barge lying alongside. But cases where only one ship is involved, such as collision by a single ship with a jetty, lock gate, or other shore installation, are not covered under the Act.

(2) The Act applies to liability not only for damage or loss to ships or to persons or property on board them but also to liability for loss or freight including passage money and hire.

(c) Principles of apportionment of liability under the 1911 Act

In English law only causative faults, faults that contributed to the collision or other casualty from which the damage or loss concerned arose, could be taken into account in apportioning collision liability under the Act. Causative fault should be considered in two ways: first, from the point of its blameworthiness or culpability—the extent of its contribution to the damage or loss concerned will not be considered; secondly, from the point to which it contributed to the damage—this consideration does not include its blameworthiness and culpability. In apportioning liability under 1911 Act culpability and the causative potency of all relevant faults must be taken into account together and weighed in the balance. This point was stated by Lord Denning in *The Koningin Juliana*[35] as follows: "the degree of fault is not to be measured by counting up the number of faults on each side. It is to be measured by assessing both their blameworthiness and their causative effect".

35. [1974] 2 Lloyd's Rep. 353, at p.356.

The exceptional cases could be where it did not contribute to the collision or other casualty itself but did contribute to the damage or loss that arose out of it.

The general principle of the common law of negligence applies also to Admiralty cases. The burden of proof is on the plaintiff to establish that the defendant was negligent and his negligence caused the damage or loss sued for. The abolition of the statutory presumption of fault rule is important here. Under this rule it was presumed that the infringement of a collision regulation rule had caused the collision. This rule was brought into being by virtue of the Merchant Shipping Act 1894. With the enactment of the 1911 Act an infringement of a collision regulation is now no more than *prima facie* evidence of negligence on the part of the offending ship. But the plaintiff must prove that the act of negligence was the effective cause of the collision or at least that it contributed to the casualty. In other words the breach of the collision regulations must be causative like any other kind of fault. Under the 1911 Act the court has to decide whether breach of the collision regulations was causative or not.

Damages which are too remote and not directly consequential upon the original act of negligence cannot be recovered. In *The Fogo*[36] the *Trentbank* was found wholly to blame for a collision with the *Fogo*. After the collision the *Trentbank* sank. The court found that the sinking was due to her failure to employ a tug for assistance. This failure was considered as a *novus actus interveniens* which broke the chain of causation. On the other hand in some cases, as in *The Calliope*, the contrast can be seen. In *The Calliope* the court found that the subsequent negligence of the plaintiffs did not break the chain of causation as this would amount to a resurrection of the last opportunity rule.[37] An early case, *Davies* v. *Mann*, was the basis for the last opportunity rule.[38] The last opportunity rule applied in collisions between ships. But the courts' view began to change gradually and they came to the view that a fairer rule to apply to collisions at sea should be whose act caused the wrong, instead of who had the last opportunity of avoiding mischief.[39] Finally the courts began to apply different rules rather than the last opportunity rule. The faults committed by the ships concerned and contribution of the faults to the loss or damage in issue are the topics to be considered.[40]

36. [1967] 2 Lloyd's Rep. 208.
37. [1970] 1 Lloyd's Rep. 84.
38. In *Davies* v. *Mann* (1842) 10 M. & W. 546, the owner of a donkey, which had been negligently left hobbled and unguarded on a highway, sued the defendant, by the negligence of whose servant in driving on a highway at too rapid speed the donkey was run over and injured. It was held that although he was negligent the donkey-owner could recover. This case established that where one party had created a negligence situation which was fixed and stabilised and another party had later also been negligent and caused loss or damage, the later party would be solely to blame if he had the opportunity to avoid the caused loss or damage. See Christopher Hill, *Maritime Law* (4th ed., 1995), pp.283–284.
39. The applicability of the last opportunity rule in collision cases was considered by the House of Lords in *The Boy Andrew* v. *St Rognvald (Owners)* [1948] A.C. 140; [1947] 2 All E.R. 350 (HL). In this case approval was given to the statement made by the Law Revision Committee in 1939 (Law Revision Committee, Command No. 6032 at 16), which said: "In truth, there is no such rule—the question, in all questions of liability for a tortious act, is, not who had the last opportunity of avoiding the mischief, but whose act caused the wrong."
40. The contribution of faults to the loss or damage was considered in *Admiralty* v. *The Camroux I (Owners)* [1958] 1 Lloyd's Rep. 378. In this case two rules applied. If the negligence of both parties to

There has been uncertainty about the tests to be applied to determine what is remote or not. In *Re Polemis*[41] the test was that if some damage could be foreseen all loss and damage consequential upon and directly related to the original negligent act would be recoverable by the innocent party. Obviously there was no clear distinction between all foreseeable and all direct consequences. This case is no longer regarded as good law following the decision in the *Wagon Mound* case. In *Overseas Tankship (UK) Ltd.* v. *Morts Dock & Engineering Co. Ltd. (The Wagon Mound) (No. 1)*[42] it was held that foreseeability should be the key word. The consequences which could not be reasonably foreseeable as directly or immediately resulting should fall into the too remote category. A plaintiff can recover damages for the negligence of the defendant only if that damage could have been foreseen by a reasonable man. It is not enough to show that the damage was a direct physical consequence of a negligent act.

In *The Fritz Thyssen*[43] it was held that where a vessel is damaged in a collision due to the negligence of another vessel and later sinks, the owners of the other vessel are liable only where the sinking is the direct result of the collision. Again in *The Edison*[44] it was held that damages which are too remote cannot be recovered. In this case a profit-earning dredger was sunk in a collision in Patras harbour. The owners claimed (1) the market price of a comparable dredger; (2) the cost of adapting a new dredger and insuring and transporting her from her moorings to Patras; (3) compensation for disturbance and loss in carrying out their contract of dredging the harbour; and (4) loss due to their inability for financial reasons to buy a substitute dredger and the resulting delay in proceeding with the work. The House of Lords held that the owners' claim under heads (1), (2) and (3) succeeded, but that under head (4) failed because such damage was too remote.

The Edison has been subject to a certain amount of reinterpretation in recent years. In *Dodd Properties (Kent) Ltd.* v. *Canterbury City Council*[45] the main issue for the Court of Appeal to decide was the date when the cost of repairs was to be assessed. The plaintiffs argued that it should be the date of the judgment, 1978, but the trial judge fixed it at 1970, the date to which he decided it was reasonable to postpone carrying out the repairs. The importance of the issue lay in the fact that inflation had vastly increased the cost of repairs between 1970 and 1978. The trial judge held, following *The Edison*, that in deciding on what was a reasonable time at which to do the repairs any impecuniosity or financial stringency had to be disregarded and he opted for the 1970 date. The Court of Appeal thought it was

---

the litigation continues until the moment of collision and the collision is the result of the continued negligence of both, then each party is to blame for the collision. If the negligent act of one party causes the other party to make a negligent mistake that he would not otherwise have made then both parties are to blame.

41. *Re Polemis and Furness Withy & Co Ltd.* [1921] 3 K.B. 560; [1921] All E.R. 40.
42. [1961] AC 388.
43. [1967] 2 Lloyd's Rep. 199. Affirmed in the Court of Appeal.
44. [1933] A.C. 449; [1933] All E.R. Rep. 144.
45. [1980] 1 All E.R. 928 (CA).

reasonable to postpone repairs until 1978. In the Court of Appeal it was stated by Donaldson L.J. that "the position of the plaintiffs in the present case seems to me to be quite different. They were not impecunious in *The Liesbosch* [in this case *The Edison* case was referred to under its alternative name *The Liesbosch*] sense of one who could not go out into the market. On the contrary they were financially able to carry out the work of reinstatement in 1970". According to Donaldson L.J. *The Liesbosch* had no application to the present case. Donaldson L.J.'s treatment of *The Edison* has been cited with apparent approval by Kerr L.J. in *Perry* v. *Sydney Phillips & Son*.[46] In this case Lord Denning M.R. was prepared to go so far as to dismiss Lord Wright's view in *The Edison* concerning non-recovery of loss due to impecuniosity as "not of general application" and "restricted to the facts of the *Liesbosch* case".[47]

In general culpability, irrespective of the extent to which it contributed to the damage or loss concerned, and causative potency, the extent to which it contributed to the damage or loss concerned, of the faults are taken into account in considering liability under the 1911 Act. In other words in apportioning liability under the 1911 Act the culpability and the causative potency of all relevant faults must be taken into account and weighed in the balance.

(i) Division of damages

Loss of or damage to ships and cargo

At common law, where two vessels were negligent damages would be equally divided regardless of the degree of fault. This rule was abolished in cases where the 1911 Maritime Conventions Act applies. Under the Act the liability of vessels at fault for damage to property and freight, where there are different degrees of blame and such different degrees can be determined, must be apportioned according to the degree to which each vessel is in fault. Section 1(1) of the Act provides as follows:

"(1) Where, by fault[48] of two or more vessels,[49] damage or loss is caused to one or more of those vessels, to their cargoes or freight, or to any property on board, the liability to make good the damage or loss shall be in proportion to the degree in which each vessel was in fault:

Provided that:

(a) if, having regard to all the circumstances of the case, it is not possible to establish different degrees of fault, the liability shall be apportioned equally; and

(b) nothing in this section shall operate so as to render any vessel liable for any loss or damage to which her fault has not contributed; and

46. [1982] 3 All E.R. 705; [1982] 1 W.L.R. 1297.

47. Similarly Donaldson L.J.'s treatment of *The Edison* was also found helpful by Peter Pain J. in *Archer* v. *Brown* where he applied a reasonable foreseeability test: [1985] Q.B. 401; [1984] 2 All E.R. 267.

48. The definition of fault is not given in the 1911 Act or the 1910 Convention.

49. For the definition of "vessel", see MSA 1894, s.742. In *The Manorbier Castle* (1922) 129 L.T. 31, 16 Asp. H.L.C. 151, a wreck and a wreck-marking vessel were held to be "vessels" within the meaning of the 1911 Act. In *Edwards* v. *Quickenden and Forester* [1939] P. 261 it was held that no craft propelled by oars is a "vessel" under the 1911 Act.

(c) nothing in this section shall affect the liability of any person under a contract of carriage or any contract, or shall be construed as imposing any liability upon any person from which he is exempted by any contract or by any provision of law, or as affecting the right of any person to limit his liability in the manner provided by law.

(2) For the purposes of this Act, the expression 'freight' includes passage money and hire, and reference to damage or loss caused by the fault of a vessel shall be construed as including references to any salvage or other expenses, consequent upon that fault, recoverable at law by way of damages."

Fault which causes or contributes to damage is relevant to the apportionment of liability. Therefore it is not simply the contribution to the collision that is relevant but the contribution to the damage. To prove fault the onus is on the party alleging negligence to prove both the breach of duty and the consequent damage. For instance in *The Margaret*,[50] although both vessels were guilty of negligence, only one of them caused the negligence. In *The Bovenkerk* it was held that the presence of the dredger, while setting the scene for the collision, was not a contributory cause as its presence was well known to the pilots of both vessels.[51] As stated above in *The Fritz Thyssen*[52] the court found that the sinking of the vessel was not caused by the collision but was due to her failure to take necessary action. This is *prima facie* evidence that the ship was lost as a result of the collision as it was badly holed by a collision and sank soon after as a result of water entering the hole so made. In such a case the defendant should try to disprove the case by showing that there was intervening negligence on board the plaintiff's ship which broke the chain of causation. Then the defendant must show that the plaintiff failed to take some precaution as a good seaman and that if the precaution had been taken the loss would probably have been avoided.

The need for a judicial approach for the establishment of degrees of fault was stressed in *The Peter Benoit*.[53] In this case a collision occurred between two vessels off the mouth of the River Tees. There was no evidence on which the blame could be apportioned with any certainty. The judicial approach to be adopted was laid down in the House of Lords[54]:

"... The conclusion that it is possible to establish different degrees of fault must be a conclusion proved by evidence, judicially arrived at, and sufficiently made out. Conjecture will not do: a general leaning in favour of one ship rather than of the other will not do: sympathy for one of the wrongdoers, too indefinite to be supported by a reasoned judgment, will not do. The question is not answered by deciding who was the first wrongdoer, nor even of necessity who was the last. The Act says, 'having regard to all the circumstances of the case'."

Attention must be paid not only to the actual time of the collision and the manoeuvres of the ship, when about to collide, but to their prior movements and

50. (1881) 6 P.D. 76.
51. [1973] 1 Lloyd's Rep. 63.
52. See n.43 *supra*.
53. (1915) 114 L.T. 147.
54. *Ibid.*, p.152.

opportunities, their acts and omissions. Matters which are only introductory, even though they preceded the collision by but a short time, are not really circumstances of the case but only its antecedents, and they should not directly affect the result.

The words "damage and loss" cover the words "damages and losses". And the court can make different apportionments of liability for different types of damage or loss resulting from the same event.[55] The provision is not limited only to cases of collisions where two or more vessels are at fault. It also applies to cases where the two or more vessels at fault are not themselves in collision.

Although the most common collision case is where two ships are at fault, sometimes casualties occur where three or more ships at fault are involved. The section applies where there are more than two ships at fault. The approach for the apportionment of liability in a case where a primary casualty involving two ships led to a secondary casualty to a third ship was considered in *The Abadesa* where the *Abadesa* and another ship were in collision in a river and, as a result, a third ship was damaged.[56]

Under this section the apportionment of liability is mandatory not discretionary, therefore the court must apportion liability unless it is impossible to do so.[57] This is followed by the provision that if the court finds it impossible to apportion different degrees of fault, it is to declare an equal distribution of fault.[58] Obviously this principle must not be used to evade responsibility for weighing the faults of each ship against the other. As stated above, the USA is not a party to the 1910 Brussels Convention but adopted a similar rule by the decision of the Supreme Court in *United States* v. *Reliable Transfer Co.*[59]

The words "fault of two or more vessels" in section 1(1) of the 1911 Act cover fault on the part of those responsible for the vessel, but are not limited to faults in navigation. They could be faults in respect of management apart from navigation. In *The Norwhale* there was an argument that the 1911 Act applied only to

55. In *The Calliope* Brandon J. said that "... The effect of sect. 1(1)(a) of the Interpretation Act, 1889, is that, unless the contrary intention appears, the words 'damage' and 'loss' in sect. 1(1) of the Act of 1911 must be construed as including the plural as well as the singular. So construing sect. 1(1), I do not see any reason why, where the facts of the case justify it, the court should not be entitled under it to hold that liability for one head of damage or loss resulting from a casualty or event should be apportioned on one basis, and that for another head of damage or loss resulting from the same casualty should be apportioned on another basis. There is certainly nothing in sect. 1(1) itself to show a contrary intention ...' See n.37 *supra*.

56. *The Miraflores and The Abadesa* [1966] 1 Lloyd's Rep. 97, [1966] P. 18 (CA), [1967] A.C. 826 (HL).

57. In *The Anneliese* [1970] 1 Lloyd's Rep. 355 (CA), it was said: "Now that section [s.1(1) of the Maritime Conventions Act], as I read it is mandatory. It does not say that the liability shall be apportioned equally unless different degrees of fault are shown. It is the other way round. It says that the Court must apportion the liability in proportion to the degree in which each vessel was at fault unless it is impossible to do so. Of course the different degrees of fault must be proved like anything else in a Court of law."

58. See *The Lucile Bloomfield* [1967] 1 Lloyd's Rep. 341.

59. 421 U.S. 397, 1975 A.H.C. 541 (1975). In this case it was stated: "When two or more parties have contributed by their fault to cause property damage in a maritime collision or stranding, liability for such damage is to be allocated among the parties proportionately to the comparative degree of their fault, and liability for such damage is to be allocated equally only when the parties are equally at fault or when it is not possible fairly to measure the comparative degree of their fault".

navigational faults. An attempt to limit the scope of the Act to cases of damage by collision or damage caused without actual collision by navigational fault failed. It was held that the Act applies also to faults in management.[60] Another important point is that the fault of one of the ship's officers or crew does not mean that the vessel is at fault under the 1911 Act. In other words the negligence of an officer on board a ship is not always to be treated as the fault of the ship.[61] He should be in charge of the navigation and his responsibilities should be connected with the operation of that vessel.

Section 1(1) applies only to apportionment of liability between vessels at fault. If ship A, without any fault, is damaged by the combined fault of ships B and C they will be jointly and severally liable to ship A for the whole of her damage. Therefore in such cases there will be no apportionment of liability and the innocent third party may recover in full against either vessel.[62]

The section applies to damage or loss caused to "vessels, to their cargoes or freight". If the damage is done to some other property by the fault of two or more vessels, i.e. where a ship collides with any other object, such as a pier or dock gate, it will not come within the section.[63]

Claims for loss of life or personal injury

Section 2 of the Maritime Conventions Act 1911 states as follows:

"Where loss of life or personal injuries are suffered by any person on board a vessel owing to the fault of that vessel and of any other vessel or vessels, the liability of the owners of the vessels shall be joint and several:

Provided that nothing in this section shall be construed as depriving any person of any right of defence on which, independently of this section, he might have relied in an action brought against him by the person injured, or any person or persons entitled to sue in respect of such loss of life, or shall affect the right of any person to limit his liability in cases to which this section relates in the manner provided by law."

Under this section claimants for personal injury or loss of life can recover their claims in full against either vessel. The liability of the vessels at fault is joint and several, so the plaintiff can sue either vessel for the full amount of his claim and

60. *Owners of Barge Norwhale* v. *The Ministry of Defence (The Norwhale)* [1975] 1 Lloyd's Rep. 610.

61. See *The Sobieski* [1949] P. 313 (CA). In this case the collision occurred between two vessels, the *SS Espérance* and *SS Sobieski*. The collision was the result of the negligence of each of them and also of a senior escort officer, L, on board a naval vessel, the *Loch Alvie*, which was escorting the *SS Sobieski*. L's negligence was in failing to transmit information of the approach, detected by radar of the *SS Espérance*. It was held that L's negligence was not the negligence of his vessel. Therefore the 1911 Act had no application to the claim of *SS Sobieski* against L.

62. See *The Cairnbahn* [1914] P. 25 (CA). In this case a collision occurred between a steamship and a barge in tow by and controlled by a tug. The tug owners had no interest in the barge. The steamship and tug were found equally to blame for the collision. The owners of the barge recovered the full amount of their damage from the steamship.

63. In *Manchester Ship Canal Co. (Alpha)* v. *Helgoy* (1924) 18 Ll.L.R. 191, a jetty belonged to the owners of vessel A and was damaged in a collision between vessels A and B. Both vessels were equally to blame. Although half of the damage to A and B was recoverable from the owners of B and A respectively, no part of the damage to the jetty was recoverable from the owners of B.

leave the shipowner to seek a proportionate recovery from the other. This provision preserves the defence available to the shipowner in such a case.

There may be cases where a plaintiff's claim is excluded or restricted by a term in a passenger ticket, contract of employment or by a provision of the law of the flag of that ship. The claimant can bring his claim against the other ship and avoid the application of such exclusion or restriction under section 2 of the 1911 Act.

On the other hand cargo owners can only claim from each vessel in the proportion in which each vessel was to blame. In other words the liability of the two vessels at fault is several and limited to the proportion of fault that that ship bears. Therefore if a cargo in one vessel is damaged by a collision with another vessel by the fault of both vessels, the cargo owner can only recover a proportion of his loss from each vessel in relation to the degree to which that vessel is to blame for the collision.

Contribution between ships at fault

Section 3(1) of the 1911 Act provides as follows:

> "Where loss of life or personal injuries are suffered by any person on board a vessel owing to the fault of that vessel and of any other vessel or vessels, and a proportion of the damage is recovered against the owners of one of the vessels which exceeds the proportion in which she was in fault, they may recover by way of contribution the amount of the excess from the owners of the other vessel or vessels, to the extent to which those vessels were respectively at fault:
>
> Provided that no amount shall be so recovered which could not, by reason of any statutory or contractual limitation of or exemption from liability, or which could not, for any other reason, have been recovered in the first instance as damages by the persons entitled to sue therefor."

Under this section each ship at fault bears only the proportion of the claim equivalent to the degree to which that ship was at fault. Sections 2 and 3 of the 1911 Act protect the interests of the plaintiff who has suffered personal injury or loss of life. If a person on board a vessel suffers personal injury or loss of life because of the fault of that vessel and any other vessels, he may recover a proportion of the damages from the owners of one of the vessels. In such a case that ship's fault may be less than the proportion recovered. The owners of the ship may recover by way of contribution the amount of the excess from the owners of the other vessel or vessels to the extent to which those vessels were respectively at fault. In cases where a plaintiff has recovered in full against one ship under section 2 the 1911 Act, that ship will not recover from the other ship more than the plaintiff could have recovered if he had made a claim directly against that other ship.[64] Under section 3 the right of contribution is limited to cases where the vessel from which contribution is claimed would have had no defence to the claims in the first instance. This may be by reason of any statutory or contractual limitation of, or exemption from, liability or for any other reason.

64. N. Meeson, *Admiralty Jurisdiction and Practice* (1993), p.213.

Limitation of actions

Section 8 of the 1911 Act provides that legal proceedings in a collision case should be commenced within two years from the arising of the cause of action.

"No action shall be maintainable to enforce any claim or lien against a vessel or her owners in respect of any damage or loss to another vessel, her cargo or freight, or any property on board her, or damages for loss of life or personal injuries suffered by any person on board her, caused by the fault of the former vessel, whether such vessel be wholly or partly in fault, or in respect of any salvage services, unless proceedings therein are commenced within two years from the date when the damage or loss or injury was caused or the salvage services were rendered, and an action shall not be maintainable under this Act to enforce any contribution in respect of an overpaid proportion of any damages for loss of life or personal injuries unless proceedings therein are commenced within one year from the date of payment:

Provided that any court having jurisdiction to deal with an action to which this section relates may, in accordance with the rules of court, extend any such period, to such extent and on such conditions as it thinks fit, and shall, if satisfied that there has not during such period been any reasonable opportunity of arresting the defendant vessel within the jurisdiction of the court, or within the territorial waters of the country to which the plaintiff's ship belongs or in which the plaintiff resides or has his principal place of business, extend any such period to an extent sufficient to give such reasonable opportunity."

This section applies only to claims in respect of damage to or loss of cargo or property or loss of life or personal injury which lie against the other vessel. It does not apply to all claims. In *The Niceto de Larrinaga*[65] the judge emphasised this principle. In this case the mother of a man killed in a collision between the two ships on 23 September 1961 issued a writ on 18 September 1964 against the owner of the ship in which her son was. The shipowner contended that the action was time barred as the two-year limit under the 1911 Act applied. It was held that the two-year limit did not apply to third party claims against the carrying vessel. The mother had brought her action within three years which was in time.

Actions to enforce contributions in respect of an overpaid proportion of damage must be commenced within one year from the date of payment. The court has discretion to extend the period in circumstances which justify it. The discretion of the court was exercised in *The Alnwick*. In this case it was held that the court's discretion should be exercised to allow the action to continue.[66] The limitation period will cease to run if a writ is issued. The original writ remains in force for 12 months, so once the writ is issued it must be served within 12 months unless the court in its discretion grants an extension of time.

(ii) Difficulties in practice in apportionment of liability

There are no general rules for the apportionment of liability in collision cases. Maritime accidents happen in so many different circumstances and the assessment of degrees of fault depends on the particular circumstances of the given accident.

65. [1965] 2 Lloyd's Rep. 134.
66. [1965] P. 357; [1965] 1 Lloyd's Rep. 69.

The courts find it difficult to assess causative potency in terms of percentages and this difficulty was expressed in the words of Lord Wright in *The MacGregor*[67]:

"[Apportionment] is a question of the degree of fault, depending on a trained and expert judgment considering all the circumstances, and it is different in essence from a mere finding of fact in the ordinary sense. It is a question not of principle, or of positive findings of fact or law, but of proportion, balance and relative emphasis, and of weighing different considerations. It involves an individual choice of discretion, as to which there may well be differences of opinion by different minds."

As was stated in *The MacGregor*[68] it is necessary to take into consideration all the circumstances of the case. But while considering the fault of each ship concerned the nature and quality of a ship's fault should be more important than the number of faults committed by that ship. This principle was stated clearly in *The Koningin Juliana*[69]:

"Apportionment of blame in the Admiralty Court is not to be arrived at by totting up and comparing the number of faults in the navigation of each vessel. If I may borrow from the poet, Wordsworth, I would say that the Admiralty Court '... rejects the role of nicely calculated less or more'.

The inquiry must be qualitative rather than quantitative. It is necessary to look at the over-all picture ..."

In some cases the serious fault of one ship will be obvious so that there is no difficulty in apportionment. In *The Sea Star* Brandon J. held that[70]:

"a) the *Sea Star* was very seriously to blame for the collision for altering course to starboard at an improper time and it was to be inferred that this improper manoeuvre was caused by previous defective look-out, or defective appreciation of the situation, or both. b) *Horta Barbosa* was also to blame for not taking proper avoiding action in time. c) As to apportionment, there was no doubt that the situation of danger was created by the fault of the *Sea Star*, and that the fault of *Horta Barbosa* lay only in her failure to react properly to the situation so created."[71]

In section 4 of the 1911 Act[72] the statutory presumption of fault is abolished, but in practice some types of fault are generally regarded as seriously culpable. For instance if a ship has or keeps excessive speed in fog it will almost always be treated as seriously culpable because such behaviour involves a direct contravention of

67. [1943] A.C. 197; (1942) 74 Ll.L.R. 82 at pp.201 and 85 respectively.
68. *Ibid.*
69. [1974] 2 Lloyd's Rep. 353 at 364, [1975] 2 Lloyd's Rep. 111 (HL).
70. See n.18 *supra*.
71. See *Liquilassie Shipping Ltd.* v. *M.V. "Nipigon Bay"* [1975] 2 Lloyd's Rep. 279.
72. S.4 of the 1911 Act states: "(1) Subsection (4) of section four hundred and nineteen of the Merchant Shipping Act, 1894 (which provides that a ship shall be deemed in fault in a case of collision where any of the collision regulations have been infringed by that ship), is hereby repealed.

(2) The failure of the master or persons in charge of a vessel to comply with the provisions of section four hundred and twenty-two of the Merchant Shipping Act, 1894 (which imposes a duty upon masters and persons in charge of vessels after a collision to stand by and assist the other vessel), shall not raise any presumption of law that the collision was caused by his wrongful act, neglect or default and accordingly subsection (2) of that section shall be repealed."

Rule 16 of the Collision Regulations.[73] In *The Savina* bad look-out and a breach of Rule 21 of the Collision Regulations combined to bring two ships into collision. It was held that the *Forest Hill*'s breach of Rule 21 was more blameworthy than the *Savina*'s bad look-out, and liability was proportioned as to 60 per cent to the *Forest Hill* and 40 per cent to the *Savina*. In this case it was pointed out that:

"A positive fault [deliberate breach of a rule] is not necessarily more blameworthy than negligent navigation. But that it may be so is apparent from taking extreme cases—grossly negligent navigation causing a collision compared with the deliberate running-down of one vessel by another . . . It is, in my view, open to an Admiralty Court . . . to attach more blame to any deliberate breach of a rule than to a fault or omission. Whether the Court should do so depends on all the circumstances—in particular, the nature of the rule and the mode of its breach, on the one hand, and the degree of negligence in navigation on the other."[74]

In *The Miraflores and The Abadesa*[75] the tanker *George Livanos* grounded while manoeuvring to avoid the tanker *Miraflores* and the tanker *Abadesa* which had collided in the River Scheldt. It was held that the *Abadesa* was two-thirds and the *Miraflores* was one-third to blame. The *George Livanos* was 50 per cent to blame. It was entitled to recover two-thirds of the remaining 50 per cent from the *Abadesa* and one-third from the *Miraflores*. The *George Livanos* appealed and the distribution of fault and liability were changed to: the *Abadesa* 45 per cent; the *Miraflores* 30 per cent; the *George Livanos* 25 per cent to blame. On appeal by the *Miraflores*, it was held by the House of Lords that the apportionment of blame should be: the *Abedesa*, two-fifths (40 per cent); the *Miraflores*, one-fifth (20 per cent); the *George Livanos* two-fifths (40 per cent).

In this case the following rule was set out by Lord Pearce: "To get a fair apportionment it is necessary to weigh the fault of each negligent party against that of each of the others. It is or may be quite misleading to substitute for a measurement of the individual fault of each contributor to the accident a measurement of the fault of one against the joint fault of the rest".[76] This case shows that in most cases it will be right to treat the fault of a ship that creates a situation of difficulty or danger in the first place as greater than that of a ship that fails to react properly to such a situation after it has been created.[77]

The general aim of the courts is to deal with the apportionment of fault in a broad way. But there is no single correct approach to the problem and judgments may differ in different courts.

(d) The Lisbon Rules

(i) Introduction

In the first draft of the 1910 Brussels Convention there was an article designed to

73. H.V. Brandon, "Apportionment of Liability in British Courts Under the Maritime Conventions Act of 1911" (1977) 51 *Tulane Law Review* 1038.
74. [1976] 2 Lloyd's Rep. 123.
75. See n.56 *supra*.
76. *The Miraflores and The Abadesa*, n.56 *supra*.
77. H.V. Brandon, p.1039.

ensure for the victim complete compensation of his loss. Under this article it was provided that the damages should cover "the expenses due to collision, the profit lost and compensation for the detention". But this article was omitted from the Convention. The difficulties in the apportionment of liability and division of damages under the 1910 Brussels Convention led the Comité Maritime International (CMI) to prepare another convention. In 1962 at the Athens Conference of the CMI an attempt was made to propose a convention on the subject, but the Assembly of the CMI adopted a timid resolution.[78] In 1983 a questionnaire was sent to the National Maritime Law Associations to find out how they deal with the problems of compensating shipowners for losses suffered as a result of a collision. Ultimately 24 National Associations replied. The CMI appointed an International Working Group to prepare a set of rules about collision damages on the basis of such documents. The first draft was considered by delegates to the CMI Conference in Lisbon in 1985. After several further meetings of the working group the Lisbon Rules received unanimous approval at the meeting of the Assembly of CMI, held at Brussels in April 1987.[79]

The CMI abandoned the idea of a convention and instead recommended Rules which could be adopted by the parties. The aim of the CMI was to create a set of rules to ensure that claims arising out of collisions would be dealt with on a uniform basis. The Rules do not have the force of law and cannot be imposed on anyone. Although the Lisbon Rules are not binding, but may simply be agreed by the parties, it is possible that in some countries where there is not already a developed system for calculation of compensatory damages in collision cases, the Rules may be adopted as a matter of law. The Rules merely seek to put forward a model which the parties can agree to follow.

One of the problems which was addressed by those who drafted the Rules was the problem of inflation, and how this should be dealt with where the parties come from countries where inflation is rampant. The Rules provide that the parties must agree on a currency of account, but if they do not do so then the damages are calculated in SDRs, which would be re-exchanged into national currencies when

78. The resolution adopted stated:

"1. The party who has sustained a damage by collision is entitled to be put as far as practicable in the same position pecuniarily as if the injury had not been suffered.

2. The victim must always mitigate the damage if he can reasonably do so.

3. The wrongdoer shall only be liable for damage which can be reasonably attributed to his wrongful act.

Considering, however, the application of those principles requires further study, [the CMI]

—Resolves that said Report be sent to the National Associations for consideration and requests these Associations to express their views thereon,

—Instructs the International Sub-committee to continue the study of the topic and report again to the next International Conference."

79. See J. Warot, "A Comment on The Lisbon Rules on Compensation for Damages in Collision Cases" (1987) 18 *Journal of Maritime Law and Commerce* 583–585. Lecture given by Patrick Griggs, Secretary and Treasurer of the British Maritime Law Association, in 1992.

the time came for payment. The Rules can be adopted on a case-by-case basis.[80] The Rules consist of three parts: first there is a set of definitions, then there is a set of basic lettered rules and a set of numbered rules. Within the Rules "rig or platform" is included under the definition of "vessel". The definition of "collision" includes an incident where there has in fact been no physical contact between two ships. Death and personal injury claimants were excluded from the definition of "claimant". These claims are going to be dealt under national legislatures.

(ii) The lettered rules

These provisions restate the general rules of law in the matter of compensation for collision damage.

Under Rule A it is stated that the Rules are "available for adoption . . . following a collision", and do not have the force of law. Rule B confirms that the Rules "do not extend to the determination of liability or affect rights of limitation of liability". Rule C says: "subject to the application of the numbered rules the Claimant shall be entitled to recover only such damages as may reasonably be considered to be the direct and immediate consequence of the collision". Under this Rule the primacy of the "Numbered Rules" is established. The remoteness of the damage will be assessed in relation to the direct and immediate consequence of the collision. Rule D says: "subject to the application of Rule C and of the numbered Rules, damages shall place the Claimant in the same financial position as he would have occupied had the collision not occurred". The principle of *restitutio in integrum* governs the calculation of collision damage. Rule E states: "the burden of proving the loss or damage sustained in accordance with these Rules shall be upon the Claimant. Damages shall not be recoverable to the extent that the person against whom the claim is made is able to show that the Claimant could have avoided or mitigated the loss or damage by the exercise of reasonable diligence". This Rule reflects the principle that the burden of proof rests on the plaintiff and the principle of mitigation. If the claimant alleges that the loss could have been avoided or minimized by the defendant the burden of proof rests on the claimant.

(iii) The numbered rules

Rules I, II and III cover total loss, and damage to vessel and property on board.

80. The Lisbon Rules might be adopted by the parties in the following form:
RE:..(0)

The parties hereto agree that their respective claims arising out of the above incident shall be assessed in accordance with the Lisbon Rules 1987.

The damages shall be calculated in

....(X)

(0) Insert names of ships involved and date/place of collision.
(X) If a currency is not specified, Rule V provides that the damages will be calculated in Special Drawing Rights (SDR).
* The Lisbon Rules, CMI Newsletter, Spring 1987, p.8.

Rule IV covers interest and Rule V deals with currency. Rule I:1 states: "In the event of a vessel being a total loss, the Claimant shall be entitled to damages equal to the cost of purchasing a similar vessel in the market at the date of the collision. Where no similar vessel is available, the Claimant shall be entitled to recover as damages the value of the vessel at the date of the collision calculated by reference to the type, age, condition, nature of operation of the vessel and any other relevant factors".

2. "Damages recoverable in the event of a total loss shall also include:

(a) Reimbursement of salvage, general average and other charges and expenses reasonably incurred as a result of the collision.

(b) Reimbursement of sums for which the Claimant has become legally liable and has paid to third parties in respect of such liability, arising out of the collision by reason of contractual, statutory or other legal obligations.

(c) Reimbursement for the net freight lost and the value of bunkers and ship's gear lost as a result of the collision.

(d) Reimbursement for the net freight lost and the cost of replacing bunkers and vessel's gear lost as a result of the collision and not included in the cost of repairs under Rule II 1.(a)."

Under Rule II in the event of a vessel being damaged but not being a total loss as defined in the Rules:

2. "Damages recoverable shall also include:

(a) Subject to reimbursement for any claim for loss of freight under Rule II 1.(d), compensation for the net loss of earnings arising from the collision. This compensation shall be assessed by establishing the gross earnings of the vessel lost during detention, calculated by reference to the vessel's earnings or by reference to the earnings of comparable vessels in the same trade and then deducting from the gross earnings the operating costs which would normally have been incurred in order to achieve the gross earnings, such as hire payable, crew and bunker costs, port disbursements and insurance.

(b) Operating costs and expenses actually incurred during detention, other than those included under Rule II 1.

3. "In the interpretation of Rule II 2., the following particular provisions will also apply:

(a) When detention occurs during the performance of a voyage charterparty and such detention does not entail cancellation of the charterparty, compensation shall be calculated by applying the average net earnings on the two voyages prior to and the two voyages subsequent to the detention.

When no reference to two prior and two subsequent voyages is possible, the net earnings on other relevant voyages or if there are no other relevant voyages on the voyage during which the collision took place shall form the basis of compensation.

If in consequence of such detention the charterparty is cancelled, and freight remains unearned compensation shall include the net freight lost.

(b) When detention occurs while the vessel is being operated on a liner service, compensation for detention shall be assessed as follows":

Rule IV 3

3. "Where under Rule V the damages are to be calculated in Special Drawing Rights (SDR), the rate of interest shall be the average London rate for three months SDR linked deposits in the period that interest runs; otherwise the rate of interest shall be ten per cent per annum."

Rule V: "Unless the parties have agreed to apply a specific currency in the calculation of their damages the following procedure shall be adopted:

– losses or expenses shall be converted from the currency in which they are incurred into Special Drawing Rights (SDR) at the rate of exchange prevailing on the day the losses or expenses were incurred.

– the final amounts due shall be calculated in SDR and the balance due shall be paid to the Claimant in the currency of his choice at the rate of exchange prevailing on the date of payment.

– where no official SDR exchange rate is quoted for the currency, conversions to and from SDR shall be made by reference to US dollars."

The existence of this Rule V is based on increasing inflation and the problems that it causes. Therefore Special Drawing Rights are referred to as a stable currency.

The Lisbon Rules should be examined with the prospect of forum shopping. In order to ascertain recoverable amounts in collision cases countries use different practices which give rise to forum shopping. In other words the claim of one ship against another may be determined in one jurisdiction while the claim of the other ship against the first ship is determined in a different jurisdiction which adopts a different practice. Therefore an international set of rules would eliminate the problem. But the Lisbon Rules should not be considered as complementary to the 1910 Brussels Convention. The Rules apply only to collision compensation claims with regard to loss and damage assessment.

The proposed compensation for damages as set out in the Lisbon Rules may be higher than the compensation awarded in some countries. But in some countries the compensation available under the Lisbon Rules could be less than what is otherwise available. Dutch law can be given as an example. Under Dutch law a time charterer of a ship can make a claim for economic losses against the owners of the other ship. Under the Lisbon Rules the application of this right is different. The Lisbon Rules also permit such a claim but they are not available to a time charterer as a right because the Lisbon Rules are adopted between two shipowners after a collision has occurred and therefore are not available to a time charterer unless they are offered to him.

CHAPTER 10

INSURANCE ASPECTS

GENERAL

A collision may cause many different kinds of liabilities. The most common types are: liability for damage to the colliding vessels, liability for damage to property carried on the colliding vessels, liability for delay and loss of use of the colliding vessels or property carried thereon, general average and salvage charges arising in relation to the vessels involved in the collision and the cargo carried thereon, loss of life and personal injury liabilities caused to crew members and/or passengers on the colliding vessels, oil pollution liabilities, wreck removal liabilities.

These liabilities are partly covered by the hull market and partly by the P&I Clubs.

HISTORY OF COLLISION LIABILITY COVER

At the beginning of the nineteenth century collision liability cover was limited to the value of the ship and freight outstanding. In 1836 the English High Court held that a marine insurance policy did not cover liability for damages to third parties. Following *De Vaux* v. *Salvador*[1] the running down clause was first introduced into a hull policy to provide cover for this uninsured liability. At the beginning the Running Down Clause was optional and its acceptance by the underwriters was not universal. It was incorporated in the Institute Time Clauses (Hulls) in 1888 when these were first published by the Institute of London Underwriters which made its use common. This clause is a good example of how a clause was initially introduced to fill a gap in the insurance cover.

At the beginning many underwriters disliked this situation as they believed that it would make masters less careful than they might otherwise have been. Between 1850 and 1854 they attempted to obtain legislative restriction on the adoption of the clause. These attempts failed and a supplementary clause known as the collision clause or as the running down clause was attached to hull policies. This clause provides limited cover against collision liabilities. The reason for this restriction was to calm the fears of the objecting underwriters, and the hull

1. (1836) 4 A. & E. 420.

insurance market felt that self-insured liability would encourage people to take greater care in the operation of their vessels. Thus, the running down clause restricts its cover to three-quarters of the liability. It leaves the assured to carry the other quarter. And it also excluded very substantial liabilities arising as a consequence of a collision, such as wreck removal, damage to cargo or property carried on the insured vessel, loss of life, personal injury or illness. This cover has been refined over the years to meet changing requirements but its basic cover is unchanged.

P&I COLLISION COVER

(a) General

P&I collision cover covers liabilities which are not recoverable under the collision liability clause (which used to be called the running down clause) contained in the hull policy of the insured vessel. Under the club rules it is assumed that the member has taken out full cover under the hull policy and then wants to fill in the gaps. Before we begin to explain the P&I collision cover the relation between the transition of shipowners' liability laws and changes of collision clause of the hull policy will be examined.[2]

(b) History of the collision liability clause in relation to P&I collision cover

(i) Establishment of Protection Clubs

As a matter of common knowledge the "running down" clause was first introduced to the hull policy following *De Vaux* v. *Salvador*.[3] The decision in this case was that an underwriter on a Lloyd's form of policy was not liable to indemnify a shipowner for the amount which he had been found liable to pay to another vessel in consequence of a collision. After a couple of years the underwriters found it necessary to include a special clause in their policies. With the introduction of the running down clause the extent of the underwriter's liability for damages paid by someone in respect of damage done in a collision is limited to three quarters of those damages. This cover does not exceed three quarters of the insured value of the ship in respect of any one collision. Therefore the shipowner was left without

2. For detailed information on transition of shipowners' liability laws and changes of P&I clubs in the UK see Takatada Imaizumi, *Transition of Shipowners' Liability Laws and Changes of P&I Clubs in the U.K.* (1992).

3. In this case the sailing ship *La Valeur* collided with the steamer *Forbes*. Both vessels were held to blame and the damages sustained were divided equally. The *Forbes* had suffered the greater damage and therefore the *La Valeur* had to make a payment in settlement of the collision liability. This payment was claimed as a particular average loss due to a peril insured against under the hull policy on the *La Valeur*. The court held that the assured could not recover under the policy because the payment of "damages" to the other ship was not proximately caused by a peril of the sea, but arose out of "an arbitrary provision of the law of notions not dictated by natural justice, nor possibly quite consistent with it".

cover in respect of one quarter of the expenses together with any excess liability above the value of the vessel and also liabilities in respect of death, personal injury and damage to fixed and floating objects falling outside the definition of a vessel. The shipowners started to look for protection for the potential liabilities. Following the decision in *De Vaux* v. *Salvador* the Protection Clubs extended their cover to include collision liabilities incurred by their members. This led to discussion about the purpose of establishing the Protection Clubs. In general it had been said that the one quarter collision liability left uncovered was the most important factor in the establishment of Protection Clubs. The contrary view was made clear by report No.109 entitled "The History and Development of Protecting and Indemnity Clubs" (1957) of the Advanced Study Group of the Insurance Institute of London (ASG Report). In this report it was said that the main purpose of establishment of Protection Clubs was not the protection of shipowners against the one quarter collision liability left uncovered by the proprietary market, but in fact to protect shipowners against loss of life and personal injury under the Merchant Shipping Act 1854, and to protect them against the excess of the aggregate of the amounts of liability for damage done to another vessel etc. and to protect them against the damage sustained by the protected vessel in a collision cover over the insured amount of the hull policy on the protected vessel.

Shipowners' liabilities at the time of establishment of Protection Clubs

As regards the circumstances of establishment of collision liability cover it is necessary to discuss the development of shipowners' liabilities at the time of establishment of Protection Clubs.[4]

LIABILITY OF SHIPOWNERS FOR LOSS OF OR DAMAGE TO CARGO CARRIED BY THEM AND FOR COLLISION DAMAGE

Since the eighteenth century the liability of British shipowners for loss of or damage to cargo carried by them has been very strict. They could not be exempted from liability unless they proved that the loss or damage had been caused by an act of God, the King's enemies, a defect in the cargo or the packaging thereof or by a general average sacrifice. There were not many collision cases, so there was no remarkable development of British law with regard to collision. When there was a collision case both vessels were held equally liable. The liability of ships was attributed without considering their fault. Obviously the courts' attitude towards collision liability seems to have changed since the eighteenth century.

In 1734 a shipowner's limited liability system was introduced for the first time. Under the Responsibilities of Shipowners Act 1734, shipowners' liability for loss of or damage to cargo carried by them which was caused by a tort committed by the crew was limited to the value of the vessel and the full amount of the freight due.

4. See Takatada Imaizumi, pp.2–21.

The scope of this Act was extended by the Responsibilities of Shipowners Act 1786. This Act applied to loss of or damage to cargo carried by ships which was caused by a tort committed by those other than their crew, and shipowners were excepted from liability for loss of or damage to cargo caused by fire.

In 1813 collision cases became a serious problem in the UK and the Responsibilities of Shipowners Act 1813 was passed. By this Act the shipowners' limited liability system was extended to apply to wash and crowding damage. The liability of the owner of the vessel which was to blame for the collision, for loss of or damage to the other vessel and the cargo on board his own and the other vessel was limited to the aggregate of the value of his own vessel and the full amount of the freight due.

In 1846 Lord Campbell's Act (Fatal Accidents Act) was passed. Under this Act[5] dependants had the right to sue for damages for the death of relatives caused by the negligence of shipowners. This Act was of great importance in three areas: first, at the time the Act was passed the liability of a shipowner for loss of life and personal injury was governed by the common law and was unlimited; secondly, when this Act was passed there were many immigrants travelling to America and Australia by British ships—in other words there was a large increase in the number of passengers travelling by sea to emigrate; thirdly, with the passing of this Act the vicarious liability principle, which means that the shipowner is liable not only for his own negligence but also for his servants, was established in the UK. Under this principle the shipowner was responsible for the negligent acts of master and crew in the course of their employment. Thus, in these circumstances Lord Campbell's Act brought enormous potential liability on the owners, and shipowners were not protected against such liability by their hull policies.

A year after Lord Campbell's Act, the Harbours, Docks and Piers Clauses Act 1847 was passed.[6] This Act gave power to port authorities to recover for damage to port works. The shipowner's liability was not dependent on his negligence or that of his employees. Obviously, like Lord Campbell's Act, this new Act placed new liability on shipowners. When it was combined with Lord Campbell's Act it amounted to a considerable potential liability. Another new Act passed during this period was the Employers' Liability Act of 1880.[7] This Act provided for payment to workmen including crew members injured in the course of their employment by their employer including shipowners.

EXTENSION OF THE LIMITED LIABILITY SYSTEM FOR LOSS OF LIFE OF OR PERSONAL INJURY TO PASSENGERS (MERCHANT SHIPPING ACT 1854)

As explained above, by the common law of England a shipowner's liability for loss of or damage to property and loss of life or personal injury was for the full amount of such loss or damage. Following the passing of the Liability Acts shipowners were

5. Fatal Accidents Act 1846 (9 & 10 Vict. c.93).
6. 10 & 11 Vict. c.27.
7. 43 & 44 Vict. c.42.

concerned about the effects of this legislation and wanted to extend limitation of liability to include loss of life and personal injury with new legislation. The Merchant Shipping Act 1854 revised and consolidated the previous shipowners' liability laws. Shipowners' liability for loss of life of or personal injury to passengers and for loss of or damage to cargo was limited to the value of the vessel and the freight due. In the case of loss of life of or personal injury to passengers, the value of the vessel and freight were not to be taken to be less than £15 per registered ton.

Merchant Shipping Act 1854, section 504, provides that:

"No owner of any sea-going ship or share therein shall, in cases where all or any of the following events occur without actual fault or privily, (that is to say,)

(1) Where any loss of life or personal injury is caused to any person being carried in such ship;
(2) Where any damage or loss is caused to any goods, merchandise, or other things whatsoever on board any such ship;
(3) Where any loss of life or personal injury is by reason of the improper navigation of such sea-going ship as aforesaid caused to any person carried in any other ship or boat;
(4) Where any loss or damage is by reason of any such improper navigation of such sea-going ship as aforesaid caused to any other ship or boat, or to any goods, merchandise or other things whatsoever, on board any other ship or boat;

Be answerable in damages to an extent beyond the value of his ship and the freight due or to grow due in respect of such ship during the voyage which at the time of the happening of any such events as aforesaid is in prosecution or contracted for, subject to the following proviso, (that is to say,) that in no case where any such liability as aforesaid is incurred in respect of loss of life or personal injury to any passenger, shall the value of any such ship and the freight thereof be taken to be less than fifteen pounds per registered ton".

This Act assumed that all ships were worth £15 per registered ton. This provision did not help shipowners very much as many ships were valued at less than that amount, so shipowners were paying claims in excess of the value of their own vessels. In other words shipowners remained potentially liable for large sums.

(ii) Liability for collision damage at the time of establishment of the first Protection Club and the 1824 Collision Clause

At the beginning of the nineteenth century collision liability became a serious problem for shipowners and in 1824 a collision clause was introduced (the 1824 Clause). The clause was as follows[8]:

"AND WE, for ourselves and each of us, do covenant and agree, that in case the said Ship shall, by accident or negligence of the Master and Crew, run down or damage any other Ship or Vessel, and the assured shall thereby become liable to pay and shall pay any Sum not exceeding the value of the said Ship or Vessel and her Freight, by or in pursuance of the Judgement of any Court of Law or Equity, or by or in pursuance of any Award made upon any Reference entered into by the Assured with the concurrence of Two of the Directors for the time being of the said Company, the Capital Stock and Funds of the said Company shall

8. Takatada Imaizumi, p.9.

and will bear and pay such proportion of Three Fourth Parts of the Sum so paid as aforesaid as the Sum of . . . Pounds hereby assured bears to the value of the said Ship or Vessel and her Freight."

Assessment of the 1824 Clause

CARGO ON THE INSURED VESSEL

It was presumed that the clause covered liability for collision damage not only to another vessel but also to cargo on the insured vessel. This clause was drawn up after the *Woodrop Sims*[9] case. Where it was held that a shipowner was not liable for any collision damage to another vessel unless there was fault on his part. Therefore it seems that the words "by accident" were specifically included in the clause to cover liability for collision damage to cargo on the insured vessel that had to be held by the assured as a common carrier under the common law, irrespective of whether there was fault on his part or not.[10]

LOSS OF LIFE AND PERSONAL INJURY

In the 1824 Collision Clause there was no express provision for collision liability for loss of life or personal injury. It has to be remembered, however, that this clause was drawn up more than 20 years before the passing of Lord Campbell's Act in 1846. At that time there were not as many passengers being carried in vessels and the shipowner was not liable to pay damages for the death of passengers at common law. Therefore shipowners were not in urgent need of covering collision liability for loss of life and personal injury. When one looks at the issue in this way it is possible to say that the draftsman of the 1824 Collision Clause did not intend to cover collision liability for loss of life or personal injury. However, within that time the number of passengers carried in vessels subsequently increased and by the passing of Lord Campbell's Act 1846 shipowners became liable for loss of life due to their negligence. Although the shipowners' limited liability system was introduced by the Merchant Shipping Act 1854 the minimum limit of liability was too high for shipowners.

Considering these circumstances shipowners wanted to change their collision liability for loss of life and personal injury. Therefore in 1860 shipowners brought the case of *Excelsior Co.* v. *Smith*[11] to cover these liabilities and won the action. But a Lloyd's underwriter disputed the decision in *Taylor* v. *Dewar*,[12] and there it was held that claims for loss of life, which the owner of a ship had been compelled to pay because his ship had collided with another ship, were not recoverable by reason of the exclusion clause in the policy. In view of the decisions in the above two cases, in

9. (1815) 2 Dodson 83; (1815) 165 E.R. 1422.
10. See Takatada Imaizumi, pp.9–10.
11. (1860) 2 L.T. 90 (SC).
12. (1864) 33 L.J. Q.B. 141, (1864) 5 B. & S. 58; 3 New Rep. 540, 10 L.T. 267, 10 Jur. N.S. 361, 12 W.R. 579; 5 Mar. L.C. 5, 122 E.R. 754.

March 1861 the Committee of Lloyd's extended the end of the Collision Clause of their marine policies. They wanted to make clear that they did not intend to cover shipowners' liability for loss of life and personal injury. The added part was as follows: "But this Agreement is in no case to be construed as extending to any sums which the Assured may become liable to pay or shall pay in respect of loss of life or personal injury to individuals, from any cause whatever".[13]

LEGAL COSTS

The next addition to the clause was made in 1868 following *Xenos* v. *Fox*.[14] In this case a ship sank following a collision with the insured vessel and the owners of the insured vessel defended the claim of another ship. Under the sue and labour clause in the policy the assured claimed for the legal costs incurred in defending the other ship's claim. The claim was rejected by the court. The court held that the legal costs were not damages recoverable under the collision liability clause, nor was it possible to recover legal costs under the sue and labour clause. First, the sue and labour clause applied only to the ordinary perils covered by the policy and did not extend to those covered by the Collision Clause, which was a supplemental agreement and did not contain any provision relating to costs. Secondly, legal costs were not incurred in order to avoid or minimise loss or damage to the insured vessel.

The unfairness of the result in *Xenos* v. *Fox*[15] was recognised by the underwriters and the position of the assured in this respect was protected for the future by the amendment of the clause to include a provision whereby underwriters would meet their due proportion of the costs, provided that they had given their prior consent to the proceedings.

REMOVAL OF WRECK EXPENSES AND SHIPOWNERS' LIABILITY FOR DAMAGE DONE TO HARBOUR, ETC.

At the time the Protection Clubs were established it was not clear whether the shipowners' liability system extended to their liability for damage done to harbour, dock, pier, etc. Under the Harbours, Docks and Piers Clauses Act 1847 it was, and it was probably after the enforcement of the Removal of Wrecks Act 1877 that the above liability and expenses of removal of wrecks came to be covered by the Protection Clubs.[16] It seems that the liability imposed by the 1847 Act was not a heavy burden on shipowners and was not covered under the limited liability system. Such liability and the expenses of removal of wrecks were also not covered under the hull policies by marine underwriters. Therefore it might be possible to say that the draftsman of the 1824 Collision Clause did not intend to cover the

13. Takatada Imaizumi, p.10.
14. (1868) 38 L.J. C.P. 351; 17 W.R. 893.
15. *Ibid.*
16. Takatada Imaizumi, p.11.

expenses of removal of wrecks which were an obstruction to harbour, dock pier, etc. and shipowners' liability for damage done to harbour, dock, pier, etc.[17]

The first Protection Club

As explained above, until the middle of the nineteenth century shipowners had to bear enormous potential liability. In particular, with the enactment of Lord Campbell's Act, they faced stricter rules of liability for death and personal injury. But shipowners could not obtain sufficient cover for such liability from marine underwriters. By the enactment of the Merchant Shipping Act 1854 limited liability was partly obtained. But, again, the minimum limit of liability under this Act was above the shipowners' expectations and the amount of potential liability that they had to bear was still enormous.

The shipowners' need for a new type of insurance to cover their potential liability was seen by some managers of Hull Clubs who reorganised their members into Protection Clubs. These clubs provided cover not only for loss of life and personal injury liability but also for excess liability which was not covered by the three-quarters Collision Clause of the hull policy. The first Protection Club, the Shipowners Mutual Protection Society, was formed in 1855.[18]

The Preface and the first three paragraphs of the Original Deed of the Shipowners Mutual Protection Society explain the reasons behind the establishment of Protection Clubs. The Society was established to protect shipowners against their liabilities for loss of life and personal injury after the Merchant Shipping Act 1854 and risk of running down other vessels not covered by their ordinary marine policies.[19] By stating that insurance could not legally be effected upon a ship or upon freight the third paragraph showed that the concept of liability interest was not established at that time. The preamble and the succeeding three paragraphs were as follows[20]:

"The Ship Owners Mutual Protection Society
Capital—one million

established for the purpose of protecting Ship Owners against the liability incurred under the 504 Sect. of the 'Merchant Shipping Act, 1854', and also the risk of running down other Vessels and Craft, not covered by the ordinary Marine Policies. This Society admits every description of Vessels, without respect to class, including those built from Iron, to the extent of £10,000 each.

17. *Ibid.*

18. That club is still in existence today, namely the Britannia Steam Ship Insurance Association. Soon it was joined by others. In 1855, the Shipowners' Protection Association (the predecessor of the present West of England Protection and Indemnity Association) and in 1860 the North of England Protecting Association (the predecessor of the present North of England Protecting and Indemnity Association) were established.

19. *The History and Development of Protecting and Indemnity Clubs*, The Advanced Study Group No. 109 of the Insurance Institute of London, p.2.

20. Takatada Imaizumi, pp.13–14. For a copy of the Deed of the Shipowners' Mutual Protection Society see The Advanced Study Group No. 109 of the Insurance Institute of London, Appendix XI, pp.17–20.

To all to whom these Presents shall come.

The several persons whose names are hereunto subscribed and Seals affixed send greeting.

WHEREAS ships and vessels very frequently receive considerable damage, and together with their freights (if any) are sometimes wholly lost, by reason of their coming in contact with other ships and vessels; and it often happens that the owners of the first-mentioned ships and vessels have not only to pay to the owners of the secondly-mentioned ships and vessels, or of their cargoes, damages amounting to the full value of the first-mentioned ships and vessels and their freight (if any), but have also to expend considerable sums of money in repairing the damage so as aforesaid occasioned to their own ships and vessels:

AND WHEREAS since the passing of the Statute 17th and 18th Vic., Cap. 104, sect. 504, the owners of ships and vessels may, and probably will, in some instances, also become liable to pay damages in consequence of loss of life, or personal injury caused to passengers, either on board their own or other ships or vessels, by the improper navigation of their ships or vessels;

AND WHEREAS an Insurance cannot legally be effected upon a ship or upon freight beyond the value of such ship or freight, consequently ship owners may be subjected to heavy losses, claims, and demands arising as aforesaid against which they cannot and will not be able to protect themselves by ordinary insurances upon their ships and freights."

The eighth paragraph shows that the Club covered liabilities and provided protection for damages which exceeded the value of the entered ship.

The eighth paragraph read as follows[21]:

"THAT in ascertaining the sums to be paid by virtue of these presents in respect of losses, claims, and demands arising as aforesaid, the Committee hereinafter mentioned shall take into account not only the damages and costs payable by the owners of ships or vessels for injuries done by their ships or vessels to other ships or vessels or their cargoes, but also the damages and costs payable in consequence of loss of life or personal injury caused to passengers as aforesaid, and also the injury done or to be sustained by their own ships or vessels owing to, or in consequence of, their coming in contact with other ships or vessels: and if aggregate amount of such damages and injury shall exceed the value of the ship doing or causing the damage or injury, and the freight (if any), the excess not exceeding the sum agreed to be secured aforesaid, shall be borne and paid by all the parties executing these presents, other than and except the owner (as such) of the particular ship doing or causing the damage or injury rateably, and in proportion to the sums in that behalf, set opposite their names and aforesaid."

Changes in the 1824 Collision Clause

When the 1824 Collision Clause was drawn up shipowners' liability was limited to the value of the vessel and her freight. With the enactment of the Merchant Shipping Amendment Act 1862 the limit of shipowners' liability changed to fixed amounts per ton of the vessel. Therefore the 1824 Collision Clause became inadequate to meet the situation after the passing of the Merchant Shipping

21. In this paragraph ¼ of collision liability was not mentioned. According to Imaizumi this para. proves that the purpose of establishment of the Protection Club was, contrary to what had been generally said, not to protect shipowners against the loss of ¼ of collision damages which were not covered by the ¾ Collision Clause of hull policy. See *ibid.*, p.14.

Amendment Act 1862. With this development the Clause was amended in accordance with the provisions of this Act.

The Original Lloyd's Clause is typical of the forms of the Collision Clause that were used by Lloyd's underwriters until 1833. The clause was as follows:

"And it is further agreed, that if the ship hereby insured shall come into collision with any other ship or vessel, and the insured shall in consequence thereof become liable to pay, and shall pay, to the persons interested in such other ship or vessel, or in the freight thereof, or in the goods or effects on board thereof, any sum or sums of money not exceeding the value of the ship hereby assured, calculated at the rate of eight pounds per ton on her registered tonnage, we will severally pay the Assured such proportion of three-fourths of the sum so paid as our respective subscriptions hereto bear to the value of the ship hereby assured, *calculated at the rate of eight pounds per ton*, or if the value hereby declared amounts to a larger sum, then to such declared value, and in cases where the liability of the ship has been contested with our consent in writing, we will also pay a like proportion of three-fourth parts of the costs thereby incurred, or paid, provided also, that this Clause shall in no case extend to any sum which the insured may become liable to pay or shall pay in respect of loss of life or personal injury to individuals for any cause whatsoever."

As can be seen from the clause £8 was to be calculated per ton on the ship's registered tonnage. Thus, this Clause is called the Limited Clause. In 1883 Lloyd's underwriters started to use the Unlimited Clause, until in 1888 the Institute Collision Clause was introduced for the first time.

Comparison between the Original Lloyd's Clause and the 1824 Collision Clause

When the 1824 Collision Clause was drawn up shipowners' liability was limited to the value of their vessel and freight. In 1862 the Merchant Shipping Amendment Act was passed. With this Act shipowners' liabilities for loss of or damage to vessels, cargoes, etc. and for loss of life or personal injury were respectively limited to £8 and £15 per ton of their vessel.[22] Under the Original Lloyd's Clause the limit of the assured's liability cover was revised according to the provisions of the Merchant Shipping Amendment Act 1862.

Under the 1824 Collision Clause liability for wash damage or crowding damage was covered. The Original Lloyd's Clause excluded these liabilities and its cover was limited to that for collision damage only. The 1824 Collision Clause covered liability for loss of or damage to cargo on the insured vessel. This liability was excluded under the Original Lloyd's Clause.

(iii) Indemnity Clubs

Until the latter half of the nineteenth century cargo claims were not a serious burden for shipowners. In 1870 a ship called the *Westenhope* which was entered in a Protecting Association was lost off the South African coast. Although the ship had

22. In the case of sailing vessels this was per registered ton; in the case of steam vessels it was per gross ton.

been loaded with cargo for Cape Town it had proceeded first to Port Elizabeth. As a result of this deviation the court held that the owners could not rely on the bill of lading exceptions in respect of the deviation to Port Elizabeth. Therefore they were liable for the value of the cargo which was lost. Its loss was not covered by the Protecting Association as the loss was not covered by its rules. This case was outlined in the ASG Report[23] as follows:

"In 1870, however, a ship called the *Westenhope* was lost off the South African coast. This ship had on board a quantity of cargo which had been carried past Port Elizabeth and was intended to be discharged on the return journey to Cape Town. The Shipowners were compelled to pay for the value of the lost cargo, the Court holding that they could not rely on the Bill of Lading exceptions in respect of such a deviation. The *Westenhope* was entered in a Protecting Association, but the Directors refused to reimburse the Shipowner on the grounds that the loss was not covered by the Rules, although a small *ex gratia* payment was eventually made".

As a result of this accident and potential liabilities to cargo interests the Steamship Owners Mutual Protection and Indemnity Association was formed at Newcastle in 1874.[24] As a consequence of the *Westenhope* case the shipowners' interests in indemnity risks were increased and the Protection Clubs which had so far not covered such risks established Indemnity Clubs to cover such risks. Protection Clubs which had already been covering such risks separated Indemnity Club from Protection Club, so therefore the foundation of the modern P&I clubs had both clubs: the Protection Club covered third party liabilities of shipowners or demise charterers due to improper navigation of their vessel and their liability for the crew as employers. The Indemnity Club covered their liabilities for cargo owners as carriers.[25] This was then the beginning of the Mutual Protection and Indemnity Clubs.

The Institute Collision Clause of 1888

After the enactment of the Merchant Shipping Act 1862 many forms of collision clause were drawn up by the Indemnity Mutual Marine Insurance Co. In 1884 the Institute of London Underwriters was established. As a result of the Institute's efforts to standardise various marine insurance clauses the first Institute Time Clauses—Hulls were drawn up in 1888. The Collision Clause included was as follows:

"And it is further agreed, that if the ship hereby insured shall come into collision with any other ship or vessel, and the assured shall in consequence thereof become liable to pay and shall pay by way of damages to any other person or persons any sum or sums not exceeding in respect of any one such collision the value of the ship hereby insured, this Company will pay the assured such proportion of three-fourths of such sum or sums so paid as its subscription hereto bears to the value of the ship hereby insured, and in cases in which the

23. The Advanced Study Group No. 109 of the Insurance Institute of London, p.5.
24. Takatada Imaizumi, pp.17–18.
25. *Ibid.*

liability of the ship has been contested with the consent in writing of this Company, the Company will also pay a like proportion of three-fourths of the costs which the assured shall thereby incur or be compelled to pay.

Provided always that this Clause shall in no case extend to any sum which the assured may become liable to pay or shall pay for removal of obstructions under statutory powers, for injury to harbours, wharves, piers, stages, and similar structures, consequent on such collision, or in respect of the cargo or engagements of the insured vessel, or for loss of life or personal injury."

The Institute Collision Clause of 1888 was amended two years later by the decision in *The Balnacraig*.[26] Following this decision the clause was extended to include the concept of "cross liabilities". In this case there was a collision where both vessels were equally to blame. The *Balnacraig* had sustained the greater damage and under the principle of single liability a balance was going to be paid by the other vessel to the owners of the *Balnacraig*. The claim was made by the owners of the *Balnacraig* on the basis of the collision clause in their policy for their liability for the damage done to the other vessel. Their liability had been set off against the other side's liability, which reduced the amount recovered by the *Balnacraig*. As the assured had not paid anything to the owner of the other vessel the claim under the policy failed.

In 1890 the above clause was extended to include the concept of "cross liabilities". The amended part was as follows:

"but when both vessels are to blame, then unless the liability of the owners of one or both of such vessels becomes limited by law, claims under this clause shall be settled on the principle of cross-liabilities as if the owners of each vessel had been compelled to pay to the owners of the other of such vessels such one-half or other proportion of the latter's damages as may have been properly allowed in ascertaining the balance or sum payable by or to the assured in consequence of such collision."

Under the amended 1888 clause three-quarters of the amount which the owner of the insured vessel was liable to pay by way of damages for loss of or damage to the other vessel and/or cargo thereon in consequence of collision was covered. The limit of this cover was up to the insured amount of the vessel. This was excluding loss of or damage to the insured vessel. The exclusions under the clause were: damages for loss of or damage to cargo on the insured vessel, expense of removal of wreck, etc., damages for injury to harbour, wharf, pier, etc. and damages for loss of life or personal injury to passengers or crew, on board either the insured vessel or another vessel.

Apart from a few small technical alterations and additions in 1918, 1952 and 1971 the Collision Liability Clause continued in much the same form for almost a century until 1983. In 1983 the present form of the Institute Collision Liability Clause was issued.

26. *The London Steamship Owners' Mutual Insurance Association* v. *The Grampian Steamship Co.* (1890) 24 Q.B.D. 32, 663 (CA), 59 L.J. Q.B. 549, 62 L.T. 784, 38 W.R. 651, 6 T.L.R. 288, 6 Asp.M.L.C. 506 (CA).

In the earlier part of the nineteenth century it was doubted whether there could properly or legally be insurance against liabilities of shipowners attributable to the negligence of themselves, their servants or agents.[27] This uncertainty on the part of shipowners and underwriters was removed by the Marine Insurance Act 1906. Section 3 of the Act provides that: "(1) Subject to the provisions of this Act, every lawful marine adventure may be the subject of a contract of marine insurance. (2) In particular there is a marine adventure where—(c) Any liability to a third party may be incurred by the owner of, or other person interested in or responsible for, insurable property, by reason of maritime perils". And section 74 of the Marine Insurance Act 1906 provides:

"Liabilities to third parties

74. Where the assured has effected an insurance in express terms against any liability to a third party, the measure of indemnity, subject to any express provision in the policy, is the amount paid or payable by him to such third party in respect of such liability."

(iv) Present P&I cover for collision liabilities

General

Under the P&I Collision Liability Rule a member is covered for liabilities he may incur as a result of a collision with another ship to the extent that these liabilities are not covered under the collision liability clause in the member's hull policy.

As explained above the Institute Running Down Clause covers loss or damage to:

- — Property on the other vessel or any other vessel.
- — Delay to or loss of use of any such other vessel or property thereon.
- — General Average or Salvage under contract of the other vessel arising out of the collision.
- — Legal and other costs.

These are all covered to the extent of three-quarters and to a maximum of three-quarters of the insured value of the vessel.

The Running Down Clause excludes the following risks:

- — Removal or disposal of obstructions, wreck or cargoes.
- — Damage to any property except other vessels or property on other vessels.
- — The cargo or other property on the insured vessel.
- — Loss of life, personal injury or illness.
- — Pollution (except of the other vessel).

Any or all of these liabilities may arise out of a collision. These liabilities are all insured by the P&I club. In other words the P&I club's cover starts where the Hull cover stops. "Club Rules assume that the member has taken out full hull cover and then attempts to fill in the gaps which exist in relation to a member's potential liabilities for collision."[28]

27. Donald O'May, *Marine Insurance Law and Policy* (1993), p.218.
28. Steven J. Hazelwood, *P&I Clubs: Law and Practice*, p.17.

The scope of the club cover (Rules of the Britannia Steam Ship Mutual Insurance Association Ltd.)

For an explanation of the scope of the cover we take the rules of the Britannia Steam Ship Mutual Insurance Association Limited (this is in no way a representation of their view or practice).[29]

RULE 19(9): LIABILITIES ARISING FROM COLLISIONS

"Liabilities which a Member may incur, together with costs and expenses incidental thereto, as a result of a collision between the Entered Ship and any other ship. The following shall be recoverable."

The Association shall cover liability or loss incurred as a result of a collision with another ship. The rule operates where two or more ships are involved in a collision, but not where a collision occurs between a ship and a fixed or floating object. Under a separate category P&I cover extends to liabilities arising from loss of or damage to fixed or floating properties. The definition of collision could be different according to the conditions in different hull policies. For instance under the Norwegian hull conditions a near collision, where there is no contact between the ships, may be treated as a collision.[30] Under English hull conditions a collision can only happen when there is physical contact between the colliding ships. However, the required physical contact need not necessarily be an actual contact between the hulls of the two vessels.

COLLISION CLAUSE (A)

"The one-fourth of the Member's liability (or such other proportion as may be applicable and agreed by the Managers) which is not covered under Clause 8 of the Institute Time Clauses Hulls 1/10/83 or under other forms of Hull Policies on the Entered Ship approved by the Managers."

One-fourth of collision liability: traditionally hull underwriters have agreed to cover only three quarters of damage caused to a third party vessel, subject to a limit of 75 per cent of the insured value of the ship. One-quarter collision liability is the part that is excluded under the running down clause. It covers liability if and to the extent that such liability is not covered under the hull policies on the ship. In other words the club does not provide cover for any liability or loss which would have been covered by the hull policies if the ship had been fully insured on standard

29. The Britannia Steam Ship Mutual Insurance Association Limited, Rules of Class 3, Protection and Indemnity, 1997.

30. The Scandinavian Hull Conditions cover contractual liability for collision with assisting harbour tugs. Under the Institute Time Clauses a distinction has to be made between collision with vessels and other property, fixed and floating objects. Under the Scandinavian system the hull insurance would cover both liability to the other vessel and to property outside the vessel. See "The Collision Aspects of P&I" *Gard News*, Issue 141, April 1996, p.12.

terms without deductible. Therefore the terms of the shipowner's hull policy reflect the extent of the liability cover under the club cover. For instance under the English hull insurance clauses a member will have collision liability cover up to only three-quarters. On the other hand under the American or Norwegian hull insurance clauses they may have cover up to four-quarters.

Some time ago the clubs which formed the International Group of P&I Clubs[31] initiated the idea of extending P&I cover to include all collision liability, not merely the one quarter covered by the hull market. Now it is for the individual club whether it wishes to continue offering the traditional one quarter collision liability cover or whether it will offer four quarters. That is why the master should be aware of the terms of the ship's hull insurance as he should be able to decide where to contact the ship's P&I club or her hull underwriters, in the case of an accident. As explained above it is possible to negotiate for the members the framework of their cover under this head of risk. As some of the P&I clubs provide cover up to four-quarters for collision liability negotiation between the club and the member avoids the possibility of having double insurance or not having full insurance cover for collision liability. If a vessel is entered for one quarter collision cover with the club, the club will repay the member the one quarter of the collision liability. If the member's vessel is entered for four quarters collision cover, the club will indemnify its member for four quarters of collision liability. For example, if ship A entered with the club for one quarter collision risks, is found to be liable for a collision with ship B, where ship B sustained US$500,000 worth of provable damage, the club will repay the member US$62,500 towards ship B's damage, subject of course to the member's deductible. The hull underwriters will pay the remaining three quarters of ship B's claim, subject to the owner's deductible.

EXCESS COLLISION LIABILITIES (B)

"That part of the Member's liability which exceeds the amount recoverable under the Hull Policies on the Entered Ship and any amount recoverable under paragraph (A) above solely by reason of such liability exceeding the valuation under the said policies."

Excess liability cover: under the terms of the Running Down Clause English hull underwriters and those writing hull risks on similar terms are not obliged to make payments in respect of collision liabilities beyond a sum representing three-quarters of the ship's insured value under the hull policies. The contribution of hull underwriters is limited by the insured value of the entered ship. Therefore hull will contribute only to the extent of three quarters or four quarters of the insured value. Where the member's liability exceeds the sum recoverable under the hull policies solely by reason of such a limit on the sum insured, and not because the hull insurers for some reason fail to pay out under the hull policies, the club will cover the excess collision liability. The excess liability cover is provided by the clubs

31. The International Group of P&I clubs exists to arrange collective insurance and reinsurance for P&I clubs. The Group represents the views of shipowner and charterer members on matters of concern to the shipping industry and provides a forum for the exchange of information.

when three-quarters of the vessel's insured value—the extent of the underwriter's liability under the RDC clause—is less than three-quarters of the vessel's liability to the other vessel involved in the collision.

In this respect a different approach was reflected in the proposed draft marine insurance clauses drawn up by the UNCTAD Working Group on International Shipping Legislation.[32] The report suggests that excess liability clauses are not necessary as additional insurance cover because the collision liability clause is a separate insurance contract and provides indemnity in addition to the basic coverage for damages. And the extent of the cover provided by the collision liability clause need not necessarily be the same as for the basic policy. The agreed value of the vessel establishes its insurable value for the maximum limit of reimbursement for damage to the hull. But this sum does not always cover the potential collision liabilities. Therefore instead of limiting the collision liability cover to the agreed value of the vessel the clause might be amended slightly to leave the limit of liability open, to be filled in by agreement. In this case there will be no need for a separate excess liabilities clause.

Obviously if the cover is decided according to present standard conditions the assured will obtain cover against the risk of excess collision liability through a P&I club. This aspect of the cover is subject to the provision that the managers may for the purpose of assessing any sum recoverable determine the proper amount for which the entered ship should have been insured for collision liability. And the club shall pay only the excess, if any, of the amount which would have been recoverable under the hull policy if the ship had been insured for the proper amount.

COLLISION LIABILITY TO CARGO (C)

"Loss of or damage to cargo carried in an Entered Ship, arising out of a collision between the Entered Ship and another ship caused by the fault both of the Entered Ship and of the other ship, for which a Member may be liable to indemnify the owner or charterer of such other ship solely by reason of responsibility for such loss or damage being determined in a country where the liability for such loss or damage is joint and several and the 'Both to Blame Collision Clause', is held invalid.

Provided always that there shall be no cover under this paragraph unless the Member shall have covered his liabilities in respect of cargo under Rule 19(17) and cover under this paragraph shall be in accordance with the provisions of Rule 19(17)."

Collision liability to cargo carried in an entered ship—Rule 19(17): an important part of the cover provided by the club is related to the liability of the shipowner under his contract of carriage to pay for any loss of or damage to cargo. Clubs' cargo cover is conditional upon the application of the Hague or Hague-Visby Rules. The cover will be given on the basis that the shipowner's contract with the owner of the cargo is on terms at least as favourable to the shipowner as the

32. See UNCTAD, "Legal and documentary aspects of the marine insurance contract" (1982) TD/BC.4/ISL/27/Rev.1, pp.29–30.

provisions of the Hague or Hague-Visby Rules. These Rules exclude claims by cargo in respect of negligent navigation of the carrying ship.

In most jurisdictions it is most unlikely that the owner of cargo in the insured vessel could succeed in a claim against the owner of that ship in a collision situation. The cargo owner may make a claim against the non-carrying vessel in accordance with her degree of blame, if any, but cannot recover either from the carrying ship or from the non-carrying ship in respect of that part of the blame attributable to the carrying ship. For instance, under the 1910 Convention a cargo owner can recover against the non-carrying vessel only to the extent to which that vessel may be at fault. Therefore the cargo owner receives a proportionate part of the damage sustained by his goods in the collision, according to the degree of fault of the non-carrying vessel. If the carrying vessel is one quarter to blame the cargo interests on that vessel can recover 75 per cent of their damage from the other vessel. But 25 per cent of the damage stays uncovered.

However, in the USA a cargo owner may recover from the non-carrying ship the whole of his loss. The only condition for this recovery is that there is some degree of blame attributable to the non-carrying ship, which is then entitled to recover against the carrying ship in respect of the carrying ship's degree of blame for the collision. The owner of the carrying ship may become obliged to pay in respect of claims for the cargo of the cargo carried on board his own ship. Under the running down clause the shipowner will not be able to recover in respect of his payment from his hull underwriters.

Club rules impose restrictions on the cover in respect of deviation from the contract voyage and of other departures from the proper carrying practice such as delivery of cargo without production by the master of the relevant bills of lading or the issue of a "clean" bill of lading for cargo which is patently damaged.

INJURY AND DEATH (D)

"The liability of a Member for the injury to, or death of, Seamen or others insofar as such liability may be covered under Rule 19(1), Rule 19(2), Rule 19(3) and Rule 19(4)."[33]

Historically, the Merchant Shipping Act of 1854 introduced the right for a legal representative of a deceased person to sue for damages against a third party, whose wrongful act, neglect or default had caused or contributed to a death. The first clubs specifically covered this Act and extended the cover to include personal injury, together with liabilities under the Running Down Clause. Where personal injury claims are concerned many countries have adopted no-fault liability. As such litigation usually allows for "contributory negligence" to be taken into account, it is possible to negotiate on it. The cover extends to crew, passengers and stevedores but it is subject to the conditions of contracts operating between the shipowners and the injured.

33. Liabilities in respect of a seaman, of passengers, of supernumeraries, of illness or injury to or death of third parties.

Officers and crew—Rule 19(1): a member's liability to his crew can be related to the law of the flag or the nationality of the crew depending on the country in which the claims are brought. It is possible for the club to cover liabilities under a contract of employment when it is extended by an agreement with the club. Therefore if there are any additional contractual liabilities they must be specifically approved by the club before they are covered. The payments under this section are made as a consequence of an illness, or injury to or death of a crew member. The clubs cover loss of life and personal injury to crew and passengers on board the entered vessel and also the entered vessel's liability to crew and passengers on board another vessel.

Passengers and others on board/near the ship—Rule 19(2), 19(3), 19(4): for the claims of the passengers the terms of the passenger ticket should have been approved by the managers. Club cover covers death, illness or injury claims including claims for loss of or damage to a passenger's baggage.

People other than passengers carried on board are known as supernumeraries. These people may be carried on board with the approval of the managers. Therefore the terms and conditions of the contract should made clear for the club.

If the member is found liable following an accident to other people who are working near the ship or may come aboard including stevedores, surveyors, customs officials, pilots, etc., the club will provide the cover.

PROPERTY DAMAGE (E)

"Liability for loss of or damage to property (other than the ship with which the Entered Ship collided, or cargo or other property in that other ship) insofar as such liability may be covered under Rule 19(10)."

The Harbour Acts of most jurisdictions impose strict liability on the owners of a vessel using the facilities if it causes damage to docks, locks and navigational marks, including buoys. In dock damage cases, it is necessary to determine whether the age and condition of the installation contributed to the damage found, in order to negotiate the claim. The owners could also be liable to third parties if damage were caused to bridges, waterside properties, underwater pipelines and cables, cranes, shoreside lifting equipment and railway equipment. All these liabilities would be covered by the club.

Damage to property—Rule 19(10): under this rule damage usually arises from physical contact by the entered ship. The club covers liability for damage to property belonging to other persons, such as docks, wharves, locks, etc.

NON-CONTACT DAMAGE TO SHIPS (F)

"Liability for losses caused to any other ship or cargo or other property therein insofar as such liability may be covered under Rule 19(11)."

Non-contact damage to ships—Rule 19(11): a member may incur liability for damage to another ship without actual contact between the two ships. Under this rule the club provides cover for liability that may arise where another vessel or her cargo is damaged as a result of the fault of the entered ship without any actual contact between the two ships. This may happen through the negligent navigation of the entered ship that causes another ship to run aground or to strike a third ship in taking action to avoid a collision with the entered ship, or by causing damage to a moored vessel by passing her closely at excessive speed.

It is possible to obtain cover for this type of liability under the cover for collision liability or as a separate complementary head of cover.

POLLUTION (G)

"Liability in respect of an escape or discharge or threatened escape or discharge of oil or any other substance insofar as such liability may be covered under Rule 19(12)."

Pollution—Rule 19(12): this rule covers a member's liability for damages and compensation as a result of an escape or discharge or threatened escape or discharge of oil or other substances. It covers any reasonable costs of preventing, minimising or cleaning up any pollution. These costs may be incurred by the member or by another party who is entitled to recover his costs from the member.

A member's liability under TOVALOP or any other such agreement was also covered, but with the termination of TOVALOP and CRISTAL on 20 February 1997 members may be parties to the international and regional conventions on oil pollution liability. But such an agreement should have been approved by the club. The member should have paid the premium or agreed to pay any additional call as may be required by the Association.

The clubs bear the pollution liability of the entered ship and that proportion of any oil pollution or clean-up liabilities caused by the other vessel as is assessed to be the responsibility of the entered ship.

WRECK REMOVAL (H)

"Liability for removal of wrecks insofar as such liability may be covered under Rule 19(13)".

Removal of wreck—Rule 19(13): the cost of removing or marking the wreck of another ship which may have sunk following a collision with the entered ship is not covered under the collision liability clause of normal hull policies and is covered by the club. Where the member is required to remove the wreck of his ship by law or by order of the port or other competent authority his liability is covered under this Rule. The club gives cover for liability which a shipowner may incur in respect of the raising, removal, destruction, lighting or marking of the wreck of his ship. From the cost of the operation will be deducted the value of the wreck or any part thereof that is recovered as a result of the removal operation.

"Liabilities which a Member may incur, together with costs and expenses incidental thereto, in respect of raising, removal, destruction, lighting or marking of:

Entered ship (A) The wreck of an Entered Ship, or of

(B) Any cargo, equipment or other property which is or was carried on board an Entered Ship.

Other ship (C) The wreck of any other ship, and of any cargo or other property which is or was on board such wreck.

PROVIDED ALWAYS THAT:

(i) the cost of raising, removal, destruction, lighting or marking were compulsory by law, or the costs thereof were legally recoverable from the Member under a contract approved by the Managers;

(ii) the value of the wreck itself and of the stores and materials, or the cargo or other property, saved shall either be credited to the Association or deducted from any recovery due from the Association;

(iii) there shall be no recovery under this Rule 19(13) if the Member shall, without the consent of the Managers in writing transfer his interest in the wreck, cargo or other property (otherwise than by abandonment) prior to the said raising, removal, destruction, lighting or marking."

Both to blame collision clause

In English law the degree of blame is apportioned according to the degree of fault of the ships concerned. Therefore in case of damage to cargo on another ship the cargo owner receives a proportionate part of the damage sustained by his goods in the collision from the non-carrying vessel according to the degree of blame attaching to the non-carrying vessel.

In the USA when both vessels are to blame liability is unlimited. The measure of liability for damages is determined as if both vessels were equally to blame. Therefore the system does not consider the actual differing degrees of blameworthiness. That is why the cargo owner can recover his full damages for injury to his goods from the non-carrying vessel although she was held responsible only to the extent of 50 per cent. In such a case the carrying vessel is indirectly obliged to pay 50 per cent of its own cargo damages. It is not possible to recover this liability from the hull underwriters because "the cargo and engagements of the insured vessel" are specifically excluded in the collision liability clause and it is not possible to recover from any P&I club. It is possible to say that US law aims to favour the cargo owner in contrast to British law that aims to protect the shipowner. The "Both to Blame Collision Clause" was introduced to restrict the liability of the owner of the carrying ship. In standard conditions the following clause appears[34]:

"This insurance is extended to indemnify the Assured against such proportion of liability under the contract of affreightment 'Both to Blame Collision Clause' as is in respect of a loss recoverable hereunder. In the event of any claim by shipowners under the said Clause the Assured agree to notify the Underwriters who shall have the right, at their own cost and expense, to defend the Assured against such claim."

In 1952 the US Supreme Court held that the both to blame collision clause was contrary to US sea carriage legislation,[35] but of course effect might be given to it in

34. Institute Cargo Clauses (A), (B), (C), cl.3.

35. In *United States of America* v. *Atlantic Mutual Insurance Co.* [1952] A.M.C. 659 the Supreme Court held that such a clause is invalid in bills of lading on the grounds of public policy.

an action on a contract containing the clause heard in any other jurisdiction. Braekhus[36] says that the clause would be void when included in a bill of lading issued by a common carrier.[37] But when the clause was used in a charterparty it was held valid[38] only if the charterer was also the owner of the cargo. If an action takes place in the courts of a state that recognises the validity of the clause it is possible to rely on it even if the bill of lading is governed by the law of the USA. The difference between English and American law can lead to forum shopping by cargo owners. If the cargo owners can sue the owners of the non-carrying vessel in the USA they will have the opportunity to recover the whole amount of their loss.

In the UK the P&I clubs advise their members to insert a both to blame collision clause in a bill of lading for shipments to or from ports in the USA. The Britannia Steam Ship Insurance Association Limited's recommended both to blame collision clause reads as follows:

"If the ship comes into collision with another ship as a result of the negligence of the other ship and any act, neglect or default of the master, mariner, pilot or the servants of the carrier in the navigation or in the management of the ship, the owners of the goods carried hereunder will indemnify the carrier against all loss or liability to the other or non-carrying ship or her owners in so far as such loss or liability represents loss of, or damage to, or any claim whatsoever of the owners of said goods, paid or payable by the other or non-carrying ship or her owners of the said goods and set-off, recouped or recovered by the other or non-carrying ship or her owners as part of their claim against the carrying ship or carrier.

The foregoing provisions shall also apply where the owners, operators, or those in charge of any ship or ships or objects other than, or in addition to, the colliding ships or objects are at fault in respect of a collision or contract."

Sister-ship clause

The aim of this clause is to protect the shipowner in respect of a collision between ships in the same ownership. When two vessels, both in the same ownership and management, collide there can be no legal liability for salvage. As a shipowner cannot sue himself the apportionment of liability is referred to an arbitrator. The existence of such a situation in the case of the collision clause was confirmed in *Simpson* v. *Thompson*.[39] If two vessels under the same ownership are in collision there is no tortious liability between the two ships as a person cannot bring an action against himself. To avoid inequity arising from the accident of ownership the sister ship clause was introduced.[40] The clause follows the ¾ths collision liability clause in the Institute clauses.

"Should the Vessel hereby insured come into collision with or receive salvage services from another vessel belonging wholly or in part to the same Owners or under the same

36. Braekbus & Rein, *Handbook on P&I Insurance* (2nd ed. 1979), p.189.
37. In the USA a common carrier is a carrier whose vessel is made generally available for the carriage of goods.
38. *American Union Transport Inc.* v. *United States of America* [1976] A.M.C. 1480.
39. (1877) 38 L.T. 1, 3 Asp. M.L.C. 567, 3 App. Cas. 279.
40. Cl. 9 of the Institute Time Clauses.

management, the Assured shall have the same rights under this insurance as they would have were the other vessel entirely the property of Owners not interested in the Vessel hereby insured; but in such cases the liability for the collision or the amount payable for the services rendered shall be referred to a sole arbitrator to be agreed upon between the Underwriters and the Assured."

Under the clause the assured shall have the same rights as they would have were the other vessel entirely the property of different owners. The damages are assessed by a single arbitrator and then the cross-liabilities can be applied. If this is not the case underwriters would just pay for the damage suffered by each vessel and the shipowner would not recover any part of his demurrage. Under the clause it is possible to claim salvage in the same way as if the vessels were in different ownership.[41]

Duty to insure the ship for her full value and on standard terms

The member is obliged to insure the ship for her full value. The insured value of the ship shall not be less than her market value. As explained above the club covers the members' reasonable or legal liability for collision damages which exceed the total amount insured under the member's hull and machinery policy and excess liability policies. If the member insures his ship with his hull underwriters for a low value the club's cover will be increased unfairly, so clubs try to avoid this happening. If the committee believes that the ship has not been insured for her full value under the hull policy the claim for excess collision liability can be reduced appropriately. This rule still exists if a member deletes the collision liability clause from his hull policy and covers the whole collision risk with the club.

For instance: (1) Collision Liability £2,000,000
Insured Value £1,800,000

Here there is under-insurance and collision liability is apportioned as follows:
Hull policy pays £1,350,000 (¾ of insured value)
P&I club pays £ 450,000 (¼ of insured value)
Assured bears £ 200,000 uninsured

(2) Collision Liability £2,000,000
Insured Value £2,200,000

There is no under-insurance and collision liability is apportioned as follows:
Hull policy pays £1,500,000 (¾ of the liability)
P&I club pays £ 500,000 (¼ of the liability)

Under the clubs' rules ships entered with the Association are required to be insured on standard terms. Therefore if a ship is insured under a hull policy that

41. In *Cargo ex "Laertes"* in the case of salvage services to a sister-ship the shipowner is precluded under English law from claiming salvage in respect of services to the ship and freight. However, the shipowner can claim salvage from the cargo: (1887) 6 Asp. M.L.C. 174; N.G. Hudson and J.C. Allen, *The Institute Clauses Handbook* (1986), p.108.

provides more limited cover than is available on standard terms, the Association will not cover the difference between the hull policy and standard terms. In other words the Association does not reimburse any loss or liability which would have been covered if the ship had been covered on standard terms.

THE HULL MARKET COVER

In practice the collision liability clause is referred as the Running Down Clause. The clause became known as the "¾ths Running Down Clause".

(a) Institute Collision Liability Clause[42]

"8. 3/4ths Collision Liability

8.1 The Underwriters agree to indemnify the Assured for three-fourths of any sum or sums paid by the Assured to any other person or persons by reason of the Assured becoming legally liable by way of damages for

8.1.1 loss of or damage to any other vessel or property on any other vessel

8. 1.2 delay to or loss of use of any such other vessel or property thereon

8.1.3 general average of, salvage of, or salvage under contract of, any such other vessel or property thereon,
where such payment by the Assured is in consequence of the Vessel hereby insured coming into collision with any other vessel.

8.2 The indemnity provided by this Clause 8 shall be in addition to the indemnity provided by the other terms and conditions of this insurance and shall be subject to the following provisions;

8.2.1 Where the insured Vessel is in collision with another vessel and both vessels are to blame then, unless the liability of one or both vessels becomes limited by law, the indemnity under this Clause 8 shall be calculated on the principle of cross-liabilities as if the respective Owners had been compelled to pay to each other such proportion of each other's damages as may have been properly allowed in ascertaining the balance or sum payable by or to the Assured in consequence of the collision.

8.2.2 In no case shall the Underwriter's total liability under Clauses 8.1 and 8.2 exceed their proportionate part of the three-fourths of the insured value of the Vessel hereby insured in respect of any one collision.

8.3 The Underwriters will also pay three-fourths of the legal costs incurred by the Assured or which the Assured may be compelled to pay in contesting liability or taking proceedings to limit liability, with the prior written consent of the Underwriters.

Exclusions

8.4 Provided always that this Clause 8 shall in no case extend to any sum which the Assured shall pay for or in respect of

42. Cl. 8 of the Institute Time Clauses.

8.4.1 removal or disposal of obstructions, wrecks, cargoes or any other thing whatsoever

8.4.2 any real or personal property or thing whatsoever except other vessels or property on other vessels

8.4.3 the cargo or other property on, or the engagements of, the insured Vessel

8.4.4 loss of life, personal injury or illness

8.4.5 pollution or contamination of any real or personal property or thing whatsoever (except other vessels with which the insured Vessel is in collision or property on such other vessels)."

(b) Analysis of the Collision Liability Clause

(i) General

The Institute Collision Liability Clause provides cover for the liability of the assured shipowner to third parties caused by collision with another vessel. The clause relates only to legal liabilities incurred by the assured in consequence of the insured being in collision with any other vessel.

The cover provided under the clause is supplementary, "in addition to the indemnity provided by the other terms and conditions of the policy" (Clause 8.2). In this respect the Collision Liability Clause shows similarity to the Institute Amended Running Down Clause 1/10/1971 (the Running Down Clause). Unless it is stated otherwise the cover provided by the clause is deemed to be supplementary to the main insurance contract. The Running Down Clause opened with the words "It is further agreed . . ." and was therefore held to be a supplementary contract.

Under the collision liability clause it is possible to claim both for the total loss of the vessel insured and collision liability to the other vessel. Therefore when one thinks of a case where the insured ship sinks in a collision with another vessel and the assured incurs liability to the other vessel consequent upon the collision there are two things the assured can claim under his policy:

1. The full sum insured in respect of the total loss of his ship.
2. His liability in consequence of the collision.

The indemnity to the assured under the Institute Collision Liability Clause is limited to three-quarters of any sum or sums paid by the assured. And in no case shall the total liability of underwriters exceed their proportionate part of three-quarters of the insured value of the vessel in respect of any one collision (Clause 8.2.2).

The indemnity under the collision liability clause operates only when the insured vessel was at fault and the assured is legally liable. In order for the clause to apply the insured vessel must be at fault and the assured must be legally liable. If the collision occurred without any relevant fault or negligence[43] of the insured vessel the collision liability clause does not apply.

43. "In maritime cases it is the act of negligence which creates the liability, not necessarily the contact of the ships": see V. Dover, p.521.

(ii) Requirements under the collision liability clause

"Paid by the Assured"

Under the Institute Amended Running Down Clause (1/10/1971) the assured had to have paid the claimant against him before he could claim to be indemnified by underwriters.[44] The collision liability clause has a similar effect.

With the words "indemnify the Assured . . . for any sum or sums paid by the Assured . . ." the underwriters' liability depends on the payment made by the assured for the damages. In *Thompson* v. *Reynolds* it was made clear that the assured can recover only what he has been required to pay in respect of his liability to third parties.[45]

Under the Third Parties (Rights Against Insurers) Act 1930, if an assured incurs a liability to a third party, and before making payment becomes bankrupt or goes into liquidation, the rights in the policy of insurance covering that liability are transferred to the third party. The question of the rights of third parties to receive the insurance money which accrued under a policy of insurance to the insured persons who were liable to the third parties has raised questions from time to time.[46] In *Re Nautilus Steam Shipping Co. Ltd.*[47] it became clear that the Third Parties Act is applicable to the collision liability clause. In this case the issue was whether the Third Parties Act 1930 applied at all, as the liquidation occurred after the enforcement of the Act but the collision occurred before the Act. In this case Maugham L.J. referred to the disputed topic under the Third Parties Act 1930. The dispute concerned the requirement that collision damages shall have been paid by the assured. There were two possibilities, a requirement in the policy that payment of damages by the assured could be a condition precedent to liability of underwriters and this would technically deprive third parties of their rights. Or such a condition may be accepted as invalid because as far as the third party is concerned section 1(3) of the Act makes clauses invalid which absolve underwriters from liability upon the assured becoming bankrupt or being wound up.[48] In *Re Allobrogia Steamship Corporation* there was substantial ground for saying that that a similar provision in a P&I club policy, making payment a condition

44. The Running Down Clause provided that "It is further agreed that if the Vessel hereby insured shall come into collision with any other vessel and *the Assured shall in consequence thereof become liable to pay and shall pay* by way of damages to any other person or persons any sum or sums in respect of such collision for . . ."

45. (1857) 26 L.J.Q.B. 93; (1857) 7 E. & B. 172; 3 Jur. N.S. 464.

46. See *Re Harrington Motor Company Ltd.* [1928] Ch. 105.

47. (1935) 52 Ll.L.R. 183.

48. Donald O'May, pp.222–224. The Third Parties (Rights Against Insurers) Act 1930, s.1(3), states that: "Where the insured has become bankrupt or where in the case of the insured being a company, a winding-up order has been made or a resolution for a voluntary winding-up has been passed, with respect to the company, no agreement made between the insurer and the insured after liability has been incurred to a third party and after the commencement of the bankruptcy or winding up, as the case may be, nor any waiver, assignment, or other disposition made by, or payment made to the insured after the commencement aforesaid shall be effective to defeat or affect the rights transferred to the third party under this Act, but those rights shall be the same as if no such agreement, waiver, assignment, disposition or payment had been made."

precedent, would not deprive third parties of their rights under the Act or those rights would be rendered invalid and inconsistent with the purpose of the Act.[49] This issue was discussed in two cases. In *The Fanti*[50] it was held that the "pay to be paid" provision in the P&I club rules was rendered invalid by section 1(3) of the Third Parties Act 1930. But in *The Padre Island*[51] the conclusion was the opposite. The two cases were heard together in the Court of Appeal[52] and in the House of Lords.[53] The decision was in favour of the P&I clubs. It was held that "pay to be paid" provisions applied throughout the lives of contracts and did not purport directly or indirectly to avoid those contracts or alter the parties' rights upon the happening of an event such as the shipowners being ordered to be wound up. Immediately before that happened the shipowners' rights under the rules were contingent upon the condition precedent that they should first have discharged the liabilities themselves and there were therefore no accrued rights to transfer where payment had not been made. In these circumstances section 1(3) did not apply and section 1(4) expressly provided that "the insurer . . . shall be under the same liability as he would have been under to the insured".[54] The principle stated by the House of Lords will apply to the collision liability clause, and as a result when the assured shipowner is wound up without payment of the collision liability or part of it rights will not be transferred to a third party under the Third Parties Act 1930.

Payment of legal costs by the assured is not necessary if they have been incurred with underwriters' prior consent in writing only. Under Clause 8.3 the insurer is liable to pay three-quarters of the legal costs incurred by the assured. Clause 8.3 was inserted following *Xenos* v. *Fox*[55] to include a provision whereby underwriters would meet their due proportion of the costs, provided they had given their prior consent to the proceedings. Legal costs are divided into two categories as costs of attack and costs of defence.[56] Costs of attack include costs incurred by the assured in legal actions to recover losses suffered by the assured in the collision. Costs of defence are legal costs incurred by the assured in defending an action brought by a third party for collision damage. In cases where these costs are incurred in relation to loss or damage which is recoverable by the assured under the policy then they are recoverable from the underwriters.[57] Costs of defence are directly related to third party liability for collision damage and are governed by the collision liability clause. Under Clause 8.3 provided underwriters have given written authority for such

49. [1979] 1 Lloyd's Rep. 190.

50. *Firma C-Trade S.A.* v. *Newcastle Protection & Indemnity Association* [1987] 2 Lloyd's Rep. 299.

51. *Socony Mobil Inc. and Others* v. *West of England Shipowners Mutual Insurance Association (London) Ltd.* [1987] 2 Lloyd's Rep. 529.

52. [1989] 1 Lloyd's Rep. 239.

53. [1990] 2 Lloyd's Rep. 191; [1990] 2 All E.R. 705.

54. Donald O'May, p.224.

55. (1868) 38 L.J. C.P. 351, 17 W.R. 893.

56. See R.H. Brown, *The Institute Time Clauses Hulls 1995, Analysis of Marine Insurance Clauses* (1996), pp.60–61; Susan Hodges, *Law of Marine Insurance* (1996), pp.319–320.

57. According to Hodges consent of the underwriters should be first obtained by the assured before he commences legal proceedings against the owners of the other vessel: Susan Hodges, pp.319–320. Brown states that costs of attack are allocated between the assured and/or the underwriters and/or the

legal proceedings the assured is entitled to recover from the underwriters three-quarters of his overall legal costs of defence. Therefore the assured will bear 25 per cent of the legal costs.

Under Clause 8.2 the underwriter's liability is in addition to any other liability incurred by underwriters under the collision liability clause. Hence, the insurer could be made liable not only for the full extent of the insured value of the damage sustained by his own vessel but he will also be liable for up to three quarters of the insured value in relation to third party liability. An additional sum of three quarters of the legal cost incurred by the assured has also to be added.

"By way of damages"

The phrase "by way of damages" means that the liability must arise by way of tort and not by way of contract or statute. The same phrase was also used in the collision liability clause. The phrase has been interpreted as being restricted to payments in respect of tortious liability. In *Furness Withy and Co. Ltd.* v. *Duder*[58] a tug was damaged by a vessel in its tow. Although the negligence was on the part of the tug the shipowners were responsible for all damage to the tug under a provision in the towage contract. The owners were not able to obtain reimbursement from their insurers as the court held[59] that the payment made to the Admiralty was not due to any tortious act by the insured vessel, but was made in consequence of the terms of the towage contract. A similar situation arose in *Hall Brothers Steamship Co.* v. *Young (The Trident)*.[60] In this case the collision was between the insured vessel and a French pilot vessel. The collision occurred due to pilot vessel's defective steering gear. There was no blame on the insured vessel. But under the local pilotage law the shipowners were obliged to pay for the repairs to the pilot boat. Therefore the shipowners had to pay for the repairs of the pilot boat but that was not in consequence of the collision but in consequence of the French law. The underwriters' liability did not extend to cover the payment.

P&I club, as the case may be; underwriters bearing the proportion of such costs that relates to a claim paid under the policy for loss of, or damage to, the insured vessel caused by the collision: R.H. Brown, p.60. In the view of Lambeth and O'May costs of attack are recoverable in full from the hull underwriters insofar as they relate to loss or damage recoverable from the hull underwriters as particular average or otherwise and prior consent of the underwriters has not been mentioned: R.J. Lambeth, *Templeman on Marine Insurance, Its Principles and Practice* (6th ed., 1986), pp.408–409; Donald O'May, p.236.

58. (1936) 55 Ll.L.R. 52, [1936] 2 All E.R. 119, [1936] 2 K.B. 461, 105 L.J. K.B. 473, 154 L.T. 663, 18 Asp. M.L.C. 623.

59. Branson J. said: "... the clause means that where in consequence of a collision there arises a legal liability upon the shipowners to pay a sum which can properly be described as damages for a tort, then the underwriters will indemnify them. The expression 'by way of damages' indicates ... a liability which arises as a matter of tort, and not as a matter of contract": [1936] 2 K.B. at p.468.

60. [1939] 1 K.B. 748, [1939] 1 All E.R. 809, 108 L.J. K.B. 313, 160 L.T. 402, 55 L.T.R 506, 19 Asp.M.L.C. 269, 44 Com.Cas. 146, 63 Ll.L.Rep. 143.

"Collision"

To establish a claim under the collision clause the insured vessel must come into collision with another vessel. The liability of the assured arises in consequence of the insured vessel "coming into collision with any other vessel". Therefore the meaning of the two words "collision" and "vessel" should become clear. Liability for damage caused by the wash of the insured ship doing an excessive speed in a confined waterway is not a collision. There are also situations where collision liability can arise without any action on the part of the insured ship.

If there is a collision with another vessel liability falls on hull underwriters under the collision clause or it will generally be recoverable under the P&I cover. Disputes that arise in practice are generally solved between the hull underwriters and Clubs on the basis of decided cases. In *The Niobe*[61] it was held that the collision of the tug with the damaged vessel must be taken to have been a collision between *The Niobe* and another vessel within the meaning of the policy. Thus, to constitute a collision the contact between the two vessels need not necessarily be an actual contact between the hulls. This case showed an exception to the rule that there must be an actual collision with the insured vessel and another vessel. In *Re Margetts* v. *Ocean Accident*[62] it was held that the tug had come into collision with a vessel within the meaning of the Running Down Clause. But, in contrast, it was found in *Bennett Steamship Co. Ltd.* v. *Hull Mutual Steamship Protecting Society* that contact between a vessel and the nets of a fishing vessel, a mile distant, did not constitute a "collision with any other ship or vessel".[63] The difference was summarised in the latter case in the Court of Appeal: "Whenever any part of the tackle of a vessel is being used in connection with the vessel, although it may be outside the ambit of the hull, as the anchor or a boat towing astern or working ahead to warp the vessel, it may just as well be said to be a part of the vessel . . . Nets, however, are not a part of the ship in that sense, nor are they things which are necessary for her to have and without which she could not prudently put to sea". Therefore the nets were held not to be part of the ship, but on the other hand, those parts which are necessary to make the ship seaworthy do not necessarily fall within the ambit of the ship.

"Vessel"

Since 1952 the Running Down Clause and the Collision Liability Clause have confined a collision to that with "any other vessel" rather than "any other ship or vessel", but a definition of a vessel is not given in the clauses. In general, the term "vessel" covers the term ship vessel. The court cases that determine whether or not

61. *McCowan* v. *Baine & Johnston* [1891] A.C. 401, 7 Asp. M.L.C. 89, [1891–4] All E.R. 343, 65 L.T. 502, 7 T.L.R. 713.

62. [1901] 2 K.B. 792, 9 Asp. M.L.C. 217, 17 T.L.R. 538, 70 L.J. K.B. 762, 85 L.T. 94, 49 W.R. 669.

63. [1913] 3 K.B. 372 (Affd. [1914] 3 K.B. 57), 19 Com. Cas. 353, 12 Asp. M.L.C. 522, 30 T.L.R. 515, 111 L.T. 489, 83 L.J. K.B. 1179.

given circumstances constitute a collision within the meaning of the hull policy generally require that the vessel with which the insured ship came into contact must be one used in navigation. In *Wells & Another* v. *Owners of Gas Float Whitton (No. 2)* it was held that a floating beacon was not constructed for navigation purposes so it was not a vessel.[64] The view in *Wells & Another* v. *Owners of Gas Float Whitton (No. 2)* was supported in *Merchants' Marine Insurance Co.* v. *North of England Protecting and Indemnity Association.*[65] In this case an insured vessel collided with a pontoon crane which was permanently moored to a river bank in a naval dockyard. It was held that a floating pontoon crane was not a vessel as it was customarily moored to the river bank and had been designed and adopted to float and to lift, not to navigate. The judge pointed out that whatever other qualities are attached to a ship or vessel, adaptability for navigation and its use for that purpose is one of the most essential elements for bringing the craft within the definition. In *Polpen Shipping* v. *Commercial Union Ass. Co.* it was held that a seaplane was not a vessel for the purposes of the collision clause in the hull policy.[66] In *European & Australian Royal Mail Co.* v. *P&O Steam Navigation Co.* it was held that a ship which had been used as a coal hulk for the past four years was no longer a "ship" but had become a mere chattel, a floating coal hulk.[67]

Another point that should be considered is whether contact with a sunken vessel constitutes a collision with another ship or vessel or not. In *Chandler* v. *Blogg*[68] the insured vessel ran into a barge, which was lying half-submerged following a recent collision with another vessel. This was held to be a collision within the terms of the policy as the barge was capable of being raised and navigated. The courts have generally taken a restrictive view of what is a ship and have laid particular emphasis on the expression "used in navigation". However, in *Pelton Steamship Co. Ltd.* v. *North of England Protecting & Indemnity Association*[69] the test of navigability was rejected. A vessel fouled a wreck upon which salvage operations were in progress. The insurers declined to pay for the liability incurred and argued that the wreck was not a ship or vessel. The assured sued the club for the whole of its liability due to the owners of the wreck. The club maintained that there had been a collision with a ship or vessel within the meaning of the Running Down Clause and that insurers could not avoid liability thereunder. Greer J. upheld the view of the club which maintained that its liability was restricted to the one-quarter not payable by the underwriters. Greer J. preferred to apply his own test "whether or not any reasonably minded owner would continue salvage operations in the hope of completely recovering the vessel by those operations and subsequent repair". He

64. R.H. Brown, *Training Notes for Brokers, The Insurance of Merchant Ships & Shipowners' Interests, Book Two* (1990), p.11.5. [1897] A.C. 337.

65. (1926) 32 Com. Cas. 165; (1926) 42 T.L.R. 724; 43 T.L.R. 107; (1926) 25 Ll.L.R. 446.

66. (1943) 74 Ll.L.R. 157; [1943] 1 All E.R. 162; [1943] K.B. 161; 112 L.J.K.B. 198; 168 L.T. 143; 59 L.R. 106. Also in *Watson* v. *R.C.A. Victor Co. Inc.* (1934) 50 Ll.L.R. 77 it was held that a seaplane is not a "ship or vessel" within the meaning of the Merchant Shipping Acts.

67. (1866) 12 Jur. (N.S.) 909, 14 L.T. 704, 14 W.R. 843.

68. [1898] 1 Q.B. 32, 67 L.J. Q.B. 336, 77 L.T. 524, 14 T.L.R. 66, 8 Asp. M.L.C. 349, 3 Com.Cas. 18.

69. (1925) 22 Ll.L.R. 510.

said, "It seems to me, that navigability cannot be the test as to whether the thing is or is not a ship or vessel. I mean navigability at the time of the collision."

In contrast in *The St. Machar* it was decided that a newly constructed, but unfinished, ship was a "vessel used in navigation" within section 742 of the Merchant Shipping Act 1894 after she was launched into the water, although she had no motive power at the stage of construction.[70] Drill ships are vessels within the collision liability clause because they are mobile and able to move under their own power, under the control of a master or crew, from one location to another. In *The Harlow*[71] a dumb barge *Harlow* was fitted with a rudder but no means of propulsion. It was used under tow for the carriage of goods on a river and was held to be a ship. The *Gas Float Whitton No. 2* was used as a floating beacon, capable of being moved and shaped like a boat. But it was held not to be a ship or vessel within the meaning of the Merchant Shipping Acts because it was not constructed for the purpose of being navigated or of conveying cargo or passengers.[72] In *Steedman* v. *Scofield*[73] Sheen J. declared that a jet ski was not a ship, partly because it was not used in "navigation" in the traditional sense. Similarly in *Curtis* v. *Wild*[74] it was held that a sailing dinghy used on a reservoir was not used in navigation as the users were merely "messing about in boats".

When the insured vessel collides with another vessel and causes her to run into a third vessel, without any intervening negligence on the part of the third vessel, the insured vessel may be held liable to pay damages to the owners of the other two vessels. This is not an exception to the rule that there must be an actual contact to constitute a collision. In such a case there is a collision between the two vessels and the damage to the third vessel is part of the damages arising in consequence of the first collision. In *William France, Fenwick* v. *Merchants' Marine Insurance Company Ltd.* the insured vessel collided with another ship which, in consequence, damaged a third ship. The insurers were held liable for three-quarters of the whole damage to both ships arising out of the first collision.[75]

Exclusions

The exclusions in 8.4.1 to 8.4.5 are similar to those in earlier versions of the Running Down Clause. The cover shall in no case extend to "removal or disposal of obstructions, wrecks, cargoes or any other thing whatsoever" (8.4.1). When a ship is sunk or stranded within the jurisdiction of an authority which is responsible for keeping the waterways clear for navigation that authority has legal power to force the shipowners, at their own expense, to remove the obstruction or the authority can remove the obstruction and pass the cost of removal to the owner of the vessel. If the underwriter has settled a total loss on the insured ship and has

70. (1939) 64 Ll.L.R. 27, (1939) 65 Ll.L.R. 119 (CA).
71. [1922] P. 175, (1922) 13 Ll.L.Rep. 311.
72. [1897] A.C. 337 (CA), [1896] P. 42, [1895] P. 301.
73. [1992] 2 Lloyd's Rep. 163.
74. [1991] 4 All E.R. 172.
75. [1915] 3 K.B. 290 (CA).

exercised his right to take over the wreck he will become the owner of the wreck and be liable to the local authority for the costs of removal of the wreck. If that is not the case he will not be responsible for liabilities attaching to the ship apart from those covered under the collision liability clause.

Under Clause 8.4.2 the cover shall in no case extend to "any real or personal property or thing whatsoever except other vessels or property on other vessels". Damages against which the assured is indemnified under the collision clause must relate only to other vessels or property on other vessels.

Claims in respect of the cargo or engagements of the insured vessel are similarly excluded. The responsibilities of a shipowner to cargo or to other property on the insured vessel is usually governed by the contract of affreightment or by a passenger's ticket. The amount paid by the assured in respect of such items is not covered under the collision clause. This exclusion also excludes any indirect liability which the carrying vessel may have towards the non-carrying vessel in respect of damages paid by it to the carrying vessel's cargo.

Claims for loss of life or personal injury have historically been taken out of collision liability cover. The clause is mainly concerned with the insurance of the subject matter rather than the insurance of personnel operating the ship or of any other person on board at the time of an incident.

Claims arising from pollution or contamination of any real or personal property or things whatsoever (except other vessels with which the insured vessel is in collision or property on such other vessels) are also excluded. The shipowner is usually covered under the P&I policy for any liabilities, losses, damages, costs and expenses which he incurs as a result of pollution or threat of pollution from his ship. The exclusion does not extend to any sum or sums by way of damages for pollution or contamination of any vessels with which the insured vessel is in collision or to property on board such vessels.

The above-explained exclusions restrict the protection afforded by the three quarters collision liability clause. Theses liabilities are usually recoverable by the shipowner from his P&I club. The application of exclusions was examined in *Chapman* v. *Fisher*.[76] In this case the owners were entitled to three quarters of the proportion of their statutory liability applying to the damage to the vessel *Hermia* and to the salvors' equipment under the Running Down Clause. They were not entitled to recover their liability relating to the increased cost of removal of the vessel which came within the exclusion with regard to the "removal of obstructions under statutory powers" as the clause in that case was worded.

(iii) Measure of indemnity under the Institute Collision Liability Clause

The wording of the Running Down Clause on measure of indemnity has been replaced by Clauses 8.2.1 and 8.2.2.

It has long been established that when two vessels are in collision and both are to

76. (1904) 20 T.L.R. 319. See R.J. Lambeth, *Templeman on Marine Insurance. Its Principles and Practice* (6th ed., 1986), p.420.

blame the court will assess the degree of blame attaching to each party and, after calculating the liability of one to the other, will require the party with higher liability to pay the difference to the other party. This practice is known as single liability. For example:

Insured vessel suffers damage	£1,000,000
Other ship suffers damage	£ 500,000
Each ship is deemed 50 per cent to blame for the collision.	
Other ship is liable for	£ 500,000
Insured ship is liable for	£ 250,000
Other ship pays	£ 250,000 (single liability)

Under the principle of cross liability the claim on underwriters is assessed as if each side actually made a payment to the other side of their respective liability. Under this principle it is not necessary to make two payments. The two liabilities can be set off against each other and the vessel with the greater liability simply pays the difference between the two liabilities to the other vessel. Between the two vessels the end result is the same whichever principle is applied unless one of the vessels claims to be able to limit its liability.

Under Article 5 of the Convention on Limitation of Liability for Maritime Claims 1976, the shipowner can limit his liability to the balance of the amount for which he is liable after claim and counterclaim have been set off against each other. If the apportionment of liability is made under the cross liability principle the vessel having the larger gross liability may limit its liability to an amount less than the other vessel's gross liability. If the case does not involve cargo loss the larger vessel will always have to pay the smaller vessel the difference between their respective limitation amounts irrespective of how much damage each vessel has suffered. In cases where both vessels can limit liability and involving cargo or other third party losses the settlement between the vessels will be on a single liability basis and the liability of the vessel with the lesser liability will be regarded as unlimited for the purposes of settlement.

The single liability principle is used between the vessel, but between the owner of the vessel and his insurers the cross liability principle is to be used. If one assumes that the claim is completely straightforward, each vessel having only one insurance policy covering all damage and liability, the only loss suffered is damage to the vessel and liability for damage to the other vessel. The result between the insurer and assured will be exactly the same whichever principle is applied. Obviously in practice there will be more complicated claims. It is now generally accepted that between the shipowner and his insurers the cross liability principle will always give the fairest result. For instance, in all cases both vessels are equally to blame and the damages are as follows[77]:

Damage to A's ship	£100,000	Damage to B's ship	£200,000
A's demurrage	£ 40,000	B's demurrage	£ 80,000
	£140,000		£280,000

77. R.J. Lambeth, pp.403–404.

At law this settlement involves only one payment, that by A to B of £70,000.

A is liable for 50 per cent of £280,000 =	£140,000
B is liable for 50 per cent of £140,000 =	£ 70,000
	£ 70,000

Single Liability Basis

It is assumed that each vessel has been fully insured.

A's underwriters would pay the amount of the damage to ship A	£100,000
and ¾ of the amount paid by A to B namely ¾ of £70,000	£ 52,000
Total payable by A's underwriters	£152,500
A also would recover the remaining ¼ of £70,000 from his P&I club	£ 17,500
A's total recovery would therefore be	£170,000
B's underwriters would pay the amount of the damage to ship B	£200,000
and receive credit for the proportion of B's recovery from A, namely £70,000	
applying to the damage to ship B, $\frac{200{,}000}{280{,}000}$ of £70,000=	£ 50,000
Net amount payable by B's underwriters	£150,000

* B's P&I club would make no payment.

Cross Liabilities Basis

A's underwriters would pay the amount of the damage to ship A	£100,000
and ¾ of ½ of B's total damages which A is assumed to have paid to B, namely ¾ of ½ of £280,000	£105,000
	£205,000
and would receive credit for ½ of the damage to ship A, which A is assumed to have recovered from B	£ 50,000
Total payable by A's underwriters	£155,000
A's P&I club would pay ¼ of ½ of B's total damages	£ 35,000
A's total recovery would therefore be	£190,000
B's underwriters would pay the amount of the damage to ship B	£200,000
and ¾ of ½ of A's total damages, namely ¾ of ½ of £140,000	£ 52,000
	£252,500
and would receive credit for ½ of the damage to ship B	£100,000
Total payable by B's underwriters	£152,500
B's P&I club would pay ¼ of ½ of A's damages	£ 17,500
B's total recovery would therefore be	£170,000

(Any deductibles in the policies have to be deducted.)

The parties can choose between alternative rules on settlement of claims, either the traditional way, as explained above, or the Norwegian solution. Under the Norwegian Plan the main point is that one should first look at the figures for a cross liability settlement without considering the limitation of liability. Then the largest gross liability is reduced by the amount by which the vessel in question actually reduced its liability. The gross liability of the vessel having the lowest liability will not be changed and the cross liability settlement on the basis of the modified figures is carried out in the normal way.

PART 3

OIL POLLUTION

CHAPTER 11

INTRODUCTION

The sea occupies 70 per cent of the earth's surface and it constitutes the world's greatest highway. "Its use is enjoyed by all nations and the continuance and safeguarding of that use is essential. Many nations, particularly island nations such as the United Kingdom",[1] are dependent upon the sea for the maintenance of their standard of living. These nations are much threatened by passing traffic as ships are destined for their ports. They are all concerned with doing whatever is reasonably possible to protect their coastlines from pollution from merchant shipping. The pollution may take many forms but in general all involve the escape from ships of substances which should not be in the sea or on the shores. The polluting substances may be oil, hazardous and noxious substances, garbage, micro-organisms carried in ballast water or they may be in other less familiar forms. The escape may be deliberate or accidental. Some discharges of polluting material are permitted under international conventions. But illegal discharges and accidental discharges are still a great problem.

The increasing seaborne transport of oil and other hazardous substances constitutes a growing pollution risk to the seas and oceans. The most common contaminant from ships is oil. It also has the highest public profile. Whether by accident or design oil pollution will exist as long as oil is moved by sea. The word "oil" refers to a wide range of materials. It includes petroleum, bunker fuels, petrol and lubricants. Vessel source pollution includes pollution originating from vessels involved in navigation and transportation and is distinct from dumping. Pollutants from various sources are covered by the term "vessel-source pollution". Oil pollution from tankers is one aspect of this pollution. A comparatively small proportion of total oil pollution comes from single accidents like the *Amoco Cadiz*, the *Exxon Valdez* and the *Torrey Canyon*. In fact such accidents are responsible for less than 10 per cent of marine pollution. Operational discharges constitute 80 per cent of vessel-source pollution by oil. Tanker accidents are publicised because they have a dramatic impact on environmentally sensitive, although small, coastal areas. In other words large spills of oil in small areas inflict serious damage and attract public interest. Therefore most public attention is focused on these

1. *Safer Ships, Cleaner Seas*, Report of Lord Donaldson's Inquiry into the Preventing of Pollution from Ships, p.12.

accidents. On 2 July 1997 the 260,000 tonne tanker *Diamond Grace* hit a reef about 12 miles south of Tokyo. About 13,400 tonnes of crude oil gushed into important fishing grounds, creating a slick that winds carried northwest towards Tokyo and China. The oil spill was one of the main news items of the day both on TV and in newspapers.

CHAPTER 12

OIL POLLUTION FROM SHIPS AND THE PUBLIC INTERNATIONAL LAW APPROACH

Pollution of the sea by oil discharged from ships creates many legal problems. Nearly all sectors of national law are involved.[1] Criminal law has to be considered in order to punish for discharge of oil into the sea, the law of torts deals with civil liability for damage arising as a result of pollution, the law of evidence is used in order to prove that pollution arose from a given ship and not from another source, criminal procedure makes the charges and form, insurance law makes sure that adequate compensation is available. But despite the complete application of national law the most complex problems still need to be solved by international law. Maritime transport is essentially international in character. Therefore pollution may happen anywhere. If it takes place within the internal or territorial waters of a state then it will be subject to that state's exclusive jurisdiction. But when it occurs on the high seas, it will be within the ambit of international law. In international law there has been increasing recognition of the need to protect the marine environment as a whole and to lay down principles and rules within that framework. International law establishes a duty on all states to co-operate in establishing rules and standards to prevent, reduce and control pollution of the marine environment.

MARINE ENVIRONMENTAL LAW

Important legal sources in international law are customary law, international conventions, general principles of law and judicial decisions. Marine environmental law is merely a specialised branch of international law; like international law it derives primarily from treaties, customs, general principles of law and judicial decisions.

Within international law marine environmental law has a short history and the most extensive source of marine environmental law is found in the conventions. International organisations, intergovernmental and non-governmental bodies have played an important role in promoting new conventions and finding new ways of contributing to the law-making process. The environmental conventional rules

1. Yoram Dinstein, "Oil Pollution by Ships and Freedom of the High Seas" (1972) 3 *Journal of Maritime Law and Commerce*, 363–374.

which are in force mainly cover vessel source pollution. The framework of marine environmental treaties developed under the auspices of the International Maritime Organisation (IMO)[2] and the United Nations Environment Programme (UNEP). Although treaties remain the most frequent method of establishing binding rules there are some practical problems with their application. The negotiation and ratification process can be too slow. The IMO treaties provide good examples of this problem. If they do not codify or generate a new custom they bind only their parties. Furthermore, as they have to satisfy so many different interests of different parties they may lay down general requirements. Then states parties to these conventions need to provide cover for missing parts by enacting national laws and by further protocols.

Customary law is quite important, specifically in laying down fundamental guiding principles in the area of liability and compensation. It is largely derived from a generalisation of frequent and uniform state practice. The law has to be practised by states and accepted as a law by them. As marine environmental law has been in existence for a short time it is possible to question whether there has been any customary law based on state practice and its acceptance as law.[3] Like treaties customary law faced some criticism. As customary law has been derived from a general frequent and uniform state practice it has been considered inappropriate for developing specific rules. But today the view on state practice has been enlarged. The United Nations Conference on the Law of the Sea (UNCLOS III) is an example of this. Under the Conference the parties are obliged to carry out their undertakings in good faith although the Conference is not binding and does not entail international responsibility. On the other hand, in order to create customary law there is no need to go through the detailed treaty-making process as its acceptance by states brings it into force sooner than any other treaty. It often provides the framework for enacting more specific national and international law such as the US Oil Pollution Act 1990.

CONCEPTS UNDERLYING THE OIL POLLUTION SYSTEM

The law of the sea has rested on two traditional principles, which are the freedom of the high seas and the sovereignty of the coastal state on the territorial sea. In relation to technological developments and the need to protect the marine environment there is a need for a balance of interests. The freedom of the seas in the past was a different freedom from today's freedom of the seas. A balance is needed between the rights and responsibilities of maritime shipping interests and coastal states. In early concepts of international law little consideration was given to marine pollution. Now it has become clear that the oceans' ability to absorb

2. See App. XI: The status of Conventions.

3. See Douglas Brubaker, *Marine Pollution and International Law, Principles and Practice* (1993), pp.58–66; Patricia Birnie, "Protection of the Marine Environment: The Public International Law Approach" in Colin M. De La Rue, *Liability For Damage To The Marine Environment* (1993), pp.1–4.

waste is not infinite. Therefore it is not possible to allow the unrestricted introduction of pollutants into the sea. Under customary international law states are required to do the necessary work to control and regulate all sources of pollution within their territory. States are obliged to promote the adoption of routing systems designed to minimise the threat of accidents which may cause pollution of the marine environment. The UNCLOS, in Article 192, embodies "a general obligation", whereby "states have the obligation to protect and preserve the marine environment", as will be explained later in this part.

In international law the protection of the marine environment is based on two principles: under the "precautionary principle" in certain cases it is not necessary to wait for conclusive scientific proof of actual or imminent harm before taking action to control harmful activities.[4] For instance under the Second and Third Declarations of the International North Sea Conferences it was recommended that where there is a threat of serious or irreversible damage, lack of scientific certainty should not be used as a reason for postponing measures to prevent environmental degradation.[5]

"The polluter pays" principle is often referred to as an economic rather than a legal principle. This principle has been developed by the OECD. Under the OECD recommendation in 1972 the principle was described as "... the polluter should bear the expenses of carrying out ... measures decided by public authorities to ensure that the environment is in an acceptable state".[6] The primary object of this principle is to stop governments subsidising industry by carrying their environmental costs. It sets out the way to distribute the costs of pollution by assuming that the polluter in a given case is readily identifiable. Obviously it is not always possible to identify the polluter in the case of discharges at sea. Although the principle attracted public criticism economic considerations motivated EC environmental policy and in 1973 the Council of Ministers adopted a programme of action on the environment which endorsed the "polluter pays" principle. Under Article 25 of the Single European Act 1986 the "polluter pays" principle provided a new legal basis for EC environmental issues.[7] Similarly under the Preamble to the IMO's 1990 Convention on Oil Pollution Preparedness, Response and Co-operation the "polluter pays" principle is referred to as a general principle of international law.[8]

4. Patricia Birnie, pp.6–7.
5. *Ibid.*
6. Recommendation c(72) 128, 26 May 1972.
7. Art. 25 of the Single European Act 1986 states that "Action by the Community relating to the environment shall be based on the principles that preventive action should be taken, that environmental damage should as a priority be rectified at source, and that the polluter should pay".
8. According to Alan E. Boyle, "the economic motivation for harmonizing environmental policy is not disregarded, but it is less prominent than in OECD recommendations": Alan E. Boyle, "Making the Polluter Pay? Alternatives to state Responsibility in the Allocation of Transboundary Environmental Costs", chapter 15 in Francesco Franciani and Tullio Scourazzi, *International Responsibility for Environmental Damage* (1991), pp.368–370.

UNITED NATIONS CONVENTION ON THE LAW OF THE SEA 1982

The object of the Third Conference on the Law of the Sea was the formulation of a comprehensive agreement on the Law of the Sea. This Conference states that one of the primary goals of the convention is the establishment of effective, prescriptive and enforceable regimes for the prevention of pollution from vessels, and in particular for preventing accidents and dealing with emergencies, ensuring the safety of operations at sea, preventing intentional and unintentional discharges, and regulating the design, construction, equipment, operation and manning of vessels.[9] It was left to the Third Conference to improve the situation by bringing together all sources of marine pollution in one convention, the United Nations Convention on the Law of the Sea (UNCLOS 1982). The Convention deals with standards of vessel-source pollution in Part XII. The most important development in the Convention is the devotion of Part XII exclusively to "Protection and Preservation of the Marine Environment".

The provisions of the Convention are not intended to form a comprehensive set of rules to cover all the problems of the marine environment. Therefore the Convention resolves only the issues on general principles and basic jurisdictional questions concerning legislation and enforcement on marine pollution. Rather than setting out detailed regulations or standards for the control of marine pollution the provisions are intended to take the form of an umbrella treaty which sets out general procedural rules on marine pollution. The wording of the Convention articles on marine pollution mainly relies on existing treaties and on the Declaration on the Human Environment and Recommendation for Action issued at the United Nations Conference, Stockholm, 1972. The Convention also includes some provisions which are new to international law.

The Stockholm Declaration and Action Plan resulted from the United Nations Conference on the Human Environment, held in Stockholm in June 1972.[10] The Declaration was brought about by the worldwide concern for the environment of the 1960s and early 1970s. The Declaration is conventional in nature but is not in fact a treaty. Marine pollution was only one aspect of the Declaration but it managed to attract the attention of governments and provided the necessary directional impetus for states to manage the environment. After the Declaration it was considered that there was not enough protection for the human environment.[11] The principles of the Declaration cover all aspects of environmental deterioration. Article 7 specifically deals with marine pollution and Article 22[12]

9. Law of the Sea Convention, Official Records of the Third United Nations Conference on the Law of the Sea U.N.Doc. A/CONF.62/122, (1982).

10. For detailed information on the Stockholm Conference of 1972 on the Human Environment see I.A. Shearer, *Starke's International Law* (11th ed., 1994), pp.364–374.

11. See Edgar Gold, *Handbook on Marine Pollution* (1985), pp.35–36; E.D. Brown, *The International Law of the Sea*, Vol I (1994), pp.337–338.

12. Art. 22 of the Stockholm Declaration and Action Plan on the Human Environment 1972 states that "States shall cooperate to develop further the international law regarding liability and

deals specifically with liability and compensation for marine pollution damage. The principle contained in Article 7[13] had considerable influence on subsequent related treaties. A so-called Action Plan provides details of the actions which have to be taken by governments and international organisations to deal with pollution. The Declaration and the Action Plan provided general public environmental awareness and reappeared in the following related treaties such as the UNCLOS 1982.

(a) The regulatory provisions

The principal provisions on marine pollution are to be found in Part XII of the Convention, entitled "Protection and Preservation of the Marine Environment". This Part has been considered as a first attempt to provide a comprehensive framework of conventional rules which cover all sources of pollution of the marine environment.[14] Article 192 prescribes the duties and measures necessary to achieve the protection and preservation of the marine environment. Under that Article "States have the obligation to protect and preserve the marine environment". Article 194 clearly states that in honouring their obligation states have a duty to use the "best practicable means at their disposal and in accordance with their capabilities". The Convention deals with six sources of marine pollution and lays down a comprehensive framework for the taking and enforcing of measures against all these different types of pollution. Under Article 207 states are obliged to deal with pollution from land-based sources by adoption of national legislation and the establishment of global and regional rules. While doing so states should consider the economic capacity of developing states and their need for economic development. Under Article 208 states are under a similar obligation as regards pollution from seabed activities. The seabed area beyond the outer limit of the continental shelf to which Part XI of the Convention applies and in which the extensive exploitation of polymetallic nodules is expected to take place in the next century is the area in which states are under a complementary obligation to adopt national legislation in relation to the activities of vessels, installations, structures and devices flying their flag, or of their registry, or operating under their

---

compensation for the victims of pollution and other environmental damage caused by activities within the jurisdiction or control of such states to areas beyond their jurisdiction".

13. Art. 7 of the Stockholm Declaration and Action Plan on the Human Environment 1972 states that "States shall take all possible steps to prevent pollution of the seas by substances that are liable to create hazards to human health, to harm living resources and marine life, damage amenities or to interfere with other legitimate uses of the sea".

14. Under Art. 1(4) of the UNCLOS 1982 "pollution of the marine environment" means "the introduction by man, directly or indirectly, of substances or energy into the marine environment, including estuaries, which results or is likely to result in such deleterious effects as harm to living resources and marine life, hazards to human health, hindrance to marine activities, including fishing and other legitimate uses of the sea, impairment of quality for use of sea water and reduction of amenities".

authority.[15] Under Article 210 states should adopt laws and regulations to prevent, reduce and control pollution of the marine environment by dumping. National legislation must ensure that dumping is not carried out without the permission of the competent authorities of the state. Dumping within the territorial sea and the exclusive economic zone or onto the continental shelf shall not be carried out without the express prior approval of the coastal state. Article 211 deals with pollution from vessels. States have an obligation to establish international rules and standards and national laws and regulations at least to the same effect. The details of this Article will be discussed later in this chapter. The final type of pollution dealt with under Section 5, Part XII is pollution from or through the atmosphere. Under Article 212 a combination of national and international law-making is provided for to prevent, reduce and control marine pollution from or through the atmosphere.

Considering the above-mentioned Articles it is possible to say that UNCLOS 1982 creates an extensive framework for the taking and enforcing of measures on these different sources of pollution. It is also quite strong in clarifying the obligations breach of which invokes state responsibility. But when it comes to a definition of liability as a result of violation the Convention is quite vague. A major criticism of the Convention provisions is on the level of standards. As they are so general it is not possible to implement them in an objective manner.[16] These general Articles do not define a state's responsibility sufficiently. In particular, the term used in Article 194(1), "best practical means available", would be adopted differently in developing and developed countries. A similar example may be found in paragraph 2 of this Article where states are to take all "measures necessary" to ensure that activities under their jurisdiction or control are so conducted as not to cause damage by pollution to other states and their environment, and that pollution arising from incidents or activities under their jurisdiction or control does not spread beyond the areas over which they exercise sovereign rights in accordance with this Convention. Under this paragraph states seem to be under a certain liability for pollution but this liability is subject to responsibility and the liability provision under Article 235. Another example of general standards is Article 194(3)(b) where a state's responsibility is defined with regard to vessel-source pollution. Paragraph 3(b) requires a state to undertake measures to control vessel-source pollution "to the fullest possible extent". A state's responsibility is not defined clearly and standards of measures are also not stated.

States are not allowed to transfer, directly or indirectly, damage or hazards from one area to another or to transport one type of pollution into another while trying

15. See E.D. Brown, Vol. I, p.342 and Art. 209(2) of the UNCLOS 1982.

16. See Art. 194(1) of the UNCLOS 1982 which allows states to use "best practicable means at their disposal and in accordance with their capabilities" to fulfil UNCLOS 1982 objectives.

to achieve the aims of Articles 192–194.[17] The need for international co-operation was stressed in Section 2, global and regional co-operation, of Part XII of the Convention. Article 197 urges "co-operation on a global basis and as appropriate, on a regional basis directly or through competent international organisations" among states in formulating international rules and standards. Under the following Articles states agree to notify the parties concerned, as well as the competent international organisations, when other states are in imminent danger of pollution. While doing so states shall develop joint "contingency plans for responding to pollution incidents in the marine environment", "promote studies and research and encourage the exchange of information data" and "establish appropriate scientific criteria for the formation and elaboration of rules". The need for technical assistance becomes important under Section 3, Part XII where states agree on the need for scientific and technical assistance to be given to developing states. In Section 4, Part XII states undertake legal responsibility to co-operate in environmental monitoring programmes.

In order to prevent pollution from dumping, seabed operations and ships[18] states are under an obligation to establish international rules and standards. "Pollution from vessels" includes pollution originating from vessels involved in navigation and transportation and is distinct from dumping. Oil pollution from tankers is one aspect of vessel-source pollution; the term includes pollutants from various sources. In order to establish a vessel-source pollution enforcement regime two different interests had to be reconciled. These were those of the maritime shipping community and of coastal states.[19] With regard to vessel-source pollution the Convention has apportioned legislative jurisdiction between flag states and coastal states. Article 211 establishes controls on pollution from vessels and includes both flag and coastal states. Article 211(1) provides that "States, acting through the competent international organisation or general diplomatic conference, shall establish international rules and standards to prevent, reduce and control pollution of the marine environment from vessels . . .". So coastal states may adopt laws and regulations for the prevention, reduction and control of marine pollution from foreign vessels within their territorial seas, including vessels exercising the right of innocent passage. However, such measures may not impede innocent passage.[20] Under this Article it is difficult to identify exactly which

17. Under Art. 195 of the UNCLOS 1982 it is stated that "In taking measures to prevent, reduce and control oil pollution of the marine environment, States shall act so as not to transfer, directly or indirectly, damage or hazards from one area to another or transform one type of pollution into another".

18. Arts. 210, 208 and 211 of the UNCLOS 1982 respectively.

19. See M.A. Stephenson, "Vessel-source Pollution Under the Law of the Sea Convention —an Analysis of the Prescriptive Standards", (1992) *University of Queensland Law Journal* 123–124.

20. UNCLOS 1982, Art. 211(4).

standards are being referred to. While establishing international rules and standards the standard applied must have the same effect as "generally recognised international rules and standards established through the competent international organisation or diplomatic conference".[21] The competent international organisation is nowhere defined or specially mentioned in the Convention.[22] Maritime states generally understand that this organisation is the IMO. It is also not clear what meaning should be attributed to "generally accepted". Accepting the IMO as the competent international organisation it is possible to argue that the requirements of the 1973/78 MARPOL, the IMO codes and other relevant international guidelines would also be incorporated. But does this mean that even if the state concerned has not ratified these Conventions it will be bound by their standards as customary law? Obviously if a state has not signed a convention it will not bind the state and states may not wish earlier conventions to be binding on them just because they are parties to the Law of the Sea Convention. If a state has not ratified the 73/78 MARPOL but ratified the Law of the Sea Convention then can the rules and standards of MARPOL be said to be "generally accepted"? Even if the IMO rules are considered to be "generally accepted" what is the meaning of the qualifying phrase, "shall at least have the same effect"?[23] And the further requirement under Article 211(1) on the establishment of international standards through "general diplomatic conference" only creates further ambiguity rather than making them clear.

Under the regulatory provisions of the Convention states are under a duty to legislate to prevent, reduce and control pollution to protect the marine environment. But although this comprehensive framework for the taking and enforcing of measures is phrased in obligatory terms the Convention does not provide a precise explanation of the detailed rules and standards and also does not refer to any international body. Considering the regulatory provisions it is possible to conclude that the Convention keeps its characteristic of an umbrella treaty and provides general guidelines for an effective oil pollution regime.

(b) The enforcement provisions

The rules set out in Part XII on setting standards for the protection and preservation of the marine environment are completed by the rules on enforcement of these standards under Articles 213–222. Traditionally anti-pollution laws can be enforced under coastal-state and flag-state jurisdiction. The Convention

21. Art. 211(2) of the UNCLOS 1982 provides that "States shall adopt laws and regulations for the prevention, reduction and control of pollution of the marine environment from vessels flying their flag or of their registry. Such laws and regulations shall at least have the same effect as that of generally accepted international rules and standards established through the competent international organization or general diplomatic conference".

22. Arts. 208(5) and 210(4) refer to international organisations in the plural.

23. See Patricia Birnie, pp.15–16, M.A. Stephenson, pp.124–126.

has continued to recognise the primacy of the flag state's jurisdiction over ships but it also recognised the increasing role of coastal and port states as they supervise and ensure conformity with international law.

A coastal state has full sovereign powers in its territorial sea. In the territorial seas coastal states have retained jurisdiction over pollution. Under this territorial jurisdiction a coastal state may enforce its pollution laws on its territory and in its ports, internal waters and territorial sea. In principle a coastal state has jurisdiction to protect and preserve the marine environment within the exclusive economic zone. With regard to land-based sources of pollution, dumping and discharges landward of the 200 mile EEZ limit the Convention employs coastal state jurisdiction.

Traditionally, a vessel on the high seas has been subject to the exclusive jurisdiction of the flag state. In other words the flag state has traditionally enforced the law on ships flying its flag or registered in its territory. Part XII of the Convention incorporates flag-state enforcement.[24] But if a foreign vessel causes pollution beyond the limits of a state's territorial jurisdiction then the flag state cannot exercise jurisdiction. This problem has been solved by Article 218 which embodies port-state jurisdiction. This Article is subject to the safeguards of flag state interests laid down in Article 228.[25] Under Article 218 port states may exercise jurisdiction and even institute proceedings over foreign vessels entering their ports and offshore terminals. This jurisdiction is subject to various conditions. In order to apply the jurisdiction, the foreign vessel either should have discharged on the high seas in violation of "applicable international rules and standards" established through competent international organisations or general diplomatic conference or should have discharged in waters, including the EEZ, subject to the jurisdiction of another state. In cases where discharge is made in waters, including the EEZ, the application of jurisdiction is possible only if requested to act by that state, the flag state or a state damaged or threatened by the discharge.[26]

Article 220 governs enforcement by coastal states. If a foreign vessel is suspected of having, during its passage through the territorial sea, violated the coastal state's anti-pollution legislation or applicable international rules relating to pollution from ships, the coastal state may, without prejudice to its general enforcement competence in the territorial sea as set out in Section 6, Part XII, of the Law of the Sea Convention, undertake physical inspection of the vessel and, where the evidence so warrants, institute legal proceedings.[27]

24. See Art. 217 of the UNCLOS 1982.
25. See Patricia Birnie, pp.16–17; E.D. Brown, Vol I. pp.342–343.
26. See Patricia Birnie, p.17.
27. UNCLOS 1982, Art. 220(2).

(c) Provisions on state responsibility

The UNCLOS 1982 does not have detailed provisions on state responsibility. In fact it is quite weak in relation to the issues of responsibility and liability, which are dealt with in Article 235. This covers both state responsibility and the civil liability of natural or juridical persons. With regard to state responsibility, Article 235(1) simply provides that "States are responsible for the fulfilment of their international obligations concerning the protection and preservation of the marine environment. They shall be liable in accordance with international law". In cases where pollution from the foreign vessel is wilful and serious, the passage of that vessel is no longer innocent. In such cases the coastal state has unrestricted enforcement jurisdiction. Under Section 7, Part XII, of the Convention the coastal states have been given the power of arrest. For this right to be used the violation should cause or threaten "major damage to the marine environment of the straits".[28]

In the matter of civil liability Article 235(2) and (3) again provides a general framework of obligations in general terms rather than detailed provisions. Paragraph (2) provides that "States shall ensure that recourse is available in accordance with their legal systems for prompt and adequate compensation or other relief in respect of damage caused by pollution of the marine environment by natural or juridical persons under their jurisdiction". Paragraph (3) provides that:

> "With the objective of assuring prompt and adequate compensation in respect of all damage caused by pollution of the marine environment, States shall co-operate in the implementation of existing international law and the further development of international law relating to responsibility and liability for the assessment of and compensation for damage and the settlement of related disputes, as well as, where appropriate, development of criteria and procedures for payment of adequate compensation, such as compulsory insurance or compensation funds."

In general the implementation of "liability in accordance with international law" and "existing international law" is left to national legal systems. Surprisingly the Convention does not have much to say on this important aspect of marine environmental law.

28. UNCLOS 1982, Art. 233.

CHAPTER 13

MARINE POLLUTION CONVENTIONS: GLOBAL

In maritime law there are rules designed to prevent pollution and there are rules designed to provide compensation for the victims of pollution. The following conventions are those which aim to prevent pollution.

INTERNATIONAL CONVENTION FOR THE PREVENTION OF POLLUTION OF THE SEA BY OIL (OILPOL 1954)

Oil is the pollutant with the longest history of international attention. In other words the first pollutant for which international control standards were set was oil. It has been on the international agenda since the 1920s. It first became the subject of an international conference in 1926, when the United States recognised the international nature of the problem and convened a governmental conference. The conference drafted a treaty to regulate intentional discharge. A similar effort was made by the United Kingdom in 1935 under the auspices of the League of Nations Committee of Experts. Both Conferences succeeded at drafting conventions but none of these conventions was signed or entered into force. Although the first attempts failed, in 1954 the International Convention for the Prevention of Pollution of the Sea by Oil was signed in London.[1]

After its efforts to achieve any signed agreements failed, the United Kingdom continued to lead the call for international regulation. Particular interest groups were the major source of pressure for action by the United Kingdom. As a result of this pressure the United Kingdom established the Committee on the Prevention of Pollution of the Sea by Oil, known as the Faulkner Committee.[2] In its 1953 report the Committee recommended an international ban on discharges above 100 parts per million (ppm) by all ships, including tankers and non-tankers, throughout the ocean. It was also suggested that until an international agreement was reached, ships registered in the United Kingdom should be banned from discharges over 100 ppm "within a wide zone around the United Kingdom".[3] The United

1. For detailed information on the International Convention for the Prevention of Pollution of the Sea by Oil 1954 (OILPOL 1954) see Ronald B. Mitchell, *International Oil Pollution At Sea*, "International Oil Pollution: History and Context", Chapter 3 (1994), pp.68–118.
2. UK Ministry of Transport, *Report of the Committee*, l, cited in Ronald B. Mitchell, p.83, n.33.
3. *Ibid.*, 33.

Kingdom government believed that the zones were inadequate and UK domestic shipping and oil interests led the government to press for international action.

In 1953 tankers transported some 250 million metric tonnes of oil by sea. They intentionally discharged some 300,000 metric tonnes of it into the ocean. The growing amount of oil transported by sea has resulted in estimates of total discharges that have ranged up to five million metric tonnes per year. When these figures are considered in relation to all ship-generated oil pollution they traditionally represent two-thirds of it. Tanker accidents and discharges from non-tankers constitute the remaining one third.[4] How does intentional discharge of oil happen? When an oil tanker has discharged its cargo a certain amount of the oil remains clinging to the tanks. This oil has to be disposed of before a new cargo can be taken on board. There are two ways for tankers to clean their tanks, and these two standard industry practices led to the mixing of this clingage with seawater. One way of cleaning tanks is for tankers to wash out their empty tanks at sea. They use seawater in high pressure cleaning machines to wash down their tanks. The second way is to fill empty cargo tanks with seawater as ballast to stabilise the tankers during their return voyages. This water, containing the residues of oil, has to be pumped out before a new cargo can be taken on board. Other ships also use seawater as ballast in their empty fuel tanks and they of course eventually pump out this seawater. In any of these ways oil may enter the sea.

In 1954 32 countries attended the governmental conference in London. Only eight of the 32 countries attending said they regarded oil pollution as a problem, and some did not want a convention at all. Therefore the fact that the convention was adopted at all was something of an achievement. The United Kingdom proposed to limit discharges throughout the ocean. It basically required all tankers to stop discharging waste oil at sea. They were not in favour of discharge being made under a zonal approach, which meant as far as possible from the shore. It was argued that tankers could eliminate oil pollution if they "refrained from cleaning their tanks and mixed oily ballast residues with new cargo oil" or practised retention on board.[5] Oil and shipping companies objected to this proposal because retention on board required lengthy port delays to discharge slops and this could easily be done while tankers were under way. Governments also opposed the proposal because they needed to provide expensive reception facilities to receive these slops. As a result there was no agreement to limit discharges throughout the oceans.

Despite the failure to agree to limit discharges throughout the oceans OILPOL 1954 was a significant convention at that time. It was the first international treaty exclusively designed to deal with the oil pollution problem. As the first convention which attempted to control marine oil pollution it served as a basis for establishing the succeeding conventions.[6] However, it was concerned only with operational oil

4. See Ronald B. Mitchell, p.72.

5. See Sonia Zaide Pritchard, *Oil Pollution Control* (1987), p.95, cited in Ronald B. Mitchell, p.84, n.42.

6. See Douglas Brubaker, pp.121–122; R.R. Churchill and A.V. Lowe, *The Law of the Sea* (2nd ed., 1988), pp.248–249.

pollution from merchant ships. There was no attempt to introduce measures concerning accidental pollution, or to deal with pollution by other substances. It prohibited intentional operational discharge of oil and oily mixtures by certain ships in specified areas of the oceans. Under Article II it prohibited intentional discharge of oil and oil mixtures by ships registered in the contracting parties or unregistered ships having the nationality of a contracting party. Tankers under 150 tonnes gross tonnage and other ships under 500 tonnes, whaling ships, ships navigating the Great Lakes of North America and naval ships were exempted.

In its original form the convention's primary role was the prohibition of discharge of oil and any oily mixture having an oil content of more than 100 parts per million within 50 miles of land and in certain special areas such as the North Sea, Baltic and Black Sea (Article III and Annex A). This prohibition did not apply if a ship other than a tanker was making for a port not equipped with reception facilities for oily residues (Article III(2)). Punishment for offences under convention provisions both in the territorial sea and the high seas must be equally severe. Under Article VII carrying ballast in oil fuel tanks must be avoided if possible. Parties are required to take all appropriate steps to provide port and ship repair facilities to receive residues and oily mixtures from ships, to provide oil-loading terminals with facilities to receive residues and oily mixtures from ships, and to provide oil-loading terminals with facilities to receive residues and oily mixtures from tankers (Article VIII). Ballast discharges have to be confined to permitted areas and all loading and discharges of oil must be recorded in an "oil record book" which is to be available for inspection at all reasonable times.

The discharge standards adopted under OILPOL 1954 were tightened up by the amendments to the convention adopted in 1962. As the results of OILPOL 1954 were not satisfactory the Advisory Committee on the Prevention of Oil Pollution of the Sea sponsored a conference in 1959. The participating countries recommended the extension of the 1954 zones and globally banning operational discharges. In 1962 IMCO sponsored another conference, and by 1962 the British government modified its 1954 proposal. Under the 1962 proposal only new tankers over 20,000 tonnes were to be banned from discharging anywhere in the oceans. Under the proposal oil-content monitors and oil–water separators were not expressly required, but the rule implicitly required new tankers to practise retention on board and monitor all discharges wherever they occurred.[7] Despite rejection by several countries including the USA, Japan, Norway and the Netherlands the proposed restriction on new tankers was adopted. Article III(c) of the 1962 Amendment prohibited the discharge of oil or oily mixture from new ships of 20,000 gross tonnes or more, i.e. those for which the building contract was placed on or after the entry into force of the 1962 amendments. The 1962 amendments entered into force in 1967.

After the 1962 amendments there was no significant increase in reception

7. The main reason for the rejection of this proposal by the USA was that devices and procedures that would have allowed a tanker to monitor its own compliance with the oil content standard did not exist at that time.

facilities. It became clear to the oil companies that expensive equipment would be required for new ships to comply with the general prohibition. The rules only resulted in increased research into alternative methods of reducing discharges. The failure of the 1954 Convention and its 1962 amendments to prevent oil discharges at sea encouraged the search for alternative ways of disposing of unwanted oil. The industry's solution was to separate the oil and water on board, discharge the water to the sea and retain the oil residues aboard. In most situations the cargo could then be loaded on top of the residues. This was called the "load on top" system.[8] By operating this system governments could save nearly all the oil which would otherwise have been discharged to sea. Many governments liked this system because it reduced the total oil discharges and the need to build reception facilities. The system was quickly adopted by 80 per cent of all tankers but not accepted by all of them because in practice not all tankers operate the system or for some reason some tanker owners do not want to use the system.[9]

In 1969 further amendments to the 1954 Convention were adopted. In general amendments to the convention, in 1962 and 1969, expanded its scope by narrowing the permitted areas for pollution discharges. The grounding of the *Torrey Canyon* in 1967 provided a major new impetus for oil pollution control. The 1969 amendment was in response to the *Torrey Canyon* disaster which had shown up some of the weaknesses in OILPOL.

The 1969 amendments entered into force on 20 January 1978. They introduced a standard which, for the first time, was based upon the load on top system. Under the new system tankers were required to discharge oil *en route* at sea. The discharge had to be accomplished through use of the load on top system. The 50 mile zones were retained. Outside the zones all tankers had to keep discharges below 60 litres per mile and the total quantity of oil discharged was limited to $^{1}/_{15,000}$ of the total cargo-carrying capacity. Ships other than tankers were obliged to discharge their oily water through a suitable operator and to retain all oily residues on board for eventual disposal ashore. The outcome of the 1969 amendments is that tankers either operate the load on top system or keep all oily residues on board for eventual disposal ashore. In 1971, two further amendments were approved by the IMO Assembly. The amendments did not enter into force but the substances have been incorporated into MARPOL 73/78.

INTERNATIONAL CONVENTION RELATING TO INTERVENTION ON THE HIGH SEAS IN CASES OF OIL POLLUTION 1969 (INTERVENTION CONVENTION 1969)

The Intervention Convention was another result of the *Torrey Canyon* disaster. The question of the existence of a right of intervention by coastal states on the high

8. For detailed information on, the "load on top" system see David W. Abecassis (ed.), *Oil Pollution From Ships International, United Kingdom and United States Law and Practice* (1985), pp.24–32.
9. For problems associated with load on top and solutions to those problems see *ibid.*, pp.27–31.

seas following a pollution casualty arose in relation to this incident.[10] When the *Torrey Canyon* ran aground on the high seas it caused major pollution damage to the British and French coastlines and related interests. The British government first tried to rescue the tanker, but salvage attempts were unsuccessful. Then the government decided to bomb the wreck in order to set it on fire. At that time the attitude of the British Government seemed to be contrary to International Law of the Sea principles. As a general rule if the vessel is in the territorial sea the coastal state may take any measures it considers appropriate.[11] Because at that stage the vessel is not on "innocent passage", as it will have stopped to enjoy a right of innocent passage it will be subject to the sovereignty of the coastal state. But if the vessel is beyond the territorial sea then the position changes. In the *Torrey Canyon* case, when the incident occurred the tanker was flying a foreign flag and was on the high seas. Therefore theoretically it was protected by the principle of exclusive jurisdiction of the flag state, the corollary to the principle of the freedom of the high seas. The attitude of the British government, the decision to bomb the wreck, was criticised in the country by the Chamber of Shipping as it created a very dangerous precedent. Outside the United Kingdom the decision was also criticised in France by the French government as the bombing of the *Torrey Canyon* resulted in further pollution of the French coast. As a result of these criticisms the British government invoked a right of self-protection under general international law principles. And doubts about the legality of its action led the United Kingdom government to refer the question to the IMO.[12] The Legal Committee of IMCO recommended an international conference in order to define the rights of coastal states in relation to pollution casualties off their shores. The international conference was held at Brussels in November 1967 and resulted in the adoption of an International Convention Relating to Intervention on the High Seas in Cases of Oil Pollution Casualties.

The Intervention Convention is a global convention, and under Article I the right of intervention of coastal states on the high seas in cases of oil pollution casualties is expressly recognised.[13] Under this key Article the right of intervention may be exercised not only with respect to an actual case of pollution but also to a threat of pollution, provided that there is "grave and imminent danger" to the parties' "coastline or related interests". The right of intervention may be exercised anywhere on the high seas. It is not limited to any kind of pollution control zone.

10. For detailed information on the right of intervention of coastal states on the high seas in cases of pollution casualties see Claude C. Emanuelli, "The Right of Intervention of Coastal States on the High Seas in Cases of Pollution Casualties" (1976) V. XXV *University of New Brunswick Law Journal* 79–96.

11. This is subject to the principle of proportionality.

12. At the meeting of IMCO in May 1967 three categories of problems were studied These were: preventive measures against oil pollution, measures to limit the extent of damage and necessary changes in international law.

13. Art. I of the Intervention Convention 1969 provides that states parties "may take such measures on the high seas as may be necessary to prevent, mitigate or eliminate grave and imminent danger to their coastline or related interests from pollution or threat of pollution of the sea by oil, following upon a maritime casualty or acts related to such a casualty, which may reasonably be expected to result in major harmful consequences".

The coastal state may also take any kinds of measures which seem to be suitable at the time of the casualty.

Obviously this Convention caused some debate because it was a real departure from the traditional international principle of freedom of the high seas. The Convention is a limitation on freedom of the high seas. Therefore the 1969 Convention grants wide powers to coastal states to act on the high seas in cases of oil pollution casualties. For the first time states other than the flag state were permitted to take preventive and mitigating action against foreign vessels in cases where there was real concern that oil pollution might result in major harmful consequences. The rights of coastal states on the high seas were subject only to a reasonable exercise of these rights and powers. Before taking measures under Article I the state concerned must, except in cases of extreme urgency, consult the flag state and other states affected by the casualty and notify anyone likely to be affected by the proposed measures. In any case, after the measures have been taken, such states and persons, as well as the IMO, must be notified (Article III).

The right of intervention is limited to a maritime casualty resulting in a case of pollution involving a privately owned ship of one of the contracting parties. Under Article II "maritime casualty" is defined as a collision, stranding or other incident of navigation or occurrence resulting in actual or threatened material damage to a ship or its cargo. Therefore the right of intervention does not apply to cases of voluntary pollution, pollution casualties resulting from the exploration or exploitation of the continental shelf or of the deep seabed, pollution casualties involving warships, state-owned vessels used on government non-commercial services or privately owned vessels of a flag state which is not party to the Convention. Under Article V measures taken by the coastal state must be proportionate to the actual or threatened damage and must not unnecessarily interfere with the rights and interests of the flag state or other states. So the coastal state must take into account "the extent and probability of imminent damage if those measures are not taken; and the likelihood of those measures being effective; and the extent of the damage which may be caused by such measures".[14]

As stated above the Convention grants wide powers to coastal states, so there is always a possibility of abuse in the exercise of these powers. In order to prevent such abuse the Convention provides for a mechanism of control over the exercise of the right of intervention. Article VIII provides a compulsory mechanism of peaceful settlement of disputes arising from the exercise of the right of intervention and sanctions against coastal states in cases of wrongful exercise of this right. According to Article VIII disputes involving measures taken in contravention of the provisions or compensation obligations must upon request be submitted to conciliation or arbitration procedures set out in the Annex to the Convention.

The Convention has been in force since 6 May 1975. In 1973 a protocol to the Convention was adopted, under which a coastal state's powers of intervention were extended to casualties causing or threatening pollution by substances other

14. Intervention Convention 1969, Art. 5(3).

than oil. The protocol entered into effect in 1983. To assist in the application of the protocol, in 1974 the IMO issued a list of substances to which it would apply. The 1991 amendments revised the list of substances drawn up in 1974, and entered into force on 30 March 1993. Following the adoption of new criteria for the selection of the list of substances attached to the 1973 Protocol, on 10 July 1996 the 1996 amendments were adopted and entered into force on 19 December 1997.

Until the 1969 Intervention Convention the question of oil pollution was merely covered by OILPOL 1954. In accordance with traditional international law OILPOL 1954 recognises the exclusive jurisdiction of the flag state over violations of the prohibition to discharge. Not being in the traditional international law framework the Intervention Convention has been criticised both for allowing too much discretion to coastal states and for limiting the rights of such states to take action.[15] Another criticism was on the scope of the convention and its protocol. As they both applied only to measures taken on the high seas, literally it means that the coastal state could not rely on the powers given to it to take action in its Exclusive Economic Zone (EEZ). As there is no other convention which gives powers of intervention in the EEZ and unless such powers derive from customary international law, the powers given under the Convention mean that a coastal state has greater powers of intervention on the high seas than in its EEZ. When the Convention was adopted the concept of the EEZ did not exist, so it would be more reasonable to consider the phrase "high seas" as the seas beyond the territorial sea.[16] It was also stated that extending the powers of the coastal state on the high seas is not necessarily the best way to reduce the danger.[17]

INTERNATIONAL CONVENTION ON THE DUMPING OF WASTES AT SEA 1972 (LDC 1972)

Dumping is the deliberate disposal into the sea from ships or aircraft of waste loaded on board for this purpose and used to be a legitimate means of disposal because of the low cost.[18] International action to control the dumping of wastes at sea began in the early 1970s. The earliest conventions covering dumping were the global Convention on the Prevention of Marine Pollution by Dumping of Wastes and Other Matter 1972 (London Dumping Convention, LDC 1972) and the regional Oslo Dumping Convention. The general pattern of the Convention is that there is a general prohibition on the dumping of wastes, subject to specified exceptions. The Convention aims effectively to control all sources of marine pollution and to prevent pollution of the sea by means of regulation of dumping into the sea of waste materials. It covers materials transported to the sea for the purpose of dumping and incineration activities. The LDC defines dumping as the deliberate disposal of waste from ships and aircraft, but excluding the disposal of

15. Edgar Gold, pp.38–39.
16. R.R. Churchill and A.V. Lowe, p.262.
17. See Yoram Dinstein, pp.371–373.
18. Douglas Brubaker, p.36.

waste incidental to the normal operation of ships and aircraft.[19] The definition was later modified to include incineration at sea.

The Convention divides wastes into three categories. Before discussing the hazardous substances listed in Annex I (the "Black List") one can trace the LDC back to the 1972 Stockholm Declaration.[20] In Stockholm the view was that waste should be considered hazardous if it contained (i) "substances that are liable to create hazards to human health, to harm living resources and marine life, to damage amenities or to interfere with other legitimate uses of the sea", or (ii) substances which were discharged "in such quantities or concentrations as to exceed the capacity of the (marine) environment to render them harmless".[21] This original and globally agreed definition of "hazardous waste" was not binding upon the subsequent diplomatic conferences dealing with marine pollution. When the London Dumping Conference convened it adopted a very general definition of "wastes"[22] but in order to define what wastes were to be considered "hazardous" it used the "list approach". The London Dumping Convention seems to have accepted that approach in the Stockholm Conference. Even certain substances on the "Black List" may be dumped at sea "if they are contained in wastes as trace amounts or are rapidly rendered harmless at sea". Under the LDC there is no clear definition and reference must be had to the "Black List" (and its exceptions) and the "Grey List" in order to determine what wastes are hazardous and whether they may be dumped at all and, if so, under what conditions.

The categories of wastes under the Convention are as follows. The first category consists of the substances listed in Annex I (the "Black List"), which include organohalogen compounds, mercury, cadmium, persistent plastics and synthetic materials, various oils taken on board to be dumped and high level radioactive wastes defined by the International Atomic Energy Agency (IAEA). There is a specific, absolute prohibition on dumping of substances listed in the "Black List" (Article IV(1)). The absolute prohibition does not apply if these substances are present only as trace contaminants.[23] The second category of wastes comprises the noxious substances listed in Annex II (the "Grey List"). These include arsenic, lead, copper, zinc, organosilicon compounds, cyanides, fluorides, pesticides, beryllium, chromium, nickel, vanadium, mixtures of these radioactive wastes not included in Annex I, and containers, scrap metal and other bulky wastes liable to

19. The LDC 1972, Art. III(1), provides that "For the purposes of this Convention:
(a) 'Dumping' means:
- (i) any deliberate disposal at sea of wastes or other matter from vessels, aircraft, platforms or other man-made structures at sea,
- (ii) any deliberate disposal at sea of vessels, aircraft, platforms or other man-made structures at sea."

20. See F.L. Wiswall, "Hazardous Waste, Dumping Convention, and Liability" In Colin M. De La Rue, pp.189–192.

21. Declaration of the UN Conference on the Human Environment, Stockholm, 16 June 1972, Principles 7 and 6 respectively.

22. LDC 1972, Art. III(4) provides that "'wastes or other matter' means material and substance of any kind, form or description".

23. The LDC 1972, Annex I, para. 9.

present a serious obstacle to fishing or navigation. The dumping of such substances requires special care. It is permitted only if a prior special permit has been obtained. The permit must be issued by the national authority and is subject to specified precautionary requirements (Article IV(1)). The parties must take full account of the recommendations of the International Atomic Energy Agency in issuing a permit for the dumping of radioactive waste. The third category consists of all wastes which are not on the Black or Grey List. This is called the "White List". The dumping of wastes on the "White List" requires a general permit (Article IV(1)(c)). In fact there is no list of substances on the White List. Annex III provides a list of factors for evaluating whether or not a permit to dump at sea should be issued. Special and general permits may be issued only after careful consideration of the factors set out in Annex III. These factors include the characteristics and composition of the subject matter, the characteristics of the dumping site, the method of deposit and the possible effects of dumping.

At the beginning of 1969 some countries began incinerating hazardous wastes at sea using specially designed ships. In 1978, at the Third Consultative Meeting of the Parties, the Convention was amended to make specific provision for incineration at sea. Under the amendments a special permit was required for the incineration of wastes listed in Annexes I and II. And such incineration had to comply with the regulations set out in the revised annexes. In order to issue special permits states had to apply the Regulations for the Control of Incineration of Wastes and Other Matters at Sea and also take into account technical guidelines. But during the 1980s opinion began to turn against incineration.[24] In 1988 the contracting parties agreed that incineration should be ended by 1994. The LDC does not mention the dumping of radioactive wastes into the sea-bed, because when the Convention was considered this was not technically possible. By the 1980s serious discussions had begun on the issue and in 1986 the contracting parties agreed that sea-bed disposal should not take place until it had been proved to be technically and environmentally acceptable. In 1990 the Consultative Meeting agreed that the disposal of radioactive wastes into sub-sea-bed repositories accessed from the sea constituted "dumping" within the terms of the Convention. In many countries the strict controls on the disposal of wastes resulted in the transboundary movement of wastes. In 1991 the Consultative Meeting agreed that no transboundary movement of wastes intended for dumping should take place without the prior informed consent of the country of destination.[25]

Under the Convention jurisdiction rests with the flag states, loading states and coastal states. Flag states have to enforce convention measures against vessels and

24. The 1980 amendments adopted on 24 September 1980 are related to those concerned with incineration and list substances which require special care when being incinerated. They entered into force on 11 March 1981.

25. On 20 February 1994 the 1993 amendments entered into force. They banned the dumping into the sea of low-level radioactive wastes and the incineration at sea of industrial wastes.

aircraft registered in the flag state or flying its flag. Loading states are under an obligation to enforce conventional measures against vessels and aircraft loading waste for dumping in their territory or territorial sea. It is quite possible that loading states will enforce the Convention regulations against vessels and aircraft flying the flag of a non-party. Under the Convention the coastal state's area of jurisdiction was not defined clearly due to compromise between coastal states and the major maritime states. This area was deliberately left to UNCLOS 1982 for definition of coastal state jurisdiction over dumping. Under Article VI(I) parties must keep a record of all dumping they permit and monitor the condition of the sea for the purposes of dumping.

A protocol to the Convention was adopted at a Conference held at IMO headquarters which ended on 8 November 1996. It is expected that the protocol will be the biggest change to the Convention since its adoption in 1972 and will represent a major change of direction that reflects the changes in attitude towards marine waste disposal that have taken place in recent years. The decision to include the changes in a protocol rather than simple amendments was made because the proposed changes are so extensive that they will, if adopted, amount to a virtual re-writing of the Convention. The protocol will supersede the original 1972 Convention.

The protocol will enter into force 30 days after ratification by 26 countries, 15 of which must be Contracting Parties to the 1972 Treaty. The protocol represents a major change in approach to the question of how to regulate the use of the sea as a depository for waste materials. Article 3 of the protocol introduces a "precautionary approach". This approach, which is set out in Article 3, requires that "appropriate preventive measures are taken when there is reason to believe that wastes or other matter introduced into the marine environment are likely to cause harm even when there is no conclusive evidence to prove a causal relation between inputs and their effects". Again Article 3 states that "the polluter should, in principle, bear the cost of pollution". It is also emphasised that Contracting Parties should ensure that the protocol should not simply result in pollution being transferred from one part of the environment to another.

Under the 1972 Convention dumping is permitted in certain conditions. The severity of these conditions changes according to the danger to the environment presented by the materials themselves. The Black List contains materials which may not be dumped at all. The approach of the protocol is more restrictive. Article 4 states that Contracting Parties "shall prohibit the dumping of any wastes or other matter with the exception of those listed in Annex I". These materials include:

1. Dredged material;
2. Sewage sludge;
3. Fish waste, or material resulting from industrial fish processing operations;
4. Vessels and platforms or other man-made structures at sea;
5. Inert, organic geological material;
6. Organic material of natural origin;

7. Bulky items primarily composing iron, steel, concrete and similar harmless materials for which the concern is physical impact, and limited to those circumstances where such wastes are generated at locations, such as small islands with isolated communities, having no practicable access to disposal options other than dumping.

The only exceptions are contained in Article 8 which permits dumping to be carried out "in cases of *force majeure* caused by stress of weather, or in any case which constitutes a danger to human life or a real threat to vessel . . .".

Article 6 states that "Contracting Parties shall not allow the export of wastes or other matter to other countries for dumping or incineration at sea".

The protocol contains three annexes and amendments to the annexes are adopted through a "tacit acceptance" procedure and will enter into force not later than 100 days after being adopted. The amendments will bind all Contracting Parties except those which have explicitly stated their non-acceptance.

MARPOL 73/78

The International Convention for the Prevention of Pollution from Ships 1973 and its 1978 Protocol entered into force in October 1983. It effectively supersedes OILPOL 1954, which with amendments proved unsatisfactory to deal with marine pollution, so that an International Conference on Marine Pollution was held in 1973. The result of the Conference was the global MARPOL 73.[26] In the Convention detailed pollution standards are set out in five annexes. These are concerned with oil (Annex I), noxious liquid substances in bulk (Annex II), harmful substances carried by sea in packaged forms (Annex III), sewage (Annex IV) and garbage (Annex V). The acceptance of Annexes I and II is obligatory for all Contracting Parties, but the acceptance of the remaining annexes, Annexes III, IV and V, is optional.

The original 1973 MARPOL Convention was intended to deal with all forms of intentional pollution of the sea from ships other than dumping. In 1977–1978 a series of very serious tanker accidents took place. This led the IMO to convene an international conference on tanker safety and pollution prevention in 1978. On the other hand due to the considerable economic cost and technical difficulties of complying with its provisions by 1978 the Convention was still a long way from receiving the necessary number of ratifications so as to enter into force. In order to speed up ratification a Protocol to the Convention was adopted at the IMO's Conference on Tanker Safety and Pollution Prevention. As a result, both MARPOL 73 and the 1978 Protocol were to be merged into one legal instrument, and the provisions of both were to be read and interpreted together. Accordingly, by accepting the 1978 Protocol, states agree and undertake also to implement all of MARPOL. Ratification of one meant that a state was a party to both. An exception

26. For general information on MARPOL 73/78 see Douglas Brubaker, pp.122–129; Edgar Gold, pp.58–74; E.D. Brown, Vol. I, pp. 380–382; R.R. Churchill and A.V. Lowe, pp.249–251.

to this general rule relates to the provisions of MARPOL concerning pollution by chemicals, Annex II, which was postponed for three years from the date of entry into force of the 1978 Protocol. The period could also be longer if so decided by a two-thirds majority of the Marine Environment Protection Committee (MEPC) of the IMO.[27] The stated aim of MARPOL is to achieve the complete elimination of international pollution of the marine environment by oil and other harmful substances from ships and to minimise the accidental discharge of such substances.

(a) Definitions

Articles 1–6 deal with prescriptive and enforcement jurisdiction. Article 2 covers definitions. "Harmful substance" includes any substance discharged into the sea which is liable to:

(a) create hazards to human health;
(b) harm living resources and marine life;
(c) damage amenities or interfere with other legitimate uses of the sea.

"Discharge" includes any release howsoever caused from a ship and includes any escape, disposal, spillage, leak, pumping, emission or emptying. The definition does not include dumping as defined by the London Dumping Convention, the release of harmful substances arising from offshore exploration, exploitation or processing, or the release of harmful substances for research into pollution abatement. "Ship" is defined as any vessel of any type whatsoever operating in the marine environment and includes hydrofoil boats, air-cushion vehicles, submersibles, floating craft and fixed and floating platforms. The definition of "incident" includes every event involving either actual or probable discharge of harmful substances into the sea.

The Convention provisions do not apply to state ships such as warships or others on non-commercial service. The Convention is intended to apply to all ships under the flag of a state which is a party to it, but it also applies to ships which are not entitled to fly the flag of a state which has accepted the Convention, but operating under the authority of a party to it (Article 3).

According to Article 4 violations of Convention requirements must be prohibited and sanctions must be established under the law of the flag state for ships everywhere. Any violation of Convention requirements within the jurisdiction of any party must be prohibited and sanctions established under national law. In the case of a violation the state must either bring proceedings or give the flag state evidence that a violation has occurred and the flag state must in turn inform the state and the IMO of the action taken. All states must hold a certificate of compliance, International Oil Pollution Prevention Certificate, with MARPOL 73/78 standards, both technical and operational. If a ship does not carry the necessary certificate and it is clear that the condition of the ship or its equipment does not comply with the necessary certification standards then the port authority

27. Annex II entered into force on 6 April 1987.

of the state party carrying out the inspection must take steps to ensure that the ship does not sail until it can proceed "without presenting an unreasonable threat of harm to the marine environment" (Article 5). Article 6 requires states parties to "cooperate in the detection of violations and the enforcement of provisions". The Article empowers the port state to inspect foreign vessels entering its ports and terminals in order to verify whether any discharge has been made contrary to the Convention. But a port state may not bring any legal action for violations outside its jurisdiction. Proceedings may only be brought by the flag state, on receipt of evidence from the port state.

(b) Oil pollution regulations

(i) Annex I

The detailed regulations dealing with oil pollution are contained in Annex I. Regulation 26 of Annex I as modified by the Protocol of 1978 applies to every oil tanker of 150 gross tons or more and every ship other than oil tanker of 400 gross tons or more.[28] Oil covered by Annex I generally includes crude oil, fuel oil, sludge, oil refuse and refined products and specifically includes approximately 45 varieties of asphalt solutions, oils, distillates, gas oil, gasoline blending stocks, gasolines, jet fuels and naphtha. The oil discharge criteria under Annex I are similar to those set out in the 1969 amendment to OILPOL 1954. But some significant additions were made. Seven new areas—the Mediterranean, the Baltic, the Black Sea, the Red Sea, the "Gulfs" area, the Gulf of Aden and the Antarctic area—have been designated as special areas. Oil tankers and other ships of 400 tons and above are not allowed to discharge oil in these areas. Apart from the cases where the oil content does not exceed 15 parts per million, discharges from ships of less than 400 tons are prohibited. In the Antarctic area oil discharge is completely prohibited. In cases of clean ballast and processed bilge water these prohibitions do not apply. The rate of discharge has been reduced from 60 to 30 litres per mile. The maximum quantity of oil discharged by new ships has also been reduced. For new ships the maximum quantity of oil which may be discharged on a ballast voyage is reduced from $1/15{,}000$ to $1/30{,}000$ of cargo capacity. Apart from persistent oils non-persistent oils and refined oils are also within the framework of these criteria. All parties to MARPOL are required to provide facilities for the reception of residues and oily mixtures at oil loading terminals, and repairs ports. All vessels over 400 GRT, including vessels other than tankers, must be equipped with oily water separating or filtering systems for the discharge of machinery space bilges. New tankers, those for which the building contracts were placed after 31 December 1975, over 70,000 dwt must be provided with segregated ballast tanks (SBT) of sufficient capacity to facilitate safe operation on ballast voyages without

28. For the purposes of MARPOL 73/78 Annex I, an oil tanker is a ship constructed or adapted primarily to carry oil in bulk in its cargo spaces and includes combination carriers and any chemical tanker as defined in Annex II when it is carrying a cargo or part cargo of oil in bulk.

the need to use cargo tanks for ballasting. Under the 1978 Protocol the rules became stricter, SBTs are required to be fitted on all new tankers over 20,000 dwt rather than above 70,000 dwt. They must also be protectively located.[29] Tankers must be fitted with the necessary equipment for the load on top (LOT) system. In the earlier discharge standards the LOT system depended upon the conscientious application of tank vessel operators. MARPOL 73 sought to reduce reliance on the human element by requiring the use of a discharge monitoring and control system.[30] The LOT procedure was made compulsory for all oil tankers of 150 tons and over. The existing tankers had three years to fit the necessary slop tanks and discharge monitoring and control systems.

Two important new regulations were added to Annex I in 1992. They entered into force on 6 July 1993. Regulation 13F applies to all new tankers ordered after 6 July 1993 whose keels were laid on or after 6 January 1994 or which are delivered on or after 6 July 1996. Under it tankers of 5,000 dwt and above must be fitted with double bottoms and double hulls extending the full length of the ship's side. The mid-deck design is permitted as an alternative. Other designs may be allowed in due course, provided that they ensure the same level of protection against pollution. Other amendments adopted in 1992 made it mandatory for existing tankers to be fitted with double hulls or an equivalent design when they reached the age of 25, but these requirements did not take effect until 6 July 1995. By 1992, concern was growing about the condition of some of the world's large number of ageing tankers and bulk carriers because the bulk of the world's tanker fleet including many of the world's largest tankers was built before MARPOL 73/78 became mandatory. So it was agreed that a solution should be found in order to improve the maintenance of these ships. Regulation 13G was therefore adopted and it deals with older tankers. The regulation took effect from 6 July 1995. It states that pre-MARPOL ships must, not later than 25 years after the date of delivery, have side or bottom protection to cover at least 30 per cent of the cargo tank area.

All ships must carry a newly designed and more comprehensive oil record book in which all operations involving oil are recorded. Any state party to MARPOL may inspect this book. The original MARPOL made provision for initial, periodical and intermediate surveys in order to ensure compliance with the Convention. As stated above an International Oil Pollution Certificate was required by tankers entering ports or offshore terminals of other parties. Under the 1978 Protocol the procedure for inspection and certification became stricter. Under Regulation 4(3)(b) provision is made for unscheduled inspections.

Under the amendment to Annex I, Regulation 26 of Annex I adopted in 1991 and in force on 4 April 1993, tankers of 150 tons and above and other ships of 400 tons and above have to carry an emergency plan. This shipboard emergency plan

29. For instance they must be located in such a way that they protect the cargo tanks in cases of stranding or collision.

30. The problem with this requirement is that in 1973 there were no commercially viable monitoring systems.

must be approved by the administration of the vessel's flag state. The requirement has immediate effect for all ships delivered on or after 4 April 1993. For all ships delivered before 4 April 1993 the requirement applies from 4 April 1995. The Convention also applies to any fixed and floating rigs and platforms, "when engaged in the exploration, exploitation and associated off-shore processing of sea-bed mineral resources". At its 32nd session in March 1992, the IMO adopted a set of guidelines for the preparation of emergency plans. These guidelines aim to assist shipowners in preparing shipboard oil pollution emergency plans that comply with Regulation 26. They also aim to assist governments in developing and enacting domestic laws which give force to and implement the regulation.

(ii) Annex II

Annex II deals with noxious liquid substances in bulk.[31] It provides that ballast water, tank washing residues or mixtures containing any noxious substances be discharged only at a reception facility unless the concentration is diluted to acceptable levels. This Annex is mandatory and it lists approximately 200 chemicals of great variety and with classification levels denoting toxicity to aquatic life. Annex II also contains provisions to minimise pollution in case of an incident. There is a requirement for ship operators to maintain a cargo record book, similar to the oil record book in Annex I, and show all loading, transferring and unloading of chemical cargo, ballasting, cleaning, transfer and discharge of ballast from cargo tanks, discharge of residues, and accidental or other discharges. The cargo record book may be inspected by any state which is a party to MARPOL. States parties to MARPOL are also required to control the design, construction and operation of chemical tankers which must comply with the Code for the Construction and Equipment of Ships carrying Dangerous Chemicals in Bulk.

The tough concentration standards and the corresponding need for reception facilities under Annex II and the fact that it was compulsory compounded the difficulties. By 1978 there was strong opposition to MARPOL 73 and there was no uniform application of any international standards. During the period from December 1976 to January 1977 a series of tanker disasters occurred, all of which were caused by human error or negligence.[32] As a result of these disasters and the need for legitimate progress to control source-vessel oil pollution the Conference on Tanker Safety and Pollution Prevention was held in London in February 1978. The 1978 Protocol implemented a more effective system for the prevention of oil

31. The substances covered by Annex II are divided into four categories on the basis of environmental criteria: Category A: high hazard to human health or aquatic life and moderately to highly toxic to aquatic life; Category B: produce tainting sea food and slightly to moderately toxic to aquatic life; Category C: substances slightly toxic to aquatic life; Category D: substances non-toxic to aquatic life which may have a nuisance effect on amenities. See Edgar Gold, pp. 67–69.

32. These accidents include the *Argo Merchant* on 15 December 1976, the *Oswego Peace* on 24 December, the *Olympic Games* on 27 December and the *Grand Zenith* on 29 December and also the *Sansinena Daphene*, *Barcola*, *Mary Ann*, *Universe Leader* and *Irenes Challenge*. See Douglas Brubaker, p.127; Akintoya A. Ayorinde, "Inconsistencies Between OPA '90 and MARPOL 73/78: What is the Effect on Legal Rights and Obligations of the United States and Other Parties to MARPOL 73/78?" *Journal of Maritime Law and Commerce*, 64–65.

pollution. The modification also allowed the entry into force of Annex II to occur after the entry into force of the Articles and Annex I.[33]

(iii) Annex III

The regulations designed to prevent pollution of the sea by harmful substances carried in packaged form, including freight containers, portable tanks or road and rail tank wagons are contained in Annex III. This Annex did not enter into force when the rest of MARPOL became effective in October 1983. Under Annex III parties to the Convention are required to issue detailed requirements on packing, marking, documentation, stowage, quantity limitations and other matters. Annex III prohibits the carriage of listed harmful substances except in accordance with the provisions of Annex II to MARPOL. Adequate packing was required in order to reduce the carriage hazard. The carriage of empty receptacles, containers, portable tanks, road tankers, rail wagons, etc. was prohibited unless they were properly cleaned. Special documentation on board was required for the substances carried. A vessel had to notify port authorities in the loading and discharge ports at least 24 hours prior to the commencement of cargo operations.

In order to assist the uniform implementation of Annex III, the IMO decided to use the International Maritime Dangerous Goods Code (IMDG). Under the Code around 600 substances are identified as marine pollutants. For the carriage of these pollutants detailed requirements have been agreed upon and incorporated into the IMDG Code.[34] Until the entry into force of the amendment the Code was concerned only with the safety aspects. It consists of five volumes and is generally accepted as the most important international guide to the carriage of dangerous substances and now marine pollutants. In 1988 the Marine Environment Protection Committee (MEPC) approved a revised text of Annex III. With this revision Annex III clearly connected to the IMDG Code more and harmful substances defined as "those which are identified as marine pollutants in the IMDG code".

(iv) Annex IV

Annex IV deals with the prevention of pollution by sewage by ships and consists of 11 regulations and an appendix. Unless fitted with an approved sewage treatment system ships are not permitted to discharge sewage within four miles of the nearest land. Between four and 12 miles from land, sewage must be comminuted and disinfected before discharge. Due to this requirement the industry was faced with technical and economic problems and delayed the acceptance of this Annex in a number of states.[35] Under Annex IV it is assumed that sewage reception facilities will be available in almost all ports. This of course was not the case at that stage.[36]

33. As stated above a two-thirds majority of the parties in the Marine Environment Protection Committee could further extend the entry date for Annex II.
34. Amendment 25–89 entered into force in 1991.
35. Edgar Gold, pp.70–71.
36. *Ibid.*

(v) Annex V

Annex V sets specified minimum distances from land for the disposal of all the principal kinds of garbage and prohibits the disposal of all plastics. It was enacted to help curtail the indiscriminate dumping of trash and plastics at sea. Under Annex V garbage is defined as including all types of victual, domestic and operational waste[37] generated during the normal operation of the vessel.

The Annex does not specifically require any equipment for the treatment of garbage.[38] The coastal states are under an obligation to provide reception facilities for garbage. This requirement caused similar problems to those for sewage. Disposal of garbage shall take place at sea as far as possible from the nearest land. The disposal into the sea of all plastic substances including synthetic ropes, synthetic fishing nets and plastic garbage bags is completely prohibited.

Under MARPOL 73/78 states parties to Annexes II, IV and V are required to provide adequate port reception facilities.

On 3 March 1996 a number of amendments to MARPOL 73/78 came into force. The amendments affect four of the Convention's five technical annexes and are all designed to improve the way in which the Convention is implemented. They make it possible for ships to be inspected when in the ports of other parties to the Convention to ensure that crews are able to carry out essential shipboard procedures relating to marine pollution prevention. The amendments are contained in resolution A.742(18), which was adopted by the IMO Assembly in November 1993. The amendments were adopted in November 1994. The Annexes affected are Annex I, which deals with oil pollution, Annex II which is concerned with pollution by noxious liquid substances, Annex III containing regulations for the prevention of pollution by harmful substances in packaged form, and Annex V which deals with garbage. The 1996 amendments, which were adopted in July 1996, entered into force, under tacit acceptance, on 1 January 1998. The amendments include more precise requirements for the sending of reports of incidents involving harmful substances.

(c) The impact of MARPOL 73/78

MARPOL 73/78 is often considered to be a very effective international instrument to combat marine pollution. On 5 and 6 February 1990 a workshop organised by the National Academy of Sciences in Washington DC[39] concluded that the entry into force of the MARPOL Convention had had a substantial positive impact on decreasing the amount of oil that entered the sea as a result of marine transportation activities.[40] Although the MARPOL Convention has been accepted

37. The definition excludes fish waste.

38. According to Gold "as reference is frequently made to comminuters or grinders" this requirement may be implied: Edgar Gold, p.72.

39. IMO Doc. MEPC 30/INF. 13 (IMO London, September 1990) cited in Gerard Peet, "The MARPOL Convention: Implementation and Effectiveness" (1992) 7 *International Journal of Estuarine and Coastal Law*, 277, n.2.

40. Gerard Peet, pp.277–295.

as the most important international treaty dealing with the prevention of pollution from ships it has not escaped criticism. First, it is said that the Convention has been too slow to enter into force. The MARPOL Convention of 1973, even in its modified form, took 10 years to come into force. Secondly, even where the Convention has entered into force, it has not been widely ratified. In case of MARPOL it is true that the ratification process took longer than was expected due to technical and economic problems. Developing states in particular appear to have had some difficulty in ratifying it. The Convention is not very easy to follow, so the Maritime Safety Committee of the IMO is engaged in the preparation of a series of guidelines which are intended to assist the implementation of the various IMO requirements. Table 2 shows that the support of the IMO has increased the ratification level of the MARPOL Convention.

In the second criticism the number of states which ratified the Convention is misleading. When one considers the number of states alone there do not seem to be many, but in fact the parties to the Convention collectively own about 90 per cent of the total world shipping tonnage. MARPOL represents all the shipping industry including the owner, operator, master, engineer, seaman, surveyor, salvor, underwriter or government official and its basic aim is to achieve the IMO's "cleaner seas and safer ships" principle.[41] But the Convention would have achieved more if it had been properly implemented. The National Academy of Sciences Report states that "there is a lack of worldwide enforcement of the

Table 2. Summary of Ratification Percentages[42]

Country groups MARPOL Annexes	I	II	III	IV	V
Developed market economy countries	85%	85%	73%	46%	77%
Socialist and former socialist countries[43]	77%	77%	46%	46%	54%
Developing countries with a 1986 *per capita* GDP above $1,500	52%	52%	24%	24%	30%
Developing countries with a 1986 *per capita* GDP between $500 and $1,500	26%	26%	21%	21%	24%
Developing countries with a 1986 *per capita* GDP below $500 or about which there is no information about the 1986 *per capita* GDP	20%	20%	11%	9%	11%

MARPOL Convention, there is a lack of worldwide efficient monitoring, there is a difficulty in identifying the source of oil spillage, and there is a lack of worldwide port states control systems".[44]

41. Edgar Gold, pp.73–74.
42. *Ibid.*, p.279.
43. States included in this group are: Albania, Bulgaria, Czechoslovakia, Hungary, Poland, Romania, Estonia, Lithuania, Russia, China, the Democratic People's Republic of Korea, Mongolia and Vietnam. With an increasing number of new states in this category it is to be expected that the ratification in this category will go down. Cited in *ibid.*
44. Gerard Peet, pp.277–278.

Under Article 11(1)(e) of the MARPOL Convention contracting parties are required to submit reports about the application of the convention to the IMO.[45] Since the entry into force of the MARPOL Convention only six contracting parties have submitted reports. More than 30 contracting parties have never submitted a report to the IMO. The other contracting states have submitted reports for one or a few years only. Unless all contracting parties comply with all the requirements of the MARPOL the Convention will not be as effective as it could be.

INTERNATIONAL CONVENTION ON OIL POLLUTION PREPAREDNESS RESPONSE AND CO-OPERATION 1990 (OPRC)

As is well known, the *Torrey Canyon* in 1967 resulted in the 1969 Civil Liability Convention, the 1969 Intervention Convention and the 1971 Fund Convention. In 1977 the *Amoco Cadiz* disaster gave rise to amendments to SOLAS 1974 and the Salvage Convention 1989. This incident also affected the process of the introduction of the MARPOL Convention 1973/78. The US Oil Pollution Act 1990 was a result of the *Exxon Valdez* incident which was the largest and most expensive US oil spill. The *Exxon Valdez* also gave rise to another international convention, the International Convention on Oil Pollution Preparedness, Response and Co-operation 1990 (OPRC).

On 19 October 1989 the International Maritime Organisation (IMO) adopted Resolution A.674 (16) on International Co-operation on Oil Pollution Preparedness and Response in response to recent severe pollution incidents resulting from tankers and offshore installations, in which the Assembly decided to convene an international conference on the problem. The actual diplomatic conference was held in London under the aegis of the IMO and adopted the OPRC.[46] To enter into force the Convention had to be accepted by 15 states. This total was achieved on 13 May 1994 and the Convention became international law on 13 May 1995.

The aim of the Convention is to increase the level of effective preparation to combat oil pollution incidents, to improve the response to such incidents and promote international co-operation in this area. In general the Convention provides a global framework for international co-operation in combating major incidents or threats of marine pollution.

The Convention recognises that in the event of a pollution incident, prompt and effective action is essential. Under Article 1(1) parties undertake individually or jointly to take all appropriate measures under its terms to prepare for and respond

45. There is a special circular letter (MEPC/Circ. 138) and the IMO has developed a special format for these reports consisting of 8 different sections of information.

46. For a general overview of the OPRC, see Ashok Mahapatra, "International Convention on Oil Pollution Preparedness, Response and Cooperation—An Overview" in 1995 International Oil Spill Conference, 27 February–2 March 1995, Long Beach, California, pp.775–778; Edgar Gold, "International Convention on Oil Pollution Preparedness, Response and Co-operation 1990", Issue 121, June 1991. *Gard News*, pp.3–4; John H. Bates, and Charles Benson, *Marine Environment Law* (1993), pp.3–180 & 3–22.

to an oil pollution incident. The Convention applies to an occurrence anywhere, whether on the high seas, in territorial waters or even in internal waters. In order to apply the incident need only pose a threat to the marine environment. Article 2 defines a ship as a vessel of any type operating in the marine environment and an offshore unit as any fixed or floating installation engaged in gas or oil operations. The Convention also applies to incidents from sea ports or oil handling facilities, which are defined as those which present a risk of oil pollution incidents, including oil terminals, pipelines, etc.

Article 3 of the Convention requires that parties to the Convention shall require ships under such parties' flags to have on board a shipboard oil pollution emergency plan as required by the IMO. Operators of offshore units under the flags of parties to the Convention are required to have similar emergency plans. Emergency plans are required for oil tankers of 150 gross tons and above and other ships of 400 gross tons and above, any fixed or floating offshore installation or structure engaged in gas or oil exploration, exploitation, production activities or loading and unloading oil and any seaport and oil handling facility that presents a risk of an oil pollution incident.

Specific oil pollution reporting procedures for vessels and aircraft are set out in Article 4. Masters or other persons in charge of ships flying a state's flag and persons in charge of offshore units under the state's jurisdiction are required to report without delay any discharge or probable discharge of oil that emanates from their ships or units to the nearest coastal state and to the coastal state within whose jurisdiction the offshore units are operating. The reports by ships, offshore units and ports are mandatory. Parties to the Convention must require ships and those in control of units or ports to make such reports. Under Article 4(1)(d) states are under an obligation to instruct their maritime inspection services, naval forces and other relevant bodies or officials to make oil pollution reports either to their competent national authority or to the nearest coastal state. Pilots of civil aircraft are requested to report incidents involving a discharge of oil or the presence of oil to the nearest coastal state.

The Convention imposes an obligation to establish a national system for responding promptly and effectively to oil pollution incidents. As a part of contingency planning each party, either individually or through co-operation with other countries and, as appropriate, oil and shipping industries, port authorities, and other relevant entities shall establish:

- a minimum level of pre-positioned oil spill response equipment, proportionate to the risk involved, and programmes for its use;
- a programme of exercises for oil pollution response organisations and training of relevant personnel;
- detailed plans and communication capabilities for responding to oil pollution incidents;
- an arrangement to co-ordinate the response to an oil pollution incident with capability to mobilise the necessary resources if necessary.

The most important provision of the Convention is Article 7, which is concerned with international co-operation in responding to a pollution incident. Parties have undertaken to co-operate and provide advisory services, technical support and equipment at the request of any other party affected or likely to be affected by a particular incident. Parties' response to a request is subject to their economic strength and the availability of relevant resources. Parties need only to comply with their undertaking "when the severity of the incident so justifies" which means that a party could refuse assistance because in its judgement the incident was not severe enough.

The Convention encourages, directly or through bodies such as the IMO, the promotion of the holding of regular international symposia. It also allocates to the IMO an important role as operator of an Oil Pollution Co-ordination Centre. Apart from this centre the IMO's Marine Environment Protection Committee established an OPRC Working Group in 1991. The Group oversees implementation of the Convention.[47]

THE INTERNATIONAL CONVENTION ON LIABILITY AND COMPENSATION FOR DAMAGE IN CONNECTION WITH THE CARRIAGE OF HAZARDOUS AND NOXIOUS SUBSTANCES BY SEA 1996 (HNS CONVENTION)

The International Convention on Liability and Compensation for Damage in Connection with the Carriage of Hazardous and Noxious Substances by Sea (HNS Convention) was adopted on 3 May 1996 at an international conference convened under the auspices of the IMO. The subject of carriage by sea of hazardous and noxious substances has been on the working programme of the IMO Legal Committee for more than 20 years. However, the international legal framework is not very well developed regarding liability for hazardous substances other than (persistent) oil. Therefore the liability aspect of the issue is mainly regulated by national maritime law and rules. Governments participating in the 1969 International Conference on Marine Pollution Damage were aware of the fact that pollution may also be caused by agents other than oil. However, after the *Torrey Canyon* incident public attention was mainly focused on oil pollution, so the problem of oil pollution had priority.

Appendix III (Calendar of major oil pollution incidents and international response) sets out major oil pollution incidents which have led to international conventions. Following the *Sea Empress* incident in February 1996 the HNS Convention had been adopted in May 1996. The *Sea Empress* released approximately 72,000 tonnes of crude oil and about 360 tonnes of heavy fuel oil in the period from 15 to 21 February 1996 when it grounded off Milford Haven in south west Wales and while it was at the jetty. The *Sea Empress* spill is the third largest

47. See David T. Edwards, "IMO's Strategy For Implementation Of the OPRS Convention", in 1995 International Oil Spill Conference, 27 February–2 March 1995, Long Beach California, pp.787–789.

tanker spill in UK waters after the *Torrey Canyon* in 1967 and the *Braer* in 1993 and among the top 20 worldwide.

The question of a convention on liability for the carriage by sea of hazardous and noxious substances was discussed at the diplomatic conference in London in 1984 but no agreement was reached. The main dispute at the conference was the relationship between the HNS Convention and the Limitation of Liability Conventions 1924, 1957 and 1976[48] and the sharing of liability between the shipowner and cargo interests. One of the most controversial issues has been the question of who should be liable to pay compensation under the new HNS Convention. Neither the industries involved in the transportation of HNS nor the chemical and gas industries were willing to accept new liabilities and the resulting financial burdens. According to shipowners and P&I Clubs any special risks created by the transport of HNS are caused by their inherent dangerous qualities and there is no need to impose on a shipowner a liability which would go beyond his traditional liability for negligence. The chemical industry has argued that shipowners are aware of the special risks involved with the transportation of the HNS and so they should know which precautions must be taken to avoid HNS accidents.[49] On the question of the link between the HNS Convention and the existing treaties on limitation of liability it was decided that the difficulties of achieving a satisfactory link were so great that the outcome of the entire conference could be threatened by a failure to do so. Thus, the HNS Convention is not linked to other treaties.

Before the 1984 Conference the Legal Committee had discussed two options: joint and several liability of the shipowner and the shipper, and a two-tier system comprising a first tier of strict shipowner's liability up to a certain limit, and a second tier of supplementary shipper's liability also up to a certain limit. Both the owner's and shipper's liability would have to be covered by compulsory insurance. The Legal Committee elaborated a draft convention on the basis of the second option, in other words it contained the solution of shared liability between shipowner and cargo interests. The shipowner has strict liability during the carriage of hazardous and noxious substances named in the Convention. This liability is linked with the obligation to maintain insurance or other financial security.

"The owner of a ship registered in a state party actually carrying hazardous and noxious substances shall be required to maintain insurance or other financial security, such as the guarantee of a bank or similar financial institution in the sums fixed by applying the limits of

48. See Peter Wetterstein, "Trends in Maritime Environment Impairment Liability" [1994] L.M.C.L.Q. 230–233.

49. Robert Cleton, "Liability and Compensation For Maritime Carriage of Hazardous And Noxious Substances (HNS) in Liability For Damage To The Marine Environment" in Colin M. De La Rue (ed.) (1993), p.175.

liability prescribed in Article 9, paragraph 1,[50] to cover liability for damage under this Convention."[51]

It is prohibited for a vessel subject to the HNS Convention to trade without a compulsory insurance certificate having been issued, in force for, and displayed on board the vessel.

The strict liability is subject to exceptions similar to those in the CLC. The defences available to the shipowner have been limited to those enumerated in Article 7(2) and (3). Article 7(5) provides that no claim for compensation for damage under the Convention or otherwise may be made against the persons mentioned in the paragraph, unless they have caused the damage by their personal intentional or reckless act. The period of liability has been defined in Article 1(9) as the period from the time when the HNS enter any part of the ship's equipment, on loading, to the time they cease to be present in any part of the ship's equipment, on discharge. If no ship's equipment is used, the period begins and ends respectively when the hazardous and noxious substances cross the ship's rail. The shipowner is entitled to limit his liability according to Article 9(1) unless it is proved that the damage resulted from the personal act or omission of the owner, committed with the intent to cause such damage, or recklessly and with knowledge that such damage would probably result.

Article 3 of the Convention contains provisions with respect to its territorial scope. It will apply to any damage caused in the territory, including the territorial sea, of a state party and to damage by contamination of the environment caused in the EEZ of a state party and to preventive measures, wherever taken, to prevent such damage. Article 3, paragraph c, contains an important extension to the scope of the Convention. By this paragraph the Convention will also apply to damage, other than damage by contamination of the environment, caused outside the territory of a state party (i.e. on the high seas), if such damage has been caused by a ship registered in a state party or, in the case of an unregistered ship, by a ship entitled to fly the flag of a state party and to measures to prevent such damage. "If an empty oil tanker registered in an HNS state collides on the high seas with a passenger ship from a non-HNS state and causes casualties among the passengers because of fire and explosion on board the tanker, the passengers and/or their dependants will have a claim under the HNS Convention against the owner of the

50. HNS Convention, Art. 9(1) states:
"The owner of a ship shall be entitled to limit liability under this Convention in respect of any one incident to an aggregate amount calculated as follows:
(a) 10 million units of account for a ship not exceeding 2,000 units of tonnage; and
(b) for a ship with a tonnage in excess thereof, the following amount in addition to that mentioned in (a):
for each unit of tonnage from 2,001 to 50,000 units of tonnage, 1,500 units of account
for each unit of tonnage in excess of 50,000 units of tonnage, 360 units of account provided, however, that this aggregate amount shall not in any event exceed 100 million units of account

51. HNS Convention, Art. 12(1).

tanker and possibly also against the HNS Scheme".[52] The definition of damage aims to avoid speculative claims based on an abstract calculation of damage caused by the impairment of the environment. Hence, compensation for impairment of the environment other than loss of profit from such impairment shall be limited to costs of reasonable measures of reinstatement actually undertaken or to be undertaken.

Chapter III of the Convention contains provisions relating to a second tier of compensation supplementary to that payable by the shipowner. The purpose of the scheme is to provide compensation for damage resulting from the carriage of HNS substances by sea to the extent that the protection afforded by the shipowner is inadequate or not available. Contributions to the second tier will be levied on persons in the contracting parties who receive a certain minimum quantity of HNS cargo during a calendar year. The tier will consist of one general account and three separate accounts for oil, liquefied natural gas (LNG) and liquefied petroleum gas (LPG). The system with separate accounts has been seen as a way to avoid cross-subsidisation between different HNS substances. The HNS Fund shall in each state party be recognised as a legal person capable under the laws of that state of assuming rights and obligations and of being a party in legal proceedings before the courts of that state. Each state party shall recognise the Director as the legal representative of the HNS Fund.[53]

The aggregate amount of compensation to be paid by the shipowner and by the Scheme in respect of any one incident is limited to a certain amount. The owner of a ship shall be entitled to limit liability under the Convention in respect of any one incident to an aggregate amount calculated as follows:

(a) 10 million units of account for a ship not exceeding 2,000 units of tonnage; and
(b) for a ship with a tonnage in excess thereof, the following amount in addition to that mentioned in (a);

for each unit of tonnage from 2,001 to 50,000 units of tonnage, 1,500 units of account;

for each unit of tonnage in excess of 50,000 units of tonnage, 360 units of account;

provided, however, that this aggregate amount shall not in any event exceed 100 million units of account.[54] The unit of account used in the Convention is the Special Drawing Right (SDR) of the International Monetary Fund.

The second tier of the system is financed primarily by purchasing HNS certificates in respect of contributing cargoes. The Fund will be made up of contributions paid by the importers of HNS materials (primarily chemical companies). There is also an initial contribution system. However, the collecting mechanism under the Convention is quite complex and will be difficult to implement.

52. Robert Cleton, pp.179–180.
53. HNS Convention, Art. 13(2).
54. HNS Convention, Art. 9.

CHAPTER 14

REGIONAL MARINE POLLUTION CONVENTIONS AND THE UNEP REGIONAL SEAS PROGRAMME

AGREEMENT FOR CO-OPERATION IN DEALING WITH POLLUTION OF THE NORTH SEA BY OIL 1969 (THE 1969 AND 1983 BONN AGREEMENTS)

The Bonn Agreement was the first regional oil pollution prevention agreement which was concluded as a result of the *Torrey Canyon* incident. Although it was originally designed as a response to vessel-source oil pollution threats it was broad enough to cover discharges from offshore installations as well.[1] The Agreement came into force on 9 August 1969. All the North Sea coastline states are parties to it and it covers the whole North Sea and the English Channel.[2] The Agreement calls on contracting states to exchange information on national contingency plans, oil spill alerting procedures and developments in oil spill response measures. Under Article 5 the ships and aircraft of states parties are to report to their flag states all large oil slicks and accidents causing or likely to cause pollution which represents a danger for one or more of the state parties' coasts or related interests. And the flag state is under an obligation to notify other affected states parties. In case of a pollution or a potential pollution situation Article 6(1) to (4) applies to regulate the initial responsibility for assessment and observation.

For the purposes of oil spill monitoring and control, the North Sea has been divided up into eight zones with supervisory responsibilities being ascribed to each of the contracting states. This division avoids duplication of effort and provides the initial framework for a single command structure to be set up. When a slick or an accident occurs in a zone for which one state is responsible that state must keep the slick under observation and report to other states an appraisal of the situation and measures taken to combat the oil. By Article 7 a state party requiring help to deal with the oil may request assistance from the other states parties. The parties

1. E.D. Brown, Vol. I, p.357.

2. Under Art. 1 the Agreement applies "whenever the presence or the prospective presence of oil polluting the sea . . . presents a grave and imminent danger to the coast or related interests of one or more contracting states". According to Abecassis, as a result of Art. 1 the Agreement does not apply to small slicks or spills as these would not present a grave danger to the coasts or other interests. But it seems that fishing grounds and offshore installations would be within "related interests" and therefore any serious danger to them would be covered under the Agreement: David W. Abecassis (ed.), pp.133–134.

requested must do their best to render such assistance as is within their power. According to Article 8 nothing in the Agreement can be interpreted to contradict any of the states' rights and duties under international law especially as related to preparedness for and fighting oil pollution. And nothing regarding the zones of responsibility can be interpreted as a precedent or argument for state sovereignty or jurisdiction.[3] It is for states parties to change the common boundaries of their zones of responsibility. Under Article 17 two or more states parties may change them.

On 13 September 1983 in Bonn, the contracting parties to the original Bonn Agreement were joined by the European Economic Community (EEC) and concluded a revised Bonn Agreement for Co-operation in Dealing with Pollution of the North Sea by Oil and Other Harmful Substances. The parties to the Bonn Agreement are Belgium, Denmark, France, Germany, the Netherlands, Norway, Sweden, the United Kingdom and the European Community. By Article 19(2) of the new Agreement, it superseded the 1969 Agreement on its entry into force. Under the new Agreement the 1969 Agreement is extended not only with regard to signatory states but also in scope as it is no longer limited to oil but also covers other harmful substances. The new Agreement also provides information on payment for assistance. The 1969 Agreement was silent on this issue. Under Article 9, in the absence of agreements regarding costs involved in clean-up operations, a party requesting assistance must pay the assisting party the cost of the help rendered. Payments do not affect the requesting state party's rights to claim compensation from a third party. The revised Bonn Agreement provided active co-operation in response to oil pollution which presented a grave and imminent danger to the coast or related interests.

The Agreement was amended in 1989. Under that amendment a provision was made for co-ordinated surveillance and an adjustment of the geographical scope of the Agreement. In 1990 France, Morocco, Portugal and Spain and the European Economic Community made similar arrangements for the neighbouring area of the Northeast Atlantic in the Lisbon Accord.[4] An international centre was established to help the states to react rapidly and effectively to incidents of pollution. And the region was divided into zones of responsibility according to the EEZs of the states parties.[5]

3. Douglas Brubaker, pp.181–182.

4. Accord of Co-operation for the Protection of the Coasts and Waters of the Northeast Atlantic against Pollution due to Hydrocarbons or Other Harmful Substances, 17 October 1990 (the Lisbon Accord), (1991) 30 ILM, 1227–40, cited in E.D. Brown, Vol. I, p.402, n.128.

5. E.D. Brown, Vol. I, p.357.

AGREEMENT BETWEEN DENMARK, FINLAND, NORWAY AND SWEDEN CONCERNING CO-OPERATION IN MEASURES TO DEAL WITH POLLUTION OF THE SEA BY OIL 1971 (THE 1971 COPENHAGEN AGREEMENT)

The Agreement came into force on 16 October 1971. The governments of Denmark, Finland, Norway and Sweden are contracting states to the agreement. It replaces an earlier Agreement on the same subject of 8 December 1967. In scope it is similar to the Bonn Agreement. In general it covers co-operation among the Nordic states in cases of oil pollution incidents. As with the 1969 Bonn Agreement, by Article 1 the parties undertake to inform each other of the sighting of any significant oil slick which may drift towards a party's territory.[6] A threatened state may request help from another and the party requested must do its best to give this help (Article 3). Despite these similarities the 1971 Copenhagen Agreement does not divide the sea into zones like the 1969 Bonn Agreement.

By Article 6 states parties are required to inform others through the competent authority of any offence against anti-oil-pollution regulations committed within the territorial or adjacent waters by a flag ship of the other state party. Article 8 requires states parties to exchange information on various topics. The state authorities must co-operate directly in the planning and co-ordination of other measures required of them under the Copenhagen Agreement. They also need to exchange information for co-operation in planning (Article 9). In addition to the obligations contained in the 1969 Bonn Agreement Article 4 places a duty on parties to maintain stocks of slick-fighting equipment. Each of the states undertakes to set up an emergency service to deal with heavy concentrations of oil at sea. Another difference from the 1971 Agreement can be seen in Article 7 where co-operation in enforcing international regulations is provided for. Article 7 requires states to assist one another in investigating offences against anti-oil-pollution regulations presumed to have been committed within the territorial or adjacent waters of states, including inspecting the ship's oil record book, official log book, engine room log, and the taking of samples.[7]

CONVENTION FOR THE PREVENTION OF MARINE POLLUTION BY DUMPING FROM SHIPS AND AIRCRAFT 1972 (OSLO DUMPING CONVENTION)

The earliest conventions covering dumping are the global London Dumping Convention and the regional Oslo Dumping Convention. The Oslo Convention is

6. Similarly, Art. 2 requires a state threatened by the presence of substantial oil pollution likely also to cause damage to other state parties to investigate the oil, the type, approximate quantity, extent, location, direction and speed and communicate these results and information on measures taken to the other states. Under Art. 5 a state is required to send the other states information concerning heavy concentrations of oil which have obliged it to take anti-pollution action, and the measures taken and the results achieved.

7. David W. Abecassis (ed), pp.134–135; Douglas Brubaker, pp.182–183.

the predecessor of the London Dumping Convention. The Oslo Convention, for the first time, addresses the deliberate disposal of wastes at sea.[8] It defines dumping as "any deliberate disposal of substances and materials into the sea by or from ships or aircraft other than [*inter alia*] the placing of substances and materials for a purpose other than the mere disposal thereof, if not contrary to the aim of this Convention". The Convention was concluded between 13 Western European states.[9] As amended in 1989; the Convention applied to a maritime area in specified parts of the Atlantic and Arctic Oceans. The format of the Oslo Convention was adopted by the London Dumping Convention and the latter only expanded the regional scope of the Convention worldwide. Dumping required the approval of the national authorities. The dumping of Black List substances was prohibited. Grey List substances could be dumped in significant quantities. And even for insignificant quantities the national authorities needed to issue a permit for the dumping.

In June 1991 the Oslo Commission Guidelines for the Disposal of Offshore Installations at Sea were adopted. The Guidelines were designed to assist contracting parties to the Oslo Convention in the management of the disposal of offshore installations and structures so as to prevent pollution in Oslo Convention waters.[10] The Guidelines fall into two parts. Part A deals with the assessment and management of disposal at sea and Part B provides guidance on the monitoring of disposal sites. During this period the Oslo Commission and the Paris Commission under the Paris Convention for the Prevention of Marine Pollution from Land-Based Sources 1974 had agreed to revise the Oslo and Paris Conventions. The Paris Convention for the Protection of the Marine Environment of the North-East Atlantic was signed on 22 September 1992. With its entry into force it will replace the Oslo Dumping Convention 1972 and also the Paris Convention on Land-Based Sources 1974. But the decisions, recommendations and other agreements adopted under the Oslo Convention will continue to apply unless they are incompatible with or terminated by the new Convention or instruments adopted under it.[11] The new Paris Convention is open for signature by the contracting parties to the earlier Paris and Oslo Conventions and other states and organisations specified in Article 25. Under Article 26 additional states and organisations may be invited to accede to the Convention. The original Paris Convention covered the North Sea and parts of the North Atlantic and Arctic Oceans. It applied to a maritime area which extended to the high seas, including that part which is superjacent to the continental shelf, the territorial sea and

8. The Oslo Convention is also the only regional convention so far to deal with the incineration of wastes at sea. In March 1984 a general prohibtion on incineration at sea was introduced. The prohibition was subject to limited exceptions; specific permits could be issued by the national authorities. But in June 1990 the Oslo Commission decided to terminate incineration at sea within the Convention area by 31 December 1991.

9. Belgium, Denmark Finland, France, Germany, Iceland, Ireland, Netherlands, Norway, Portugal, Spain, Sweden and the United Kingdom.

10. Oslo Commission Guidelines, Introduction, para. 1.

11. E.D. Brown, Vol. I, pp. 371–372.

internal waters up to the freshwater limit. States parties were required to combat marine pollution from land based sources. The Paris Convention 1992 describes the Oslo and earlier Paris Conventions as not adequately controlling some of many sources of pollution. That is why it was considered "justifiable to replace them with the ... new Paris Convention, which addresses all sources of pollution of the marine environment and the adverse effects of human activities upon it, takes into account the precautionary principle and strengthens regional co-operation". The 1992 Paris Convention is similar in geographical scope to the 1974 Convention but extends to "all sources of pollution".

CONVENTION ON THE PROTECTION OF THE ENVIRONMENT BETWEEN DENMARK, FINLAND, NORWAY AND SWEDEN 1974 (STOCKHOLM CONVENTION)

The aim of this Convention is to give effect to wider environmental principles as enunciated by the 1972 Stockholm Conference on the Human Environment. It attempts to protect and improve the environment through co-operation between the Nordic States. Under the co-operation aimed for, the Convention attempts to ensure that activities under the jurisdiction of one state do not cause damage to the environment of other states. Under Article 13 the scope of the Convention is extended to the continental shelf areas of contracting states. It covers a wide range of pollutants. By Article 1 the discharge of gas into watercourses, lakes or the sea is included within the frame of environmentally harmful activities. Any person who is affected or may be affected by a nuisance caused by environmentally harmful activities in another contracting state has the right to bring before a court or administrative authority of that state the question of the permissibility of such activities, including the question of measures to prevent damage, and claims for compensation. The Convention has become a precedent for regional approaches for pollution prevention and enforcement.[12]

THE HELSINKI CONVENTION ON THE PROTECTION OF THE MARINE ENVIRONMENT OF THE BALTIC SEA AREA 1974 AND THE HELSINKI CONVENTION 1992 (HELSINKI CONVENTION)

The Helsinki Convention follows the United Nations Environment Programme's "Regional Seas" principles. These principles foresaw action plans for various vulnerable maritime regions throughout the world.[13] The Baltic Sea area was not originally designated as a UNEP regional sea but it nevertheless met all the criteria.

12. Edgar Gold, p.40.
13. *Ibid.*

The Helsinki Convention 1974 is the first Convention on the Protection of the Marine Environment of the Baltic Sea Area. In order to achieve this object it establishes a legal framework to ensure that states parties either individually or jointly take all appropriate legislative, administrative or other relevant measures to prevent and abate pollution. The states parties to the Convention also agree to take all appropriate measures to control and minimise land-based pollution of the marine environment of the Baltic Sea Area. Under Article 2(2) pollution of the sea caused by discharges from land reaching the sea airborne is included in the definition of land-based pollution.[14]

The new Helsinki Convention, signed on 9 April 1992, the Helsinki Convention on the Protection of the Marine Environment of the Baltic Sea Area, is designed to "embody developments in international environmental policy and environmental law" and to modernise the legal regime for the protection of the Baltic environment. The 1992 Convention covers similar undertakings but also requires parties to implement detailed procedures and measures set out in Annex IV. It undertakes to apply the "precautionary principle" and the "polluter pays principle" and to promote the use of "Best Environmental Practice" and "Best Available Technology". The new Convention prohibits the deliberate disposal at sea of fixed or floating platforms and other man-made structures. The governing body of the Convention is the Helsinki Commission–Baltic Marine Environment Protection Commission also known as HELCOM. The present contracting parties to HELCOM are Estonia, European Union, Finland, Latvia, Lithuania, Poland, Russia and Sweden. The 1974 Helsinki Convention will be replaced by the new Helsinki Convention 1992[15] once it enters into force.

The prevention of pollution from ships is dealt with by the Helsinki Conventions 1974 and 1992 and a wide range of measures is prescribed to protect the Baltic from pollution by deliberate, negligent or accidental release of oil and other substances. The conventions deal with liability and compensation for pollution damage in very general terms.

THE UNEP REGIONAL SEAS PROGRAMME AND CONVENTION FOR THE PROTECTION OF THE MEDITERRANEAN SEA AGAINST POLLUTION 1976 (BARCELONA CONVENTION)

Under the auspices of the United Nations Environment Programme (UNEP) the Regional Seas Programme was initiated in 1974. The aim of the programme is to endorse a regional approach to the control of marine pollution and the management of marine coastal resources and to encourage the promotion of

14. Under Article 6(8) and Annex II, states parties are bound to "endeavour to use best practicable means in order to minimise the airborne pollution of the Baltic Sea Area by noxious substances".

15. "Convention on the Protection of the Marine Environment of the Baltic Sea Area." 9 April 1992, LOS Bull. No. 22, January 1993, p.54, RMC II.7.325.

sub-regional co-operation to enhance national capabilities in marine emergency preparedness and response. Under the UNEP's Regional Seas Programme a number of regional seas are identified as requiring particular environmental attention. The programme currently covers thirteen areas where regional action plans are operative or under development[16]:

The Barcelona Convention (adopted 1976) for the Mediterranean States;
The Kuwait Convention (adopted 1978) for the Gulf States;
The Abidjan Convention (adopted 1981) for West and Central African States;
The Lima Convention (adopted 1981) for the South-East Pacific States;
The East Asian Seas agreement (adopted 1981);
The Jeddah Convention (adopted 1982) for Red Sea and Gulf of Aden States;
The Cartagena Convention (adopted 1983) for the Wider Caribbean States;
The Nairobi Convention (adopted 1985) for the East African States;
The Noumea Convention (adopted 1986) for the South Pacific States;
The Bucharest Convention (adopted 1992) for the Black Sea States;
The Northwest Pacific agreement (adopted 1994);
The South Asian States agreement (adopted 1995);
The South West Atlantic agreement (currently being formulated).

The substantive aspect of any regional programme is outlined in an action plan. In order to clarify the substance and scope of the most suitable action plan for the region the governments are consulted through a series of meetings. Although action plans are specifically adapted to the needs of each region, in general their structures are the same. Each action plan usually has the following parts: Environmental Assessment, Environmental Management, Environmental Legislation, Institutional Arrangements and Financial Arrangements.

In the 1970s the Mediterranean was considered to be a region requiring urgent attention. In 1975 the Mediterranean Action Plan was adopted. Following the approval of this plan by all Mediterranean coastal states, except Albania, the Convention for the Protection of the Mediterranean Sea Against Pollution and two protocols were adopted in Barcelona in February 1976. The Protocols to the Convention covered "Pollution by Dumping from Ships and Aircraft" and "Co-operation in Combating Pollution by Oil and Other Substances". The importance of the Barcelona Convention is that it is the first convention adopted under the auspices of the UNEP Regional Seas Programme and it became a model for the UNEP's other action plans. The Convention and Procotols entered into force on 12 February 1978. The signatories to the Convention are Albania, Algeria, Bosnia Herzegovina, Croatia, Cyprus, Egypt, France, Greece, Israel, Italy, Lebanon, Libya, Malta, Morocco, Slovenia, Spain, Syria, Tunisia, Turkey, and the European Union. The Convention has been established as a framework and under this framework the parties undertake general obligations to each other. By Article 3 parties will enter into bilateral, regional and subregional agreements

16. ITOPF. "Country Profiles. A Summary of Oil Spill Response Arrangements and Resources", *Worldwide*, March 1997, p.16.

for the protection of the marine environment of the Mediterranean Sea. These agreements shall be in conformity with the Barcelona Convention and international law. The Convention refers to pollution from ships in very general terms. Under Article 6 "the contracting parties shall take all measures in conformity with international law to prevent, abate and combat pollution of the Mediterranean Sea Area caused by discharges from ships and to ensure the effective implementation in that area of the rules which are generally recognised at the international level relating to the control of this type of pollution".

There are no provisions for dividing the area into zones of responsibility, but Article 8 provides specifically for co-operation in taking measures to deal with pollution emergencies. Under Article 8 of the Convention the contracting parties are required to take all appropriate measures to prevent, abate and combat oil pollution of the Convention area caused by discharges from rivers, coastal establishments or outfalls, or emanating from any other land-based sources within their territories. Article 8 of the Barcelona Convention is further developed by the "Athens Protocol". Under this Protocol, Protocol 3 for the Protection of the Mediterranean Sea against Pollution from Land-Based Sources 1980, the scope of application extended both to the sea and to inland waters up to the freshwater limit.[17] Contracting states are committed to promoting co-operation and diffusion of information within the Mediterranean region through the Regional Marine Response Centre for the Mediterranean Sea (REMPEC). The centre was established in Malta in 1976.

THE KUWAIT REGIONAL CONVENTION FOR CO-OPERATION ON THE PROTECTION OF THE MARINE ENVIRONMENT FROM POLLUTION 1978 AND THE PROTOCOL THERETO CONCERNING CO-OPERATION IN COMBATING POLLUTION (KUWAIT CONVENTION)

The Kuwait Convention and the Protocol thereto Concerning Co-operation were adopted at a Conference in Kuwait on 23 April 1978. Upon ratification by all parties it entered into force on 1 July 1979. The water covered by the Convention is the Arabian Gulf and the parties are the coastal states thereof: Bahrain, Iran, Iraq, Kuwait, Oman, Qatar, Saudi Arabia and the United Arab Emirates.

As with the 1976 Barcelona Convention for the Protection of the Mediterranean Sea Against Pollution this Convention and its Protocol were adopted under the auspices of the UNEP Regional Seas Programme. Therefore the form and many of the essential provisions of the Kuwait Convention are very similar to the Barcelona Convention. And as with Article 8 of the Barcelona Convention, it provides protection for the region's marine environment from all sources of marine pollution.[18] The terms that are provided in Article XII of the Protocol for warning

17. E.D. Brown, Vol. I., pp. 347–348.

18. According to Abecassis, in a number of significant respects the Kuwait Convention is more developed than the Barcelona Convention. For instance Art. IX(a) of the Kuwait Convention gives the

other parties and others to monitor and combat spills are very similar to those of the 1976 Protocol. With regard to provisions on notification of spills both for states and their ships and aircraft the Kuwait Convention and its Protocol are very similar to the 1976 Protocol and therefore do not make any significant change to the 1976 Convention for the Protection of the Mediterranean Sea Against Pollution.[19] The obligations assumed under the Convention have been further elaborated in two Protocols, the Protocol Concerning Regional Co-operation in Combating Pollution by Oil and Other Harmful Substances in Cases of Emergency 1978 and a second Protocol mainly concerned with pollution from offshore platforms, the Protocol on Pollution Resulting from Exploration and Exploitation of the Continental Shelf. The Convention has been supplemented by the Protocol for the Protection of the Marine Environment[20] against Pollution from Land-Based Sources in 1990. The Regional Organisation for the Protection of the Marine Environment (ROPME) was established and eventually based in Kuwait in order to implement the Kuwait Action Plan and its accompanying protocols.

CONVENTION FOR CO-OPERATION IN THE PROTECTION OF THE MARINE AND COASTAL ENVIRONMENT OF THE WEST AND CENTRAL AFRICAN REGION 1981 (ABIDJAN CONVENTION)

The Convention with its Protocol[21] on the Combating of Pollution in Cases of Emergency was adopted in 1981 and it covers the marine and coastal environment of the West and Central African region. Eleven coastal states of the West and Central African region, namely, Benin, Congo Brazzaville, Côte d'Ivoire, Gabon, Gambia, Ghana, Liberia, Mauritania, Nigeria, Senegal and Togo, are signatories to the Abidjan Convention.

CONVENTION FOR THE PROTECTION AND DEVELOPMENT OF THE MARINE ENVIRONMENT OF THE WIDER CARIBBEAN REGION 1981 (JAMAICA CONVENTION)

The Convention and the Protocol on Combating Oil Spills were adopted in Jamaica in 1981. They were developed from the Action Plan adopted in Jamaica in

---

obligation of parties further elaboration in dealing with the taking of necessary measures to deal with emergencies. The obligation to harmonise policy for the purpose of co-operation in preventing and combating pollution of the marine environment is given more detailed elaboration in Art. XII of the Protocol: see David W. Abecassis (ed.), pp. 138–139.

19. *Ibid.*

20. Kuwait Region.

21. Abidjan Protocol concerning Co-operation in Combating Pollution in Cases of Emergency.

1981. The format of the Convention is similar to that of the Barcelona, Kuwait and Abidjan Conventions.[22]

CONVENTION FOR THE PROTECTION OF THE MARINE ENVIRONMENT AND COASTAL AREAS OF THE SOUTH-EAST PACIFIC 1981 (LIMA CONVENTION)

The Lima Convention is based on the Action Plan adopted for the region in Lima in 1981. The Convention covers the marine environment of the South-East Pacific and includes a further agreement which deals with "Regional Co-operation in Combating Pollution by Hydrocarbons and other Harmful Substances in Cases of Emergency". The other Protocol which was adopted in 1983 deals with a similar topic, "Regional Co-operation in Combating Pollution in Cases of Emergency".[23] The Convention was supplemented by the Protocol for the Protection of the South-East Pacific against Pollution from Land Based Sources in 1983. The five South-Eastern Pacific states, namely, Chile, Colombia, Ecuador, Panama and Peru, are signatories to the Convention.

REGIONAL CONVENTION FOR THE CONSERVATION OF THE MARINE ENVIRONMENT OF THE RED SEA AND THE GULF OF ADEN 1982 (JEDDAH CONVENTION)

The Jeddah Convention covers the Red Sea and Gulf of Aden region. The Convention and the Protocol on "Regional Co-operation in Combating Marine Pollution by Oil and Other Harmful Substances in Cases of Emergency" were adopted in 1982 by seven states of the Red Sea/Gulf of Aden region.[24] The parties to the Convention are, Jordan, Palestine, Saudi Arabia, Somalia, Sudan and the Yemen. As with other pollution conventions concluded under the UNEP Regional Seas Programme this Convention deals also with the problem of pollution caused by dumping. Under Article V of the Convention the parties agree to take all appropriate measures to prevent, abate and combat pollution caused by dumping and to ensure effective compliance with generally recognised international rules in relevant international conventions. For the conservation of the marine environment and coastal areas of the Red Sea and Gulf of Aden an action plan has been adopted. A Regional Organisation for the Conservation of the Red Sea and Gulf of Aden Environment has been established under the auspices of the Arab League Educational, Cultural and Scientific Organisation based in Jeddah.

22. Edgar Gold, p.42.
23. *Ibid.*
24. *Ibid.*

CONVENTION FOR THE PROTECTION AND DEVELOPMENT OF THE MARINE ENVIRONMENT OF THE WIDER CARIBBEAN 1983 (CARTAGENA CONVENTION)

The Cartagena Convention and its Protocol concerning Co-operation in Combating Oil Spills were adopted in 1983 at a Conference in Cartagena de Indias, Colombia. The parties to the Convention are Barbados, Colombia, the EEC, France, Grenada, Guatemala, Honduras, Jamaica, Mexico, the Netherlands, Nicaragua, Panama, St Lucia, the United Kingdom, the USA and Venezuela. In addition Antigua and Barbuda, Cuba and Trinidad and Tobago are signatories to the Protocol, which encourages co-operation in combating oil spills. The Convention and its Protocol cover the marine environment of the Wider Caribbean. As part of maintaining the means of responding to oil spill incidents the Convention contains a specific obligation for the parties to develop and promote contingency plans, individually and jointly, for pollution response.[25] The Convention also includes an Article on "airborne pollution". In order to provide a framework for the Convention and in particular the supplementary protocol, a Caribbean Islands OPRC Plan has been developed with the assistance of the IMO.

CONVENTION FOR THE PROTECTION, MANAGEMENT AND DEVELOPMENT OF THE MARINE ENVIRONMENT OF THE EAST AFRICAN REGION 1985 (NAIROBI CONVENTION)

The Convention covers the marine and coastal environment of the Eastern African Region and was adopted in 1985. As with the Cartagena Convention it refers to "airborne pollution" as well. The EEC, France, Madagascar, the Seychelles and Somalia are parties. The Comoros are signatories to supplementary protocols concerning the protection of wild flora and fauna and co-operation in combating spills at sea.

CONVENTION FOR THE PROTECTION OF THE NATURAL RESOURCES AND ENVIRONMENT OF THE SOUTH PACIFIC REGION 1986 (NOUMEA CONVENTION)

The Convention and related Protocols cover the natural resources and environment of the South Pacific Region and were adopted in 1986. Australia, the Cook Islands, the Federated States of Micronesia, France, the Marshall Islands, Nauru, New Zealand, Palau, Papua New Guinea, Tuvalu, the United Kingdom, the USA and Western Samoa are parties. As with the Cartagena and Nairobi Conventions this Convention requires all parties to take all appropriate measures to prevent,

25. Art. 11 of the Cartagena Convention and Art. 3 of its Protocol.

reduce and control pollution of the convention area resulting from discharges into the atmosphere from activities under their jurisdiction. The Convention has been supplemented by a specialised protocol on dumping, the Noumea Protocol for the Prevention of Pollution of the South Pacific Region by Dumping, 25 November 1986.

CONVENTION ON THE PROTECTION OF THE BLACK SEA AGAINST POLLUTION 1992 (BUCHAREST CONVENTION)

The Convention was adopted in 1992 and covers the Black Sea. It has been supplemented by the Protocols on the Protection of the Black Sea Marine Environment against Pollution by Dumping and on the Protection of the Black Sea Marine Environment against Pollution From Land-Based Sources in 1992. With regard to "airborne pollution" the parties, by Article XII, agree "to adopt laws and regulations and take individual or agreed measures to prevent, reduce and control pollution of the marine environment of the Black Sea from or through the atmosphere, applicable to the airspace above their territories and to vessels flying their flag or vessels and aircraft registered in their territory". The Convention includes undertakings to make provision for liability and compensation for pollution damage, but only in very general terms. Bulgaria, Georgia, Romania, Russia, Turkey and Ukraine are parties.

PROTOCOL ON ENVIRONMENTAL PROTECTION TO THE ANTARCTIC TREATY 1991 (MADRID ENVIRONMENTAL PROTOCOL)

The Protocol[26] was adopted in 1991 to supplement the Antarctic Treaty,[27] so it is consistent with other components of the Treaty. It represents one of the most comprehensive multilateral environmental protocols. It was opened for signature in Madrid, Spain, by the Antarctic Treaty Consultative Parties. It aims to prevent marine pollution in the Antarctic Treaty Area. Five principal sources of pollution affecting the Antarctic marine environment can be identified. These are land-based pollution, ship-based pollution, dumping, pollution caused by exploration and exploitation of the seabed in offshore areas and marine pollution from the air. The Madrid Protocol aims to regulate all these pollution resources. It contains certain fundamental rules that provide basic obligations, set specific provisions and fix acceptable standards for activities in Antarctic marine areas. The Protocol explicitly bans drilling activities offshore Antarctica that might create pollution in the circumpolar marine environment.

26. For detailed information on the 1991 Madrid Environmental Protocol see Christopher C. Joyner, "The 1991 Madrid Environmental Protection Protocol" (1996) 20 *Marine Policy*, 183–197.
27. The Antarctic Treaty entered into force in 1961.

Two institutions are provided in order to put into effect the legal principles set out in the Protocol. The Antarctic Treaty Consultative Party Meetings (ATCMs) will make decisions for implementing the Protocol regime. The Committee for Environmental Protection (CEP) will give advice and formulate recommendations to parties regarding implementation of the Protocol and its annexes. But this body has no decision-making authority. Under Article 13 parties are obliged to take appropriate measures to ensure compliance with the anti-pollution provisions of the Protocol.

Attached to the Protocol are five annexes. Annex I sets out procedures for environmental impact assessment.[28] Annex II restates the need for conservation of Antarctic fauna and flora. It also updates the Agreed Measures that apply most directly to land and ice shelves of the continent.[29] Under Annex III[30] waste is classified into five main groups. With regard to marine pollution Article 5 of Annex III pertains to disposal of wastes at sea. Under certain conditions[31] sewage and domestic liquid wastes may be discharged directly into the sea.

Annex IV directly refers to the "Prevention of Marine Pollution", and it is specifically linked to the International Convention for the Prevention on Pollution from Ships, MARPOL 73/78. Annex IV deals with discharges from ships, particularly discharge of oil, noxious liquids, garbage and sewage. Under Article 2 the Annex applies "with respect to each party, to ships entitled to fly its flag and to any other ship engaged in or supporting its Antarctic operations, while operating in the Antarctic Treaty area". Under Article 3 any discharge of oil into the sea is prohibited except in cases permitted under Annex I of MARPOL 73/78. "The discharge of any noxious liquid substance, and any other chemicals or other substances, in quantities or concentrations that are harmful to the marine environment" is also prohibited under Article 4. With a direct link to MARPOL 73/78 parties are obliged to "eliminate all discharge of untreated sewage . . . within 12 nautical miles of land or ice shelves". Beyond 12 nautical miles any sewage discharge should be made "at a moderate speed, and where practicable, while the ship is en-route at a speed of no less than 4 knots". And this prohibition applies "except where it would unduly impair Antarctic operations", but the terms "unduly", "impair" and "Antarctic operations" have not been defined under Annex IV.[32] By Annex IV contracting parties' ships are under an obligation to have "sufficient capacity" on board for the retention of garbage while within the Antarctic Treaty area. They must also ensure that their ships entering Antarctic

28. Annex I to the Protocol on Environmental Protection to the Antarctic Treaty: Environmental Impact Assessment, XI ATSCM/2, 21 June 1991.

29. Annex II to the Protocol on Environmental Protection to the Antarctic Treaty: Conservation of Antarctic Fauna and Flora, XI ATSCM/2, 21 June 1991.

30. Annex III to the Protocol on Environmental Protection to the Antarctic Treaty: Waste Disposal and Waste Management, XI ATSCM/2, 21 June 1991.

31. Three conditions have to be met: (a) the capacity of the marine environment to assimilate such wastes is taken into account; (b) the discharge is made wherever practicable under conditions for initial dilution and rapid dispersal; (c) large quantities of these wastes are treated by maceration: Christopher C. Joyner, p.193.

32. *Ibid.*, p.194.

waters are fitted with tanks of sufficient capacity for the retention of all sludge, dirty ballast, tank washing water, oily residues and garbage from all ships. Under Article 2 of Annex IV enforcement of compliance with Annex IV is left to each contracting party to exercise over ships flying its own flag or supporting that government's Antarctic operations. If Protocol parties are also parties to MARPOL 73/78 then they are explicitly bound to the rights and obligations of MARPOL 73/78 notwithstanding any provisions in Annex IV.

Annex V, as adopted following the proposals of the United Kingdom and the United States, gives an integrated approach to the creation and management of protected areas in the Antarctic.[33] This Annex expands the future scope of the Antarctic protected area system. With regard to marine pollution law "any area, including any marine area" can be a candidate for designation as either an Antarctic Specially Protected Area or an Antarctic Specially Managed Area.

THE NORTH WEST PACIFIC AGREEMENT 1994

The agreement establishes a framework for regional co-operation. Parties to it are China, Japan, the Democratic People's Republic of Korea, the Republic of Korea and Russia.

THE SOUTH ASIAN SEAS AGREEMENT 1995

Parties to the Agreement are Bangladesh, India, the Maldives, Pakistan and Sri Lanka. An action plan for the South Asian seas has been adopted.

A number of bilateral and trilateral agreements have also been created. They are formal and informal in character. These agreements are[34]:

Countries	*Areas Covered*
Argentina and Uruguay	Rio Plata
Australia and Indonesia	Timor Sea
Australia and New Zealand	Tasman Sea
Australia and Papua New Guinea	Torres Strait
Bermuda and the USA	North Atlantic (Bermudan Waters)
Brunei, Darussalam and Malaysia	South China Sea and Brunei Bay
Canada and Denmark	Baffin Bay, Davis Strait, etc.
Canada and the USA	Great Lakes, St. Lawrence Seaway, etc.
Colombia and Venezuela	Gulf of Venezuela and Southern Caribbean Sea

33. *Ibid.*, p.196.
34. ITOPF, March 1997, pp.20–21.

Cyprus, Egypt and Israel	South-East Mediterranean
Denmark and Germany	Wadden Sea, Southern Baltic Sea
Djibouti, Somalia and Yemen	Gulf of Aden
Estonia, Finland and the Russian Federation	Southern Baltic Sea
France and Italy	Tyrrhenian Sea, Ligurian Sea (Medi-plan)
France, Italy and Monaco	Ligurian Sea (RAMOGE)
France and Spain	Bay of Biscay, North-West Mediterranean
France and the UK	English Channel (Manche Plan)
Germany and The Netherlands	Wadden Sea, South-East North Sea
Germany and Sweden	South-Western Baltic Sea
Greece and Italy	Ionian Sea
Ireland and the UK	Irish Sea
Indonesia, Malaysia and Singapore	Straits of Malacca and Singapore
Indonesia, Malaysia and the Philippines	Sulawesi Sea
Indonesia and Malaysia	Lombok-Macassar Straits
Japan and South Korea	Sea of Japan
Japan and the USA	North Pacific
Mexico and the USA	Mexican Gulf
Netherlands Antilles and Venezuela	Southern Caribbean Sea
Norway and the UK	North Sea
Norway and the Russian Federation	Barents Sea
Russian Federation and the USA	Bering Strait, Chukchi Sea
Trinidad and Tobago and Venezuela	Gulf of Paria and Southern Caribbean Sea

CHAPTER 15

LIABILITY AND COMPENSATION FOR MARINE POLLUTION

IN GENERAL

Shipowners are liable for damage caused by the negligence of themselves, their servants and agents. But in order to encourage shipowning this liability has been limited. By the 1924 International Convention for the Unification of Certain Rules Relating to the Limitation of Liability of Owners of Sea-going Vessels and the 1957 International Convention Relating to the Limitation of the Liability of Owners of Sea-going Ships the right to limit in respect of claims for oil pollution damage was provided. When the 1976 International Conference on the Limitation of Liability for Maritime Claims took place in London, the 1969 Civil Liability Convention had already entered into force, the 1971 Fund Convention had been adopted and oil pollution had already been accepted as a special case. Therefore under Article 3(b) of the 1976 Convention on Limitation of Liability for Maritime Claims "claims for oil pollution damage within the meaning of the International Convention on Civil Liability for Oil Pollution Damage, or of any amendment or Protocol thereto" are excluded. Under this definition if a claim for loss or damage is one described in the 1969 Convention as oil pollution damage the 1976 Convention will not apply to that claim. The exclusion does not consider whether or not the 1969 Convention is in force in the territory where the loss or damage was suffered. Thus, if a state which is a party to the 1976 Convention is also party to the 1969 Convention then the oil pollution claims will be regulated in accordance with the 1969 Convention. But if a state party to the 1976 Convention is not a party to the 1969 Convention then the state may regulate oil pollution claims as it pleases. For instance if a claim arises for pollution damage caused by a laden tanker which caused oil pollution damage in a state party to only the 1976 Convention how will the state regulate the limitation of such claims? If the state were also party to the 1969 Convention then it would be possible to limit the claims under the 1969 Convention as damage caused by laden tankers is pollution damage within the meaning of the 1969 Convention. If the damage is caused by spills of non-persistent oil then it would be possible to limit the claim under the 1976 Convention. Because the claim is not a claim for oil pollution damage "within the meaning of" the 1969 Convention or the 1992 Convention therefore it is not excluded from the application of the 1976 Convention.[1]

Consequently, whether claims excluded by Article 3(b) of the 1976 Convention are subject to any limitation depends upon other rules applicable according to international convention or national law. When a state becomes a party to the 1976 Convention it must determine whether the claims for oil pollution damage excluded by Article 3(b) thereof should be subject to limitation and, if so, adopt the national legislation required to provide for an appropriate limit of liability for such claims.[2] Considering claims which are excluded under the 1969 Convention it may be more practical for a state, when ratifying the 1976 Convention, to adopt legislation providing special limits of liability for claims not governed by the 1969 Convention. If the state is not a party to the 1969 Convention then such limit applies to all claims excluded by the 1976 Convention Article 3(b).[3]

THE INTERNATIONAL CONVENTION ON CIVIL LIABILITY FOR OIL POLLUTION DAMAGE 1969 (1969 CLC)

(a) Introduction

The devastating impact of the *Torrey Canyon* spill[4] on the Breton coast in March 1967 highlighted the need for a new international regime in two major areas: the rights of coastal states to protect themselves from pollution damage and civil liability for oil pollution damage. After the *Torrey Canyon* disaster the International Maritime Consultative Organisation (IMCO, now IMO) decided to develop international policies in these areas. Following deliberations in the IMO's Legal Committee, in November 1969 the IMO sponsored the International Legal Conference on Marine Pollution Damage for the purpose of adopting two new conventions. The first dealt with the right to intervene and the second addressed liability and compensation for pollution damage. The first concern gave rise to the International Convention Relating to Intervention on the High Seas in Cases of Oil Pollution 1969 and the second, as the private law counterpart of the public law Intervention Convention, to the International Convention on Civil Liability for Oil Pollution Damage 1969. In its preamble the purpose of the convention was defined as "... the need to ensure that adequate compensation is available to persons who suffer damage caused by pollution resulting from the escape or discharge of oil from ships, and to adopt uniform international rules and procedures for determining questions of liability and providing adequate compensation in such cases". During the conference most delegates agreed with the

1. International Maritime Organisation, *Oil Pollution Liability and Compensation, Course + Compendium, Model Course 3.16* (1991), p.270.
2. See Erling Selvig, "The 1976 Limitation Convention and Oil Pollution Damage" [1979] L.M.C.L.Q. 21–25.
3. *Ibid.*, p.25.
4. See App. IV, Largest Tanker Spills (Source: Lloyd's Maritime Information Services Ltd.).

purpose of the Convention. But many delegates considered these two objectives as conflicting. The compensation issue involved some difficult questions, such as who should be liable for pollution damage? What is the basis of this liability? What is the limit of liability? These questions had a variety of answers according to different delegates. Liability for pollution damage could be on the shipowner, cargo owner and the oil industry, or on all of these. Major shipping nations like France, Germany or the USA favoured strict liability on the shipowner alone because they believed that if they brought any liability burden on the cargo owners or oil industries their oil companies would have major financial burdens. Many coastal states were in favour of putting the burden on both the shipping and oil industries. Other countries argued that "the notion of liability based on fault was a principle of law from which there should be no departure without valid reason".[5]

Limits of liability have caused some disagreements between the nations. The UK was against high limits of liability as it believed that, having a large P&I Club industry, high limits would dry up the funds which were available in the world insurance market. The US delegation was mainly concerned with environmental issues. The parties were not agreed on the precise definition of polluter but despite this disagreement in the final convention draft the delegation applied the principle of "the polluter pays".

(b) Main principles of the 1969 CLC

The aim of the 1969 CLC was to ensure that adequate compensation was available to persons who suffered damage caused by oil pollution as a result of discharge or escape from ships. The convention also aimed to standardise the international rules and procedures on oil pollution liability and compensation.

(i) The scope of the Convention

Article II describes the geographical scope of the Convention: "This convention shall apply exclusively to pollution damage caused on the territory including the territorial sea of a Contracting State and to preventive measures taken to prevent or minimise such damage".

Under this Article the domicile, residence or nationality of the defendant is not relevant. The only criterion is territorial. For instance if a ship causes oil pollution on the shores of state A, which is a party to the Convention, and state B, which is not a party to the Convention, then the Convention will apply only to the damages caused in state A. But this includes any pollution damage suffered in state A by nationals of any state. In the case of preventive measures the Convention will apply to reasonable preventive measures[6] taken anywhere in state A, irrespective of the

5. Dagmar Schmidt Etkin, *Oil Spill Intelligence Report, The Financial Costs of Oil Spills* (1994), pp.146–147.

6. Under 1969 CLC Art. 1(7) preventive measures "means any reasonable measures taken by any person after an incident has occurred to prevent or minimise pollution damage".

persons' nationality who took them.[7] If the action is taken outside the territorial sea of a contracting state it is not covered by the convention.

The Convention applies to seagoing vessels of any type carrying oil in bulk as cargo. It defines ship in Article 1(1) as "any sea-going vessel and any sea-borne craft of any type whatsoever, actually carrying oil in bulk as cargo". Under this definition tankers and combination carriers laden with bulk cargo are covered, but when they are carrying oil only as slops they are excluded. Vessels in ballast are also not covered. In *The Olympic Bravery* the vessel went aground off Ushant, France, in 1976 and the damage was caused by bunker oil escaping from it. The damage was not covered under the Convention, as the vessel was not carrying any oil at the time of the incident. But if the vessel had been carrying oil in bulk as cargo the Convention would have applied to oil pollution caused by any oil emanating from it.[8]

Dry cargo ships, river and lake vessels and seaborne offshore installations are excluded from the definition of a ship in the Convention.

(ii) Pollutants

By Article 1(5) oil means "any persistent oil such as crude oil, fuel oil, heavy diesel oil, lubricating oil and whale oil, whether carried on board a ship as cargo or in the bunkers of such a ship". Under the definition it is not clear whether slop and bilge oils are included or not. According to Abecassis the answer depends on whether the clause at the end of the definition is interpreted restrictively or illustratively. If the definition is interpreted as excluding them it would be unnecessarily arbitrary and a better interpretation would be to include them within the definition.[9] The Convention also does not define the meaning of persistent. Under the International Oil Pollution Fund's practical study[10] the definition of non-persistent oil is given so all other oils excluded under this definition are persistent. Dirty ballast is not covered under the Convention. Polluting substances other than oil, like hazardous and noxious chemicals are also excluded.

(iii) Damages

The Convention defines pollution damage as "loss or damage caused outside the ship carrying oil by contamination resulting from the escape or discharge of oil from the ship, wherever such escape or discharge may occur, and includes the costs of preventive measures and further loss or damage caused by preventive measures".[11] Article 3(1) imposes liability on the shipowner only for pollution

7. David W. Abecassis, (ed.), p.204; International Maritime Organisation, p.288.
8. *Ibid.*, p.195, *ibid.*, p.279.
9. *Ibid.*, p.196, *ibid.*, p.280.
10. A Non-technical Guide to the Nature and Definition of Persistent Oil, attached to FUND/A.4/11. "Non-persistent oil is oil which, at the time of shipment, consists predominantly of non-residual fractions and of which more than 50 per cent by volume distils at a temperature of 340 degrees Centigrade when tested by the ASTM Method D 86/78 or any subsequent revision thereof".
11. The 1969 CLC, Art. 1(6).

damage. The Convention covers the damage resulting from the escape or discharge of oil as well as the costs of prevention measures to mitigate the damage and further loss or damage caused by such measures. The question of what is reasonable and how far the cost of a measure is to be considered in deciding whether it is reasonable is not defined in the Convention. The answers to these questions are left to the interpretation of national law.

The Convention covers both incidental and intentional discharges, but the discharge must have been from the ship. In the case of pure threat situations the Convention will not apply. In case of the *Tarpenbeck*, the tanker *Tarpenbeck* loaded with lubricating oil collided with the Royal Fleet Auxiliary *Sir Geraint* in England. As a result of the collision neither was cargo spilled nor were the cargo tanks of the *Tarpenbeck* damaged, but some non-persistent light diesel oil from the bunker tanks escaped to sea. The tanker was towed to a sheltered bay and the cargo was pumped out. The dispute in this case was whether any cargo escaped during the pumping operation. The United Kingdom Government and some local authorities incurred costs in taking certain preventive measures. When these costs were added to the owner's costs of towing away the ship and pumping her out the total exceeded the ship's limitation amount under the 1969 CLC. The International Oil Pollution Compensation Fund has refused to pay these claims, as there was no spill of persistent oil. Neither of the conventions, the 1969 CLC and the 1971 FC, applied.[12]

In order to apply the 1969 CLC the oil must escape as a result of the incident. Article 1(8) defines an incident as "any occurrence or series of occurrences having the same origin, which causes pollution damage". In the case of a collision where both ships spill oil the spill of oil from each ship is an incident.

In general the 1969 CLC covers a wide range of damages but most of the issues are left to the national law for interpretation. Actions must be brought in the courts of the contracting state in which the damage occurred, within three years from the date of the damage, but not later than six years after the date of the incident.[13]

(iv) Liability

The Convention abandons the traditional concept of liability based on fault and instead imposes on shipowners strict liability for oil pollution damage. Under Article III(1)[14] it is the owner of the ship who is liable. The Convention places no liability upon any other person. The liability of the owner[15] does not depend on his

12. IOPC Fund Documents: Annual Reports 1982 and 1983, FUND/EXC.4/3;FUND/EXC.7/2, cited in David W. Abecassis (ed.), p.198, n.9; International Maritime Organisation, p.282.

13. The 1969 CLC, Art. VIII.

14. The 1969 CLC, Art. III(1) provides that "Except as provided in paragraphs 2 and 3 of this Article, the owner of the ship at the time of the incident, or where the incident consists of a series of occurrences at the time of the first such occurrence, shall be liable for any pollution damage caused by oil which has escaped or been discharged from the ship as a result of the incident".

15. Under Art. I(3) of the 1969 CLC owner has been defined as "the person or persons registered as the owner of the ship or, in the absence of registration, the person or persons owning the ship. However, in the case of a ship owned by a State and operated by a company which in that state is registered as the ship's operator, 'owner' shall mean such company".

residence or domicile or on the state in which his ship is registered. Thus the liability is placed upon the shipowner whether or not the state of his ship's flag is a contracting state.

With regard to channel liability to the owner Article III(4) provides that "No claim for compensation for pollution damage shall be made against the owner otherwise than in accordance with this Convention. No claim for pollution damage under this Convention or otherwise may be made against the servants or agents of the owner". Therefore if an incident is covered by the Convention it is not possible to bring a claim against the owner under general principles of law. If the damage suffered is pollution damage within the scope of the Convention but exempted from liability then there will be no remedy against the owner. It may only be possible to bring claims against the ship's managers, bareboat charterers, builders or other person under general principles of law such as common law or civil code claims. The second part of Article III(4) states that "No claim for pollution damage under this Convention or otherwise may be made against the servants and agents of the owner". The Article does not offer any protection to salvors, charterers and others. If there is a claim against such people it is possible to bring it under general principles of national law while the claim against the owner will be brought under the Convention. In *The Tanio* the bareboat charterers, the technical managers, the classification society and the ship repairer were sued under French law while the registered owner was sued under the Convention.[16] Obviously states, under their national law, may make provisions for the protection of persons who are not protected under the Convention.

Joint and several liability has been referred to under Article V. In cases where there is an escape or discharge of oil from two or more ships as defined in the Convention and the resulting pollution damage is not reasonably separable then the owners of all the ships concerned shall be jointly and severally liable.[17]

(v) Compulsory insurance

Under the Convention the owner of a tanker carrying more than 2,000 tonnes of persistent oil as cargo is obliged to maintain insurance to cover his liability. There is a right of direct action against the insurer. The Convention does not require an owner to be insured beyond the amount to which he could limit liability under Article V(1). Every ship to which the Convention applies must, under Article VII(2), be issued with "a certificate attesting that insurance or other financial security is in force in accordance with the provisions of the Convention. The Certificate shall be issued or certified by the appropriate authority of the State of the ship's registry after determining that the requirements of part (1) of Article VII have been complied with".

16. IOPC Fund Document FUND/EXC.9/3.

17. The 1969 CLC Art. IV states that "When oil has escaped or been discharged from two or more ships, and pollution damage results therefrom, the owners of all the ships concerned, unless exonerated under Article III, shall be jointly and severally liable for all such damage which is not reasonably separable".

According to Article VII(6) the state of registry shall determine the conditions of issue and validity of the certificate and certificates issued or certified under the authority of a contracting state shall be accepted by other contracting states as having the same force as certificates issued or certified by them.[18] The certificates issued by a contracting state shall be regarded by other contracting states as having the same force as certificates issued or certified by them. If a contracting state believes that the insurer or guarantor named in the certificate is not financially capable of meeting the obligations imposed by the Convention it may at any time request a consultation with the state of a ship's registry.[19] States parties must ensure that ships lacking the necessary certificates are not allowed to trade. In case of change of ownership the insurance cover with the new owner stays at least for three months.[20]

The obligation to maintain a financial security certificate also applies to ships which are registered in a non-contracting state but wish to trade to a contracting state. Therefore if any tanker, even one registered in a non-contracting state, wishes to trade to a contracting state, it must have a certificate. States parties must ensure that all ships in their ports, wherever registered, must be insured or secured in accordance with the requirements of the Convention.[21]

If insurance or other financial security is not maintained in respect of a ship owned by a contracting state then the ship must carry a certificate that the ship is owned by the state and that its liability is covered up to the Article V(1) limit.[22]

In countries where the shipowner limits his liability to the value of the ship and freight, following the casualty the shipowner gets the hull insurance and very little or nothing is left for the claimant. In order to make sure that the claimant benefits from the compulsory insurance under the Convention Article VII(9) provides that "Any sums provided by insurance or by other financial security maintained in accordance with paragraph 1 of this Article shall be available exclusively for the satisfaction of claims under this Convention". But the Article does cover hull insurance.

By Article VII(8) proceedings may be brought directly against the insurer or other person providing financial security for the owner's liability for pollution damage but the insurer may limit his liability to the Article V(1) amount. An insurer's right to limit is absolute. Even in the case of actual fault or privity of the assured owner the assured may limit his liability. In any event the insurer has the right to set up a limitation fund.[23] Apart from the bankruptcy or winding-up of the owner the insurer has all the defences which the owner could have invoked.

18. The 1969 CLC, Art. VII(7).
19. *Ibid.*
20. The 1969 CLC, Art. VII(5).
21. The 1969 CLC, Art. VII(11).
22. *Ibid.*
23. The 1969 CLC, Art. V(11), provides that: "The insurer or other person providing financial security shall be entitled to constitute a fund in accordance with this Article on the same conditions and having the same effect as if it were constituted by the owner. Such a fund may be constituted even in the event of the actual fault or privity of the owner but its constitution shall in that case not prejudice the rights of any claimant against the owner".

Furthermore, the insurer may avail himself of the defence that the pollution damage has resulted from the wilful misconduct of the owner himself, but the insurer shall not avail himself of any other defence which he might have been entitled to invoke in proceedings brought by the owner against him.[24]

(vi) Exemptions from liability

Under the 1969 CLC the shipowner is strictly liable subject to various exceptions which the shipowner must prove. By Article III(2):

> "No liability for pollution damage shall attach to the owner if he proves that the damage:
> (a) resulted from an act of war, hostilities, civil war, insurrection or a natural phenomenon of an exceptional, inevitable and irresistible character, or
> (b) was wholly caused by an act or omission done with intent to cause damage by a third party, or
> (c) was wholly caused by the negligence or other wrongful act of any Government or other authority responsible for the maintenance of lights or other navigational aids in the exercise of that function."

In sub-paragraph (a) the common law defence of inevitable incident, to prove that the accident could not have been avoided by the exercise of ordinary care and maritime skill, is not sufficient. With regard to the term "irresistible" used in sub-paragraph (a) the defendant must show that in no circumstances could anyone have avoided the accident.[25] The exceptions in sub-paragraphs (b) and (c) were accepted under the requirement of the United Kingdom delegation. According to the delegation the demand of compulsory insurance under the Convention could only be supplied if those exceptions in sub-paragraphs (b) and (c) were adopted.

(vii) Limitation of liability

If there is no actual fault or privity by the shipowner he is entitled to limit his liability in respect of any one incident to an aggregate amount of 2,000 francs for each ton of the ship's tonnage. However, this aggregate amount shall not in any event exceed 210 million francs. The franc referred to is the gold franc and it shall be converted into the national currency of the state in which the fund is being set up on the basis of the official value of that currency on the date of constitution of the fund.[26] The basis for the conversion is left substantially to national law. Under Article V(1) it is only the owner who may limit liability, and it is only his liability he may limit under Article V. If some person other than the registered owner has become legally liable under the general principles of law oil pollution plaintiffs can recover more than the Article V limited amount. When the gold franc was adopted

24. The 1969 CLC, Art. VII(8).
25. M. Forster, "Civil Liability of Shipowners for Oil Pollution" [1973] J.B.L. 23 at p.26, cited in David W. Abecassis (ed), p.205, n.28.
26. The 1969 CLC, Art. V(9).

at the 1969 CLC it was hoped that it would ensure a measure of uniformity in the real value of the limits of liability. But the world currency crisis resulted, with the flotation of the major currencies on the international markets. As a result the efficacy and conversion of the gold franc at an official rate was destroyed. In November 1976 a Conference held in London adopted a Protocol to the 1969 CLC. The Protocol established that the Poincaré Franc values set out in the Convention should be replaced by Special Drawing Right (SDR) as defined by the International Monetary Fund (IMF). This Protocol entered into force in 1981. Article II of the 1976 Protocol replaces the amounts of 2,000 francs and 210 million francs with 133 SDRs and 14 million SDRs respectively. The stated amounts will be converted into the national currency of the state in which the fund is being set up on the date of setting up of the limitation fund. If the state is a member of the IMF then it applies the same method of valuation as applied by the IMF. If the state is not a member of the IMF then it is free to determine the method itself.

In order to gain the benefit of limitation of liability the owner, under Article V(3), must set up a fund for the total sum representing the limit of his liability with the court or other competent authority of any one of the contracting states in which the action has been brought under the Convention, either by depositing the sum or by producing a guarantee acceptable to the court. If the right to limit is granted, under Article V(4), the fund shall be distributed among the claimants in proportion to the amounts of their established claims. With regard to the benefit of interest accruing on the limitation fund between the date of its establishment and of its distribution the Convention is silent. The solution to this problem is left to national law.

When the owner has set up a fund in accordance with Article V and is entitled to limit his liability the fund becomes the sole source of satisfaction of claims under the Convention. After the establishment of the fund a person having a claim for pollution damage arising out of that incident shall not be entitled to exercise any right against any other assets of the owner in respect of such claim and any arrested property must be released.[27] Claims for damage other than pollution damage suffered in contracting states are unaffected by the establishment of a limitation fund. Similarly all claims suffered in non-contracting states are also not affected by the establishment of the fund under the 1969 CLC. The shipowner may limit his liability in all contracting states to the single amount and he does not have to limit liability separately in each state in which an action is brought.

27. The 1969 CLC, Art. VI.

THE INTERNATIONAL CONVENTION ON THE ESTABLISHMENT OF AN INTERNATIONAL FUND FOR COMPENSATION FOR OIL POLLUTION DAMAGE 1971 (1971 FC)

(a) Introduction

At the 1969 Diplomatic Conference which adopted the 1969 CLC it was already recognised that the regime of compensation established by the Civil Liability Convention was insufficient. During the Conference there had been a long debate on the main issues of strict liability and who was to bear that liability, the shipowner, the cargo owner or both of them. The majority of delegates were in favour of strict liability for the ship which was combined with liability for the cargo interests in the form of a fund. A Working Group was set up to examine the question of liability based on an international fund. But at the time the group produced its report it seemed obvious that the establishment of such a fund was not possible. So the Conference adopted a resolution requesting IMO to elaborate a suitable convention based on the existence of an international fund. The resolution was based on two principles: first, victims should be fully and adequately compensated under a system based upon the principle of strict liability and, secondly, the fund should in principle relieve the shipowner of the additional financial burden imposed by the 1969 CLC.[28] The International Convention on the Establishment of an International Fund for Compensation for Oil Pollution Damage 1971 (1971 FC), setting up a scheme called the International Oil Pollution Compensation Fund (the IOPC Fund), was adopted on 18 December, 1971. The 1971 FC entered into force on 16 October 1978. Membership of the 1971 FC is open only to states that are parties to the 1969 CLC.[29] Thus a state can become a party to the Liability Convention but not to the Fund Convention, both Conventions or neither and rely on its own national law. Before termination of TOVALOP and CRISTAL, becoming party to these agreements was another option.

The main function of the 1971 FC was to ensure the availability of adequate compensation to persons who suffer injury from an oil pollution incident resulting from the escape or discharge of oil from ships and to indemnify the shipowner for a portion of his liability under the 1969 CLC. The Fund Convention also ensured that oil cargo interests bear a proportion of the economic responsibility for oil pollution damage to relieve the shipping industry of total responsibility.

28. David W. Abecassis (ed.), p.253; International Maritime Organisation, p.340.

29. See App. V. Contracting states to the 1969 CLC and 1971 Fund Conventions. (Source: The International Tanker Owners Pollution Federation Limited.)

(b) Main principles of the 1971 FC

(i) The scope of the Convention

Under Article 3(1), compensation under Article 4 applies exclusively to pollution damage caused on the territory including the territorial sea of a contracting state and to measures taken to prevent or minimise such damage. The only limitation in this Article is territorial in nature, so the benefits of the Convention are conferred upon those in contracting states irrespective of the flag of the ship causing the damage or the domicile or residence of the owner.

(ii) Supplementary compensation

The Fund Convention supplements the 1969 CLC and indemnifies the parties for part of their liability under the Civil Liability Convention. The Fund Convention requires that a party to the Fund Convention must also be a party to the Civil Liability Convention. In order to receive indemnities under the Fund Convention the flag state of the ship must be a party to the Convention.

The IOPC Fund pays compensation to any person suffering oil pollution damage in an IOPC Fund member state. The person seeking compensation under the Fund Convention should be unable to obtain full compensation under the Civil Liability Convention for one of the following reasons[30]:

(a) no liability for pollution damage arises under the Civil Liability Convention because the shipowner invokes one of the exemptions from liability under that Convention;

(b) the shipowner is financially incapable of meeting his obligations under the Civil Liability Convention and he has inadequate insurance to satisfy the claims for compensation for pollution damage;

(c) the damage exceeds the shipowner's liability under the Civil Liability Convention (most common reason).

If the pollution damage[31] has been suffered within the geographical scope of Article 4 and the claimant has been unable to obtain full and adequate compensation for it for one of the stated reasons then the IOPC Fund will be liable to him. But this liability is subject to the exceptions. Article 4(2) of the Fund Convention exempts the IOPC Fund from liability if:

(a) it proves that the pollution damage resulted from an act of war, hostilities, civil war or insurrection or was caused by oil which had escaped or been discharged from a warship or other ship owned or operated by a State and used, at the time of the incident, only on Government non-commercial service; or

30. The 1971 FC, Art. 4.

31. Under Art. 1(2) of the 1971 FC "ship", "person", "owner", "oil", "pollution damage", "preventive measures", "incident" and "organisation" have the same meaning as in Art. I "of the Liability Convention, provided however that, for the purposes of these terms, 'oil' shall be confined to persistent hydrocarbon mineral oils".

(b) the claimant cannot prove that the damage resulted from an incident involving one or more ships.

In 1995 the 1971 Fund was informed of an oil spill which had occurred in 1994 in Morocco. The Moroccan authorities claimed compensation for clean-up costs but did not give any indication as to the source of the spill; they stated that the oil could only have come from the sea as a result of the escape of ballast water, the cleaning of tanks or accidental pollution. In December 1995 the 1971 Fund informed the Moroccan authorities that it had not been established that the oil originated from a ship as defined in the 1971 Fund Convention and for that reason the 1971 Fund could not accept the claim for compensation.[32]

Apart from specific exemptions in Article 4(2), Article 4(3) provides that:

"If the Fund proves that the pollution damage resulted wholly or partially either from an act or omission done with intent to cause damage by the person who suffered the damage or from the negligence of that person, the Fund may be exonerated wholly or partially from its obligation to pay compensation to such person provided, however, that there shall be no such exoneration with regard to such preventive measures which are compensated under paragraph 1. The Fund shall in any event be exonerated to the extent that the shipowner may have been exonerated under Article III, paragraph 3, of the Liability Convention."[33]

The wording of Article 4(3) has caused some problems in practice. The first sentence of the Article says that preventive measures are always recoverable, but according to the following sentence "in any event" preventive measures are not always recoverable.[34] The inconsistency in Article 4(3) has been removed by a change made under the 1984 Protocol to the 1971 FC.

(iii) Compensation under Article 4

Originally Article 4(4)(a) provided compensation to a maximum amount in respect of any one incident of 450 million francs. The Article 4 limit is set by the Convention at a figure which must be between 450 million and 900 million gold francs. The figure started at 450 million. By Article 4(6), the Assembly of the IOPC Fund may, having regard to the experience of incidents which have occurred and in particular the amount of damage resulting therefrom and to changes in monetary values, decide that the amount of 450 million francs shall be changed. But the amended sum shall in no case exceed 900 million francs or be lower than 450 million francs. The change shall apply to incidents which occur after the date of the decision effecting the change. In order to make such a change a three-quarters majority of those present is required.[35]

Following the *Amoco Cadiz* incident in France in March 1978, the French

32. International Oil Pollution Compensation Fund, *Annual Report 1996*, pp.77–78.

33. Art. III(3) of the 1969 CLC provides that "If the owner proves that the pollution damage resulted wholly or partially either from an act or omission done with intent to cause damage by the person who suffered the damage or from the negligence of that person, the owner may be exonerated wholly or partially from his liability to such person".

34. David W. Abecassis (ed.), pp.259–260; International Maritime Organisation, pp. 346–347.

35. The 1971 FC, Arts. 32(c) and 33(1).

government proposed to the Assembly of the IOPC Fund that it increase the limit to 900 million gold francs. The proposal was not accepted completely and with effect from 20 April 1979 the limit was raised to 675 million gold francs. In October 1981 the French government again proposed to increase the limit up to 900 million gold francs. As it failed to attract the required majority the proposal was again refused. Under the 1984 FC Protocol the system for revising the Article 4 limit was completely changed.

The amount of compensation, 675 million gold francs, set out under the 1971 FC includes that actually paid under the 1969 CLC. Part of the compensation paid under the Civil Liability Convention may have been the subject of a claim by the shipowner under Article 5 of the Fund Convention. Therefore the amount received from the IOPC Fund is not affected by whether the IOPC Fund has paid a claim to the shipowner under Article 5 of the Fund Convention or not.

The amount of compensation recovered under the IOPC Fund is calculated by subtracting from 675 million francs the amount of compensation actually recovered under the Civil Liability Convention. Under Article 4(4)(b) where the pollution damage has been caused by a natural phenomenon of an exceptional, inevitable and irresistible character the aggregate amount of compensation payable by the Fund shall not exceed the limit of 675 million francs. Therefore if the phenomenon has caused more than one tanker to cause pollution damage the Article 4(4)(b) limit applies to all of them irrespective of how many ships spill oil.

As with the 1969 CLC in 1976 a Protocol to the 1971 FC was adopted. Under the Protocol the gold franc was replaced as the monetary unit by the Special Drawing Right of the International Monetary Fund (SDR). The 1976 Protocol to the Fund Convention follows the 1976 Protocol to the Liability Convention. It expresses the gold franc amounts in SDRs on the basis of 15 gold francs to 1 SDR. Pursuant to the 1976 FC Protocol 450 and 900 million francs become respectively 30 and 60 million SDRs and 675 million francs were 45 million SDRs. Although the 1976 Protocol to the Civil Liability Convention entered into force in 1981, the 1976 Protocol to the Fund Convention entered into force in 1994. This led to an unusual situation because the Civil Liability Convention limit could be expressed and calculated by reference to the SDR but the compensation limit under the Fund Convention was still calculated by reference to the gold franc. However, on the basis of the word "official" included in the text of the Civil Liability Convention to rule out the application of the free market value of gold, the IOPC Fund had taken it for granted that the conversion of the gold franc into national currency should be made on the basis of the SDR.[36] Obviously the reason for such an assumption was to make the unit of account in the two conventions the same.

36. Mans Jacobsson, "The International Conventions on Liability and Compensation for Oil Pollution Damage and the Activities of the International Oil Pollution Compensation Fund", in Colin M. De La Rue (ed.), pp.43–44.

1984 PROTOCOLS TO THE CIVIL LIABILITY AND FUND CONVENTIONS

The International Conference on Liability and Compensation for Damage in Connection with the Carriage of Certain Substances by Sea held in London from 30 April to 25 May 1984 adopted two protocols:

(a) the Protocol of 1984 to amend the International Convention on Civil Liability for Pollution Damage 1969; and

(b) the Protocol of 1984 to amend the International Convention on the Establishment of an International Fund for Compensation for Oil Pollution Damage 1971.

The 1984 Protocols provide higher limits of compensation and a wider scope of application than the Conventions in their original version. As the required number of ratifications have not been obtained the 1984 Protocols have not come into force.

Under the 1984 Civil Liability Convention Protocol, Art. 11, the Convention is called the International Convention on Civil Liability for Oil Pollution Damage 1984 (1984 CLC).

The 1984 CLC, Article 2(1), redefined "ship" to include combination carriers and unladen vessels not clean of oil residues.[37] Under the new definition the Convention covers unladen tankers carrying slops or those carrying only bunkers. Combination carriers which can carry oil in bulk as cargo are now covered not only during their voyage of carriage but also on any voyage following such carriage until the slops are fully discharged. Dry cargo ships which can carry oil in bulk in their deep tanks are to be treated as combination carriers, and so are covered by the Convention. River and lake vessels are still excluded.

Under Article 2(2) of the 1984 CLC the definition of oil has to a great extent been clarified by the complementary inclusion of "hydrocarbon mineral". But a definition of "persistent" has still not been provided.

The definition of pollution incident was broadened to include the threat of pollution. Under this new definition the framework of the Convention was expanded to include the time before an escape or discharge of oil actually occurs. The preventive measures taken before an escape will not be recoverable unless the threat of causing pollution damage is grave and imminent.[38]

Article 4(2) provides protection for a wider range of persons than those mentioned in the 1969 CLC. The geographical scope of the 1969 CLC has been

37. The 1984 CLC Protocol, Art. 1(1) defines ship as "any sea-going vessel and sea-borne craft of any type whatsoever constructed or adapted for the carriage of oil in bulk as cargo, provided that a ship capable of carrying oil and other cargoes shall be regarded as a ship only when it is actually carrying oil in bulk as cargo and during the voyage following such carriage unless it is proved that it has no residues of such carriage of oil in bulk on board".

38. Under Art. 2(4) of the 1984 CLC Protocol, incident means "any occurrence, or series of occurrences having the same origin, which causes pollution damage or creates a grave and imminent threat of causing such damage".

extended to the exclusive economic zone (EEZ) established under the UNConvention on the Law of the Sea 1982. Under Article 3 of the 1984 CLC certain claims for loss or damage will be allowed which were not valid under the 1969 CLC. Loss or damage caused by contamination outside the territorial sea but inside the 200 mile zone will be covered by the new text. Measures taken anywhere to prevent or minimise such damage are also within the ambit of the new text.

The 1984 CLC extended the definition of pollution damage to include compensation for impairment of the environment as long as this is limited to the costs of reasonable measures of reinstatement actually undertaken or to be undertaken. In other words, the costs incurred on reasonable measures to restore a contaminated environment are covered under the new pollution damage definition.[39]

Under the 1984 CLC a shipowner is entitled to new limits of liability, depending on the size of the ship. Article 6 of the 1984 Protocol replaces Article V(1) of the 1969 Liability Convention with the following:

> "1. The owner of a ship shall be entitled to limit his liability under this Convention in respect of any one incident to an aggregate amount calculated as follows:
>
> (a) 3 million units of account for a ship not exceeding 5,000 units of tonnage;
>
> (b) for a ship with a tonnage in excess thereof, for each additional unit of tonnage, 420 units of account in addition to the amount mentioned in subparagraph (a); provided, however that this aggregate amount shall not in any event exceed 59.7 units of account."

The units of account that are used in the 1984 CLC are the same as those in the 1976 Protocol to the 1969 CLC. The unit of tonnage is changed to the gross tonnage calculated in accordance with the International Convention on Tonnage Measurement of Ship 1969. With this change the 1969 CLC comes in line with the 1976 Convention on Limitation of Liability for Maritime Claims.

Under the 1969 CLC the owner was not entitled to limit his liability if the incident occurred as a result of the actual fault or privity of the owner. By the 1984 CLC the owner shall not be entitled to limit his liability if it is proved that the pollution damage resulted from his personal act or omission committed with the intent to cause such damage, or recklessly and with knowledge that such damage would probably result.

Only the states parties to the 1984 Liability Convention may become parties to the 1984 Fund Convention.

The 1984 FC establishes the International Oil Pollution Compensation Fund 1984 (IOPC Fund 1984). The purpose of the IOPC Fund 1984 is to provide compensation for pollution damage to the extent that the protection afforded by

39. Under Art. 2(3) of the 1984 CLC Protocol, pollution damage means "a) loss or damage caused outside the ship by contamination resulting from the escape or discharge of oil from ship, wherever such escape or discharge may occur, provided that compensation for impairment of the environment other than loss of profit from such impairment shall be limited to costs of reasonable measures of reinstatement actually undertaken or to be undertaken; b) the costs of preventive measures and further loss or damage caused by preventive measures".

the 1984 Civil Liability Convention is inadequate. The new Fund has a virtually identical internal structure and organisation to the existing IOPC Fund but it dissolves the Executive Committee to streamline the Fund organisation.

The 1984 Fund Convention leaves Article 4(1) unamended apart from redefining some words. It retains the provision that the IOPC Fund will compensate for damage where exoneration applies under the CLC, where the damage exceeds CLC liability limits or where the shipowner is insolvent. As the references in Article 4(1) to the Liability Convention are replaced by references to the 1984 Liability Convention, the claimants must show that they have been unable to obtain full and adequate compensation for their loss under the terms of the 1984 Liability Convention. In other words, the 1984 Fund Convention applies revised definitions and geographical designations of the Civil Liability Convention to the Fund Convention.

The main and most important purpose of the 1984 Conference was to amend the limits of liability for both Conventions. The system of revision of limits contained in Article 4(6) of the Fund Convention was completely changed by the 1984 Protocol. Article 33 of the 1984 Protocol replaces the old system with a new one which is almost identical to that of the 1984 Liability Convention.

As explained above the 1967 *Torrey Canyon* incident produced a great improvement in the international law relating to oil pollution. Apart from this development it also affected the practice of the oil and tanker industries and their attitudes towards to the oil pollution problem. After the *Torrey Canyon* incident the oil industry felt that waiting for the entry into force of the CLC and FC was not enough. Two schemes were voluntarily introduced by the oil industry to fill the gap somehow until the 1969 CLC and 1971 FC came into force. TOVALOP and CRISTAL[40] have the same purposes: they both aim, first, to remove voluntarily some of the legal uncertainties that arise from an oil spill from a tanker at sea, secondly, to encourage quick action for cleaning up or mitigating the effects of an actual or threatened oil spill, and thirdly, to provide adequate compensation for the victims of an oil spill. This includes the reimbursement of the reasonable costs incurred by owners or governments.[41] TOVALOP stands for "Tanker Owners Voluntary Agreement Concerning Oil Pollution" and CRISTAL stands for "The Contract Regarding An Interim Supplement To Tanker Liability For Oil Pollution".

Although TOVALOP and CRISTAL expired on 20 February 1997 these Conventions will be examined as they were part of the international maritime scene for a very long period and International Tanker Owners Pollution

40. For detailed information on TOVALOP and CRISTAL see *TOVALOP & CRISTAL—A Guide to Oil Spill Compensation* which can be obtained from The International Tanker Owners' Pollution Federation Limited; Ian C. White, "The Voluntary Oil Spill Compensation Agreements—TOVALOP and CRISTAL", in Colin M. De La Rue (ed.), pp.57–69; Ian C. White, "The International Oil Spill Compensation Regimes (CLC, Fund Convention, TOVALOP, and CRISTAL) and Recent Claims Experience", in Marine Cargo & Hull Liability Conference, Sydney, 1–2 June 1994, pp.2–18.

41. The Britannia Steam Ship Insurance Association Limited, *TOVALOP and CRISTAL Explanatory Notes* (Revised ed.), February 1994, p.3.

Federation's (ITOPF) existing members will remain members after 20 February 1997, unless any member notifies the ITOPF in writing that they wish to resign.

THE VOLUNTARY OIL SPILL COMPENSATION SCHEMES

(a) TOVALOP

(i) In general

TOVALOP is entered into by tanker owners and bareboat charterers and originated from the determination of tanker owners to take constructive action to mitigate and to provide compensation for damage by oil pollution from tankers. TOVALOP came into effect on 6 October 1969. At that time owners of 50 per cent of the world's tanker tonnage became parties to the agreement. TOVALOP has been amended on a number of occasions and, with effect from 20 February 1987 a Supplement (TOVALOP Supplement) was added to the existing Agreement which was known as the Standing Agreement. The TOVALOP Supplement can apply only when a participating tanker involved in an incident is carrying a cargo owned by a party to CRISTAL.

TOVALOP applies when a participating tanker spills, or threatens to spill, persistent oil (whether it be cargo or fuel). In the case of such a spill or spill threat the owner or bareboat charterer will take appropriate action and will reimburse governments and others who incur reasonable costs in responding to the incident or who suffer pollution damage. Measures taken in response to the incident include attempts to eliminate a grave and imminent threat of escape or discharge of oil from a tanker and actions to prevent or minimise pollution damage, for example by using booms, skimmers, dispersants and other clean-up techniques.

In the Standing Agreement and the Supplement pollution damage has been defined to cover loss or damage caused by contamination resulting from the escape or discharge of oil, and by preventive measures. The TOVALOP Supplement specifically covers proven economic loss sustained as a direct result of the oil contamination but not as a consequence of physical damage to property. Since 20 February 1990, under the TOVALOP Supplement it is possible to be reimbursed for reasonable costs actually incurred in taking reasonable and necessary measures to restore or replace natural resources damaged as a direct result of an incident. Claims based on theoretical scientific or economic assessments and other forms of damage to natural resources that are not commercially exploited are not covered under the Supplement.

TOVALOP is administered by the International Tanker Owners Pollution Federation Limited (the Federation). Each party to TOVALOP is also a member of the Federation. An important function of the Federation from its early days was the provision of a technical advice service to its members. The Technical

Department was established to reduce the cost of oil spills to the tanker industry, to clean up any spill quickly and efficiently and minimise the damage to the marine environment. Over the years the department has expanded to meet the requirements of members, both tanker owners and their P&I clubs.[42]

TOVALOP is described as a voluntary agreement, but each party undertakes to comply with the terms and conditions of the Agreement without exception. The Agreement must apply to all of shipowners' tankers. Each party to TOVALOP is required to notify the Federation immediately if he fails to comply with or to perform or observe any of the general conditions set out in the Agreement. The result of such a failure is that the party concerned automatically ceases to be a party to the Agreement.

In general TOVALOP is subject to termination upon various contingencies and individual parties may withdraw at any time upon six months' prior written notice. Originally TOVALOP was to expire when the CLC had been ratified by a majority of maritime nations but on two occasions it was extended for further periods. The latest extension was until 20 February 1997.

(ii) Limits of financial responsibility

Under the TOVALOP Standing Agreement the maximum compensation for all claims arising out of any one incident is US$160 per limitation ton or US$16.8 million, whichever is the less. As explained above in May 1984 Protocols to the Liability and Fund Conventions were agreed. The Protocols substantially increased the limits of compensation and coverage of the two Conventions. In November 1992 in order to facilitate their prompt and widespread ratification by governments the entry-into-force provisions of these Protocols were revised. The Protocols are called the 1992 Protocols. Following the agreement between tanker owners and oil companies in 1986, the increased limits of compensation are available to claimants through voluntary agreements. With effect from 20 February 1994, the limits of financial responsibility under the TOVALOP Supplement[43] are expressed in SDRs and they are identical to the limits of liability in the 1992 Protocol to the Liability Convention. The new limits are:

—For a tanker of up to 5,000 gross tons, a maximum of SDR 3 million.

—For a tanker of over 5,000 gross tons, SDR 3 million plus SDR 420 for each gross ton in excess of 5,000 gross tons, up to a maximum of SDR 59.7 million.

If the tanker owner is required to establish a limitation fund under the terms of the Civil Liability Convention the applicable date for conversion of SDRs into national currency is the date that the limitation fund is established. Otherwise the

42. The first incidents attended by the Technical Department were in the Mediterranean in 1972. These were followed by the *Metula* in the Magellan Straits in 1974, the *Jakob Maersk* off Portugal in 1975, the *Venoil/Venpet* collision of South Africa in 1977, the *Amoco Cadiz* off Northern France and the *Esso Bernicia* in the Shetlands in 1978.

43. The TOVALOP Supplement can only apply where the cargo on board the tanker at the time of the incident was owned by a party to CRISTAL.

TOVALOP Supplement establishes the conversion date as the date of any payment.

(iii) Submission of claims

Under TOVALOP the settlement of any valid claims is the responsibility of the tanker owner or bareboat charterer who is a party to TOVALOP. The Federation itself is never liable for the settlement of claims and there is no TOVALOP fund money. Anyone with a claim under the Agreement must notify the participating owner or bareboat charterer in writing within two years of the date of the incident. As the responsibility for settling claims under TOVALOP falls to the participating tanker owner or bareboat charterer each party should be financially capable of meeting its maximum potential financial responsibility under the TOVALOP Agreement and its Supplement. Normally this is done by arranging oil pollution insurance with one of the P&I clubs. After an applicant's insurance arrangements have been found satisfactory the International Tanker Owners Pollution Federation Ltd. will issue TOVALOP certificates in respect of the entered vessel. The TOVALOP certificate is not a certificate of financial security. It only shows that the named tanker owner or bareboat charterer and vessel satisfied the entry requirements at the date of issue.

If there is a dispute between the claimants under the Agreement and the TOVALOP party the Agreement provides for arbitration. The person making the claim has three years from the date of the incident in which to commence the arbitration procedure.

(b) CRISTAL

(i) In general

CRISTAL was formed as a voluntary system for compensating states and persons while they were waiting for the entry into force of the Fund Convention. At the same time it aimed to provide compensation supplemental to that available from tanker owners and bareboat charterers under TOVALOP. Therefore many of the provisions and definitions and also the overall scope of the two voluntary agreements are complementary. CRISTAL came into effect in April 1971 and like TOVALOP it has been amended on a number of occasions. Like TOVALOP it was amended with effect from 20 February 1987. The Amendment in 1987 substantially increased the amount of compensation available to those affected by oil pollution from tankers who would not be fully compensated under the terms of the TOVALOP Supplement. Under the more recent amendments the definition of pollution damage was changed and made identical to the revised definition in the TOVALOP Supplement.

CRISTAL does not apply to incidents which are covered under the Fund Convention. It specifically begins to apply when other regimes such as the Civil Liability Convention or TOVALOP are inadequate. Participation in TOVALOP is compulsory for coverage under CRISTAL and affected cargo must be owned by an oil company participant. Therefore in order to apply CRISTAL to an incident two conditions have to be satisfied[44]: first, the incident must involve an escape or discharge of persistent oil, or threat thereof, from a sea-going tanker[45] carrying a cargo that is owned or deemed to be owned at the time of the incident by a party to CRISTAL. Ownership is fairly broadly interpreted and so may cover situations where a party to CRISTAL does not actually posses legal title to the oil cargo.[46] If a non-CRISTAL party owning the cargo has agreed to sell it to a party to CRISTAL the cargo will be considered owned by a party to CRISTAL even if legal title to the oil is still with the non-CRISTAL party. If a CRISTAL party has transferred title to a non-party, again the CRISTAL party may be considered to be the owner of the cargo. Further, if a CRISTAL party or one of its affiliates is carrying a cargo owned by a non-party to CRISTAL it they will be considered as the owner of that cargo. In the last two cases the election has to be made in writing to CRISTAL prior to any accident occurring. In cases where the actual title to the oil is with a non-party and the shipment is contracted to go to or from a terminal in which a CRISTAL party has ownership or other interests, and an incident occurs or pollution damage is caused within 250 nautical miles of that terminal then the shipment of oil will also be considered to be owned by a party.

The second condition for the application of CRISTAL is that the tanker owner or bareboat charterer must first pay compensation up to the applicable limit calculated in accordance with the TOVALOP Supplement. In other words, in the first instance the claimant must seek recovery from the owner of the tanker involved in the incident up to the limit stated in the TOVALOP Supplement. The tanker does not have to be actually entered in TOVALOP. After meeting the necessary requirements if the limited exceptions[47] do not also apply then it is possible to apply for compensation under the terms of the CRISTAL contract. As with TOVALOP the CRISTAL contract provides for reimbursement of reasonable costs incurred by a tanker owner or any other person in taking threat removal measures, preventive measures or through having sustained pollution damage.

44. TOVALOP & CRISTAL, pp.6–7.

45. In order for CRISTAL to apply the tanker must be a sea-going vessel or seaborne craft of any type, designed and constructed for carrying oil in bulk as cargo. It must also be actually carrying oil in bulk as cargo at the time of the incident. The tanker need not be at sea at the time of the incident, but must be capable of sea-going or being seaborne: CRISTAL, *Memorandum of Explanation of the Contract Regarding a Supplement to Tanker Liability for Oil Pollution*, Revised 20 February 1994.

46. Contract Regarding a Supplement to Tanker Liability for Oil Pollution, as amended 20 February 1994, Cl. V. Ownership of Shipments.

47. If the incident (a) resulted from an act of war, hostilities, civil war, insurrection or a natural phenomenon of an exceptional, inevitable and irresistible character, or (b) was wholly caused by an act or omission done with intent to cause damage by a third party, or (c) was wholly caused by the negligence or other wrongful act of any Government or other authority responsible for the maintenance of lights or other navigational aids in the exercise of that function. In addition, if the pollution damage, or the taking of Threat-Removal Measures, resulted wholly or partly either from an act or omission done with intent to cause damage by, or from the negligence of, the Person who sustained the Pollution Damage or who took the Threat-Removal Measures, then any compensation otherwise payable to that person will be denied or reduced proportionately.

The reimbursement mechanism does not affect the claimants directly. It is mainly designed to ensure that payments made by tanker owners and CRISTAL members are consistent with their respective limits of financial responsibility under the TOVALOP Supplement and CRISTAL contract.

Any company engaged in the production, refining, marketing, storing or trading of oil or which receives oil in bulk for its own consumption or use can become a party to the CRISTAL contract.[48] As with TOVALOP, although CRISTAL is a voluntary agreement parties to the contract will be bound by the terms of it and agree to become members of CRISTAL Limited and to abide by its bye-laws, rules and directives.

Under TOVALOP or under the CLC a tanker owner's obligations are normally met through individual insurance arrangements. CRISTAL itself maintains a fund of money from which claims are met. Depending upon the quantities of crude and fuel oil received by sea parties to the contract are called upon to make contributions to the fund. The amount and frequency of calls depend upon claims settlements. With TOVALOP, CRISTAL expired on 20 February 1997.

(ii) Limits of financial responsibility

Under the CRISTAL contract the maximum limit of financial responsibility is determined by the gross tonnage of the tanker. In all cases, the stated amounts include compensation that would be payable by the tanker owner under the TOVALOP Supplement. As with the TOVALOP Supplement as from 20 February 1994, limits of liability in CRISTAL are expressed in terms of SDRs rather than US dollars. The limits are:

—For a tanker of up to 5,000 gross tons, a maximum of SDR 32 million;

—For a tanker of over 5,000 gross tons, SDR 32 million plus SDR 652 for each gross ton, in excess of 5,000 gross tons up to a maximum of SDR 120 million.

(iii) Submission of claims

The detailed claims procedure is set out in the Rules of CRISTAL Limited. Notice of claim must be given to CRISTAL Limited within two years of the incident giving rise to the claim. The directors of CRISTAL Limited, a Bermudan company in which the parties are shareholders, make the final decisions on the admissibility of claims and the payment of compensation.

INTERACTION BETWEEN THE VOLUNTARY AND LEGAL REGIMES

The common feature of the international system of compensation created by the Liability and Fund Conventions and the two voluntary agreements of TOVALOP

48. TOVALOP & CRISTAL, p.8.

and CRISTAL is that all the regimes are based on the principle of strict liability. Under this principle the Conventions apply regardless of fault. In case of an incident the owner has the primary liability. If the amount of compensation provided by the owner is not sufficient to meet all valid claims then it is possible to obtain supplementary compensation from the international funds established under the terms of the Fund Convention and CRISTAL. This system of compensation consists of two steps. In the first step the individual tanker owner whose vessel causes a spill is responsible. The second step is the responsibility of the cargo owner, which starts where the tanker owner's applicable limit of liability is exceeded. Where there is an incident in which the cargo owner is involved he has no direct liability for paying compensation.

The TOVALOP Standing Agreement applies only to incidents where no liability is imposed under the terms of the Liability Convention or its Protocol. But the TOVALOP Supplement and CRISTAL apply worldwide. They can apply whenever a participating tanker involved in an incident is carrying a cargo owned by a party to CRISTAL. If the tanker owner suffers a spill in a state which is a party to the CLC while carrying a cargo owned by a party to CRISTAL claimants should first bring their claims under the CLC before seeking any additional compensation under the voluntary agreements. Similarly under the TOVALOP Supplement if a tanker owner suffers a spill in a state which is party to the CLC or which has its own applicable domestic legislation the owner, as a matter of priority, is required to pay the necessary amounts under the CLC or domestic legislation to fulfil his legal liabilities. The payment of compensation by a tanker owner to fulfil his legal obligations goes towards meeting a party's financial obligations under the TOVALOP Supplement and is not additional thereto.

Where a tanker carrying a cargo owned by a party to CRISTAL suffers an oil spill in a state which is a party to both the CLC and Fund Conventions, the oil company contributors to the IOPC Fund will be required to contribute to the settlement of valid claims that exceed the tanker owner's legal liability under the CLC. In such a situation the IOPC Fund contributors relieve the tanker owner of the additional financial responsibility which he agreed to under the TOVALOP Supplement. In order to make sure that CRISTAL members in Fund states do not face unreasonable contributions to settlements made under CRISTAL and the Fund Convention, the TOVALOP Supplement and CRISTAL contract provide a reimbursement mechanism. Through the reimbursement mechanism the tanker owner is required to reimburse the parties to CRISTAL that have contributed to the Fund for claims up to the applicable TOVALOP Supplement limit. This mechanism also recognises the fact that contributors to the Fund settlement would otherwise have relieved the tanker owner of his financial responsibility under the TOVALOP Supplement. Such a reimbursement would not be made to non-CRISTAL contributors to the IOPC Fund settlement.

The reimbursement mechanism does not affect the normal functioning of the Liability and Fund Conventions. In other words, it does not affect the claimants in states which are parties to both Conventions. It is designed to ensure that payments made by tanker owners and CRISTAL members are consistent with their respective limits of financial responsibility under the TOVALOP Supplement and CRISTAL contract. By this mechanism claimants in states where the CLC and Fund Conventions apply are not excluded from the improved levels of compensation available under the voluntary agreements.

THE TERMINATION OF TOVALOP AND CRISTAL

With the rapid growth in acceptance of the Civil Liability and Fund Conventions by maritime states and the known entry into force in May 1996 it became clear that both voluntary agreements had served their intended purpose and it no longer made sense to have two sets of regimes working in parallel. In November 1995 the Board of Directors of ITOPF and CRISTAL decided not to renew these Conventions when their present terms expired at 12:00 hours on 20 February 1997.[49] The Board of Directors was also concerned that the continued existence of TOVALOP and CRISTAL could slow progress towards widespread ratification of the Protocols to Civil Liability and Fund Conventions.

With the expiry of voluntary agreements claimants who suffer oil spill damage as a result of a tanker incident will need to rely on the provisions of international compensation schemes. In countries where international compensation conventions are not ratified, tanker owners, their P&I insurers and cargo owners will face unlimited liability. Therefore it is hoped that lack of voluntary agreements will encourage the ratification of the 1992 Protocols. After 20 February 1997 the TOVALOP certificates will disappear. Port authorities, terminal operators and others who in the past have required to see TOVALOP certificates before allowing tankers to load or discharge should therefore have amended their procedures. Where appropriate, they should look instead to the Civil Liability Convention certificates issued by flag states.

Tankers trading to signatory countries will be required to hold a Civil Liability Convention certificate, but it is unclear how port state inspectors from countries that have not ratified the Convention will treat a ship lacking a valid TOVALOP certificate. According to some P&I clubs, ships trading to signatory countries have been asked to carry a valid Civil Liability certificate as well as a P&I club certificate of entry and a letter reminding non-Civil Liability Convention countries that the TOVALOP Agreement is no longer in force.

Tanker owners who participated in TOVALOP gave charterers the right, in certain circumstances, to take measures at an owner's expense in response to a spill or threat of a spill from a tanker. This clause is no longer relevant. The fundamental implication of termination of voluntary agreements is that claimants who suffer oil spill damage as a result of a tanker accident will in the future have to

49. Ian C. White, "Recent Developments in Compensation and Related Pollution Response Issues", in The International Two-Day Forum on Oil Pollution 1997, London, 17–18 February 1997, p.2.

rely on the provisions of the international compensation conventions. Therefore all maritime states at risk from tanker spills should consider ratification of the 1992 Protocols to the Civil Liability and Fund Conventions (the 1992 Conventions), otherwise tanker owners, cargo owners and P&I insurers could face an uncertain liability and compensation situation in the event of a spill in a country that has not ratified any of the international conventions.

TOVALOP has played the role of universal baseline liability and compensation regime for persistent oil discharges. The agreement ensured that shipowners had liability insurance. They could immediately respond to a spill and promptly pay costs and damages. Therefore even in case of a dispute regarding the application of various international or domestic legal regimes TOVALOP acted as a guarantee that a shipowner would respond without delay. Hence, the absence of TOVALOP and CRISTAL may generate some problems in incidents involving a non-party tanker or occurring in a non-party's waters. With TOVALOP and CRISTAL shipowners were contractually obliged to respond and pay claims under the agreements. Without these voluntary agreements choice of law will become a problem again. One may think that some countries, although they are not parties to international conventions, have strong domestic liability and compensation laws which will solve the problem. For example the USA has, under OPA 1990, a strict domestic liability and compensation law which includes financial responsibility requirements. But application of this law is limited. It does not apply to US tankers when they are overseas. Therefore if a US registered tanker has a spill on the high seas or in the waters of a non-party state then the choice of law problem remains the same.

The domestic liability and compensation laws of countries do not really solve the problems that arise in the absence of TOVALOP and CRISTAL. Hence, ratification of international conventions is vital.

Under the arrangements made by the International Tanker Owners Pollution Federation directors and P&I clubs the existing 3,900 tanker owner and bareboat charter members of the Federation will remain members after 20 February 1997. If an owner or charterer notifies the Federation in writing that he wishes to resign then his membership will cease. It has been indicated by the P&I Clubs that in future clubs will enter all their new tanker owners and bareboat charterers into the Federation unless they specifically state that they do not wish to become members of the Federation.

THE 1992 PROTOCOLS TO THE CIVIL LIABILITY AND FUND CONVENTIONS

During the 1984 IMO Conference the Protocols to the 1984 CLC and FC 1984 were structured in such a way that without the participation of the USA their criteria for entry into force would be difficult to meet. Following the *Exxon Valdez* oil spill in Prince William Sound, Alaska, on 24 March 1989 many previous protocol supporters believed that liability for oil pollution should not be limited. The US public also put pressure on the government to make polluters suffer unlimited liability for the damage they caused. In the resulting enactment of its own Oil Pollution Act 1990 the USA preserved the state's prerogatives and rejected the international solutions embodied in the 1984 Protocols to the CLC and FC. The fact that the 1984 Protocols would never enter into force was recognised and a Diplomatic Conference was held in November 1992 under the auspices of IMO. The Conference considered draft protocols containing the substantive provisions of the 1984 Protocols but with lower entry-into-force conditions. The IMO Conference adopted two protocols to amend the 1969 Civil Liability Convention and the 1971 Fund Convention. The Protocols provide higher limits of compensation and a wider scope of application than the Conventions in their original versions. As it had become clear that the 1984 Protocols would not obtain the required number of ratifications to enter into force the 1992 Protocols contained lower entry-into-force provisions. But they contained the same substantive provisions as the two Protocols adopted in 1984.

As a result of the coming into force of the 1992 Protocol, there will now be two funds: the 1971 Fund and the 1992 Fund. It was decided by the 1971 Fund and 1992 Fund Assemblies that the 1971 Fund and 1992 Fund should have a joint Secretariat and that the 1971 Fund Secretariat should, for the time being, also administer the 1992 Fund.

(a) Entry into force

The 1992 Protocols require ratification by four states with not fewer than one million units of gross tanker tonnage.[50] The entry into force requirements for the 1992 Fund Convention Protocol stipulate ratification by states representing together at least 450 million metric tons, which is 132.3 billion gallons. Under the 1984 Protocols the required limit was 600 million metric tons, which was 176.4 billion gallons. The 1992 Protocols entered into force on 30 May 1996.[51]

When the total quantity of contributing oil received in the states parties to the Protocol to the Fund Convention reaches 750 million tonnes the 1969 CLC and the 1971 FC will be denunciated.[52] When the Netherlands deposited an instrument of accession to the 1992 Protocol on 15 November 1996, the requirements for the compulsory denunciation of the 1969 and 1971 Conventions

50. The 1984 Protocols required ratification by 6 states with not fewer than one million units of gross tanker tonnage.

51. See Art. VI. States Party to the 1992 Protocols to the CLC and the FC. (Source: International Tanker Owners Pollution Federation.)

52. In its instrument of ratification Spain made a declaration that the 1992 Protocol to the Fund Convention would not enter into force until 18 months after the total quantity of contributing oil has reached 750 million tonnes. No corresponding declaration was made pursuant to the Protocol to the Civil Liability Convention: see International Oil Pollution Compensation Fund, *Annual Report* (1995) p.27.

were fulfilled. Consequently, the transitional period for the 1992 Conventions will cease on 15 May 1998. During the transitional period, states may belong both to the old régime governed by the 1969 and 1971 Conventions and the new régime governed by the 1992 Protocols to the 1969 and 1971 Conventions. During the transitional period compensation is payable first by the shipowner or his insurer under the 1969 Civil Liability Convention, then by the 1971 Fund, then by the shipowner and his insurer under the 1992 Civil Liability Convention and finally by the 1992 Fund under the 1992 Fund Convention.[53] During the transitional period the differences in the limits of compensation between the two sets of Conventions may cause difficulties in states which are members of both the 1971 Fund and the 1992 Fund. As the scope of application of the 1992 Conventions is wider than that of the 1969 and 1971 Conventions in certain respects complications may arise.

After 16 May 1998 it will not be possible for a state to belong to both regimes. States which have deposited instruments of ratification, acceptance, approval or accession in respect of the 1992 Fund Protocol (whether or not the Protocol is in force for the state in question) are obliged to deposit the instruments of denunciations of the 1969 Civil Liability Convention and the 1971 Fund Convention by 15 May 1997.[54]

In the United Kingdom the Merchant Shipping (Liability and Compensation for Oil Pollution Damage) (Transitional Provisions) Order 1996, which came into force on 30 May 1996, makes transitional provision for the period following entry into force of the 1992 Protocols to the 1969 Civil Liability Convention and the 1971 Fund Convention. The effect of the Order is that if an incident occurs during the transitional period, compensation may, in principle, be available under both the original Conventions and the 1992 Conventions. Compensation could be paid successively from four "distinct" tiers:

(1) The shipowner would initially meet the costs of compensation payments up to the limit of liability determined by the 1969 CLC.

(2) If this limit is not sufficient, the 1971 IOPC Fund would pay additional compensation up to a total amount of 60 million SDRs.

(3) If this amount is not sufficient either, then the shipowner would (unless the flag state is a party to the 1969 CLC only) be liable for the balance under the 1992 CLC. However, the total compensation paid by the shipowner would not exceed the limits of liability under the 1992 CLC. The reason for this restriction is that any compensation paid under the 1969 CLC would be considered to have also been paid under the 1992 CLC.

(4) The 1992 IOPC Fund would only pay compensation if the shipowner had paid compensation up to 1992 CLC liability limits and the 1971 IOPC Fund had also paid the maximum amount of compensation under the 1971 Fund Convention.

53. International Oil Pollution Compensation Funds, *Claims Manual* (5th ed, December 1996), p.13.

54. International Oil Pollution Compensation Fund, *Annual Report* (1996), p.107.

It is also provided under the Order that ships registered in states party to the 1969 CLC need only have insurance up to 1969 CLC limits rather than 1992 CLC limits.

On 30 December 1997 the Merchant Shipping (Liability and Compensation for Oil Pollution Damage) (Transitional Provisions) (Overseas Territories) Order 1997 came into force. This Order gives effect to the rules set out in the 1992 Protocols in the Territories listed in Article 2 of the Order by extending to these Territories the provisions of the Merchant Shipping (Liability and Compensation for Oil Pollution Damage) (Transitional Provisions) Order 1996.

(b) Main amendments

The 1992 CLC Protocol[55] introduces a special liability limit for small vessels and increases the liability of larger vessels substantially. The revised limits will be:

—for small ships under 5,000 units of gross tanker tonnage, 3 million SDR;

—for ships with tonnages between 5,000 and 140,000 units of gross tonnage, 3 million SDR plus 420 SDR for each additional unit of tonnage;

—for ships exceeding 140,000 gross tonnage, 59.7 million SDR.

The 1992 Protocol to the Fund Convention[56] increases the limit of compensation payable by the IOPC Fund to 135 million SDR. This amount includes the compensation payable by the shipowner under the 1992 Protocol to the Civil Liability Convention. If the amount of oil which is received in three of the contracting states whose combined quantity of contributing oil received during a given year in their respective territories exceeds 600 million tonnes then the total compensation payable under both the 1992 CLC and Fund Conventions will increase to 200 million SDR. However, this aggregate amount shall not in any event exceed 59.7 million units of account, as opposed to the 14 million in the 1969 Convention. The unit of account is the SDR.

The 1992 Protocols bring in a simplified procedure for increasing the limitation amounts in the two Conventions. Under the 1992 Fund Convention if the signing parties agree on the need for further increase on the limitation amounts they may increase them. The 1992 IOPC Fund follows the method of the 1984 Protocol and it does not indemnify the shipowner for part of its liability under the CLC.[57]

By the 1992 Protocols, the geographical scope of application of the CLC and the FC[58] is extended to the EEZ established under the United Nations Convention on the Law of the Sea 1982. Under Article II of the CLC 1992 the application of the Convention has been extended to pollution damage caused in the EEZ of a contracting state, or, where none has been established, in the area beyond the

55. See App. VII. International Convention on Civil Liability for Oil Pollution Damage 1992 (CLC 1992).

56. See. App. VIII. International Convention on the Establishment of an International Fund for Compensation for Oil Pollution Damage, 1992 (FC 1992).

57. Under the 1971 Fund Convention the IOPC Fund indemnifies the shipowner for part of his liability under the CLC.

58. The 1969 CLC and the 1972 FC applied only to pollution damage in the territory or territorial sea of a contracting state.

territorial sea determined by the coastal state extending not more than 200 miles from territorial sea baselines.

The coverage provided under the 1992 Protocols has been extended to pollution damage caused by spills of persistent oil from unladen tankers. Therefore a spill from a tanker occurring during any voyage following a laden voyage will be covered under the 1992 CLC unless it is proved that the tanker had no residues of the bulk cargo from the previous voyage on board at the time of the spill. Hence, 1992 CLC can apply where there is a spillage of bunker oil even though the ship is proceeding in ballast. The definition of oil excludes non-persistent products as well as vegetable and animal oils and fats.[59] As long as there is grave and imminent danger of pollution damage the preventive measures are covered under the 1992 Protocols even when no spill of oil occurs.

Pollution damage means loss or damage caused outside the ship by contamination resulting from the escape or discharge of oil from the ship, wherever such escape or discharge may occur, provided that compensation for the impairment of the environment, other than loss of profit from such impairment shall be limited to costs of reasonable measures of reinstatement actually undertaken or to be undertaken and/or the costs of preventive measures, and further loss or damage caused by preventive measures. The new definition of pollution damage under the 1992 Protocols retains the basic wording of the present definition. In the concept of pollution damage the liability of the shipowner is limited as to compensation for impairment of the environment to the cost of reasonable measures or reinstatement actually undertaken or to be undertaken. The new definition of pollution damage removes any implications of contingent value of the affected environment. "Incident" is amended to mean not only an occurrence actually causing pollution but also a grave and imminent threat of causing such damage.

The 1992 Protocol to the Fund Convention sets a cap on contributions to the IOPC Fund payable by oil receivers in any given state. This cap was fixed by the Conference at 27.5 per cent of the total annual contributions to the IOPC Fund.[60] The capping system will cease to apply when the total quantity of contributing oil received during a calendar year in all member states of the new Fund set up under the 1992 Protocol exceeds 750 million tonnes, or on the expiry of a period of five years from the entry into force of the 1992 Protocols to the Fund Convention (i.e. 30 May 2001), whichever is the earlier. The 1992 Fund Convention will enter into force in Ireland, South Korea and Spain on 16 May 1998. With these countries' contributions to the IOPC Fund, the total quantity of contributing oil will exceed 750 million tonnes and the capping of contributions to the 1992 Fund will cease to apply in 1998 rather than 2001.

59. Under the 1992 Protocols oil is defined as any persistent hydrocarbon mineral oil such as crude oil, fuel oil, heavy diesel oil and lubricating oil, whether carried on board a ship as cargo or in the bunkers of such a ship.

60. This issue was raised by the Japanese delegation which drew attention to the fact that Japanese contributors' share of the annual contribution to the IOPC Fund in 1980 was 44% and in 1990, 27%. It would be difficult for the Japanese Government to ratify the 1984 or 1992 Protocol to the Fund Convention.

The 1992 Protocol offers wider scope of protection. Under the 1969 CLC only servants or agents of the shipowner were protected. Under the 1992 Protocol no claim for compensation for pollution damage under the Convention or otherwise may be made against the servants or agents of the owner, pilots or other persons performing services for the ship, charterers (including bareboat charterers), salvors, or any person taking preventive measures, or any claim for compensation for pollution damage be made against any servants or agents of a charterer, salvor or any person taking preventive measures. The 1992 Protocol to the CLC channels more of the liability to the shipowner by exempting other groups from this exposure. By widening the categories of exempt persons it becomes clear that claims for pollution damage can be brought against the shipowner alone.

The registered owner has a right to limit liability both under the 1969 CLC and the 1992 Protocol. The limit in respect of any given ship is much higher under the 1992 Protocol. The maximum amount under the 1992 Protocol is SDR 59.7 million as opposed to SDR 14 million under the 1969 CLC. Under the 1969 CLC it is possible for claimants to break the right to limit if the registered owner has been guilty of "actual fault or privity". This is the test which was applied under the 1957 Limitation Convention. Under the 1992 Protocol in order to break the right to limit the claimant must prove that the pollution damage was caused by the personal act or omission of the registered owner, committed with the intent to cause pollution damage, or recklessly and with knowledge that such damage would probably result. Therefore under the 1992 CLC Protocol it will be rare that a claimant will be able to break the right to limit.

THE MERCHANT SHIPPING ACT 1995 AND THE MERCHANT SHIPPING AND MARITIME SECURITY ACT 1997

In England the 1969 CLC was ratified and brought into effect by the Merchant Shipping (Oil Pollution Act) 1971. The 1971 Act was supplemented by the Merchant Shipping Act 1974. The 1974 Act ratified and implemented into English law the 1971 FC. In general the system of compensation for oil pollution damage from ships is set out in the Merchant Shipping Acts 1971 and 1974. On the basis of the system provided under the 1969 and 1971 Conventions the 1974 Merchant Shipping Act established a compensation fund contributed to by oil cargo owners and divided the responsibility for oil pollution damage between shipowners and cargo owners. Certain sums which were not recoverable under the 1971 Act became recoverable under the 1974 Act. Some changes to the method of calculating the limits of liability under the 1969 and 1971 Conventions were made by Protocols in 1976. On August 24 1994 Japan deposited its instrument of ratification and the 1976 Protocol entered into force on 22 November 1994. The Oil Pollution Act 1971 was amended by the Merchant Shipping Acts 1979 and 1984 in order to include modifications to the Civil Liability Convention.

Major changes to the CLC and the FC were agreed under the 1984 Protocols and the Merchant Shipping Act 1988, section 34 and Schedule 4, prospectively amended the Merchant Shipping Act 1971 to take account of the changes. The 1984 Protocols themselves allowed reference to the amended Conventions to be by their 1984 dates rather than their 1969 or 1971 dates. Therefore section 34(1)(a) and (b) of the Merchant Shipping Act 1988 referred to the Civil Liability Convention 1984 as the 1984 Liability Convention and to the Fund Convention 1984 as the 1984 Fund Convention. As a result of the United States' decision not to ratify the 1984 Protocols but to implement its own pollution legislation the Protocols have not entered into force internationally. Consequently, the prospective changes made to the 1971 Act by the Merchant Shipping Act 1988 were formally repealed by the Merchant Shipping (Registration) Act 1993, Schedule 4, paragraph 18.

In 1992 an international compromise produced two more Protocols to the 1969 CLC and the 1971 FC. On 29 September 1994 the UK deposited its instrument of ratification to the 1992 Protocols. Section 5 of the Merchant Shipping (Salvage and Pollution) Act 1994 conferred a power on the government to implement the 1984 Conventions by Order in Council following UK ratification. Following the exercise of implementing power under section 5 of the Merchant Shipping Act 1994 consequential amendments to the Merchant Shipping Act 1971 were enacted ready to be brought into force too. Section 6 of the 1994 Act extended strict liability to oil pollution emanating from vessels other than oil tankers. Section 7 modified the operation of subrogation rights acquired by the Fund following payment by it.

All the above legislation was repealed and consolidated by the Merchant Shipping Act 1995.[61] The Act came into force on 1 January 1996. However, the most recent development in UK oil pollution legislation is the Merchant Shipping and Maritime Security Act 1997 which amends the Merchant Shipping Act 1995. The Act makes changes to merchant shipping and maritime security legislation. Under the new Act the Secretary of State has more power to intervene in relation to ships which may cause oil pollution in UK waters and to cover cases of pollution by substances other than oil.

Part VI of the 1995 Act deals with prevention of pollution and Chapter I is on pollution generally. Section 128 provides the government with the power to give effect to the various conventions, protocols and international agreements mentioned in section 128(1). Section 5 of the 1997 Act inserts a new chapter into Part VI of the 1995 Act, following Chapter I. The inserted Chapter IA is on waste reception facilities at harbours. Under this chapter it may be possible for a harbour authority to impose charges by levying ship, passenger and goods dues, for the purpose of recovering the whole or a part of the costs of the provision by or on behalf of the authority of waste reception facilities at the harbour.[62] If a person

61. For detailed information on the Merchant Shipping Act 1995, Chapters III and IV, see, Nevil Phillips, *Merchant Shipping Act 1995. An Annotated Guide* (1996), pp.116–139.
62. Merchant Shipping and Maritime Security Act 1997, Chapter IA, s.130(C)(1).

contravenes a requirement under the regulations he is guilty of an offence. Under section 6 of the 1997 Act the Secretary of State is given the right of indemnification in connection with counter-pollution measures: "Where under Section 293(1) of the 1995 Act the Secretary of State agrees that another person shall take any measures to prevent, reduce or minimise the effects of marine pollution, he may agree to indemnify that other person in respect of liabilities incurred by that person in connection with the taking of measures."[63]

Part VI, Chapter II, of the 1995 Act is on oil pollution. This chapter mainly restates or re-enacts the Prevention of Oil Pollution Act 1971. Section 7 of the 1997 Act increases the penalty provided for under Chapter II, section 131, from £50,000 to £250,000. Similarly section 7 of the 1997 Act increases the fine on summary conviction of an offence of discharging oil into certain United Kingdom waters from £50,000 to £250,000. The security required for release of a ship in a case where the same offence is committed has been raised from £55,000 to £255,000.

Part VI, Chapter III, of the 1995 Act deals with liability for oil pollution. In this Chapter the Liability Convention means the International Convention on Civil Liability for Oil Pollution Damage 1992; a Liability Convention country means a country in respect of which the Liability Convention is in force and the Liability Convention state means a state which is a party to the Convention.[64] In this Chapter section 154 deals with liability for oil pollution in the case of other ships and extends oil pollution liability to bunker oil from tankers. Limitation of liability was covered in section 157. In this section limitation is possible provided that the person seeking to claim against the party seeking to limit is unable to show that the damage resulted from a personal act or omission which was committed with intent to cause such damage and with knowledge that any such damage would probably result.[65] Personal acts or omissions were dealt with by the section in a strict way. Therefore the actions of a ship's manager or charterer will not affect the right of the owner to limit. But if the owner delegates responsibility to another that delegation does not absolve him of the duty to ensure that a safe system of work is employed on board.[66]

After section 100B of the 1995 Act, which is inserted by section 1 of the 1997 Act, section 100C, on power to require ships to be moved, was inserted. The powers conferred by section 100C shall be exercisable where a ship in United Kingdom waters is not a qualifying foreign ship or is such a ship but appears to the Secretary of State to be exercising neither the right of innocent passage nor the right of transit passage through straits used for international navigation. Section 100D sets out offences in relation to section 100C.

Part VI, Chapter IV, of the 1995 Act covers the International Oil Pollution

63. Merchant Shipping and Maritime Security Act 1997, s.6.

64. Merchant Shipping Act 1995, s.152. This section restates s. 19 of the Merchant Shipping (Oil Pollution) Act 1971 as amended.

65. Nevil Phillips, p.121.

66. *Ibid.*

Compensation Fund. In this Chapter the Liability Convention has the same meaning as in Chapter III. The CLC 1992, and the Fund Convention means the International Convention on the Establishment of an International Fund for Compensation for Oil Pollution Damage 1992. The Fund means the International Fund established by the Fund Convention and a Fund Convention country means a country in respect of which the Fund Convention is in force.[67] Section 175 provides compensation from the Fund where the liability of or amount recovered from a shipowner under the 1969 Civil Liability Convention and sections 153 and 154 of the 1995 Act is insufficient to meet the damage which has been suffered. Section 179 deals with subrogation. This section re-enacts section 8 of the 1974 Merchant Shipping Act as amended by the 1994 Merchant Shipping Act. The 1974 Act did not give a right of recourse where a Fund paid a claim before the shipowner had been found liable and limited its liability. The wording of section 8(3) only gave a right against third parties where the shipowner had paid the appropriate sum. By the 1994 Act section 8 of the Merchant Shipping Act 1974 is amended to allow the Fund to acquire by subrogation rights which the victim has. Under the new wording, if the IOPC Fund settles the claim including those parts for which the shipowner would be liable under the 1969 CLC, the Fund is entitled to replace that first tier of liability from the shipowner by subrogation.[68]

After Part VI, Chapter V of the 1995 Act, a new chapter on carriage of hazardous and noxious substances has been inserted. In this Chapter the Convention means the International Convention on Liability and Compensation for Damage in Connection with the Carriage of Hazardous and Noxious Substances by Sea 1996 (HNS Convention). This section gives the government the power to give effect to the HNS Convention.

Part VII of the 1995 Act deals with liability of shipowners and others. Section 16 of the 1997 Act inserts a section after Part VII, section 192, of the 1995 Act. This new section covers regulations requiring insurance or security. Section 16 of the 1997 Act gives the Secretary of State the power to make regulations concerning compulsory insurance and security. These regulations shall not apply in relation to a qualifying foreign ship while it is exercising the right of innocent passage or the right of transit passage through straits used for international navigation, any warship or any ship for the time being used by the government of any state for other than commercial purposes.

The Oil Pollution (Compulsory Insurance) Regulations 1997 came into force on 1 September 1997. They revoke and replace the Oil Pollution (Compulsory Insurance) Regulations 1981, consequentially on the United Kingdom becoming a party to the International Convention on Civil Liability for Oil Pollution Damage 1992.

The Regulations define "oil" for the purposes of section 163 of the Merchant

67. Merchant Shipping Act 1995, s.172. This section substantially re-enacts s.1 of the Merchant Shipping Act 1974.

68. Current Law Statutes, Merchant Shipping (Salvage and Pollution) Act 1994, Chapter 28, p.28–24.

Shipping Act 1995 and provide for the cancellation of certificates issued by the Secretary of State under section 164 and add to the Merchant Shipping (Fees) Regulations 1996 the fee for an application for and issue of such a certificate.

THE IOPC FUND AND THE ADMISSIBILITY OF CLAIMS

(a) In general

The International Oil Pollution Compensation Fund (IOPC Fund) was set up in October 1978. The Fund is a worldwide intergovernmental organisation and it provides compensation for those suffering pollution damage. The basic function of the Fund is to provide supplementary compensation to oil pollution victims in Fund member states who cannot obtain full compensation for damage under the Civil Liability Convention. The Fund operates within the framework of the 1969 Civil Liability Convention and the 1971 Fund Convention. The aim of the Fund[69] is to settle claims out of court to provide compensation for the claimants as soon as possible. But the claimants also have the right to take their claims to the competent national court.

The Fund is governed by an Assembly consisting of representatives of the governments of all member states. The Assembly elects an Executive Committee of 15 member states. The Director of the IOPC Fund has the authority to make final settlement of claims within certain limits. If those limits are exceeded then the Director has to submit the claim settlements to the IOPC Fund's Executive Committee for decision. Therefore the main function of the Committee is to approve settlements of claims for compensation in cases where the Director of the IOPC Fund is not authorised to make such settlements. The Director heads a Secretariat, which administers the IOPC Fund and deals with claims for compensation.

Anyone who has suffered pollution damage in a Fund member state may make a claim against the Fund. Claimants should bring their action against the Fund within three years of the date on which the damage occurred or make a formal notification to the Fund of a court action against the shipowner or his insurer within that three-year period. In cases where the damage occurs after an incident takes place the action must be brought within six years of the date of the incident.

(b) Financing the IOPC Fund

The IOPC Fund is financed by persons who have in the relevant calendar year received more than 150,000 tonnes of crude oil or heavy fuel oil in ports or terminal installations in an IOPC Fund member state by sea transport. Every year a member state is required to submit a report to the Director of the IOPC Fund

69. For further information about the IOPC Fund see International Oil Pollution Compensation Fund, *Annual Report 1995 and Annual Report 1996*; International Oil Pollution Compensation Fund, *Claims Manual* (5th ed., December 1996); International Oil Pollution Compensation Fund, *General Information on Liability and Compensation for Oil Pollution Damage* (July 1995).

including the name and address and quantity of oil received by any person who is liable to contribute to the Fund. The levy of contributions is based on these reports.

The individual contributors have to make their contributions direct to the Fund. The only exception to direct payment is where the governments voluntarily accept such responsibility. There are two types of contributions. An initial contribution is payable when a state becomes a member of the IOPC Fund. This contribution is a fixed amount per tonne of contributing oil received during the past year in which the Fund Convention entered into force for the state in question. Annual contributions are made to meet the anticipated payments of compensation and indemnification by the IOPC Fund and the estimated administrative expenses of the Fund during the following year.[70] The 1971 Fund has initial and annual contributions, the 1992 Fund has only annual contributions.

As the payments made by the Fund in respect of claims for compensation for oil pollution damage change from year to year the level of annual contributions to the Fund changes from year to year, as illustrated in Table 3.

Table 3. 1996 Annual Contributions to the General Fund[71]

Year	General Fund £	Major Claims Fund £	Total Levy £
1979	750,000	0	750,000
1980	800,000	9,200,000	10,000,000
1981	500,000	0	500,000
1982	600,000	260,000	860,000
1983	1,000,000	23,106,000	24,106,000
1984	0	0	0
1985	1,500,000	0	1,500,000
1986	1,800,000	0	1,800,000
1987	800,000	400,000	1,200,000
1988	2,900,000	90,000	2,990,000
1989	1,600,000	3,200,000	4,800,000
1990	500,000	0	500,000
1991	5,000,000	21,700,000	26,700,000
1992	0	10,950,000	10,950,000
1993	8,000,000	70,000,000	78,000,000
1994	6,000,000	34,000,000	40,000,000
1995	6,000,000	37,000,000	43,000,000

(c) Admissibility of claims

In general the IOPC Fund is able to pay compensation only in respect of claims which fulfil the criteria for admissibility for compensation under the Civil Liability

70. The IOPC Fund, *Annual Report 1996*, p.27.
71. *Ibid.*, p.31.

and the Fund Conventions. The Fund can only accept claims which fall within the definitions of pollution damage and preventive measures set out under these Conventions. In order to benefit from the compensation system which is provided under the Conventions a uniform interpretation of the definitions are necessary.

(i) Definition of pollution damage

The 1969 CLC and the 1971 FC provide a common definition of pollution damage but leave its interpretation to the *lex fori*. By Article I.6, of the 1969 Convention "pollution damage" means loss or damage caused outside the ship carrying oil by contamination resulting from the escape or discharge of oil from the ship, wherever such escape or discharge may occur, and includes the cost of preventive measures and further loss or damage caused by the preventive measures.

The definition in Article I.6 refers to the concept of "preventive measures" defined as follows (Article I.7): "preventive measures" means any reasonable measures taken by any person after an incident has occurred to prevent or minimise pollution damage. The definition of preventive measures refers to an incident which is defined as "any occurrence, or series of occurrences having the same origin, which causes pollution damage". Under the Fund Convention 1971 the definitions of pollution damage, preventive measures and incident are included in Article 1(2) by reference to the 1969 CLC.

By the definition provided in the 1969 CLC the wording of pollution damage seems to cover damage to property and consequential losses. Although they are very rare in the pollution context loss of life and personal injury are also covered. Damage caused by preventive measures and clean-up operations is also within the scope of the definition. But the wording of the 1969 CLC does not provide any guidance on whether pure economic losses are covered.

Damage to property does not normally give rise to any questions of principle in respect of admissibility of claims. All legal systems recognise compensation for reduction in value or for the necessary cost of repair. In general persons whose property has been contaminated or damaged may suffer loss as a consequence. For instance, a fisherman may not be able to use his boat for some time as it has been polluted by oil and has to be cleaned. In general consequential loss claims do not cause fundamental legal problems because nearly all legal systems seem to accept in principle that compensation should be granted for such losses. In other words, as the claimant has at the same time suffered damage to property most legal systems recognise in principle claims for compensation of this kind. But when the problem is about pure economic loss recovery the issue becomes more complex. In some cases persons may suffer economic loss as a result of oil pollution without any damage being caused to their property. For instance, if a fishing ban needs to be imposed and fishermen suffer loss of income as a result, or if a substantial spill outside a commercial port and fishing area may also pollute an adjacent tourist area or the pollution may affect hoteliers in the vicinity of the affected tourist resorts. Obviously in such a case the hoteliers have no proprietary interest in the affected coastline. As a result of a pollution incident tourists may refrain from going to an area where the coast has been polluted and as a result the hotel, restaurant and shop owners in the area may suffer economically. In most cases the courts have taken a restrictive attitude to claims for pure economic loss because it is difficult to find a criterion for the admissibility of such claims. In most legal systems a claim for compensation is generally accepted only if it arises out of damage to a defined and recognised right such as a right of property or a right of possession. Therefore if damage is suffered by someone as a result of loss of use of the environment due to pollution that damage is normally not considered to be damage to an individual's recognised right within the context of recognised pollution damage.[72] Apart from in common law system countries the admissibility of pure economic loss claims are not very clear. The criteria of foreseeability and remoteness or the requirement that the economic loss should be a direct result of the defendant's action have been applied in some countries. In some other countries there is a requirement for a direct link of causation between the damage and the defendant's action and the damage must be certain and quantifiable in monetary terms.[73]

As stated above the problem of pure economic loss is not dealt with in the definition of pollution damage in the 1969 Civil Liability Convention. Thus, the question whether and to what extent pure economic loss is recoverable has to be decided by the national courts of the contracting states. They will decide the issue on the basis of their national law and by interpreting the definition of pollution damage in the Civil Liability Convention. The definition of pollution damage in the 1984 Protocol gave the Conference great difficulty. It was unable to agree on it, so it set up a working group to elaborate a new definition. In the working group the majority were agreed that the new definition should cover property damage and personal injury, loss of use or exploitation by people who normally use or exploit the polluted environment, costs of reinstatement of the environment and cost of preventive measures. At the same time there was an urge to disallow speculative or theoretical damages. As a result the Conference adopted the following definition: "pollution damage" means:

(a) loss or damage caused outside the ship carrying oil by contamination resulting from escape or discharge of oil from the ship, wherever such escape or discharge may occur, provided that compensation for impairment of the environment other than loss of profit from such impairment shall be limited to costs of reasonable measures of reinstatement actually undertaken or to be undertaken;

(b) the costs of preventive measures and further loss or damage caused by preventive measures.

The concept of pollution damage under the 1984 CLC included:

—loss of life, personal injury or damage to property of the environment;

72. FUND/WGR.7/2, p.8.
73. *Ibid.*, p.9.

—loss of profit or income; and
—costs of reinstatement or preventive measures.

Within the context of the pollution damage definition the main discussion was how to limit compensation for impairment of the marine environment.[74] In order to benefit from recovery the damage involved in the definition should be actual and not hypothetical or speculative. With regard to the cost of reinstatement the principle established by international practice is that compensation must, as far as possible, clear all the consequences of the illegal act and re-establish the situation which would have existed if that act had not been committed.[75] In practice the costs of preventive measures has always been the most important kind of damage in oil pollution cases. That is why the costs of preventive measures are the only defined element of compensatable damage under the 1969 CLC.

The new definition of pollution damage under the 1984 Protocol to the Civil Liability Convention clarifies the legal situation with respect to pure economic loss. It is possible to recover for loss of profit from impairment of the environment provided that one can prove that one has suffered loss of profit as a result of such impairment. The 1984 Protocol definition follows the IOPC Fund's approach under which shopkeepers, restaurateurs and hoteliers who obtain their income from tourists at seaside resorts are able to recover loss of income caused by the decrease in tourism as a result of oil pollution. Fishermen are also able to obtain compensation for loss of income actually suffered as a result of the impairment of the environment. But even under the definition provided by the 1984 Protocols it is not clear how far the right to compensation extends in the above-mentioned cases. So the solution of the problem is left to national courts which will have to find and develop a limit for the right of compensation for loss of profit.

Under the 1992 Protocols the revised definition of pollution damage means: "loss or damage caused outside the ship by contamination resulting from the escape or discharge of oil from the ship, wherever such escape or discharge may occur, provided that compensation for impairment of the environment other than loss of profit from such impairment shall be limited to costs of reasonable measures of reinstatement actually undertaken or to be undertaken; the cost of preventive measures and further loss or damage caused by preventive measures". Under this definition loss of profit may be recovered when it results from impairment of the environment. Under the 1992 Protocols the pollution damage must be caused by contamination resulting from the escape of oil. The definition of pollution damage in the 1969 CLC does not expressly require loss or damage to be directly caused by contamination. Under the industry schemes, TOVALOP and CRISTAL, the corresponding definition required economic loss to be actually sustained as a direct result of contamination. In civil jurisdictions directness of causation has been used for controlling remoter claims. At the 1984 Diplomatic Conference the

74. For the discussion on the definition of pollution damage under the 1969 CLC see Z. Brodecki, "New definition of pollution damage" [1985] L.M.C.L.Q. 382–391.

75. *Chorzow Factory* case, *Germany* v. *Poland* (1928) P.C.I.J. Reports, Series A No. 17, pp.47–48, cited in *ibid.*, p.389, n.29.

Legal Committee proposed the phrase "sustained as a direct result of contamination". One of the delegates of the CMI stressed that this phrase plays a very important role in the definition of pollution damage. It was also mentioned that the phrase could be important within the context of theories of causality, which do not use the test of direct consequences or a similar test.[76] Due to disagreement among the delegates over its true effect the phrase was omitted from the 1984 Protocol to the Civil Liability Convention. Therefore it is not very clear whether the concept of directness will on its own be enough to provide certainty on the issue. As with the 1984 Protocols the 1992 Protocols leave the issue to the national courts.[77]

(ii) Damage to the marine environment

Oil pollution incidents may also cause damage to the marine environment. As that does not have a direct market value it is very difficult to assess this kind of damage. In a claim for damage to the marine environment there will be technical, scientific and legal problems. It is very difficult to obtain detailed knowledge of the resource before the spill. Public authorities need to take measures to restore the environment as nearly as possible to its pre-spill condition. It will be very expensive to take such measures.

The definition of pollution damage under the 1969 CLC has no specific provision concerning damage to the natural resources of the marine environment. The definition also does not deal with the question of costs of restoration of the marine environment. Under the definition adopted in the 1984 Protocol the costs incurred in restoring the environment after a pollution incident are recoverable. In other words, compensation for restoration of the marine environment is expressly limited to costs of reasonable measures of reinstatement actually undertaken or to be undertaken.

But the definition makes it clear that non-economic damage to the marine environment is not to be compensated. Compensation based on a theoretical calculation of damage caused by oil to the environment without actual proof of the cost of reinstatement is excluded. With this exclusion the definition comes into line with the resolution adopted in 1980 by the IOPC Fund's Assembly which states that: "the assessment of compensation to be paid by the International Oil Pollution Compensation Fund is not to be made on the basis of an abstract quantification of damage calculated with theoretical models".[78] This approach was reinforced in the definition of pollution damage adopted under the 1992 Protocols, which contains the provision that "compensation for impairment of the environment ... shall be limited to costs of reasonable measures of reinstatement actually undertaken or to be undertaken ...".

76. Z. Brodecki, p.388.

77. For detailed information on economic loss and environmental damage see Colin M. De La Rue, "Oil Pollution From Ships, Problems of Economic Loss And Environment Damage" in Conference, Managing Risks in Shipping Operations. Managing Spill Liabilities: Claims & Compensation, 15 December 1995, pp.1–22.

78. IOPC Fund Resolution No. 3, FUND/A/ES.1/13, para. 11(a).

Under the 1984 and the 1992 Protocols only costs of reasonable measures are recoverable. The costs have to be reasonable from an objective point of view. They must be proportionate to the results achieved or to the results that may reasonably be expected. Obviously the pollution damage definition in the CLC has clarified certain issues regarding compensation for pure economic loss and the costs of reinstatement of the environment. But there are still certain problems which need to be resolved for instance.[79] The quantification of damages for the purpose of liability for loss or destruction of natural resources is one of the problems. If there is a need to place a monetary value on natural objects how are these natural objects to be measured? The 1984 and 1992 Protocols to the CLC do not specify who is entitled to claim the costs incurred in restoring the environment after the pollution incident. The claim may be filed by private citizens or the authorities. The issue has to be resolved by the national law and will lead to different solutions.[80]

(iii) P&I clubs and admissibility of claims

As P&I clubs provide the insurance for shipowners under the Civil Liability Convention there has always been a continuing link between the clubs and the IOPC Fund. Traditionally P&I clubs in the International Group accepted reasonableness as a key element. Claims for compensation for pollution damage following an incident to which the CLC applies may be settled and paid subject to the limits and other applicable terms of the Convention, provided always that those claims are reasonable.[81] The P&I clubs support the view of the IOPC Fund that a uniform interpretation of the definition of "pollution damage" is essential for the proper functioning of the regime of compensation established under the CLC and FC. The clubs agree that losses which flow generally from an incident, as opposed to contamination resulting from the escape or discharge of oil, do not fall within the definition of "pollution damage". The clubs also recognise the need to draw a line between those claims for pure economic loss which are recoverable and those which are not admissible. But they urge a cautious and restrictive approach to the criteria for "preventive measures" when considering economic loss. The clubs support the decisions of the Executive Committee in relation to environmental damage but emphasise that an environmental damage claim is not payable when the claim is not quantifiable by reference to an actual economic loss.

(d) Settlement of claims in the IOPC Fund

(i) General criteria

The IOPC Fund's policy on the admissibility of claims for compensation has been established by the governments of Fund member states. As each has its own

79. See Peter Wetterstein, "Trends in Maritime Environmental Impairment Liability" [1994] L.M.C.L.Q. 237–240.
80. *Ibid.*, p.240.
81. FUND/WGR.7/18, p.1.

characteristics, claims are considered in the light of the particular circumstances of the case. Therefore the criteria adopted by the IOPC Fund allow a certain degree of flexibility.[82] In general a claim for compensation can be accepted by the IOPC Fund only to the extent that it meets the criteria laid down in the Civil Liability and the Fund Conventions. Although the Fund prefers to proceed on a case-by-case basis different types of claim, specifically under the *Haven*, *Aegean Sea* and *Braer* casualties, showed the necessity for a clearer criterion in deciding which claims are to be accepted as eligible for compensation. In 1993 an Intersessional Working Group was set up by a mandate given by the Assembly to examine the general criteria for the admissibility of claims for compensation for "pollution damage" and for "preventive measures" within the scope of the 1969 CLC and the 1971 FC and the 1992 Protocols thereto. The Working Group would also work on claims relating to "pure economic loss" and "preventive measures".[83] During the examination of certain general issues relating to the legal regime of compensation established under the Conventions it was stated that in considering the subsequent practice of the parties regarding the interpretation and application of the provisions of the 1969 CLC relating to the admissibility of claims, reference might be made to the decisions of the courts and other competent authorities in the states parties, as well as the relevant decisions of the appropriate IOPC Fund bodies.[84] During the Working Group's research it was again stated that a uniform interpretation of the definition of pollution damage is essential for the proper functioning of the regime of compensation established by the Civil Liability Convention and the Fund Convention.[85] The definition of pollution damage in the Civil Liability and the Fund Conventions generally states that pollution damage is damage caused by contamination. As the definition is not entirely clear the IOPC Fund has developed certain principles with regard to the meaning of this definition.

Since the first cases involving the organisation the IOPC Fund applied the following general criteria for the admissibility of claims:

—any expense/loss must actually have been incurred;

—there must be a link between any expense/loss and the incident;

—any expense must relate to measures which were deemed reasonable and justifiable;

—a claimant's loss or damage is admissible only if and to the extent that it could be considered to have been caused by contamination;

—there must be a link of causation between the loss or damage covered by the claim and the contamination caused by the spill;

—a claimant is entitled to compensation only if he has suffered a quantifiable economic loss;

82. IOPC Fund, *General Information on Liability and Compensation for Oil Pollution Damage*, pp.11–12.
83. FUND/WGR.7/1.
84. FUND/WGR.7/2, p.6.
85. FUND/A.11/20, para. 5.5.

—the claimant has to prove the quantum of his loss or damage.

If one examines the past practice of the IOPC Fund in respect of the admissibility of claims[86] it becomes clear that it has acquired considerable experience with regard to the admissibility of claims for compensation. The Assembly and the Executive Committee have also taken important decisions.

(ii) Clean-up operations and property damage

In cases where measures taken to combat an oil spill cause damage to roads, piers and embankments the cost of the resulting necessary repairs is admissible. In *The Tanio* road repair costs were accepted to the extent that the roads in question had actually been affected. Reasonable costs of disposing of the oil and oil debris collected during clean-up operations are also admissible. Claims for the cost of cleaning or repairing property which has been contaminated by oil are accepted by the IOPC Fund.[87] If it is not possible to clean or repair the property then replacement costs are accepted with a reduction for wear and tear. Claims for damage to property without escape of oil into the sea were examined in the *Tsubame Maru No. 58* and *Tsubame Maru No. 16* cases. In these cases oil escaped from a ship and caused damage without any oil being spilled into the sea. The Executive Committee considered that although no oil had escaped into the sea, the damage should nevertheless be covered by the definition of pollution damage.[88]

Apart from claims relating to damage to property the Fund also dealt with personal injury claims.[89] With regard to personal injury claims, although the Committee did not take a decision on them it took note of the Director's view that, in the light of the discussions at the 1969 Conference, CLC claims for personal injury were in principle admissible, provided that the injury was caused by contamination. In such a personal injury claim it is the defendant who has to prove that the alleged damage was caused by contamination and he has to prove the amount of the loss or damage sustained.

(iii) Consequential loss and pure economic loss

In cases of claims for consequential losses the Fund takes a positive view and accepts claims for losses incurred by the owners or users of property which has

86. The IOPC Fund's past practice in respect of the admissibility of claims was reviewed in document FUND/WGR.7/3.

87. In *Tsunehisa Maru No. 8* claim for damage to the unfinished paintwork of a vessel under construction in a contaminated dock is accepted. In *Showa Maru* claims for cleaning or replacing seaweed farm nets, in *Vistabella* claims for cleaning yachts, in *Qued Gueterini* claims for cleaning the water intakes of power stations and in the *Aegean Sea* and *Braer* claims for the cleaning of buildings affected by dispersed droplets of oil in wind-blown spray were accepted by the IOPC Fund. See FUND/WGR.7/3, p.4.

88. *Ibid.*

89. See *The Braer*, UK 1993.

been contaminated as a result of an oil spill.[90] An important group of claims are those relating to pure economic loss where loss of income is suffered by persons whose losses do not derive from the fact that they are the owners or users of contaminated property. Claims for pure economic loss are admissible only if they are for loss or damage caused by contamination. Therefore in such claims the starting point is the pollution, not the incident itself. To qualify for compensation for pure economic loss the loss must have been suffered by persons who depend directly on earnings from coastal state or sea-related activities, such as hoteliers at seaside resorts or fishermen. And there must be a reasonable degree of proximity between the contamination and the loss or damage sustained by the claimant. While considering whether the criterion of reasonable proximity is fulfilled the Fund takes into account certain elements such as[91]:

—the geographic proximity between the claimant's activity and the contamination;

—the degree to which a claimant was economically dependent on an affected resource;

—the extent to which a claimant had alternative resources of supply or business opportunities;

—the extent to which a claimant's business formed an integral part of the economic activity within the area affected by the spill.

On an examination of the IOPC Fund's case practice it becomes clear that the establishment of the amount of the loss sustained in related cases is not very easy. The assessment is not based on budgeted figures. It is based on the actual financial results of the individual claimant for appropriate periods in the years before the incident. In some cases it may also be necessary to make a comparison with the claimant's income for an appropriate period after the incident. The Fund takes into account the particular circumstances of the claimant and considers any evidence presented.[92]

The IOPC Fund also dealt with financial and banking difficulties encountered by claimants. In the *Tanio* and *Braer* cases claims submitted for bank charges and similar costs incurred as a result of overdrafts and loans which became necessary as a result of loss of income following the incident were accepted as valid claims for compensation. But claims for loans to resolve cash flow difficulties resulting from the incident were rejected by the IOPC Fund as it is not supposed to take on the role of claimant's banker and it cannot provide funds to resolve cash flow problems.[93]

In some incidents such as the *Aegean Sea* claims for loss of income were submitted by near-shore cultivators of mussels, salmon, oysters and scallops. The

90. In *The Volgoneft 263* the claim of a fisherman for loss of income as a result of his nets being contaminated was accepted. Similarly in *The Tanio* a hotel/restaurant owner's claim for two days' lost income while the establishment was closed during the removal of oil from the approach road was accepted as consequential loss claim. FUND/WGR.7/3, p.5.

91. The IOPC Fund, *Claims Manual*, p.27.

92. *Ibid.*, p.28.

93. FUND/WGR.7/3, p.6.

Executive Committee accepted these on the basis that these activities were carried out within the area contaminated by oil. After the *Braer* incident oil contaminated several salmon farms and an exclusion zone was established by the United Kingdom government under the Food and Environment Protection Act 1985. Salmon farmers' claims for compensation for loss resulting from the destruction of their fish were accepted as losses relating to damage to property.[94] In the *Haven* incident claims for loss of income were submitted by the operators of beach facilities located along the stretch of coast affected by the incident. On the Executive Committee's decision the claimant's loss of income as a result of the reduction in tourism was considered as damage caused by contamination. While taking this decision the Committee considered the extent to which the reduction had been caused by the *Haven* incident and that the operators had suffered an infringement of their recognised legal right, which was to operate the facility on the beach.[95]

Claims relating to preventive measures are admissible to the extent that the operations are reasonable and the amount claimed is reasonable. Following the *Patmos* case the Executive Committee considered the relationship between salvage operations and preventive measures as defined in the Civil Liability Convention. The Committee took the view that operations could be considered as falling within the definition of preventive measures only if the primary purpose was to prevent pollution damage. If the operations had another purpose, like hull or cargo, they would not be covered by this definition and therefore would not be admissible. The primary purpose test was applied to claims submitted in respect of the removal of the stranded *Rio Orinoco* and her cargo. Following the *Agip Abruzzo* incident certain activities were undertaken both for the purpose of preventing pollution and for the salvage of ship and cargo. Similarly in the *Portfield* case the Executive Committee accepted that the measures had a dual purpose. In such cases it was necessary to consider how the costs of these operations should be distributed between salvage and pollution prevention. The Director took the view that such distribution would have to be made on the basis of an assessment of the facts in the individual case.[96]

(iv) Environmental damage

As stated above the 1984 Protocol to the Civil Liability Convention contains an amended wording of the definition of pollution damage. Under the 1984 Protocol definition compensation for impairment of the environment, other than loss of profit from such impairment, had to be limited to the costs of reasonable measures of reinstatement actually undertaken or to be undertaken. The 1992 Protocols to the Civil Liability Convention contain the same definition of pollution damage as

94. *Ibid.*, p.7.
95. *Ibid.*, p.12.
96. In the *Agip Abruzzo* case the costs were apportioned between pollution prevention and other activities. In the *Portfield* case two-thirds of the costs were attributed to preventive measures and one-third to salvage: *ibid.*, pp.16–17.

in the 1984 Protocol. But contrary to the development of international Conventions increasing attention has been given to the use of abstract quantification methods and theoretical methods.[97]

The first *Antonio Gramsci* incident was the first case in which the question of the admissibility of claims for damage to the marine environment was dealt with by the IOPC Fund. The incident occurred in Russia in 1979. The Russian government made a claim of an abstract nature for compensation for ecological damage within the scope of the Civil Liability Convention. The amount claimed had been calculated according to a mathematical formula in accordance with the Soviet legislation. According to this calculation the assessment of the damage was linked to the quantity of oil collected in Russian territorial waters. The IOPC Fund Assembly took the position that claims for non-economic environmental damage should not be accepted. In 1980 the Assembly unanimously adopted a resolution stating that "the assessment of compensation to be paid by the International Oil Pollution Compensation Fund is not to be made on the basis of an abstract quantification of damage calculated in accordance with theoretical models".[98] In 1981, following the adoption of this resolution,[99] an Intersessional Working Group was set up by the Assembly to examine the question whether, and if so to what extent, claims for environmental damage were admissible under the Civil Liability and the Fund Convention. The Working Group reached the conclusion that compensation could be granted only if a claimant who had a legal right to claim under national law had suffered a quantifiable economic loss.

Following the second *Antonio Gramsci* incident a similar claim relating to the damage to the marine environment was made in 1987. The amount claimed had again been claimed according to Russian legislation. Referring to the IOPC Fund's Resolution No.3 the Executive Committee objected to this claim as the claimant had not suffered any quantifiable economic loss.[100]

The Civil Liability and Fund Conventions were originally devised to ensure that claimants who were injured by an oil spill were left no worse off economically than had the event not occurred. Therefore claims for impairment of the environment are accepted only if the claimant has sustained an economic loss which can be quantified in monetary terms. In the *Patmos* incident a claim relating to damage to the marine environment was submitted by the Italian government, but the kind of damage which had allegedly been caused or any explanation of the basis on which the amount claimed had been calculated was not provided. The claim was ejected on the basis of Resolution No.3. Following the rejection of the claim by the Italian court of first instance the Italian government appealed. In the appeal the Italian government argued that the claim related to actual damage to the marine environment and to actual economic loss suffered by the tourist industry and

97. FUND/WGR.7/9/3 explains the complexity of the marine environment and addresses the consequent difficulty of establishing what constitutes "damage".

98. FUND/WGR.7/4, p.2.

99. IOPC Fund Resolution No.3.

100. FUND/WGR.7/4, pp.2–3.

fishermen. On the basis of this argument the Italian government maintained that the claim was not in contravention of the interpretation of the definition of pollution damage adopted by the Assembly in Resolution No.3. With regard to the economic loss which had allegedly been suffered by the tourist industry and fishermen the Committee expressed the view that compensation could only be claimed by the individual having suffered the damage who, in addition, had to prove the amount of the economic loss sustained.[101]

Similarly in the *Haven* incident[102] the Italian government presented a claim relating to damage to the marine environment. The amount of claim quantified by the claimant was £340 million and related to restoration of phanerogams, wreck removal and damage restored by the natural recovery of the resources (sea and atmosphere). There were also several important items on the claim where the quantification was left to the court to decide on the basis of equity. The IOPC Fund maintained the position that claims relating to unquantifiable elements of damage to the environment could not be admitted. In its interpretation of the Civil Liability Convention and the Fund Convention, the IOPC Fund Assembly excluded the assessment of compensation for damage to the marine environment on the basis of theoretical models. The judge held that the 1969 CLC and the 1971 FC did not exclude environmental damage. He took the view that the environmental damage could not be quantified according to a commercial or economic evaluation. During the 1984 Diplomatic Conference deliberations were based on the policy of the IOPC Fund and the intention was to codify the IOPC Fund's interpretation of "pollution damage" in the original text of the Civil Liability Convention. Therefore the 1984 Protocol excluded non-quantifiable elements of damage to the environment and limited compensation for impairment of the environment, other than loss of profit from such impairment, to the actual cost of reinstatement.[103]

Under the 1992 Protocols to the Civil Liability Convention it becomes clear that claims for damage to the marine environment *per se* are not admissible, but that reasonable costs incurred in reinstating the marine environment after a pollution incident are in principle admissible under the Conventions. It is submitted that the IOPC Fund should maintain its position that claims relating to the impairment of the environment should be accepted only if the claimant has sustained a quantifiable economic loss, and that loss must be such that it can be quantified in monetary terms. In order to be accepted as an admissible claim under the 1992 Protocol to the Civil Liability Convention compensation for reinstatement of the marine environment must be limited to costs of measures actually taken or to be undertaken. And only costs of reasonable measures are recoverable. As a result it is possible to say that the entry into force of the 1992 Protocols would not necessitate any change in the IOPC Fund's position as described above.

101. *Ibid.*, p.3.
102. Italy, 11 April 1991. See the IOPC Fund, *Annual Report, 1996*, pp.39–49. See also Seki, 30 March 1994, *ibid.*, p.73.
103. FUND/WGR.7/4, pp.4–5.

CMI GUIDELINES ON OIL POLLUTION DAMAGE

Apart from the IOPC Fund's work on oil pollution damage, following the introduction of the OPA'90 the Comité Maritime International was encouraged to take up the subject of oil pollution damage and consider the scope for establishing acceptable international norms in this field.

As is common knowledge, the scope of compensation set out in OPA is much more extensive than the concept of "pollution damage" as defined in the Conventions including the 1992 Protocols. Therefore OPA raises the question whether the international system should be accepted in future years as providing adequate solutions to oil pollution damage, or whether the approach of the United States is to be regarded as an advanced model.[104]

An International Sub-Committee (ISC) and Working Group were established under the chairmanship of the Secretary-General of the CMI. The Working Group was charged with the duty of circulating an initial report and questionnaire to explore the position in national laws. As a result of this work the CMI Working Group prepared a draft paper on the admissibility and assessment of claims for pollution damage. The draft was debated in detail and adopted at the 35th CMI Conference held in Sydney on 2–8 October 1994. The guidelines are concerned with the admissibility and assessment of claims for oil pollution damage. They do not deal in any way with claims for personal injury.

The guidelines consist of three parts: general; economic loss; preventive measures, clean-up and restoration. Under economic loss the CMI considered the question of how to draw a reasonably clear boundary line between admissible and inadmissible claims for economic loss. While considering this question the CMI examined the position in different legal systems. Attention was paid to the principles adopted in the courts of various common law and civil law jurisdictions. Under the guidelines "economic loss" comprises both consequential loss and pure economic loss. In principle compensation is payable for consequential loss. Pure economic loss must be caused by the contamination itself. It is not sufficient for a causal connection to be shown between the loss and the incident that caused the escape or discharge of the oil from the vessel involved in the incident.[105] Under Part III of the guidelines compensation for impairment of the environment, other than loss of profit, shall be limited to the cost of reasonable measures of reinstatement actually undertaken or to be undertaken. If the claim is made on the basis of an abstract quantification of damage calculated in accordance with theoretical models compensation is not payable. In order to be accepted as admissible claims the cost of preventive measures, including clean-up and disposal, should be reasonable in the particular circumstances.

The object of the CMI is the unification of maritime law. As it is a private organisation its guidelines do not pretend to have any force. They do not alter legal rights in any way. They are mainly intended to promote a consistent approach in cases of doubt about what the relevant legal rights might be.[106]

104. FUND/WGR.7/17, pp.1–2.
105. *CMI Guidelines on Oil Pollution Damage*, Part II: Economic Loss, Sydney, 8 October 1994, p. 5.
106. *Ibid.*, p. 4.

CHAPTER 16

THE US OIL POLLUTION ACT OF 1990

INTRODUCTION

On 23 March 1989 the *Exxon Valdez* grounded and spilled approximately 250,000 barrels, or 11 million gallons, of crude oil into the pristine waters of Prince William Sound, Alaska. This incident is probably not among the top 20 oil spills. The wreck of the *Torrey Canyon* spilled three times as much oil. The grounding of the *Amoco Cadiz* off Brittany led to a spill of 68 million gallons, six times the amount in Alaska. But the *Exxon Valdez* spill was the largest oil spill in United States history emanating from a vessel. The incident attracted an enormous amount of world-wide media attention and became one of the most famous environmental disasters of all time.[1] Obviously the *Exxon Valdez* caused widespread environmental damage in Alaska and put a heavy financial burden on Exxon, one of the world's largest corporations. But it mainly affected the marine transport of oil and changed the way American society, government, media and the industry will deal with oil pollution in the future. For instance before the *Exxon Valdez* there was no criminal penalty resulting from a marine incident as a consequence of errors in navigation or management of the vessel. And the prosecution of its master Joseph Hazelwood was the first instance in US history of a crew member of a vessel involved in an oil spill being criminally charged. It was also the second time in US history that a crew member was criminally charged for an error in navigation and management.[2] The *Exxon Valdez* disaster was a ship source pollution and it was also a human-error caused accident. The incident made it clear that whenever oil is transported in ships and there are human and mechanical elements involved accidents will occur regardless of the number of precautions taken.

1. For detailed information on the effects of the *Exxon Valdez* incident on US oil pollution legislation see, Michael G. Chalos, "A Perspective on the *Exxon Valdez* and the Consequences for Vessel Oil Pollution Liability in the United States", in Liverpool & London P&I, 2nd International Conference, Malta, 24–26 December 1990, pp.1–8.

2. Michael G. Chalos, p.2. The state of Alaska charged Captain Hazelwood with three misdemeanour counts and one felony count. The felony count was related to the damage caused to the property of others, such as damage to the beaches and wildlife in the area. It carried a maximum sentence of five years and a fine of $50,000. The three misdemeanour counts included: (1) reckless endangerment to human life, carrying a maximum penalty of one year and a $5,000 fine; (2) operating the vessel while intoxicated, carrying a maximum sentence of one year and a $5,000 fine; and (3) negligent discharge of oil, carrying a maximum sentence of 90 days and a $1,000 fine.

PRE-EXISTING LAW

The *Exxon Valdez* incident also demonstrated that United States, including both Federal and state, civil statutory schemes regarding oil pollution liability are not complete and are quite disorganised. Federal legislation addressing oil pollution is not a recent development in the USA. Although it started with the passage of the Oil Pollution Act of 1924 it has never been very straightforward. The *Exxon Valdez* incident is a very good example of the chaos in US oil pollution legislation. It showed that a major oil spill could involve both federal and state laws. Claimants brought actions under state and federal law. Following the incident the Exxon Corporation was charged with violating[3] the Clean Water Act, the Refuse Act, the Migratory Bird Treaty Act, the Ports and Waterways Safety Act and the Dangerous Cargo Act. Civilly, Exxon is being sued under the Clean Water Act, the Trans-Alaska Pipeline Authorisation Act, several Alaskan state strict liability statutes and general maritime law. All these statutes have different liability limits and elements of proof of negligence; the result was great confusion and the resulting liability problems showed the need to link federal general maritime law, federal statutory law, state common law and state statutory law.

Under Federal law prior to OPA's enactment, four major statutes provided for oil pollution liability and compensation: the Federal Water Pollution Control Act (Clean Water Act), the Deepwater Act, the Outer Continental Shelf Lands Act Amendments and the Trans-Alaska Pipeline Authorisation Act.[4] All these Acts apply to various geographic regions and activities and they contain different funding, liability and administrative provisions.

(a) The Federal Water Pollution Control Act

The Federal Water Pollution Control Act (FWPCA), also known as the Clean Water Act,[5] was the principal item of oil pollution legislation prior to the entry into force of the Oil Pollution Act 1990 (OPA). The Act, as the broadest statute, covers tankers and inland barges and applies to spills in the inland navigable waters, the territorial sea, the contiguous zone and the exclusive economic zone. It prohibits the discharge of any oil or hazardous substances in or on navigable waters of the United States, its shorelines or its contiguous zone. The prohibition is subject to certain limited exceptions. The Act imposes civil liabilities on the owner, operator, or person in charge of the vessel or onshore or offshore facility which discharged polluting substances unless it could be proven that the discharge was caused solely by an act of God, an act of war, negligence on the part of the US Government or an

3. *Ibid.*, pp.3–4.

4. For pre-existing US oil pollution legislation see Antonio J. Rodriguez and Paul A.C. Jaffe, "The Oil Pollution Act of 1990" (1990) 15 *Tulane Maritime Law*, 1–11; Edgar Gold, "Marine Pollution Liability After Exxon Valdez: The U.S. All-Or-Nothing Lottery!" (1991) 22 *Journal of Maritime Law and Commerce* 434–439; Gordon W. Paulsen, "US Oil Pollution Legislation Status as of mid-July 1990", 4/90, July/August, *BIMCO Bulletin*, pp.4–5.

5. 33 U.S.C. §1321.

act or omission of a third party without regard to whether any such act or omission was or was not negligent, or any combination of the foregoing, in which case there was no liability.[6] The party responsible under the FWPCA was also subject to criminal penalties under the same Act and also under the Refuse Act of 1899.[7] These penalties are on a strict liability basis.

The person in charge of a vessel or onshore or offshore facility has to report any discharges of prohibited amounts or his failure will constitute an offence. The primary duty to clean up the pollution is on the owner or operator of a vessel or onshore or offshore facility which discharges the oil or hazardous substance. If a responsible party fails to clean up he will be strictly liable to the United States government for the actual costs incurred in the removal of the oil or hazardous substance. If the shipowner proves that the incident did not occur with his privity or knowledge he may limit his liability to the value of the vessel and freight pending at the time of the accident. If the discharger is a vessel the limit is based upon its tonnage; if it is an onshore or offshore facility the limitation amount is fixed. If the discharge is the result of wilful negligence or wilful misconduct within the privity or knowledge of the owner, the owner or operator shall be liable for full removal costs.[8]

The interpretation of the FWPCA with regard to the costs of clean-up efforts undertaken by the responsible party may discourage the owners and operators of discharging units to clean up the pollution voluntarily. The reason for this is that when governments are involved in clean-up operations which are initially started by the party responsible, the party responsible cannot credit the money he has spent against the amount later sought by the government for reimbursement of its expenditure. And in most cases the governments' efforts are generally more expensive than those undertaken by private industry. States, individuals or foreign governments may recover their actual clean-up costs even when these are in excess of the limits set out in the Act. But when the source of the discharge is a vessel these claims are subject to the Limitation of Liability Act.

Under the FWPCA the US government can proceed direct against a third party who alone causes a discharge. A third party may benefit from the same defences as are available to an owner or operator. If a party responsible alleges that the spill was caused solely by a third party and the government proceeds against that party responsible the Act permits him to pay the removal costs and become subrogated to the government's claim against the third party.[9]

The FWPCA section 1321(p), requires vessels trading to the USA to have on board certificates of financial responsibility. These certificates indicate that the vessel is able to meet the financial requirements of the Act. In order to obtain such a

6. *Ibid.*, §1321(f).

7. 33 U.S.C. §407 (1988). The Refuse Act imposed the fines under 33 U.S.C. §411, which were between $500 and $2,500.

8. U.S.C. §1321(f)(4) defines costs of removal as including "any costs or expenses incurred by the Federal Government or any state government in the restoration or replacement of natural resources damaged or destroyed as a result of a discharge of oil or hazardous substance . . .".

9. 33 U.S.C. §1321(g); Antonio J. Rodriguez, and Paul A.C. Jaffe, pp.5–6.

certificate the owner or operator must furnish evidence of insurance, a surety bond, or other evidence of financial responsibility, or must qualify as self-insurer. As the requirements of these certificates were considered to be generally equivalent to the Civil Liability Convention requirements they have generally been issued by the vessel's P&I insurer.

The Act does not cover the question of civil liability in cases of marine-source oil pollution damage. Thus liability for marine-source oil pollution damage has to be established under tort principles.[10] Under general maritime law, a plaintiff must establish culpable negligence in order to recover damages resulting from a discharge.

In section 1370 of the Act it is made clear that the Federal Act does not pre-empt any of the 50 states or political subdivisions thereof from enacting any legislation which they consider necessary to control pollution within their respective territories, including "higher pollution control standards than those established" by the Federal Act. In general the Act provides for onerous liability for government clean-up costs to fall on an owner or operator of a vessel involved in a pollution incident. And the states of the United States may enact stricter liabilities or impose civil or criminal penalties.

The following three statutes, the Outer Continental Shelf Lands Act Amendments 1978, the Trans-Alaska Pipeline Authorisation Act and the Deepwater Port Act, were passed by the US Congress to deal with pollution in certain areas or resulting from certain activities.

(b) Outer Continental Shelf Lands Act Amendments of 1978

The Outer Continental Shelf Lands Act Amendments (OCSLA) regulate the discharge of oil from any offshore facility or vessel on the Outer Continental Shelf, so it is particularly relevant to oil drilling and production in the Gulf of Mexico. It provides joint, several and strict liability. Limitation of liability is based on the tonnage of the vessel, and defences are similar to those under the FWPCA. The Act does not pre-empt other federal and state laws which impose liability, but if a party brings an action under the OCSLA it is not possible for the same damages or removal costs to be recovered under other federal or state laws.[11]

If the incident is caused "primarily by wilful misconduct or gross negligence, within the privity or knowledge of the owner or operator" or where it was caused primarily by breach of certain federal regulations within such knowledge or privity,[12] the OCSLA limits did not apply. Under the Act any statutorily permitted claimant has a right of direct action against the insurer of the owner or operator of the facility or the vessel.[13] The liability of the insurer cannot exceed policy limits and the insurer may invoke all the defences available to the insured. He may also

10. Edgar Gold (1991), pp.435–436. Gold describes this as the inherent weakness in the US approach under the Clean Water Act.
11. Antonio J. Rodriguez, and Paul A.C. Jaffe, p.8.
12. 43 U.S.C. §1814 (b).
13. 43 U.S.C. §1815(c)(1).

rely on the defence that the incident was caused by the wilful misconduct of the insured. But the insurer cannot invoke policy defences that may be available in claims brought by the insured.

(c) Trans-Alaska Pipeline Authorisation Act

The Trans-Alaska Pipeline Authorisation Act (TAPAA) applies, *inter alia*, to vessels loaded with oil transported through the Trans-Alaska Pipeline. The oil should be destined for ports or terminal facilities under United States jurisdiction.[14] The Act pre-empts inconsistent provisions in the Clean Water Act.[15] In *Re Glacier Bay* it was held that "the Act completely repeals the Limitation of Liability Act, not only with respect to the vessel owner's liability under the TAPAA, but for claims in excess of the TAPAA limit asserted under applicable state law or general maritime law".[16]

As with other oil pollution acts liability under the TAPAA is joint and several and has limited defences, and it provides for a maximum limit of exposure which is applicable in all cases.[17] But unlike the OCSLA this Act does not provide for a direct action remedy against insurers.

(d) Deepwater Port Act

The Act applies to oil handling facilities off the coast of the United States and beyond the territorial sea.[18] It prohibits any discharge of oil in a safety zone or at a deepwater port. Discharging oil creates civil penalty for the owner or operator of such a vessel or licensee of the port. The Act does not apply to facilities used exclusively for the production, transportation and storage of oil produced from Outer Continental Shelf mineral leases. It applies to facilities used to tranship oil from tankers to pipelines. The Act imposes strict, joint and several liability for the owners and operators of the vessel. The owner or operator may escape liability where the discharge was caused solely by act of war, negligence of the United States government in establishing and maintaining aids to navigation, and if the damage was caused solely by the negligence of the injured party.

THE LEGISLATIVE STARTING POINT

Over the years the United States Congress considered several comprehensive oil spill measures to combine all state and federal oil spill liability laws into a uniform

14. 43 U.S.C. §1653(c)(7); Antonio J. Rodriguez and Paul A.C. Jaffe, p.9.
15. In the view of Antonio J. Rodriguez and Paul A.C. Jaffe, according to the language of the statute it appears that the extent of liability imposed by the FWPCA was in addition to the limit of exposure imposed by TAPAA. See p.9.
16. 741 F. Supp. 800, 1990 AMC 1614 (D. Alaska 1990), cited in Antonio J. Rodriguez and Paul A.C. Jaffe, p.9, n.58.
17. 43 U.S.C. §1653(c)(3).
18. 33 U.S.C. §1502(10).

national programme but never enacted a comprehensive oil spill system of liability and compensation. According to the early oil spill reform supporters the system created as a combination of federal law and state laws was inadequate. Under this system it was not possible to provide adequate compensation for victims, responsible parties were under conflicting liability and taxing requirements, most of the clean-ups were paid by general taxpayers and clean-up funds were not big enough to cover major oil spills.[19]

The Comprehensive Environmental Response, Compensation and Liability Act of 1980 (CERCLA, also known as the Superfund) was developed by the 96th United States Congress.[20] The Act was originally developed to deal with hazardous substances and oil spills. But later the House agreed to pass a Superfund bill that dealt only with hazardous substances and not oil spills. The Superfund appears to pre-empt all laws inconsistent with it, including the Clean Water Act and the Oil Pollution Act. The Act is very similar to the other US pollution legislation examined above.

Many shipowners believe that the OPA covers all legislative concerns regarding vessel pollution. But CERCLA explicitly holds "the owner and operator of a vessel or a facility" liable for violating its provisions. It imposes strict liability on the owner or operator of a vessel, and on a person who transports hazardous substances to a treatment facility. CERCLA also allows direct action "against any guarantor providing evidence of financial responsibility for such vessel". The scope of the Act is quite wide. It covers removal and remedial costs incurred by the United States, a state, or any other person and also damages for injury to or destruction or loss of natural resources. Liability is limited on the basis of vessel type, nature of the cargo and tonnage. The Act's jurisdiction specifically extends to "navigable waters, the waters of the contiguous zone, and the ocean waters for which the natural resources are under the exclusive management authority of the United States under the Magnuson Fishery Conservation and Management Act". CERCLA explicitly excludes petroleum and natural gas from the definition of "hazardous substances", so the maritime industry dismisses the statute at this point. It is believed that the OPA alone regulates a spill from a vessel because the petroleum exclusion clearly makes CERCLA inapplicable. But the OPA defines oil in a way that accommodates the ambiguity of whether petroleum spills are regulated by OPA or CERCLA. OPA defines oil to mean "any kind or in any form, including, but not limited to, petroleum, fuel oil, sludge, oil refuse, and oil mixed with wastes other than dredged spoil, but does not include petroleum, including crude oil or any fraction thereof which is specifically listed or designated as a hazardous substance under [CERCLA] and which is subject to the provisions of that Act"[21] and the Environmental Protection Agency has listed all petroleum

19. Benjamin J. Grumbles, "Federal Oil Spill Legislation in the Wake of the Exxon Valdez", MTS Journal, Vol. 24, N.4, p.5.
20. 42 U.S.C. §§9601 *et seq.*
21. OPA, S.1001(23).

constituents as hazardous. It is also possible that through contamination petroleum may contain hazardous substances which are not natural to it.[22]

The 1984 Protocols to the CLC and FC were on the US political agenda for a long time. The Reagan administration played a leading role in the negotiations and strongly supported the Protocols. The Bush administration introduced bills, S. 1066 and H.R. 232, which were to incorporate the Protocols and tighten up domestic control over the transportation of oil and hazardous substances. But neither was enacted. Until the *Exxon Valdez* incident Congresses did not pay enough attention to oil spill bills. But following the incident congressional debate on comprehensive federal legislation changed completely. For the first time in almost 20 years, both Houses, the Senate and the House of Representatives, passed similar comprehensive reform bills. H.R. 1465 was introduced on 16 March 1989 and passed the House on 9 November 1989. On 4 August 1989 the Senate unanimously passed S. 686. S. 686 and H.R 1465 were the most significant developments.

Despite the similarities between the bills certain issues were separating House and Senate conferees. At the beginning these issues included international protocols, double hulls, cargo owner liability, state and industry roles in oil-spill response and clean-up efforts. Later other issues also caused disagreement. These were on liability and damage, penalties and other enforcement tools, Alaska-specific items, research and development, size and use of the fund.[23] A lot of work has been done to reconcile the differences between the Senate and House bills. Conferees from the House of Representatives and the Senate have met many times to resolve the disputes between H.R. 1465 and S. 686.[24]

In their first and second meetings, conferees on H.R. 1465 argued extensively over the Protocols. Section 3002(a) provided that "nothing in this title shall constitute a ratification of either the Civil Liability Convention or the Fund Convention". Despite section 3002(a) section 3003 recognises the IOPCF as a legal entity. Thus the bill provided the opportunity for the United States to ratify the protocols to the 1969 CLC and 1971 FC.

In its original form the bill established a liability system for vessel owners who discharged oil. Despite the strict liability on the spiller there was a liability limit for a tanker of $500 per gross ton, not to exceed $60 million. The bill provided a government-administered Oil Spill Trust Fund of $300 million, financed by a 1.3 per cent per barrel tax on oil produced in or imported into the USA. More importantly, this federal law, introduced as the Comprehensive Oil Spill Liability and Compensation Act, would pre-empt contradictory state laws on oil spill liability.

On 4 August 1989, the Senate passed its bill after substantially increasing

22. Sharon Elliot, "The Relevance of the United States 'Superfund Law' to Vessel Owners and Operators", *Gard News*, Issue 129, April 1993, pp.12–14.
23. Benjamin J. Grumbles, p.8.
24. No bill can become law unless passed by both the Senate and the House of Representatives and either signed by the President or, if vetoed, passed in a second vote by Congress.

liability limits for spillers. Similarly H.R. 1465 was passed in November 1989, but only after members had made significant changes on the floor. The Members of Congress wanted a Trust Fund of $1 billion. The tax has risen to five cents per barrel. The limits of liability have been altered substantially. The Senate version sets liability for a tanker at $1,000 per gross tonne. The House limit has risen to $1,200 per gross tonne. This increase brought with it the new concept of "secondary cargo owner liability". Within the rise the owner or operator of the vessel is responsible for $600 per gross tonne but if damages exceed this amount the then cargo owner is liable for up to $600 per gross tonne. Under both bills there was a need to obtain evidence of financial responsibility to meet the maximum liability established in the legislation, and the limitation of liability could be exceeded if the spiller was guilty of gross negligence, wilful misconduct, or violated certain operating standards.[25] Pre-emption of state laws was one of the most important fundamental differences between the House and Senate bills. Traditionally the sponsors of the House bill were in favour of pre-emption. They were in favour of a single federal system which was believed to provide fairness to injured parties and at the same time would promote financial stability and certainty in international shipping. But during the House debate this view was not supported. As a result both House and Senate bills now say that nothing in them shall be "construed or interpreted as pre-empting any state ... from imposing any additional liability or requirements with respect to the discharge of oil ... within such state". The Congress preferred to allow states to continue to develop their separate laws relating to maritime spills.[26] With regard to international conventions on oil pollution the House bill attempts to reconcile the US law with the international scheme. The House, seeking to win support from protocol opponents, included "joinder" provision in Title III.[27] This provision aimed at protecting against pre-emption of state laws. The concept of "joinder" provides that in the event of a protocol spill which affects the USA the US Oil Spill Trust Fund shall indemnify and defend the spiller. If the sum of liability is over the amounts stated under the CLC the Fund agrees to pay it. A state's law is not pre-empted, but this does not make much difference because any liability beyond CLC amounts is paid by the US Fund, not the spiller. Thus, when a vessel trades in US waters its financial exposure is no more than elsewhere. It was believed by the joinder provision proponents that this would help to ensure compensation without pre-emption. The implementation of the protocols was rejected by the opponents. The Senate had no joinder language. According to the critics of Title III the liability regime of the protocols offered less competition than under the new federal regime and also pre-empted state laws to the extent that such laws imposed

25. Edmund B. Welch, p.3.

26. For detailed information on interaction between the state laws and the OPA see Steven R. Swanson, "Federalism, the Admiralty, and Oil Spills", (1996) 27 *Journal of Maritime Law and Commerce* 379–416.

27. Under §3002(b) the federal oil spill fund would be joined in law suits over protocol covered oil spills. The federal fund would pay victims costs and damages recoverable under state law but not covered, or exceeded limits, under the protocols.

unlimited liability on spillers.[28] With regard to double hulls the conferees were in agreement: all new tankers should have double hulls and the Secretary of Transport should not be able to waive the requirement even if more efficient or environmentally protective measures ever came to light.[29] Both bills required double hulls on new tankers serving the US regardless of flag. The resulting compromise required double hulls on all new tankers and barges operating in waters subject to US jurisdiction.

On 18 August 1990 President Bush signed into law the Oil Pollution Act of 1990, after the measure had been unanimously approved by both the House and the Senate.

A REVIEW OF THE OIL POLLUTION ACT OF 1990 (OPA)

The OPA sets out a comprehensive new regime concerning oil pollution prevention, removal, liability, damages and penalties. It is applicable to incidents that occur after 18 August 1990.[30] The Act differs substantially from the original bill (H.R. 1465) and from former comprehensive proposals. It establishes a comprehensive oil spill liability response and compensation programme consolidating several federal liability provisions into one statute. It amends a number of other federal statutes, including the Federal Water Pollution Control Act and prescribes preventive measures and penalties. The OPA, while comprehensive in scope, neither pre-empts state law nor implements the international protocols. It maintains the rights of the states to legislate in the area of pollution control.

(a) Liability

(i) In general

The Act begins with lots of definitions. Section 1001 contains definitions of the terms in the Act. These are not substantially different form those contained in the Federal Water Pollution Control Act, 33 U.S.C. section 1321. Under Section 1001 of the OPA, a "responsible party" for a vessel or facility is liable for removal costs and a wide range of damages if oil is, or is threatened to be, discharged into US navigable waters, adjoining shorelines or the exclusive economic zone which extends the limit to 200 miles. Liability will be incurred for discharge of oil from a vessel or facility in the navigable waters of the United States, up to three miles offshore, and in the United States exclusive economic zone. The Act also applies to any waters over which the United States exerts sovereignty including the waters of Puerto Rico, Guam, American Samoa, the United States Virgin Islands and the

28. Benjamin H. Grumbles, p.11.
29. *Ibid.*, p.8.
30. For general information on OPA see Bessemer A.F. Clark, "The U.S. Oil Pollution Act of 1990" [1991] L.M.C.L.Q. 247–256; Thomas J. Wagner, "The Oil Pollution Act of 1990: An Analysis" (1990) *Journal of Maritime Law and Commerce* 569–587; John E. Noyes, "The US Oil Pollution Act of 1990" (1992) *International Journal of Estuarine and Coastal Law* 43–56.

Marianas. Under the OPA claims may be lodged in United States courts by foreign governments or private claimants when a shoreline of a foreign country has been fouled by oil spilled from a vessel in United States waters, by a vessel carrying oil between two ports and the United States or by a tanker moving Alaskan oil to the United States East and Gulf coast by way of the Trans-Panam pipeline. Under this provision parts of Mexico, Canada and many of the Caribbean islands come within the scope of the Act.

The Act applies only to discharges of "oil of any kind or in any form, including, but not limited to, petroleum, fuel oil, sludge, oil refuse and oil mixed wastes other than dredged spoil ...". Liability on the party responsible is joint, several and strict. The party responsible for a vessel is defined as "any person owning, operating, or chartering by demise, the vessel". The Act also applies to the owners and operators of off-shore and on-shore facilities[31] which are involved in the handling of oil.[32] Under the Act the owner, demise charterer and operator/manager are jointly and severally liable. A "vessel" is defined as "every description of watercraft or other artificial contrivance used, or capable of being used, as a means of transportation on water other than a public vessel". Under this definition apart from tankers the Act covers not only dry cargo vessels but also pleasure yachts.

The owner/operator terminology of the Act, §1001(26), is repeated from the FWPCA.[33] Neither the FWPCA nor the OPA explains the meaning of the term "owner" or "operator". As a result of the inherent ambiguity of the meaning of these terms it is not clear whether mortgagees and individual investor/managers of a vessel-owning corporation will be considered "owners" under the Act or not. If direct day-to-day control and involvement in activities like manning and supplying a vessel are accepted as key points in order to establish liability then management companies responsible for a vessel's performance of a voyage would be responsible as "operators". Banks and lenders that repossess vessels and afterwards involve themselves in their operation may be considered "owners" under the Act. In *The Amoco Cadiz*, the parent oil company, Standard Oil Co. of Indiana, was held to be in effective operational control of, and hence responsible for, the vessel.[34] Proposals made at the legislative stage that liability should also fall on the cargo owner have not been incorporated in the Act. But as the Act expressly permits states to impose more encompassing liability provisions in some cases, under state laws, the cargo owner may be included in the definition of the responsible party.

31. A facility as defined in the Act may include such things as pipelines, drilling equipment and storage areas: 33 U.S.C. §2701(9).

32. In case of an onshore facility and an offshore facility "owner or operator" means any person owning or operating such onshore or offshore facility and in the case of any abandoned offshore facility, it means the person who owned or operated such facility immediately prior to such abandonment": OPA, S.1001(26).

33. The definition of "owner or operator" is different from that contained in CERCLA. Under CERCLA the definition specifically excludes a person "who, without participating in the management of a vessel or facility, holds indicia of ownership primarily to protect his security interest in the vessel or facility".

34. *The Amoco Cadiz* [1984] 2 Lloyd's Rep. 304.

(ii) Limits of liability

The Act increases the limits of liability of the FWPCA eightfold. Under the Act, the liability of tanker owners or operators for removal costs and recoverable damages is limited to:

For a tank vessel over 300 gross tonnes but not to exceed 3,000 gross tonnes: the greater of $2,000,000 or $1,200 per gross tonne.
For a tank vessel over 3,000 gross tonnes: the greater of $10,000,000 or $1,200 per gross tonne.
For a vessel other than a tank vessel over 300 gross tonnes: the greater of $500,000 or $600 per gross tonne.
For an offshore facility except deepwater port: the total of all removal costs plus $75,000,000.
For any onshore facility and a deepwater port: a maximum of $350,000,000.

Additional amounts are also applicable under the CERCLA.

For any vessel over 300 gross tonnes carrying hazardous substances as cargo, the greater of $5,000,000 or $300 per gross tonne.
For any other vessel over 300 gross tonnes, the greater of $500,000 or $300 per gross tonne.

The underlying basis of liability in the Act is in line with the international regime reflected in the 1969 Civil Liability Convention and its protocols. It is strict liability for removal costs and damages, with right to limitation. But it is far easier to break the carrier's right to limit liability under the OPA than it is under general international law. Under the OPA a responsible party's ability to limit liability is severely restricted by certain exceptions. Liability cannot be limited if the spill was proximately caused by:

(a) gross negligence or wilful misconduct, or
(b) the violation of an applicable federal safety, construction, or operating regulation by the party responsible, its agents or employees or any person acting pursuant to a contractual relationship with the party responsible.[35]

Under the latter exception it will be very easy to lose the ability to limit liability. Most accidents resulting in a spill also involve a breach of federal regulation. Simple negligence which violates a particular regulation in the event of a spill will subject a party responsible to unlimited liability.

Liability will also be unlimited if the responsible party fails to:

(a) report an incident;
(b) co-operate with officials in the clean-up process; or
(c) comply with an order under sub-section (c) or (e) of section 311 of the FWPCA or the Intervention on the High Seas Act, which deals with Presidential powers to order removal.[36]

35. OPA, S.1004(c)(1).
36. OPA, S.1004(c)(2).

In order to benefit from limitation of liability the party responsible is under an obligation to notify the appropriate authorities of the discharge of oil or the probable discharge of oil "immediately". But the definition of "immediately" is very vague and it will cause considerable misunderstanding. Avoiding these restrictions on the right to limit is not the only problem a party responsible will face. Section 1018 of the OPA provides that nothing in the new Federal Act shall affect, or be construed or interpreted as pre-empting the authority of any state from imposing any additional liability or requirements with respect to:

(a) the discharge of oil or other pollution by oil within such state; or

(b) any removal activities in connection with such a discharge.

Section 1018 also provides that the 1851 Limitation Act, the pre-existing federal limitation of liability statute, shall similarly have no effect on such state laws.[37] In many cases there will be state legislation co-existing with the federal Act and the owner will probably face unlimited liability under state laws.

In addition, removal costs incurred by the US Government or any state or local government in connection with a discharge or substantial threat of a discharge of oil from any Outer Continental Shelf facility or from a vessel carrying oil as cargo from such a facility is not subject to limitation. Interest on claims is also not subject to limitation. Apart from liability for removal costs and damages, an owner or operator or guarantor is also liable to the claimant for interest on the amount paid in settlement of a claim starting 30 days after the claim is presented and ending on the date the claim is paid. If a guarantor makes an offer within 60 days of the date the claim is presented, he will be protected against an interest charge, provided that his offer is equal to or greater than the amount finally paid in satisfaction of the claim. Obviously the time limit, 30 days or 90 days, provided under Section 1005 is too short to assess all the factors and to make a claim. Therefore in practice these deadlines will hardly be met and interest will be an additional liability.[38]

(iii) Defences to liability

A party responsible is exonerated from liability for removal costs or damages if the party responsible establishes by a preponderance of the evidence that the sole cause of a discharge or threatened discharge was:

(a) an act of God[39];

(b) an act of war;

(c) an act or omission of a third party;

37. The effect of the Limitation of Liability Act on federal law was discussed in the *Sammi Superstar* litigation and claims on the base of Limitation of Liability Act were rejected. In this case, in California, it was claimed that the language of Section 2702 of the OPA constitutes an express repeal of the Limitation Act with regard to maritime oil spills and the judge issued a proposed order favouring this interpretation: Dogmar Schmidt Etkin, pp.211–214.

38. Bessemer A.F. Clark, p.252.

39. Under S.1001(1) "act of God" means an unanticipated grave natural disaster or other natural phenomenon of an exceptional, inevitable, and irresistible character the effects of which could not have been prevented or avoided by the exercise of due care or foresight.

(d) any combination of (a), (b) and (c).

The sole cause requirement restricts the application of defences to liability. They will not be available to a party responsible whose slightest negligence contributes to a discharge. It will be very difficult to establish a defence to liability under the Act. Even if a defence is available, there is a danger of losing it in certain circumstances, such as failure to report the incident, failure to co-operate in the clean-up, or failure without sufficient cause to comply with certain federal orders. Contributory negligence of a claimant is available as a defence in actions for recoverable damages only if the party responsible can prove that the "gross negligence or wilful misconduct" of the claimant caused the discharge.

With regard to the defence that a spill was solely caused by the acts or omissions of a third party the party responsible has to prove that he took due care in handling the oil and took precautions against the foreseeable consequences of the acts or omissions of third parties. Under this defence to liability the third party cannot be an employee or agent or one who acts in a contractual relationship with a party responsible. Under this section a pilot or a tugboat operator cannot be a "third party".

(b) Damages

Under the OPA, parties responsible are subject to liability for the cost of removing oil spills and for damages caused by the discharge or threat of discharge of oil products. Removal costs for which parties responsible are liable include all removal costs incurred by the United States, a state, or an Indian tribe and any removal costs incurred by any person for acts taken by the person which are consistent with the National Contingency Plan.

The range of damages under the Act is wider than the damages recoverable under the FWPCA. Under Section 1002(b)(2) of the OPA, apart from clean-up and restoration costs, a party responsible may be liable for six categories of recoverable damages. These are as follows:

(1) Natural resources: includes damages for injury to, destruction of, loss of, or loss of use of "land, fish, wildlife, biota, air, water, ground water, drinking water supplies and other such resources". Under natural resources damage the reasonable cost of assessing such damage are included. Under the Act natural resources are given importance and a separate section.[40] In this category claims can be brought by the Federal government, by state governments or Indian tribes where they control resources, and in some cases by foreign governments.

(2) Real or personal property: includes economic loss. It is recoverable by a claimant who owns or leases the property. This heading constitutes a departure from the general law. Before the OPA these damages were recoverable in limited circumstances according to general maritime law. But under the OPA pure economic losses will be recoverable even in cases where

40. OPA, S.1006.

there is no claim for physical loss. For instance a fisherman whose stocks are destroyed by a spill can sue, without establishing fault, for his lost earnings.
(3) Subsistence use of natural resources, recoverable by any claimant who uses natural resources which have been injured, destroyed, or lost. Claimants do not have to be owners or managers of the resources.
(4) Revenues: governmental claimants may recover for net loss of taxes and other fees resulting from injury to real or personal property or natural resources.
(5) Profits and earning capacity covers loss of profits, impairment of earning capacity for damage to real property, personal property or natural resources. It is recoverable by any claimant who need not be the owner of the damaged resources.
(6) Public services: governmental claimants may recover the net costs of providing additional services such as fire and health prevention during removal activities.

The protection of natural resources is a major part of the Act. One of the most significant aspects of the liability provisions of the Act relates to natural resources damage. Section 1006 of the Act specifically deals with natural resources damages and provides that the party responsible will be liable to the federal government, state government, Indian tribe or foreign government owning, managing or controlling the affected natural resources. The Act calls for the appointment of trustees to represent those entities in presenting claims for, and recovering for, damage to their respective natural resources. The trustees for each entity assess the natural resource damages under Section 1002(b)(2)(A) and "develop and implement a plan for the restoration, rehabilitation, replacement, or acquisition of the equivalent, of the natural resources under their trusteeship".

The measure of natural resource damages is the sum of:

(a) the cost of restoring, rehabilitating, replacing, or acquiring the equivalent of, the damaged natural resources;
(b) the diminution in value of those natural resources pending restoration; plus
(c) the reasonable costs of assessing those damages.

To assist in damage assessment the OPA establishes a structure for evaluating natural resources damages incurred in OPA-covered oil spills. Two federal government agencies, the Department of the Interior (DOI) and the National Oceanic and Atmospheric Administration (NOAA) of the Department of Commerce, have been empowered to develop regulations which set out the manner in which that damage to the natural resources is assessed. The DOI is responsible for natural resource damage assessments (NDRA) under CERCLA. NOAA is responsible for NDRA under OPA. Both agencies support expanding the scope of natural resource damage, thus increasing potential liability. The agencies have endorsed speculative damage assessment measures, generic modelling, formulae and other non-site-specific damage formulations. One particular focus has been contingent valuation methodology (CVM). CVM is a process which attempts to measure non-use or passive uses of natural resources, normally through formulae, models and opinion polls.

The effective date of final rule of NOAA for natural resource damage assessment (NRDA) is 5 February 1996. The final rule stated that natural resource damage assessments "are not intended to replace response actions, which have as their primary purpose the protection of human health, but to supplement them, by providing a process for restoring natural resources and services injured as a result of an incident involving oil".[41] The purpose of the final rule is to provide a framework for conducting sound natural resource damage assessments that achieve restoration under the OPA. Under the rule a natural resource damage claim is based on restoration plans developed with input from the public and parties responsible. Parties responsible are allowed to implement trustee-approved and monitored restoration plans. The final rule is quite important because trustee determinations of natural resource damages made in accordance with the final rule will have the effect of a rebuttable presumption in an administrative or judicial proceeding under the Act. In any court case the government's figures will be accepted unless the responsible party assumes the burden of demonstrating the actual extent of damages.

(c) Financial responsibility

Under the OPA owners and operators of vessels are required to establish and maintain evidence of financial responsibility[42] sufficient to meet their potential liabilities under the Act for discharges or threatened discharges of oil.[43] The US Coast Guard is mandated to promulgate regulations which will implement the financial responsibility requirements under the OPA. Since the passing of the Act[44] shipowners and their liability insurers have grown increasingly concerned about the new regulations being developed by the Coast Guard for Certificates of Financial Responsibility (COFR).

On 26 September 1991 the US Coast Guard issued a Notice of Proposed Rulemaking (NPRM) concerning the establishment and maintenance of financial responsibility. The NPRM consolidates financial responsibility requirements under both the OPA and CERCLA. It develops consolidated methods for evidencing financial responsibility under both Acts. Under the NPRM a single unified Certificate of Financial Responsibility replaces separate certificates under the OPA, the Federal Water Pollution Control Act, the Outer Continental Shelf Lands Act Amendments, the Trans-Alaska Pipeline Authorisation Act and the

41. National Oceanic and Atmospheric Administration, Natural Resource Damage Assessments, *Final Rule*, 15 CFR Part 1990, January 5, 1996, p.440.

42. For detailed information on Certificates of Financial Responsibility see Leonard F. Alcantara and Mary A. Cox, "OPA 90 Certificates of Financial Responsibility" (1992) 23 *Journal of Maritime Law and Commerce* 369–386; David Bovet, pp.245–247; "Oil Pollution Act of 1990", Coast Guard Notice on Financial Responsibility, BIMCO Bulletin 6/91, November/December, pp.37–38.

43. OPA, S.1016.

44. OPA repealed 33 U.S.C., the FWPCA, §1321(p) and enacted, in its place, 33 U.S.C. §2716.

Deepwater Port Act. According to the NPRM, evidence of financial responsibility must be established and maintained in the total applicable amount. In other words, the party responsible for any vessel over 300 gross tonnes, other than a dumb barge not carrying oil, which uses any place subject to United States jurisdiction, or for any vessel using the waters of the US exclusive economic zone[45] to tranship or lighter oil destined for the USA must establish and maintain evidence of financial responsibility sufficient to meet the maximum amount of his liability as limited under Section 1004. Additional amounts are also applicable under the CERCLA. Applicable amount under the CERCLA is, for any vessel over 300 gross tonnes carrying hazardous substances as cargo, the greater of $5,000,000 or $300 per gross tonne. For any other vessel over 300 gross tonnes it is the greater of $500,000 or $300 per gross tonne. If the responsible party has more than one vessel he must meet the maximum amount of liability applicable to the largest vessel. But the financial responsibility must be applicable to each vessel owned or operated.

On 1 July 1994 the Coast Guard published the interim rule implementing the vessel financial responsibility provisions of the OPA and CERCLA. That rule, for most intents and purposes, became the final rule on 7 March 1996. The new COFR requirement comes into effect on several dates: for tankers, on 28 December 1994; for self-propelled tank vessels on 1 July 1995; and for tank barges, and mobile offshore drilling units and for all other vessels (that is passenger vessels, dry cargo vessels, etc.), upon existing COFR expiration, over the three years from 28 December 1994. When actually operating on site as an offshore facility, a mobile offshore drilling unit is exposed to tank vessel liability with respect to discharges of oil on or above the surface of the water. Accordingly, a non-self-propelled mobile offshore drilling unit is considered by the Coast Guard to be a non-self-propelled tank vessel when operating as an offshore facility. The financial responsibility implementation date with respect to self-propelled tank vessels was 1 July 1995. If a mobile offshore drilling unit is tied up at a shoreside dock or otherwise not operating as an offshore facility, the Coast Guard does not require that unit to demonstrate tank vessel financial responsibility during that period. However, on and after 1 July 1995, before that mobile offshore drilling unit may operate as an offshore facility, it must demonstrate financial responsibility because it is subject to tank vessel limits. If a mobile offshore drilling unit remains out of work and it holds an unexpired pre OPA 90/CERCLA COFR the unit would not be required to comply with the final rule until 28 December 1997, or at the time its pre OPA 90/CERCLA COFR expires, whichever is earlier. The OPA offers six methods of establishing financial responsibility: insurance, surety bonds, letters of credit, guarantees, self-insurance and other evidence. A certificate of financial responsibility will be a precondition of entry to or departure from the USA. Failure to provide evidence of financial responsibility may result in severe sanctions. The entry of a vessel into US navigable waters may be denied. If the vessel is already in navigable waters it may be seized or forfeited.

45. For COFR purposes the waters of the EEZ begin immediately after the three-mile territorial sea, i.e. waters seaward of the three-mile territorial sea are waters of the EEZ.

The Act allows direct claims against the guarantor to be made by any party affected. A claim for liability which may be established under Section 1002 may be asserted directly against a guarantor providing evidence of financial responsibility. Thus the guarantor is directly liable for clean-up costs and damages. He is entitled to invoke all rights and defences which could have been available to the owner or operator under the Act; any defence authorised under Section 1016(e); and the defence that the incident was caused by the wilful misconduct of the owner or operator. The guarantor's liability is limited to the amount of the financial responsibility provided for. The COFR guaranty ensures that, in the event of a spill, the responsible party has sufficient assets to pay for removal and damage costs up to the limit of liability. If they do not, the guarantor promises to pay up to the responsible party's limit of liability. Obviously the requirement under the OPA that a provider of financial responsibility be a guarantor, and allowing direct claims by all claimants against the guarantor, caused many problems in practice. These problems will be discussed later in the chapter.

(d) Oil Spill Liability Trust Fund

The OPA established a $1 billion Oil Spill Liability Trust Fund (OSLTF) to pay for expeditious oil removal and unrecovered damage. The Fund has a power to spend up to $1 billion per incident. Within that overall limit, the amount available for damage to natural resources is $500 million.

OSLTF administration was delegated to the Coast Guard by the Executive Order and on 20 February 1991 the National Pollution Funds Center (NPFC) was commissioned to perform this function as an independent headquarters unit reporting directly to the Coast Guard Chief of Staff.[46] The NPFC is the fiduciary agent for the Oil Spill Liability Trust Fund and the portion of Superfund[47] accessible to the US Coast Guard. The NPFC executes programmes to accomplish five principal objectives: to provide funds for removal actions, to provide funding to initiate Natural Resource Damage Assessments, to compensate claims, to recover costs from responsible parties and to issue vessel certificates of financial responsibility.

The Congress created the OSLTF in 1986 but did not pass legislation to authorise the use of money or the collection of revenue to maintain it. After the *Exxon Valdez* incident and the passage of OPA 90 that authorisation was granted. Under the OPA the funds supporting certain prior federal oil pollution laws are consolidated.[48] By consolidating the previous funds and revenue collection the

46. US Coast Guard National Pollution Funds Center, *FY94 Annual Report* (July 1995), p.4. See also US Coast Guard National Pollution Funds Center, *FY95 Annual Report* (March 1996); US Coast Guard National Pollution Funds Center (April 1996); US Coast Guard National Pollution Funds Center, *Claimant's Information Guide* (September 1993).

47. Superfund was established under the CERCLA for response and remedying of hazardous waste incidents.

48. The consolidated oil pollution laws are: The Federal Water Pollution Control Act, the Deepwater Port Act, the Trans-Alaska Pipeline Authorisation Act and the Outer Continental Shelf Lands Act.

Fund increased to $1 billion.[49] The OSLTF has four revenue sources. The primary source is an annual levy of five cents per barrel on oil transported in US waters. Since 1 January 1990 the Fund has been receiving this annual levy from the oil industry on petroleum produced in or imported to the United States. The tax was suspended on 1 July 1993 and reinstated on 1 July 1994. But on 31 December 1994 it ceased again due to the "sunset" provision in the law.[50] The five cent tax contributed 33 per cent, or $92 million, of the revenues the OSLTF received in financial year 1995. The interest on the Fund principal from US Treasury investments is the second source of Fund revenue. Most of the unused balance in the Fund accrues interest in US Treasury investments. This source accounted for almost $53.1 million, or 19 per cent of the Fund's revenue in financial year 1995. Cost recoveries from parties responsible are a third source. Those responsible for oil spills are liable for costs and damages. NPFC bills parties responsible to recover costs expended by the Fund. Moneys recovered are used to replenish the Fund. Cost recovery from parties responsible accounted for about $10.4 million, or 4 per cent of the Fund's revenue in the 1995 financial year. The fines and civil penalties paid by parties responsible are deposited into the Fund. In the 1995 financial year these payments accounted for $6.2 million, or 2 per cent of the Fund's revenue.[51]

The OSLTF mainly provides funds for oil pollution removal activities and compensation for damage where an oil discharge affects or threatens the surface waters of the United States. In general, claims for all costs and damages resulting from an oil pollution incident must be presented first to the party responsible or its guarantor. In other words, claimants must first seek compensation for removal costs and damages from the party responsible or guarantor.[52] If the claim is not voluntarily paid by the party responsible, the claimant is left with a choice. He may present the claim to the Fund or sue the party responsible. But if the claimant sues the party responsible, he must refrain from seeking recovery from the Fund while the action is on foot. The NPFC cannot certify or approve any claim which is the subject of pending litigation. The OSLTF may accept claims in order to facilitate prompt and efficient settlement of third party claims. Uncompensated damages arising from non-OPA liabilities are not recoverable from the OSLTF. In order to submit claims to the OSLTF certain conditions have to be fulfilled. A claim for uncompensated removal costs or damages should be presented to the party responsible or guarantor first and, if the claim is denied, full and adequate compensation is not available or the claim is not settled by payment by any person within 90 days, then the claim may be submitted to the OSLTF. The starting date

49. US Coast Guard, *FY95 Annual Report*, p.10. The Fund takes in amounts remaining in the Deepwater Port Fund, the Offshore Oil Pollution Compensation Fund and the Trans-Alaska Liability Fund. The Trans-Alaska Pipeline Liability Fund is being transferred to the OSLTF in yearly increments. In financial year 1995 $119 million was transferred to the OSLTF.

50. The tax, when authorised by Congress, is suspended when the Fund reaches one billion dollars but may be reinstated if the Fund falls below that. The US Coast Guard Publication, p.6.

51. US Coast Guard, *FY95 Annual Report*, pp.10–11.

52. The *Claimant's Information Guide* prepared by the US Coast Guard provides assistance in the preparation and presentation of claims: US Coast Guard National Pollution Funds Center, *Claimant's Information Guide* (September 1993).

for the 90 day limit is either the date on which the claim was presented or the date of advertising for claims by the party responsible or NPFC,[53] whichever is later.[54]

In certain cases a claim for removal costs and damages may be presented first to the OSLTF. In the *Claimant's Information Guide* these cases are set out as follows: if the NPFC has advertised or otherwise directed claimants to submit their claims to the OSLTF, by a party responsible who may assert a claim under Section 1008 of the OPA[55] (33 U.S.C. 2708), by the governor of a state for removal costs incurred by that state, or by a United States claimant for removal costs and damages resulting from a discharge or substantial threat of discharge from a foreign offshore unit under Section 1012(a) of the OPA (33 U.S.C. 2712).[56]

The OSLTF covers certain categories of losses. These are set out under Section 1012 of the OPA[57]:

(i) the payment of removal costs by the Federal Authorities or a Governor or State Official in the case of a state clean-up;
(ii) payment of costs for assessing natural resource damage;
(iii) removal costs and damages resulting from oil spilt from a foreign offshore unit;
(iv) payment of claims for uncompensated removal costs and damage;
(v) payment of federal administrative, operational and personnel costs and expenses in connection with the Act.

Claims must be submitted to the Fund within certain time limits or the NPFC will not consider the claim. Claims for damages must be made within three years after the date on which the injury and its connection with the incident were reasonably discoverable with the exercise of due care. In the case of natural resource damages[58] the time limit starts on the same date or within three years of the date of completion of the natural resources damage assessment under Section 1006(e) of the OPA. Removal costs claims must be made within six years after the date of completion of all removal actions for the incident.[59]

Although the Fund mainly aims to be quicker than the courts in compensating claims it is not very clear whether it can speed up the process. In most cases claims, such as natural damage claims, will be quite complicated and assessment of those damages will take a long time. Payment of any claim or obligation by the Fund under the OPA entitles the US government to acquire by subrogation all the rights

53. In the case of a damage due to an oil spill the source of the spill may be designated or not. If the source of the spill is designated the spiller advertises for claims and claimants submit claims to the spiller. If the spiller denies the claim or fails to settle it within 90 days then the claimant may submit it to the OSLTF. In cases where the source of the spill cannot be designated the NPFC advertises for claims and then claimants may submit claims to the OSLTF.

54. US Coast Guard, *Claimant's Information Guide*, pp.7–8.

55. S.1008 permits a party responsible for a vessel or facility from which oil is discharged to make a claim for removal costs and damages for the Fund but only if the party responsible shows that it is entitled to a defence to liability under S.1003 of the Act or is entitled to limitation of liability under S.1004.

56. *Ibid.*, p.8.

57. Bessemer A.F. Clark, p.251.

58. Under S.1002(b)(2)(A) of the OPA.

59. US Coast Guard, *Claimant's Information Guide*, p.7.

of the claimant or state to recover from the party responsible.[60] After paying removal costs and damages, or only costs or damages, the Fund will obtain the rights of the compensated person against the party responsible by subrogation. The Fund becomes subrogated to the rights of claimants of a state who received them to recover from the party responsible.[61]

(e) Double hull requirement

One of the most publicised aspects of the OPA 90 was the introduction of the double hull requirement for new tankers over 5,000 gross tons. Following the *Exxon Valdez* incident a special National Research Committee was established to investigate the preparedness for, and response to, the incident. In its report the Committee presented several alternatives of tanker design that might be used to prevent big oil spills in the future. The report also stated that the primary cause of serious incidents was groundings. Although large spills comprised only 3 per cent of incidents, more than 95 per cent of accidental spillage is caused by these large spills. On the basis of these facts the Committee concluded that the most cost-effective design[62] for the prevention of major oil spills was a double-skinned hull.[63] As a result of this recommendation Congress included the double hull requirement in the OPA. The Act set out a comprehensive scheme eventually to require tankers carrying oil in US waters or the exclusive economic zone to have double hulls.[64]

In general, the Act institutes a gradual, 25-year phase out of single-hulled vessels carrying "oil in bulk as cargo or cargo residue" based upon the age and tonnage of the vessel. All such tankers operating in US waters must be equipped with double hulls no later than 2015. The only vessels excepted from the interim deadlines are those discharging at the Louisiana Offshore Oil Port and those lightering in specified zones at least 60 miles off the US coasts. Within those zones single hull tankers will be able to discharge oil. In such circumstances, the double hull construction requirement does not come into play until 1 January 2015. The Act phases out existing non-double-hulled vessels beginning in 1995. If a vessel is less than 5,000 gross tons it will be exempted from the double hull requirement, provided that it is equipped with approved "double containment" systems. By the year 2010, all vessels over 5,000 gross tons with single hulls will be prohibited from operating. By the year 2015 all vessels over 5,000 gross tons with double bottoms or double sides will be prohibited from operating without double hulls. A vessel for which a building contract or contract for major conversion was placed before 30

60. OPA, S.1015.
61. OPA, S.1012(f).
62. For detailed information on the financial side of double-hulls see R. Scott Brown and Ian Savage. "The Economics of Double-hulled Tankers" (1996) 237 *Maritime Policy Management* 167–175.
63. A double hull consists of two skins of steel separated by a small space known as a cofferdam. It is believed that in the event of a collision or grounding, the first skin will absorb the impact while the second skin remains intact: *ibid.*, p.1.
64. *Ibid.*

June 1990, and that is delivered under that contract before 1 January 1994, must install a double hull by a certain date, which varies according to the age and size of the vessel. The phase-out schedule of vessels over 5,000 tons is shown in Table 4. By Sections 4101–4118 the Coast Guard is also authorised to issue regulations concerning various aspects of ship construction and equipment, crew hours and navigation. Furthermore the Coast Guard is also authorised to set requirements for pressure and over-fill gauges on oil cargo tanks, plating thickness and inspection requirements for vessels more than 30 years old, and maximum tanker crew hours. A crew member cannot work more than 15 hours in any 24-hour period and not more than 36 hours in a 72-hour period.

Table 4. Phase Out Timetable for Single Hull, Double Bottom and Double Sided Vessels

Size of Vessel Gross Tonnes	*Age of Single Hull Vessel*	*Age of Double Bottom, Doubled Sided Vessel on Phase Out Date*	*Date after which Vessels of designated age must cease Operation*
5,000–15,000	40 or older	45 or older	1 Jan. 1995
5,000–15,000	39 or older	44 or older	1 Jan. 1996
5,000–15,000	38 or older	43 or older	1 Jan. 1997
5,000–15,000	37 or older	42 or older	1 Jan. 1998
5,000–15,000	36 or older	41 or older	1 Jan. 1999
5,000–15,000	35 or older	40 or older	1 Jan. 2000
5,000–15,000	25 or older	30 or older	1 Jan. 2005
15,000–30,000	40 or older	45 or older	1 Jan. 1995
15,000–30,000	38 or older	43 or older	1 Jan. 1996
15,000–30,000	36 or older	41 or older	1 Jan. 1997
15,000–30,000	34 or older	39 or older	1 Jan. 1998
15,000–30,000	32 or older	37 or older	1 Jan. 1999
15,000–30,000	30 or older	35 or older	1 Jan. 2000
15,000–30,000	29 or older	34 or older	1 Jan. 2001
15,000–30,000	28 or older	33 or older	1 Jan. 2002
15,000–30,000	27 or older	32 or older	1 Jan. 2003
15,000–30,000	26 or older	31 or older	1 Jan. 2004
15,000–30,000	25 or older	30 or older	1 Jan. 2005
30,000 or greater	28 or older	33 or older	1 Jan. 1995
30,000 or greater	27 or older	32 or older	1 Jan. 1996
30,000 or greater	26 or older	31 or older	1 Jan. 1997
30,000 or greater	25 or older	30 or older	1 Jan. 1998
30,000 or greater	24 or older	29 or older	1 Jan. 1999
30,000 or greater	23 or older	28 or older	1 Jan. 2000

(f) Prevention, removal and other provisions

The OPA aims to modify existing legislation on prevention of spills and removal of oil. For instance the Act specifically tries to identify potential and current licence holders with alcohol and drug problems and provides for comprehensive licensing procedures of sea-going personnel to ensure that unqualified persons do not obtain or continue to hold merchant mariner licences.[65]

The Merchant Marine office cannot issue a licence unless the applicant's National Driver Register (NDR) information is submitted to the USCG. The NDR is a central databank which receives reports on individuals with convictions for alcohol or drug related traffic violations, accidents or traffic incidents involving speeding or recklessness. The Transport service may also review an applicant's criminal record through the FBI. Licences are valid for five years. Existing licence-holders must undergo drug testing on a periodic and random basis, or with reasonable cause, or after an accident. There is also a provision for suspension, at least temporarily, of licences if the individual performs a safety sensitive function and there is probable cause to conclude that the individual has performed the function in violation of law or federal regulation as to the use of alcohol or a dangerous drug or has been convicted of an offence that would prevent the issue or renewal of the documents and/or if the individual has been convicted of driving while intoxicated.

This Section of the Act is a result of charges brought against the master and crew members of the *Exxon Valdez*. One of the charges brought against Captain Hazelwood was of operating the vessel while intoxicated, carrying a maximum sentence of one year and a $5,000 fine. Furthermore it was brought up in evidence that Captain Hazelwood had a drink-drive conviction some years earlier. Charges were also brought against certain crew members who were drinking ashore just before the ship sailed.

As a result of this Section tanker charterparties include a "D and A Policy Clause" which puts the onus on the owner to carry out drugs and alcohol tests not just in the USA but world-wide. A typical "D and A Clause" provides as follows:

"Drug and Alcohol Policy. Owner warrants that it has a policy on Drug and Alcohol Abuse (Policy) applicable to the vessel which meets or exceeds the standards in the Oil Companies International Marine Forum Guidelines For the Control of Drugs And Alcohol Onboard Ship. Under the Policy, alcohol impairment shall be defined as a blood alcohol of 40 mg/100 ml or greater; the appropriate seafarers to be tested shall be all Vessel officers and the drug/alcohol testing and screening shall include unannounced testing in addition to routine medical examinations. An objective of the Policy should be that the frequency of the unannounced testing be adequate to act as an effective abuse deterrent, and that all officers be tested at least once a year through a combined program of unannounced testing and routine medical examinations. Owner further warrants that the Policy will remain in effect during the term of this Charter and that Owner shall exercise due diligence to ensure that

65. OPA, S.4101 (amending 46 U.S.C. §7101).

the Policy is complied with. It is understood that an actual impairment or any test finding of impairment shall not in and of itself meant that the Owner has failed to exercise due diligence."

Under Section 4106 of the Act the United States government is allowed to evaluate the manning, training, qualifications and other standards of a foreign country that issues documents for any tank vessel that operates in the exclusive economic zone of the United States where there is an accident that results in death, serious injury or harm to the marine environment. Foreign tankers must have equivalent manning, training, qualification and watchkeeping standards to those in the USA. If the Secretary of Transport determines that a country has failed to maintain or enforce standards at least equivalent to the United States standards or international standards accepted by the United States, he is required to prohibit vessels which are issued with documentation by such country from entering the United States until such time as the Secretary determines that those standards have been established and are being enforced. If the owner or operator of the vessel establishes, to the satisfaction of the Secretary, that the vessel is not unsafe or a threat to the marine environment the Secretary may make exceptions to this prohibition. The interesting point under manning standards for foreign tank vessels is that compliance with the requirements of international standards is not enough to fulfil the USA's manning standards: the international standards must be accepted by the USA. It is Congress's view that international standards are not always acceptable.

The OPA also amends the FWPCA regarding the ability of the federal government to undertake clean-up efforts. Section 4201 of the Act establishes a general requirement under new Section 311(c)(1) of the FWPCA for the President to act in accordance with the National Contingency Plan and to ensure that effective and immediate removal action is undertaken. Obviously it is not only the President who has to act immediately in accordance with the Plan. The owner or operator has to respond immediately and must also act in accordance with the National Contingency Plan. This Section is a direct response to the public criticism which followed the *Exxon Valdez* spill. After that incident the appropriate roles of the discharger and the federal, state and local governments in the removal process were the subject of comment. In the *Exxon Valdez* incident a lot of time was lost before any prevention or clean-up measures were initiated. If a spill presents a substantial threat to the public health or welfare of the USA the President must direct all response activities. In such a case the clean-up should be federalised.

The OPA provides an important exemption[66] from liability for persons giving care, assistance or advice consistent with the National Contingency Plan or as directed by the President. Apart from the party responsible any person involved in the removal of oil or hazardous substances consistently with or pursuant to the

66. This exemption is sometimes referred to as the "Good Samaritan" exemption.

National Contingency Plan is exempt from liability for removal costs and damages, except with respect to personal injury or wrongful death.[67] The exemption is not applicable if the assisting person is guilty of wilful misconduct. The party responsible will be liable instead of an assisting person to the extent that he is exempted from liability for removal costs and damages. The exemption provided under this Section is considered quite important as it encourages salvors and clean-up operators to assist at oil spills.

The National Contingency Plan was originally adopted pursuant to the FWPCA. The OPA amends the FWPCA by expanding the scope of the National Contingency Plan. The plan is designed to "provide for efficient, co-ordinated and effective action to minimise damage from oil and hazardous substance discharges, including containment, dispersal, and removal of oil and hazardous substances".[68] The National Contingency Plan must establish procedures for a worst case discharge of oil and include a fish and wildlife response plan.[69] Section 4202 directs the President to prepare and publish a National Contingency Plan for the removal of oil and hazardous substances. The Act establishes a national planning and response system and owners and operators of tank vessels and facilities are under an obligation to prepare and submit individual vessel response plans to the President. The individual response plan must identify an individual having full authority to implement removal actions, identify and ensure by contract that private personnel and equipment are available to remove a "worst case" discharge and describe the training to be provided to personnel on a vessel to mitigate or prevent a discharge of oil.[70] After submitting his plan the owner may have permission to trade for up to two years without approval of his plan but in order to do so he must show that he has concluded a commercial contract designed to achieve a result similar to the proposed plan. The owner's compliance with the response plan does not provide a defence to his liabilities and fines imposed by the OPA, but failure to comply with it means a fine of up to $25,000 per day of violation.

(g) Penalties

(i) Criminal penalties and immunity

Section 4301 increases penalties under the FWPCA for discharge of oil or hazardous substances and for other violations of the FWPCA. The amendment provides stricter criminal liabilities for failure to notify the appropriate Federal agency of a discharge. Any person in charge of a vessel or an onshore or offshore

67. OPA. S.4201.
68. OPA, S.4201(b).
69. *Ibid.*
70. The Notice of Proposed Rulemaking published on 19 June 1992 and the Navigation and Vessel Inspection Circular 8–92 (NAVIC 8–92) dated 15 September 1992 provide the specific requirements for vessel response plans: see Cynthia A. Hudson, "Vessel Response Plans Under OPA 1990", BIMCO Bulletin, 6/92, November/December, pp.10–13.

facility shall immediately notify the appropriate agency of the United States government of such discharge. The notification must be made as soon as he has knowledge of any discharge of oil or a hazardous substance from such vessel or facility. Otherwise the fine will be $250,000 for an individual and $500,000 for an organisation or up to five years' imprisonment or both.[71] Any violations of the FWPCA prohibitions on discharge of oil are subject to criminal penalties.[72] Penalties for negligent violations are $2,500–$25,000 and/or one year's imprisonment. For knowing violations penalties are $5,000–$50,000 and/or three years' imprisonment. The penalties for knowing endangerment[73] go up to $250,000 and 15 years' imprisonment.

(ii) Administrative and civil penalties

An owner, operator, or person in charge of any vessel, onshore or offshore facility from which oil or a hazardous substance is discharged or who fails or refuses to comply with any regulation issued under the FWPCA may be assessed for a Class I or Class II civil penalty. The Secretary of Transportation or the Administrator of the Environmental Protection Agency has authority to decide whether a violation should have a Class I or Class II penalty. Class I civil penalties are $10,000 per violation, not to exceed $25,000, and Class II penalties are 10,000 per violation, not to exceed $125,000.[74] The civil penalty for discharges of oil or hazardous substances is $25,000 per day of violation or $1,000 per barrel of oil. In the case of gross negligence or wilful misconduct, the penalty is to be not less than $100,000 but no more than $3,000 per barrel of oil.[75] If the owner fails to undertake removal action under order of the President he will face a penalty of $25,000 per day of violation or three times the costs incurred by the Fund as a result of such failure. The Secretary of Transportation is also authorised to assess a penalty, up to $25,000 per day, for violation of financial responsibility requirements. The Secretary is also authorised to seek a judicial order to compel compliance, including an order terminating operations.

The Department of Justice (DOJ) filed the first two judicial settlements under the OPA with a US district court in Washington State on 10 February 1993.[76] The DOJ brought the first complaint against Texaco Refining and Marketing Inc. (Texaco) for a 210,000 gallon pipeline oil spill in Anacortes, Washington, on 22

71. The previous penalty was $10,000 or up to one year in prison or both.
72. OPA, S.4301(c).
73. Knowing endangerment means that a person is placed in imminent danger of death or serious physical injury.
74. OPA, S.4301(b).
75. Under the FWPCA, 33 U.S.C. §1321(b), civil penalties for discharges of oil or hazardous substances were no more than $50,000. In the case of wilful negligence or wilful misconduct within the privity and knowledge of the owner, operator, or persons in charge the penalty was not more than $250,000.
76. Dagmar Schmidt Etkin, pp.113–115.

February 1991 and four additional spills. The result was a discharge of 400 gallons of oil at the same location. The second complaint was against US Oil & Refining Company (US Oil) and involved a 600,000 gallon pipeline oil spill in Tacoma, Washington, on 6 January 1991. Under the Texaco settlement Texaco agreed to continue spill clean-up at a cost of $8 million, to install and maintain state-of-the-art spill prevention equipment at a cost of approximately $800,000, to pay the government's removal costs of $125,000 and pay a civil penalty of $480,000 to the US Government. Texaco had also paid a $20,000 civil penalty to Washington State.

Under the US Oil settlement the company agreed to complete the clean-up of the spill for the cost of approximately $4 million, to install and maintain state-of-the art spill prevention clean-up equipment at a cost of $800,000, to pay the government's removal costs of almost $60,000 and to pay a civil penalty of $425,000 to the US government.

(h) Reactions to OPA

(i) Relationship with other laws

Since its entry into force the industry has been discussing various aspects of the OPA. The OPA has had an irreversible effect on the tanker trades as whole. One of the greatest disappointments of the OPA is the failure of Congress to make provision for the USA to adopt the International Civil Liability Conventions and their protocols. Section 3001 of the OPA states that "It is the sense of the Congress that it is in the best interests of the United States to participate in an international oil pollution liability and compensation regime that is at least as effective as federal and state laws in preventing incidents and in guaranteeing full and adequate prompt compensation for damages resulting from incidents." Despite this section, in practice ratification of international conventions does not seem to be possible. If the amount of the vessel owner's liability under the OPA would not have exceeded the amount imposed by the 1992 Protocol to the CLC then ratification would have been possible. Higher limits of liability under the OPA and its potential for unlimited liability under state law put the international conventions in a position where they will never be considered as effective as the OPA and make their ratification impossible.[77] So, in reality, the OPA rejects the United States' participation in international regimes related to oil pollution liability.

It is commonly known that oil pollution from ships is an international problem and needs an international solution. In most cases the participation of the United States is crucial for the success of international legal efforts. During the negotiation of the 1984 Protocols leadership from the United States led the international community to increase liability limits and international shipping standards. Higher limits were adopted to secure United States ratification. But despite its active role in the negotiation of the 1984 Protocols the USA did not ratify them.

77. Antonio J. Rodriguez and Paul A.C. Jaffe, pp.24–25.

Obviously the USA's approach towards international conventions and the OPA's failure to promote a uniform national and international approach through its relationship with domestic and international laws will weaken the influence of the United States in the negotiation of future maritime treaties. From the point of view of shipowners and insurance clubs the diplomatic isolation of the USA may not be terribly important. What is serious are the difficulties introduced by the Act itself.

The OPA made an important change in the application of state law to pollution incidents. Before the OPA, under the "supremacy clause" of the US Constitution, federal pollution statutes automatically overrode state laws when their provisions were in conflict. Therefore it was possible for vessel owners to invoke the lower liability limits specified under the federal Limitation of Liability Act rather than those provided under state pollution laws. Furthermore under the Limitation of Liability Act, in some circumstances, a shipowner was permitted to limit the amount of his liability to the post-casualty value of his vessel plus the freight earned on the casualty voyage. Under the OPA such a defence is no longer available for the shipowner. Section 1018 of the OPA does not allow parties responsible to limit their liability under the Limitation of Liability Act for clean-up costs and damage incurred by states and their political subdivisions.[78] As a consequence of Section 1018 a responsible party's liability to each state affected by a spill is determined by that state's law.[79] When a vessel is carrying oil for use in the United States it may be subject to a multiplicity of rules and liabilities as it moves from the jurisdiction of one state to another. First, the vessel is subject to the OPA. Then it is subject to the laws of the state where the accident occurs. The vessel may also be subject to the laws of other states where physical damage occurs, or where economic damage occurs without any physical loss.[80] The potential problem under state laws is that the OPA now permits states "to impose additional liability or additional requirements or to impose, or to determine the amount of, any fine or penalty for any violation of law".[81] States do not have the fear of pre-emption by federal law, so they may impose additional, more stringent standards with respect to liability. For instance there are 15 coastal states in which liability for all vessels is unlimited for both clean-up costs and damages following a discharge.[82] In some states liability is

78. S.1018(c) of the OPA provides:

"Nothing in this Act, or the Act of March 3, 1851 (the Limitation of Liability Act) shall in any way affect, or be construed to affect, the authority of the United States or any state or political subdivision thereof

(1) to impose additional liability or additional requirements; or

(2) to impose, or to determine the amount of, any fine or penalty (whether civil or criminal in nature) for any violation of law;

relating to the discharge, or substantial threat of a discharge, of oil."

79. Nicholas J. Healy, Gordon W. Paulsen and Matthew A. Marion, "The United States Oil Pollution Act of 1990", BIMCO Bulletin, 5/90, September/October, p.32.

80. See Steven R. Swanson, "Federalism, the Admiralty, and Oil Spills" (1996) 27 *Journal of Maritime Law and Commerce* pp.379–416.

81. OPA, S.1018(c).

82. States with unlimited liability for both clean-up costs and damage are: Alabama, Alaska, California, Connecticut, Georgia, Maine, Maryland, Massachusetts, Mississippi, New Hampshire, Oregon, Pennsylvania, Rhode Island, South Carolina and Washington.

unlimited for damages but limited for clean-up costs. In other states it is the reverse. For instance in Florida liability is unlimited for damages but limited for clean-up costs. In Virginia and New Jersey it is the reverse.[83] Congress had the opportunity to clear up the confusion under federal and state laws when it passed the OPA but it just made things more complicated by failing to promote a uniform national and international approach.

In 1995 Intertanko filed a lawsuit in the US Federal Court in Seattle challenging the Washington State Office of Marine Safety's new oil spill prevention rules.[84] The rules came into effect on 7 July 1995 and are intended to provide the "best available protection" of Washington's marine environment. Intertanko's lawsuit challenges the state's right to develop oil spill prevention standards that are more stringent than federal or international requirements. The US District Court issued a decision on 18 November 1996 dismissing the entire case. The judge ruled that Section 1018 of the Oil Pollution Act 1990 effectively gives states the authority to enact wide-ranging measures to prevent oil spills, even if those measures intrude into areas governed by international treaties and federal laws and regulations.

The judge rules that the Washington State oil/prevention statutes and regulations do not conflict with any international or federal standards for vessels and are constitutionally valid in that they do not violate the Commerce Clause or Foreign Affairs Clause of the US Constitution.

Because these laws/regulations are designed to protect Washington's "delicate and valuable marine resources" the judge rules that they are deemed to be a legitimate exercise of the state's police powers. Intertanko filed an appeal on 11 December 1996 with the Ninth Circuit US Court of Appeals from the order of the US District Court which dismissed its lawsuit against the State of Washington.

(ii) Industry's concern

The whole future of liability and compensation for oil spills has been thrown into question by the OPA as under the legislation a tanker owner can face unlimited liabilities and damage claims in the case of an oil spill in US territorial waters, even where there is no negligence on his part. After the entry into force of the OPA some oil companies modified their trading patterns. Shell-owned and operated vessels no longer trade crude or bulk products in US waters except to the Louisiana Offshore Oil Port. "It was considered that the risk was too high for the rewards likely to be gained".[85]

The concerns of the cargo/tanker owners can be summarised as follows: by Section 1016 of the OPA financial security is to be provided on the basis of the largest ship in a fleet, the other vessels being covered under this umbrella without

83. See Amy M. Stolls, "Vessel Liability and Financial Responsibility in the US Coastal States", in 1995 International Oil Spill Conference, 27 February–2 March 1995, California, p.255.

84. *Intertanko* v. *Mike Lowry, Governor of the State of Washington, et al.*, US District Court, Seattle, August 1995.

85. Shell Briefing Service Ltd., *The Tanker Industry in the 1990s*, No.5 (1991), p.9.

the provision of further security. Read on its own, the wording of Section 1016 appears to be easy to comply with. However, when it is read in conjunction with Appendix B of the Proposed Rule the requirement becomes too complex for tanker owners to comply with. The wording in Appendix B of the Proposed Rule suggests that security would be required for every vessel, calculated separately, with an owner or operator being required to aggregate the separate security requirements for all his vessels in order to arrive at the total applicable amount for his fleet. Under such a condition for a deep-sea tanker fleet the industry would have to put up evidence of financial security of some $200 billion. But if aggregation of security is not what is intended the system will be based on security of $1,500 per gross ton being provided for the largest ship in the fleet. But even this amount still represents a very large burden for the shipowning and operating community. For instance for a ship of 100,000 dwt/57,000 grt the required financial security would be $85 million.

The OPA categorises those persons who can be held directly liable: an owner, a demise charterer, an operator. The first two need no explanation, but "operator" is nowhere adequately or accurately defined. Could the charterers assume that they are immune from being sued under the OPA? As Hill[86] discusses in detail charterers can find themselves liable by way of indemnity through the terms and conditions of the charterparty. In the event of a failure to produce a Certificate of Financial Responsibility the vessel's entry to the port concerned will be denied. Tanker owners and dry cargo carriers will seek to place responsibility on a charterer for all financial losses resulting from failure to provide a COFR. Therefore charterers should not sign a charterparty which includes a clause making the charterer indemnify the owners for any or all consequences of a failure to produce a COFR.[87] In charterparties there is also a common warranty for safe port or safe berth. In breach of this provision the charterer will be strictly liable. If a tanker or dry cargo vessel is put alongside an unsafe berth or directed to an unsafe port and damaged as a result of an oil spill then the pollution damage results directly from the unsafe berth. As a consequence, the charterer will be liable to indemnify the owner for any liability he may have to the victims of the pollution.[88]

Under the OPA a qualified individual (QI) has to be identified. The law states, "identify the Qualified Individual having the full authority to implement removal actions, and require immediate communications between that individual and the appropriate federal official and the persons providing personnel and equipment". According to this requirement a new individual will be one of the multitude of people assembled at a major oil spill.[89] The QI has to be given full written authority from the owner to act on his behalf, which means that the QI is authorised to spend on behalf of the owner. The meaning of "full written authority" has not been

86. Christopher Hill, "Oil Pollution", *P&I International*, V.9, No.10, October 1995, pp.194–196.
87. *Ibid.*, pp.194–195.
88. *Ibid.*, p.194.
89. William O. Gray, "OPA 90—The Independent Tanker Owners' Perspective", BIMCO Bulletin, 6/92, November/December, pp.5–9.

explained anywhere in the regulations. It is also not clear where the money which the QI is authorised to spend would come from.[90]

The OPA places on tanker operators the responsibility for dealing with the oil pollution wherever it happens, on the water or on the beach. This is in contrast to the general philosophy adopted under the laws of other countries where the local governments are best equipped to direct clean-up activities and then ask the owners to pay for legitimate costs up to a certain limit.

The double hull requirement is another concern for the industry. All new tankers ordered after June 1990 which call at US ports must have double hulls. The requirement will be extended to all tankers calling at US ports by the year 2015. The effectiveness of double-hulled vessels in preventing oil spills has been much debated in the industry for a long time. It is estimated that 52.7 million dwt of vessels will be subject to the requirements of the OPA for double hulls.[91] According to the National Research Council the cargo carrying capacity of a double-hulled vessel is 2.6 per cent less than a comparable single-hulled vessel, which means that there will need to be 2.7 per cent more vessels, voyages and vessel miles to carry the same amount of cargo.[92] Obviously the cost of converting a vessel to a double-hulled one will not be cheap, depending on the vessel's condition. However, now, international regulations, MARPOL 73/78, require all new building tankers to be of a double hull or equivalent design.

(iii) Insurance

Since the entry into force of the OPA the inadequacy of insurance cover has remained the critical issue. Under the OPA every shipowner faces potential costs of $2–4 billion against maximum insurance coverage of $500 million to $1 billion. Following the financial problems there is difficulty in dealing with the legal procedures which are likely to follow a major oil spill. Under the Act the owner is also faced with criminal liability and under US environmental criminal law the illegal spillage does not have to be deliberate or even reckless before criminal prosecution can result. As a result a senior executive who is thousands of miles away from the site of the incident will be criminally liable for a mistake made by one of his subordinates. So, shipowners have multiple exposure to different types of liability in respect of oil pollution and there is no effective limit under federal law on the liability of the shipowner. The states being permitted to legislate on shipowners' liability the position of the shipowner becomes more complicated.

The P&I clubs provide the insurance certificates required by the 1969 CLC and, before its termination, the TOVALOP agreement. They also provide insurance certificates which are issued by the US Coast Guard under the FWPCA.

90. *Ibid.*, p.6.
91. R. Scott Brown and Ian Savage, p.169.
92. National Research Council (1991), *Tanker Spills: Prevention by Design* (Washington D.C.: NRC Committee on tank vessel design, National Academy Press), cited in R. Scott Brown and Ian Savage, p.170, n.1.

Certificates of financial responsibility issued under the terms of the FWPCA are currently in force for more than 14,000 ships, based on P&I insurance.[93] Since the *Exxon Valdez* spill there have been many spills in the USA including the incidents involving the *World Prodigy*, the *President Rivera*, the *American Trader*, the *Mega Borg* and the *Sammi Superstars*. The financial consequences of these incidents were mainly in excess of the FWPCA guarantee limits and liabilities. In these incidents the clubs acted as insurers and they paid the necessary compensation under the terms of their cover.

During the legislative process of the OPA the P&I clubs made it clear that they were not prepared to issue certificates for amounts higher than those in the 1984 Protocols. The reasons behind the rejection of the OPA were political and commercial. Politically the clubs feared that the effect of issuing certificates to meet specific US limits could in the future result in regional or national demands elsewhere. Commercially, the clubs were not prepared to act as guarantors under the Act and they did not want to be vulnerable to direct action from courts for damage caused by their members. As the principal provider of financial security throughout the world the International Group prefers liability for oil pollution to be governed in all countries by the internationally agreed conventions. Obviously by adopting the OPA the USA makes this impossible.

The fears of P&I clubs about direct action were realised in a Louisiana Supreme Court Ruling on 18 October 1993. In the case[94] an engineer on a tug was injured while serving on it. At the time of the accident the tug was in Texan territorial waters. The engineer sued the tug owner, Gulf International Marine and Indemnity, and the owner's P&I club, American Steamship Owners' Mutual Protection and Indemnity Association, and sought compensation for his injuries. The US District Court for the Eastern District of Louisiana ruled that the P&I club was not liable under the Louisiana direct action statute because the accident occurred outside the state and the insurance policy was not written in Louisiana. The engineer appealed. The court's decision was affirmed on different grounds. The Court of Appeals concluded that the P&I policy was an "ocean marine" policy and it was excluded from the direct action statute. However, in its ruling the Supreme Court argued that the direct action statute did apply to marine P&I insurance. The court stated that "the Louisiana Direct Action statute permits an injured party to maintain a direct action against a marine protection and indemnity insurer".[95]

There are different reasons for the clubs not to act as guarantor under the OPA. Under the Act the only defences that could be used by an OPA guarantor would be those the party responsible could use, and the additional defence of wilful misconduct on the part of the party responsible. It is not possible to employ contractual limitations, terms or conditions in the policy as a defence to claims made by those affected by an oil spill. Under the OPA it is quite possible to be faced

93. David Bovet, p.245.
94. Dagmar Schmidt Etkin, p.142.
95. Oil Spill Intelligence Report, 2 December 1993. Cited *ibid.*

with liabilities in excess of the stated certificate limit. It is far easier to break the carrier's right to limit under the OPA than it is under general international law.

As explained above the OPA establishes liability limits of $1,200 per gross ton, with a minimum of $10 million, for tank vessels and $600 per gross ton for other vessels. The Coast Guard Regulation states that there is a need for a $300 per gross ton security under the CERCLA. The shipowner or operator is liable for clean-up costs, third party damage, and the natural resource damage up to stated limits. But these limits may change under state laws because, under state laws, apart from the financial responsibility requirement under the OPA, some form of evidence of financial responsibility may be required. In some states the required financial responsibility is in an amount equal to the OPA amounts. In five coastal states the P&I clubs have agreed to allow shipowners to use a club membership as evidence of financial responsibility because these states do not require the clubs to be guarantors. These states are California, Florida, Texas, Virginia and Washington.[96] Table 5 illustrates the coastal states' oil spill financial responsibility and liability requirements for vessels.

Table 5. Oil Spill Financial Responsibility and Liability Requirements for Vessels: US Coastal States[97]

State	*Persons liable*	*Liability limits*	*Financial responsibility requirement?*	*Amount of financial responsibility*	*Methods of proof OK with P&I Clubs?*
Alabama	Any person who discharges oil	Unlimited	No	N/A[98]	N/A

96. Amy M. Stolls, pp.255–256.
97. Adapted from Amy M. Stolls, pp.257–260. Source: US Oil Spill Law Report.
98. N/A = Not Applicable.

State	*Persons liable*	*Liability limits*	*Financial responsibility requirement?*	*Amount of financial responsibility*	*Methods of proof OK with P&I Clubs?*
Alaska	Vessel owners and operators and cargo owners	Unlimited	Yes, for vessels that carry bulk oil as cargo	a) $337.50 per incident for each barrel of storage capacity or $112.5 million, whichever is greater, for vessels carrying crude oil; or b) $112.50 per incident for each barrel of storage capacity or $1.125 million, whichever is greater, up to a limit of $39.375 million, for carrying non crude oil	No. The statute allows a club membership to be used as "other proof of financial responsibility", but it also includes a direct action clause. Although the development may waive this direct action clause if certain criteria are met, the statute still requires the clubs to sign on as guarantors, which they refuse to do.
California	The "owner or transporter of oil or a person or entity accepting responsibility for the oil" or "the owner, operator, or the lessee of, or person who charters by demise" the vessel, or anyone accepting responsibility for the vessel	Unlimited	Yes, for owners or operators of tank vessels, barges, and/or owners of oil	$500 million until 1 July 1995; $750 million from 1 July 1995, to 1 January 2000; and $1 billion after 1 January 2000	Yes. The statute does not include a direct action clause.
Connecticut	"Any person, firm or corporation that directly or indirectly causes pollution"	Unlimited	No	N/A	N/A

State	*Persons liable*	*Liability limits*	*Financial responsibility requirement?*	*Amount of financial responsibility*	*Methods of proof OK with P&I Clubs?*
Delaware	The owner or operator of a vessel	$150 per gross ton for non-tank vessels; $300 per gross ton or $250,000, whichever is greater, up to a limit of $30 million for tank vessels	Yes, for all vessels over 300 gross tons (except a non-self-propelled barge that does not carry oil as cargo)—however, the statute specifies that such requirements be "in accordance with regulations" that currently do not exist. So this requirement is not enforced	$150 per gross ton for non-tank vessels; $300 per gross ton or $250,000, whichever is greater, up to a limit of $30 million for tank vessels	No. The statute includes a direct action clause.
Florida	"Any vessels, or its agents or servants, who permits a discharge." The owner of a pollutant transported as cargo on any vessel suffering a discharge within state waters" is liable for all clean-up costs not paid by the owner or operator	OPA limits for clean-up costs; unlimited for damages	Yes, for vessels that carry oil as cargo	OPA '90 amounts	Yes. The statute allows proof of club membership as evidence of insurance, i.e. the clubs may include a caveat in their policies that says they are not guarantors.
Georgia	Any person who causes a discharge "intentionally or negligently"	Unlimited	No	N/A	N/A

State	*Persons liable*	*Liability limits*	*Financial responsibility requirement?*	*Amount of financial responsibility*	*Methods of proof OK with P&I Clubs?*
Hawaii	Any person who violates the statute	$700 million for barges that may carry more than 60,000 barrels of oil; unlimited for all other vessels	No	N/A	N/A
Louisiana	A vessel owner or operator or any other person "who causes, allows, or permits an unauthorised discharge . . . or threatened unauthorised discharge"	Limits consistent with OPA levels	Yes but only for a federal certificate of financial responsibility	OPA '90 amounts	No. The statute requires evidence of financial responsibility "in accordance with OPA 90"
Maine	"Any person, vessel, licensee, agent, or servant" who "permits or suffers a prohibited discharge"	Unlimited	No	N/A	N/A
Maryland	Any person discharging or permitting the discharge	Unlimited	Yes, in the form of a bond, for vessels carrying 25 barrels or more of oil	$500 per gross ton	N/A
Massachusetts	A vessel owner or operator or "any person who otherwise caused or is legally responsible for" a discharge	Unlimited	No (not under chapter 21E)	N/A	N/A
Mississippi	"Any person" causing pollution	Unlimited	No	N/A	N/A

State	*Persons liable*	*Liability limits*	*Financial responsibility requirement?*	*Amount of financial responsibility*	*Methods of proof OK with P&I Clubs?*
New Hampshire	An "operator or other person" who directly or indirectly causes a discharge	Unlimited	No	N/A	N/A
New Jersey	Any person who discharges or is in any way responsible for a discharge	Unlimited for clean-up costs; $150 per gross ton for damage	Yes, but only for a federal certificate of financial responsibility (The rules are different, however, for vessels involved in vessel-to-vessel transfers)	OPA '90 levels	No. The statute requires a federal certificate of financial responsibility.
New York	Any person who discharges	OPA '90 limits	Yes	OPA '90 amounts	No. The statute accepts only such evidence that is in accordance with OPA '90.
North Carolina	Any person who discharges or causes to be discharged	OPA '90 limits	No	N/A	N/A
Oregon	"Any person owning oil or having control over oil that enters the waters of the state"	Unlimited	Yes	OPA '90 amounts	No. The statute requires compliance with federal financial responsibility requirements.
Pennsylvania	Any person who discharges	Unlimited	No	N/A	N/A
Rhode Island	Any person who discharges, causes to be discharged, or "permits" a discharge	Unlimited	Yes	OPA '90 amounts	No. The statute requires a valid COFR under OPA '90.

State	*Persons liable*	*Liability limits*	*Financial responsibility requirement?*	*Amount of financial responsibility*	*Methods of proof OK with P&I Clubs?*
South Carolina	Any person who discharges or causes a discharge	Unlimited	No	N/A	N/A
Texas	A vessel owner or operator, or "any other person who causes, allows, or permits" a discharge	OPA '90 limits for natural resource damages; limits for clean-up costs and other damages of $1 million for vessels of 300 gross tons or less that do not carry oil as cargo, $5 million for vessels of 8,000 gross tons or less, or $600 per gross ton for vessels greater than 8,000 gross tons	Yes. However, regulations have not been written so these requirements are not yet enforced	OPA '90 amounts for those vessels subject to OPA	Yes. The statute does not require P&I clubs to be guarantors.
Virginia	Any person discharging, or causing or "permitting" a discharge	Unlimited for clean-up costs; $500 per gross ton or $10 million, whichever is greater, for damages	Yes	$500 per gross ton	Yes. The statute allows club membership to be used as evidence of self-insurance. Shipowners must simply furnish a "marine insurance broker's certificate" rather than a guarantee.

State	*Persons liable*	*Liability limits*	*Financial responsibility requirement?*	*Amount of financial responsibility*	*Methods of proof OK with P&I Clubs?*
Washington	Cargo and vessel owners; any person who discharges	Unlimited	Yes, for vessels that carry oil as cargo or as fuel. However, regulations have not been written, so these requirements are not yet enforced	The greater of $2 million or $3,000 per barrel total capacity for barges carrying persistent oil, or the greater of $2 million or $1,500 per barrel total capacity for barges carrying non-persistent oil. Small tank barge owners and operators may meet the lower limits only if they have an oil spill prevention plan on file with the Washington Office of Maritime Safety that has not been disapproved or voluntarily withdrawn; $600 per gross ton or $500,000, whichever is greater, for a cargo or passenger vessel that carries oil as fuel; and $150 per gross ton or $1 million, whichever is greater, for inland fuel barges	Yes. The statute does not require P&I clubs to be guarantors.

In general, the clubs will continue to supply indemnity cover for vessel pollution liability in the United States and worldwide. In practice it is possible to obtain $500 million P&I cover for oil pollution and in addition $200 million. Most shipowners currently carry $700 million pollution cover. But as clubs refuse to act as guarantors under the OPA some alternative methods have been developed. As explained above, the OPA offers six methods to establish financial responsibility: insurance, surety bonds, letters of credit, guarantees, self-insurance and other evidence. Whichever method is used it must permit direct action against the insurer or other guarantor by any claimant who seeks to recover costs and damages under either or both the OPA and CERCLA (see Appendix IX. The Britannia Steam Ship Insurance Association Limited Financial Responsibility Clause in Respect of Oil Pollution).

The interim rule published on 1 July 1994 did not accept the central request of many shipowners and oil companies that the Notice of Proposed Rulemaking[99] be modified to allow existing P&I insurance to continue to serve as the basis of financial responsibility. The financial responsibility must satisfy the requirements of the OPA and the CERCLA which govern liability for pollution and damage caused by the escape of hazardous material. By June 1996, 1,839 self-propelled tank vessels, 3,978 tank barges, and 5,823 other vessels have complied with the rule. 37 per cent of these 11,640 vessels are covered by self-insurance and financial guarantee, 62 per cent by insurance guarantees and about 1 per cent by surety bond guarantees.[100]

Under the insurance scheme the following insurance companies have been approved to provide evidence of financial responsibility (water pollution) under OPA/CERCLA[101]:

- — Arvak Ltd., c/o Aon Risk Services (Bermuda)
- — Environmental Pollution Group, Inc.
- — HIH Casualty and General Insurance Ltd.
- — Ocean Marine Mutual Insurance Association Ltd., c/o Ocean P&I Services Ltd.
- — Puget Sound Underwriters, Inc.
- — Scandinavian Reinsurance Company Ltd., c/o Terra Nova Underwriting Agency Ltd.
- — The Shipowners Insurance and Guaranty Company Ltd. (SIGCo.)
- — Shoreline Mutual (Bermuda) Ltd.
- — Sphere Drake Insurance plc
- — Stockton Reinsurance Ltd. (The First Line Program)
- — The Water Quality Insurance Syndicate (WQIS).

99. A Notice of Proposed Rulemaking published by the US Coast Guard on 26 September 1991.

100. Daniel F. Sheeman, *Economic Impact of Vessel Certificates of Financial Responsibility* (Department of Transportation, US Coast Guard).

101. Source: The National Pollution Funds Center, Vessel Certification Division, 4200 Wilson Blvd Suite 1000, Arlington VA 22203–1804.

Surety bonds

In certain cases surety bonds have been used to handle lower limits under the FWPCA but they are not very suitable for the requirements of the OPA. The surety onds are mainly available to companies which are strongly capitalised. Companies which are not financially very strong will be unable to qualify for surety bonds. Within those unqualified companies the majority of the world's tanker and bulker operators will be included. In practice surety bonds will not be appropriate for shipowners because a bonding company requires one-for-one security against the bond. And in most cases shipowners have already mortgaged their vessels. Furthermore the bonding company will require funds in the United States subject to US jurisdiction which will not be applicable in practice as most tanker owners are domiciled overseas. In order to fulfil this requirement the owners will need to obtain letters of credit from their banks which will be very costly for them and will possibly call for offsetting funds in credit in the bank, and a confirming letter of credit from a US bank. As a result, surety bonds will be useful only for well capitalised shipowners and will leave the smaller operators out of their scope.

Self-insurance

The self-insurance alternative is impractical for most shipowners. Because the Coast Guard's regulatory test for self-insurance requires shipowners to show assets located within the USA which exceed, by the amount of financial guarantee required, the total of all liabilities worldwide. The Interim Final Rule of the Coast Guard requires both working capital and net worth equal to or greater than the financial responsibility amount for the largest vessel in the applicant's fleet. Working capital and net worth are calculated on the basis of assets located only in the United States, set against liabilities located anywhere in the world. When the international aspect of shipping has been considered it becomes clear for most shipowners that it is not possible to meet the test. It is commonly acknowledged that most vessel owners are domiciled outside the USA and they have hardly any assets in the USA. It may be possible for only certain US flag shipowners to be able to self-insure, as they will have US-based assets. But even with the US flag shipowners it may not be possible to comply with the Coast Guard's requirements as the net worth of many shipowners is far less than their requirements.

Financial guarantees

Financial guarantors are under the same obligations as self-insurers. If the guarantee covers more than one fleet of vessels, the guarantor must have working capital and net worth equal to the cumulative total of financial responsibility. Under this requirement the worldwide application of a financial guarantee becomes impossible.

Other evidence of financial responsibility

Other evidence of financial responsibility allows only a guarantee that satisfies the stated requirements applicable to insurers, the providers of surety bonds and financial guarantees. Any other evidence provider is a statutory "guarantor" subject to all rights and obligations of a guarantor.

PART 4

LIMITATION OF LIABILITY

CHAPTER 17

THE ORIGINS OF LIMITATION OF SHIPOWNERS' LIABILITY

INTRODUCTION

The right of global limitation of liability was conceived to serve the needs of commerce and is peculiar to maritime law. The basic idea behind it was to encourage shipowners to carry on their business and capitalists to invest their money in the maritime sector in spite of the horrendous perils of the sea. Obviously under the technical conditions of the old days the result of a serious maritime disaster was very different from today. In most cases the disasters would very likely have resulted in the end of the shipowner's business. The simplest way of protecting the owner against the hazards of a maritime adventure was to exempt him from liability for certain claims or for those arising from certain causes. Under the global limitation of liability the owner's liability is limited for *all* claims subject to limitation, which arise on a certain occasion and during a certain time or in connection with a certain venture.

THE DEVELOPMENT OF LIMITATION

The origin of the practice that limits a shipowner's liability to the value of his vessel is not clear.[1] The *contrat de commande*, which seems to be the base of the limited liability in mercantile matters, originated before the twelfth century. Under this method the party who had advanced the goods or funds was not personally liable for contracts.[2] And in a trading venture a merchant could limit his liability to the extent of the goods or funds which were entrusted to another party for use in that particular venture. In medieval commercial codes the *contrat de commande* is indicated by detailed rules.

It appears that limitation of shipowner's liability first developed in Italy and then spread to Spain and France. The Amalphitan Table is the earliest extant evidence of the shipowner's right to limit his liability. This commercial code was written for

1. For detailed information on the origins and development of limitation of shipowners' liability, see J. James Donovan, "Admiralty Law Institute: Symposium on Limitation of Liability. The Origins and Development of Limitation of Shipowners' Liability", (1979) 53 *Tulane Law Review*, 999–1045.

2. *The Rebecca*, 20 F. Cas. 373, 378 (D. Me. 1831) (No. 11, 619), cited in J. James Donovan, p.1001, n.10.

the republic of Amalphia (Italy) in about the eleventh century. Later the Code of Valencia and its contemporary, the *Consulato del Mare* of Barcelona, were compiled under the direction of Peter IV of Aragon.[3] Both these codes have express limitation provisions. Under the terms of the *Consulato del Mare* owners' and part owners' liability was limited only to the extent of their respective shares in the ship itself for debts incurred by the master in obtaining ship's necessaries or for cargo damage arising from improper loading or from unseaworthiness.

As stated above the reason behind the limitation was mainly economic. It was not only shipowners who wanted to live in the certainty of liability; creditors also needed certainty. Therefore the limitation system appears to have been developed to encourage the investment of risk capital and the privilege of limitation reserved for investors by limiting their personal liability.

Following the commercial revolution of the sixteenth and seventeenth centuries the privilege of shipowner's limited liability was adopted in almost all the continental maritime jurisdictions such as the Atlantic coast trading communities, the North Sea and the Baltic communities, but interestingly not in England.[4]

In the Hanseatic Ordinances of 1614 and 1644 the liability of a shipowner was limited to the value of his vessel and the proceeds of the sale of the vessel were to be the extent of the satisfaction of all claims. Under different European statutes it was possible to find examples of limitation of liability of shipowners. For instance under the Statutes of Hamburg (1603) and the Maritime Codes of Charles II of Sweden (1667) in the case of the abandonment of a ship by creditors the shipowner's other property was protected from those claims unless agreed otherwise.[5]

Within these different statutes there was also an attempt to codify and systematise the international maritime law. The Marine Ordinance of Louis XIV was the first codification and was compiled under the direction of Minister Colbert in 1681. The basic principle behind this codification was to secure the benefits of the existing marine and commercial codes and legal decisions for French maritime and commercial interests. The Ordinance of 1681 was incorporated into the French *Code de Commerce* of 1807 (the *Code Napoléon*). As a result of these incorporations it formed part of the maritime law of several European and Latin American countries. Neither England nor America adopted the Ordinance but accepted it as evidence of general maritime law if it did not contradict judicial and statutory authorities of these countries.

3. *Ibid.*; *Thompson* v. *The Catharina*, 23 F. Cas. 1028, 1029 (D. Pa. 1795) (No. 13, 949), cited in *ibid.*, n.17.

4. D.R. Owen, "The Origins and Development of Marine Collision Law" (1979) 51 *Tulane Law Review*, 759, 764.

5. J. James Donovan, p.1003.

CHAPTER 18

INTERNATIONAL VARIATIONS OF LIMITATION OF LIABILITY

The basic principle of shipowners' limitation of liability is to hold the shipowner liable in principle but to reduce this liability by limiting his total exposure. Under the first developed systems of maritime community practically all liabilities, contractual or non-contractual, were subject to limitation as long as they arose during the maritime adventure. The only exception to this rule was where they were not incurred through a personal act of the owner. In all systems the rules on limitation of liability show similarities but there are differences in their judicial application.[1]

LIMITATION OF LIABILITY FOR PROPERTY CLAIMS

Under the following two systems limitation is by way of surrendering the venture assets or their value.

(a) The abandonment and maritime lien systems

These systems originated in the Middle Ages but developed fully in the seventeenth century. Both systems were named after the country where they were fully developed. The abandonment system originated in the Romanish countries of Southern Europe and developed in France; that is why it was also called the French system. Similarly the maritime lien system was developed in the Germanic countries of Northern Europe and was perfected in Germany, so it was also called the German system.

The concept of these two systems was identical. Both systems accepted the voyage as the limitation unit and liabilities subject to limitation were those that might arise during the venture. The surrendered assets of the shipowner were the

1. For detailed information on the different systems of limitation of liability see Alex Rein, "International Variations on Concepts of Limitation of Liability." (1979) 53 *Tulane Law Review* 1259–1276.

ship, its appurtenances and the freight earned in the venture. Claimants were satisfied in accordance with the priority of the respective liens attaching to their claims. Both systems aimed to encourage seafarers to invest in maritime ventures, so they reduced the liability to the value of the venture assets. There were also some differences between the two systems. Under the abandonment system the owner was personally liable. But the owner could always divest himself of all liability by declaring the venture bankrupt. Under the maritime lien system there was no personal liability on the owner. Under this system claimants could enforce their liens at once or keep their claims on the assets and wait until a better opportunity for realisation arose.

However, under both systems the security of claims depended on the shipowner. If the owner started new ventures without any additional security for existing claims new liens would arise and would take over the old ones.[2] Obviously every system shows the characteristics of the period in which it is created. During the development of these two systems marine insurance was just beginning to develop and in modern terms liability insurance did not exist. Therefore, instead of trying to hold the shipowner liable without limitation, it was easier to give the claimants priority rights over the assets of the venture. The main danger in both systems was the possibility of the fund becoming negligible or nil when the assets were lost. Owing to changes in the prevailing conditions shipowners began to take precautions to protect themselves mainly by insuring themselves against the risk of liability. But the interesting point is that despite the changes in the surrounding circumstances many countries retained the abandonment system as their internal law.[3]

(b) The American system

The American limitation system was introduced in 1851 by the Limited Liability Act 1851.[4] Until the introduction of this Act the shipowner's liability was not limited. The American Limitation Act was mainly based on the abandonment system. But within the application of this system the owner could divest himself of further liability by transferring his interest in the vessel and freight to a court-appointed trustee. Under this system the limitation unit is the voyage and the scope of claims is quite wide as set out in the statute. As far as property claims are concerned the Liability Act has not changed much. In other words nothing much has been done to adapt the American system to modern conditions.

Under the following two systems liability was limited to an amount of money calculated on the basis of the ship's tonnage:

2. Alex Rein, pp.1261–1263.

3. For instance France ratified both the 1924 and 1957 Limitation Conventions but kept the abandonment system as internal French law until 1967. Germany kept the system until it adopted the 1957 Convention. The system is also still in force in Italy. See Alex Rein, p.1263, n.3.

4. Limited Liability Act 1851, ch. 43,9 Stat. 635 (amended and codified at 46 U.S.C. §§181–183, 184–188 (1976)).

(c) The British tonnage system

England did not take part in the development of continental systems. The British system was not adopted by countries outside the British Commonwealth. In general other countries adopted the abandonment system. The British system was fundamentally different from the Continental systems on two points: under the British system limitation could be invoked only with respect to claims arising out of wrongful acts committed by the owner's servants in the course of their service to the ship. And the limitation unit was any "distinct occasion" giving rise to liability.

(d) The 1957 Limitation Convention

The 1957 Convention is a revised version of the British tonnage system. Under the Convention limitation is restricted to liability for damage or infringement of rights and wreck liability. The limitation unit is again the "distinct occasion" and the fund is set up exclusively on the tonnage of the ship. In order to find a compromise both to states that wanted a higher limit and those that did not, the measure based on the average value of British ships was reinstated. Despite the compromise the USA and Russia have not ratified the Convention.

The following three systems are called the option systems. These provide a choice between several traditional systems in the particular case.

(e) The Belgian system of 1908

In the international shipping community there are many systems for limitation and differences between these systems cause conflicts on the choice of law, which will be discussed later in the chapter. In 1908 the CMI introduced a new system as a result of efforts to achieve some sort of an international agreement. The Belgian system so introduced was a mixture of existing systems. The shipowner was given three choices to effect limitation. These were a mixture of existing systems. The shipowner could abandon the ship and freight as in the French system or surrender the value of ship and freight at the end of the voyage as in the American system or as a last choice apply a variation of the British system and pay 200 francs per tonne for the satisfaction of all claims, property and personal, relating to the voyage.

(f) The 1924 Limitation Convention

The 1924 Convention shows similarities to the Belgian system of 1908. It also presents a compromise solution. The limitation unit is the accident and the limitation fund is constituted on the basis of the value of the ship after the accident plus 10 per cent of its value at the commencement of the voyage. Apart from claims for damage the owner can also limit his liability for salvage, contribution to general average and obligations under contracts entered into by the master for the purpose of preserving the ship or continuing the voyage. If the Convention is considered strictly under the option system it is possible to say that the convention system has double limitation for certain claims, as the limitation fund is based on the value of the ship and the freight or the aggregate of £8 per ton, whichever is the lower.

(g) Other option systems

The Greek system is offered as an example of the abandonment system with an option. Under the Greek system the owner has an option. He may either abandon the ship and freight or deposit 30 per cent of the ship's value at the commencement of the voyage for personal claims with an additional 30 per cent for property claims and personal claims not satisfied for the first fund.

LIMITATION OF LIABILITY FOR PERSONAL CLAIMS

Britain was one of the countries which provided different liability rules for personal claims. Under the 1854 Merchant Shipping Act the limitation fund was required to be not less than £15 per ton in case of a death or personal injury claim. In such a case there was always a possibility of the value of ship and freight reducing. The amount of £15 per ton was required in particular for personal injury claims. This model was adopted by the 1924 Convention, but under the Convention a special fund of £8 per ton was distributed between the personal injury claimants.

The American system also showed similarities to the abandonment and the maritime lien systems where there is only one limitation fund for all claims subject to limitation. But in 1936 the American system also adopted the British model. Under the American system all claims share in the distribution of the value of ship and freight at the end of the voyage, but for personal claims the fund cannot be less than $60 per gross ton for each distinct occasion.

The 1957 Convention provided different limits for loss of life and personal injury claims either alone or together with property claims (3,100 francs) and 1,000 francs for property claims alone. If the occurrence gives rise to both personal and property claims the maximum fund is 3,100 francs per ton and 2,100 francs are reserved exclusively for personal claims.

Under the 1976 Limitation Convention separate limitation funds are provided for the two categories of claim.

CHAPTER 19

BRITISH CONCEPTS OF LIMITATION OF LIABILITY

HISTORICAL BACKGROUND

English Admiralty law was originally based on the Rules of Oleron. Under these rules there was no provision on limitation of liability of a shipowner—in fact such a subject was not even mentioned. The Rules were included in the *Black Book of the Admiralty* and this book was used as a reference in the English Admiralty courts.[1] During this period the common law doctrine of insurer liability of common carriers was applied. As a consequence of this policy the English shipping industry was in an economically disadvantaged position for more than 50 years.

(a) Common carrier's liability at common law

It is possible to find principles governing the liability of carriers under common law. Before the application of charterparties and modern bills of lading common law had principles on the liability of carriers independently of contract. In other words these principles show the implied obligations of a shipowner apart from any express contract. Common carriers are those who are willing to carry the goods of any person who could pay their charges as long as there is enough space available. The carrier is absolutely responsible for the safety of the goods while they remain in his hands as carrier. And, further, if the goods are lost or damaged while they are in the carrier's care the carrier is liable as an insurer.

The reasoning for such a strict liability was first given in *Coggs* v. *Bernard*[2] by Lord Holt:

> "... First, if it be to a person of the first sort [a public employment] and he is to have a reward, he is bound to answer for the goods at all events. And this is the case of the common carrier, common hoyman, master of a ship, etc.: The law charges this person thus entrusted, to carry goods against all events, but acts of God, and the enemies of the King ...."

Following this case the same reason was given in *Forward* v. *Pittard*[3] where the common carrier was liable for all due care and diligence and he was liable also on

1. J. James Donovan, pp.1005–1007.
2. (1703) 2 Ld. Raym. 909, 1 Smith's L.C., 13th ed., 175, 186.
3. (1785) 1 T.R. 27; 99 E.R. 953 (KB).

his contract for any negligence. But under the common law doctrine he had a further degree of responsibility—he was in the nature of an insurer. As in *Coggs* v. *Bernard* it was laid down that the carrier is liable for every accident, except by act of God or the King's enemies.[4]

(b) Sea carrier's liability at common law

In cases where a sea carrier was a common carrier it was clear that the same principles of liability applied as to the common carrier by land. But if a sea carrier was not a common carrier then a problem arose about liability. What standard of liability did the common law impose on a sea carrier in such a case? For instance if a land carrier was not a common carrier his liability was held to be no more than a bailee's. And a bailee of goods is not treated as an insurer of them. He is only supposed to do the best he can. So in the case of a sea carrier was his liability based on the liability of a bailee? In other words how can a sea carrier's liability be defined?

In *Liver Alkali* v. *Johnson*[5] in an action against the defendant by the plaintiffs for not safely and securely carrying certain goods the question was raised whether the defendant was liable as a bailee for hire or whether he had the more extended liability of a common carrier. To ascertain liability as a bailee or a common carrier is quite important as the limits of liability change according to this definition. Under the liability of a bailee the defendant is supposed to take proper care of the goods but is not responsible for their loss. If he has the more extended liability of a common carrier he is supposed to carry the goods safely against all events but acts of God and the enemies of the Queen. The defendant was a barge owner and let out his vessels for the conveyance of goods to any customers who applied to him. Blackburn J. said that "this describes the ordinary employment of a lighterman, and both on authority and principle a person who exercises this business and employment does, in the absence of something to limit his liability, *incur the liability of a common carrier*, in respect of the goods he carries". So the Court of Exchequer Chamber held, affirming the judgment of the Court of Exchequer, that the defendant was a common carrier and liable as such. In this case Brett J. differed from the majority and held that the defendant was not a common carrier. But he held him liable on a special custom of the realm attaching to all carriers by sea and stated that "by the law of England all carriers by sea are subject to the liability which by that law undoubtedly attaches to the common carrier whether by sea or by land . . .".[6]

In this case a class of public carriers by water, such as lightermen, were held to carry goods subject to the liabilities of common carriers. Shipowners who let their

4. It is also possible to see the same rule and reason on the common carrier's liability in *Riley* v. *Horne* (1828) 5 Bing. 217, 220.
5. (1874) L.R. 9 Ex. 338 (Ex. Ch.).
6. See A.D. Hudges, *Casebook on Carriage of Goods by Sea* (1994), pp.5–8. Martin Dockray, *Cases and Materials on the Carriage of Goods by Sea* (1987), pp.466–469.

ships under charter were also included in this class of carriers. But there was still a distinction between them and common carriers. The distinction was based on liability because this class of carrier was not liable to an action for refusing to accept goods for carriage as are common carriers.[7]

Similar arguments on shipowners' liability were also made in *Nugent* v. *Smith.*[8] In this case the plaintiff shipped his horses without any bill of lading, and during the voyage one of the animals died due to heavy weather conditions. The defendant was insistent that the storm, which was the primary, and to a certain extent the proximate, cause of the loss, must be taken to have been an act of God. Therefore the question which the court had to answer was whether the loss was occasioned by the act of God. The judgment of the Common Pleas Division was in favour of the plaintiff and Brett J. reiterated his views in this case and also referred to the authority of Roman law which he described as the origin of the English law of bailment. But Cockburn J. dissented from his theory and the general doctrine based on it in the Court of Appeal. He stated that Roman law made no distinction between the "act of God" and inevitable accident arising from other causes. But it exempted the carrier whenever the loss resulted from unforeseen and unavoidable accident. Then the court held that a loss caused by a combination of weather and inherent vice was sufficient to excuse the shipowner. In *Paterson Steamship Ltd.* v. *Canadian Cooperative Wheat Producers*[9] the shipowner's liability was discussed under common law and the court held that "under common law the shipowner is called an insurer, that is he was absolutely responsible for delivering in like order and condition at the destination the goods bailed to him for carriage. He could avoid liability for loss or damage only by showing that the loss was due to the act if God or the King's enemies."[10] One of the important points that follows from this case was on excepted perils. A carrier could avoid liability by stipulating that the loss was due to one of the excepted perils. But an inquiry would still remain whether or not the loss was due to negligence or unseaworthiness. If it was, the exception did not avail the carrier.[11]

(c) Common law limitations of liability

(i) Excepted perils

Under common law the liability of carriers has always been limited by the principle that certain risks are excluded. These may be defined as follows:

"Act of God" is the first exception to the strict rule of common law. The

7. Common carriers should undertake to carry the goods (dangerous or unreasonably heavy and bulky ones are excluded) which are first offered to them. Otherwise they are liable to an action for refusing to accept the goods.

8. (1876) 1 C.P.D. 423 (CA).

9. [1934] A.C. 538 (PC).

10. *Ibid.*, Lord Wright, p.544.

11. In the words of Willes J., "the exception in the bill of lading ... only exempts him [the shipowner] from the absolute liability of a common carrier, and not from the consequences of the want of reasonable skill, diligence and care ..." in *Notara* v. *Henderson* (1872) L.R. 7 Q.B. 225, 235.

shipowner is not responsible for loss or damage resulting from "acts of God". There has been a great deal of discussion on the definition of the term.

In *Forward* v. *Pittard* a statement was made on human action and its contribution to the result. In the words of Lord Mansfield, ". . . I consider it [the act of God] to mean something in opposition to the act of man; for everything is the Act of God that happens by His permission; everything, by His knowledge . . . In this case it does not appear but that the fire arose from some act of man, for it is expressly stated not to have happened by lightning". So the first essential feature of the definition is that the human action must not have contributed to produce the result. An accident caused by some act done in navigating the vessel does not come within the exception. For instance in *Siordet* v. *Hall* the boiler of the vessel was filled the night before it was due to sail. There was a hard frost and water in a pipe froze during the night. The pipe burst and when the water thawed it caused damage to the cargo. The carrier wanted to rely on the exception clause for act of God. But the Court of Common Pleas held that the carrier could not rely on the defence as the damage had been occasioned by the negligence of the defendants.[12]

It is not clear whether all causes other than those attributable to human intervention are to be regarded as natural causes.[13] In the case of animal transport the death or sickness of an animal may fall within the definition of "act of God" if the defect or disease which has caused the loss has definitely not been caused by the act of man.[14]

In *Nugent* v. *Smith* the following definition was given, ". . . for any accident as to which he [a common carrier] can show that it is due to natural causes directly and exclusively, without human intervention, and that it could not have been prevented by any amount of foresight and pains and care reasonably to be expected from him",[15] and it was elaborated with the emphasis on the standard of care required of the carrier, ". . . it must be such a direct, and violent, and sudden, and irresistible act of Nature as could not by any amount of liability have been foreseen, or if foreseen, could not by any amount of human care and skill have been resisted".[16] The other essential feature of the definition is that it must have been an event which the shipowner could not have avoided or guarded against with reasonable care. The shipowner has to show that it was not possible to avoid or guard against the causes with reasonable care and precaution. That the causes of

12. (1828) 6 L.J. C.P. 137 (CP), Best C.J. said: "the question in this case, is whether the damage which is the subject of this action, has been occasioned by the Act of God, or by negligence of the defendants. The evidence clearly showed, that the latter was the immediate cause; for unless the boiler had been filled with water, the frost could not have so affected the pipe; or, if it could, the bursting of the pipe would not have done the damage, unless there had been water in it. We, therefore, think, that there was negligence on the part of the defendants, in allowing the water to be placed in the boiler so long before it was absolutely necessary."

13. See *Dale* v. *Hall* (1750) 1 Wils. 281.

14. In *Williams* v. *Lloyd* it was held that the bailee of a horse was excused from redelivering it when it had become sick and died without any neglect on his part, on the ground that this happened by the act of God: (1628) W. Jones Rep. 179.

15. See A.D. Hudges, pp.12–14.

16. *Ibid.*

the casualty should be independent of human action does not necessarily mean that they should be extraordinary. The important point is that the results of the event should be within the exception.

"The King's Enemies"—this exception covers the enemies of the British Crown and also the action of any forces engaged in hostilities with the carrier's state. The shipowner is supposed to be careful to avoid the acts of these enemies but he is not liable for losses occasioned by them.[17]

"Inherent vice"—a carrier is not liable if loss or damage results from an inherent quality, or defect, of the thing carried. In *Gould* v. *South Eastern and Chatham Railway Co.*[18] it was held that the implied liability of a common carrier did not include responsibility for ordinary wear and tear in transit, ordinary loss or deterioration in quality or quantity, such as evaporation or loss or damage through insufficiency of packing. But if the loss of or defect in the goods has arisen from adventitious causes introduced by the carrier the exception does not apply. The carrier is supposed to show adequate care for the goods, or he cannot rely on the exception by saying that the loss was due to an inherent weakness in the goods. The degree of care required from the shipowner will be ascertained from the extent of his knowledge and means of the knowledge of the character of the goods.

(ii) Contractual limitation of liability

There are also certain exceptions under contractual limitation of liability.

"Perils of the sea"—for this exception to apply there must be a peril or accident. And this peril must be a sea peril which cannot be avoided by taking reasonable care. Perils of the sea are "peculiar to carrying on business on the sea".[19] In *The Xantho* it was held that perils of the sea includes a collision. The bill of lading contained an exception for "perils of the sea". The vessel came into collision with another ship and damage to the goods resulted. The shipowners relied on the exception. The House of Lords held that the collision was a "peril of the sea" within the meaning of the exception. The words "peril of the sea" had the same meaning in a policy of insurance and in a bill of lading. Consequently the shipowners were not liable for the loss. In this case the meaning of the term was discussed in detail.

> "The term does not cover every accident or casualty which may happen to the subject matter of the insurance on the sea . . . For example the term does not protect against the natural and inevitable action of the winds and waves, which results in what may be described as wear and tear. There must be some casualty, something which could not be foreseen as one of the necessary incidents of the adventure."[20]

This case also gave rise to discussion on the interpretation of the term "perils of the

17. See *Russell* v. *Niemann* (1864) 17 C.B.N.S. 163. 84 E.R. 548.
18. [1920] 2 K.B. 186 (CA).
19. *Hamilton* v. *Pandorf* (1886) 17 Q.B.D. 670, 675, *per* Lord Esher M.R.
20. *Wilson, Sons & Co.* v. *The Owners of the Cargo of the "Xantho"* (1887) 57 L.T. 701, *per* Lord Herschell, p.703.

sea" in a contract of carriage and in a policy of marine insurance. In a marine policy the *causa proxima* is considered alone, but in a contract of carriage it is possible to go behind the *causa proxima* and consider the real or efficient cause of the incident. Therefore the shipowner is not protected by the exception of the perils of the sea in every case in which he would be entitled to recover on his insurance policy. According to Lord Herschell the words do not differ in meaning between the two instruments. The reason for the different interpretation is the general scope and purpose of the contract of carriage excluding the operation of the exception in certain cases.[21]

Pirates have long been regarded as a peril of the sea.[22] If a rat eats part of a ship and the sea enters through the hole and harms the cargo, then the damage is covered by the exception "perils of the sea".[23] But if the sea enters through a hole barratrously bored in the ship by the crew it is apparently not a peril of the sea.[24]

"Strikes"—in *The Laga* the vessel was chartered on terms which provided that time lost through strikes which prevented or delayed discharge would not count against laytime. In a French port the stevedores refused to unload her and all other ships thereby aiming to assist French miners who were on strike. It was held that the sympathetic strike of the stevedores was a "strike" within the meaning of the charterparty. It does not merely mean a general concerted refusal by workmen to work in consequence of an alleged grievance.[25] In *The New Horizon* the charterparty provided that time should not count during a strike or lockout of any class of workmen essential to the discharge. At the berth dockworkers customarily worked round the clock on shifts if requested, but they were not legally obliged to do so. When the *New Horizon* arrived crane and sucker drivers were refusing shift work. The charterers maintained that the resulting periods when no work was done were excluded from the calculation of laytime because they were periods of strike. In his statement Lord Denning M.R. also said that there could be a strike even though the workmen were not in breach of contract.[26]

The problems raised by strikes are quite important in the case of charterparties as both the shipowners and charterers are reluctant to pay the cost of delay.

"Robbers, thieves"—in a case of robbery there is a taking by force and the carrier cannot reasonably resist. In a case of theft there is also a taking, but by persons from outside the ship and not belonging to it.

"Barratry"—the Marine Insurance Act defines barratry to include "every wrongful act wilfully committed by the master or crew to the prejudice of the owner, or as the case may be, the charterer".[27] And within the scope of the Act barratry is expressly identified as a marine peril. The Act defines a marine peril as

21. *Ibid.*
22. *Pickering* v. *Barkley* (1648) Style 132.
23. *Hamilton* v. *Pandorf*, n.19 *supra*.
24. *The Chasca* (1875) L.R. 4 A. & E. 446.
25. *J. Vermaas' Scheepvaartbedrijf N.V.* v. *Association Technique de l'Importation Charbonniere* [1966] 1 Lloyd's Rep. 582 (QBD).
26. [1975] 2 Lloyd's Rep. 314 (CA).
27. The Marine Insurance Act 1906, Sched. 1, Rule 11.

"perils consequent on or incidental to the navigation of the sea, that is to say, perils of the seas, war perils, pirates, rovers, thieves, captures, seizures, restraints, and detainments of princes and peoples, jettisons, barratry, and any other perils, either of like kind or which may be designated by the policy".[28] Therefore in order to find out whether a barratrous act is a covered peril the marine insurance policy has to determine whether the barratry as a peril is specifically addressed by the policy.

It is also possible to find definitions of barratry in various cases. In *Earle* v. *Rowcroft*, Lord Ellenborough described barratry as "a fraudulent breach of duty by the master, in respect to his owners; or in other words, a breach of duty in respect to his owners, with a criminal intent, or *ex maleficio*, is barratry".[29] After the Marine Insurance Act 1906 the need for criminal intent became unnecessary within the definition. In *Compania Naviera Bachi* v. *Henry Hosegood & Co. Ltd.* it was stated that "every deliberate act of wrongdoing by the master or crew against the ship or cargo, without the authority of the shipowner is barratrous.[30] "For the purpose of barratry the commission of crime is not necessary. It must be a wilful act deliberately done, and to the prejudice of the owners. It is not necessary that the persons doing it should desire to injure the owners if, in fact there, is an intention to do an act which will cause injury, even if the act be done to the benefit of persons who are guilty of barratry."[31]

"Restraint of Princes, Rulers, People"—in *Nobel's Explosives* v. *Jenkins* the goods were shipped by the defendant's steamer under a bill of lading by which the goods were to be delivered "at Yokohama or so near thereunto as she may safely get". The bill of lading contained the usual exception of "restraint of princes and rulers or people". The same day the vessel arrived in Hong Kong war was declared between China and Japan. The defendants landed explosives at Hong Kong and continued the voyage to Yokohama with the remaining cargo. The plaintiffs brought proceedings for breach of contract to carry the goods to Yokohama. The main ground of defence was the exception in the bill of lading of "restraint of princes, rulers or people". It was argued for the plaintiffs that the clause did not apply unless there was direct and specific action upon the goods by sovereign authority. But in his statement Mathew J. made it clear that apart from the terms of the bill of lading the conduct of the captain would be justified by reference to the duty imposed upon him to take reasonable care of goods entrusted to him. Whether he had discharged that duty had to depend upon the circumstances of each case, and in this case, if the goods had been carried forward, there was every reason to believe that the ship would have been detained and the goods of the plaintiffs confiscated. Therefore in refusing to carry the goods further the captain acted reasonably and prudently, and delivery of the goods at Yokohama was prevented by restraint of princes and rulers within the meaning of the exception.[32]

28. The Marine Insurance Act 1906, s. 3(2)(c).
29. (1806) 8 East 126 at p.138.
30. [1938] 2 All E.R. 189.
31. *Mentz, Decker & Co.* v. *Maritime Insurance Co.* (1909) 15 Com. Cas. 17, at p.24, *per* Hamilton J.
32. [1896] 2 Q.B. 326.

In *Watts, Watts & Co. Ltd.* v. *Mitsui & Co. Ltd.* the importance of the circumstances of each case was stressed again. "It would be useless to try to fix by definition the precise imminence of peril which would make the restraint a present fact as contrasted with a future fear. The circumstances in each particular case must be considered."[33]

"Negligence"—it is possible for the shipowner to rely on an excepted peril even if he has been negligent, as long as "negligence" itself is one of the excepted perils. In *Blackburn* v. *Liverpool, Brazil and River Plate Steam Navigation Co.*[34] sugar was stored in a tank at the bottom of the ship. The engineer negligently let salt water into the tank. The shippers argued that they were entitled to recover damages because the shipowners were bound by their contract to carry and deliver the cargo safely and they failed to do so. But the shipowners claimed that the obligation to carry and deliver the cargo safely was subject to the exceptions contained in the bill of lading. It contained an exception clause in respect of "loss or damage resulting from any of the following perils (whether arising from the negligence, default or error in the judgment of the ... engineers ... or otherwise howsoever) ...". In this case the duties of a shipowner under a contract of carriage by sea were discussed in relation to a negligence clause.[35] Under a carriage contract a shipowner is under an absolute duty to deliver the goods safely. This absolute duty is subject to the exception of perils of the sea and any other exceptions in the bill of lading. But even these excepted perils are restricted by due care and diligence to deliver the goods safely. If the goods are not carried safely but the loss is due to one of the excepted perils then the owner is not liable unless he used due care and diligence. Therefore the exception does not absolve him from the obligation to use due care and diligence. But if the contract contains a negligence clause the result is different. In such a case the shipowner is exempt from liability for loss occasioned by the perils of the seas although it is caused by negligence. And there is no need to consider liability for want of due care and diligence. In this case the question was whether the loss was occasioned by a "peril of the seas or navigation". If that was the case the owner was protected by the exception, even though the loss was due to the negligence of his servants.

Consequently, carriers may protect themselves from liability for loss or damage arising from their or their servants' negligence by a negligence exemption clause. In such a case the only difficulty may be to convince the court.[36]

In *The Emmanuel C*[37] the ship was let on a time charter. The charterparty contained an exception clause for "errors of navigation". By reason of negligent navigation the vessel grounded and became a constructive total loss. The

33. [1917] A.C. 227 (HL), *per* Lord Dunedin. See also *Ciampa and Others* v. *British India Steam Navigation Co. Ltd.* [1915] 2 K.B. 774.

34. *Blackburn* v. *Liverpool, Brazil and River Plate Steam Navigation Co.* (1902) 1 K.B. 290.

35. *Ibid.*, Walton, J. p.293.

36. See *Tor Line AB* v. *Alltrans Group of Canada Ltd. (The TFL Prosperity)* [1984] 1 Lloyd's Rep. 123 (HL).

37. *Industrie Chimiche Italia Centrale SpA* v. *Nea Ninemia Shipping (The Emmanuel C)*, [1983] 1 Lloyd's Rep. 310.

charterers claimed damages. This case raised an important question on the limits of the exception clause. Does it cover the owners against liability for negligent errors of navigation or only the errors occurring without negligence on the part of the owners, their servants or agents? In order to examine the exemption clause Bingham J. referred to a series of cases and in them May L.J. provided a summary of important points on exemption clauses.[38] In his view the court must give effect to the exemption if it clearly excludes liability for negligence. In cases where the clause does not make such an exclusion expressly but its wording clearly implies such a meaning then the intention of the parties becomes important. That has to be considered in the light of relevant circumstances. Thus the court has to examine all the relevant circumstances and then reach a conclusion. In relation to related circumstances of the case it was held that the shipowners were not protected by the exception; first, none of the events specified in the clause clearly indicated negligence. On the contrary some were plainly inconsistent with it. Secondly, in a similar case,[39] an English court held that when it was the result of negligence the stated peril did not exempt the owner from liability.

(iii) Burden of proof

Traditionally when cargo owners prove loss the carrier has to prove that the loss was caused by the excepted peril or event. Then it is for the cargo owner to show that the cause of the loss was the negligence of the carrier.

In *The Glendarroch* in defence of the action of the plaintiffs for non-delivery the defendants relied on the exception of perils of the sea. The court ruled that the onus was on the defendants to show not merely a peril of the sea but a peril of the sea not occasioned by their negligence. The defendants appealed. The Court of Appeal held that once the plaintiffs had proved the contract and the defendants had proved the loss by peril of the sea, the burden of proving that the shipowner was not entitled to the benefit of the exception on the ground of negligence was on the party alleging it. In this case the burden was on the plaintiffs.[40]

DEVELOPMENTS UNDER BRITISH LAW

In England the history of the limitation of liability rule goes back to 1733.[41] In 1733 a shipowner was found to be personally liable for the loss of a cargo of bullion taken on board in Portugal and subsequently stolen by the master. The shipowner's liability was for the full value of the cargo.[42] Following the decision of the court which held the owners personally liable for losses occasioned by acts of

38. *Ibid.*, Bingham J., p.311.
39. *Re Polemis and Another* v. *Furness, Withy & Co. Ltd.* (1921) 8 Ll.L.R. 351, [1921] 3 K.B. 560.
40. *The Glendarroch* [1894] P. 226 (CA).
41. For detailed information on the British concept of limitation of liability see Michael Thomas, "British Concepts of Limitation of Liability", (1979) 53 *Tulane Law Review*, 1205–1259; James J. Donovan, pp.1007–1010.
42. *Boucher* v. *Lawson*, 95 E.R. 53 (KB 1733).

embezzlement of the master the shipowners petitioned Parliament for relief from the decision.[43] They complained that "when they became owners of ships, they did not apprehend themselves exposed to such hazard, or liable as owners to any greater loss than that of the ships and freight; and of the insupportable and unreasonable hardships to which our laws in this case subject them; and to which no owners of ships are exposed in other trading nations".[44] The petitioners also added that Parliament should realise "that, unless some provision be made for their relief, trade and navigation will be greatly discouraged, since owners of ships find themselves, without any fault on their part, exposed to ruin".[45]

As a result of this petition Parliament passed the Responsibility of Shipowners Act in 1734.[46] With this Act shipowners were absolved of liability for those acts of the master or crew which caused loss of or damage to cargo. The limitation was up to the value of the ship, its equipment, and the freight which was to be earned on that particular voyage. The main considerations of Parliament in enacting such a statute were to increase the number of ships in its merchant fleet and to encourage the participation of investors and merchants in shipping and trade.[47] But the protection offered by the statute was not very wide; in fact in nature it was extremely limited. The limit provided by the Act was retained to the value of the vessel proceeded against and the freight she was earning or under contract to earn. In the case of a time charter hire which would not begin to be earned until six months after the loss or damage occurred was not included. In cases where there was more than one incident causing loss or damage during a particular voyage, a separate fund of that amount was available for each separate incident. This principle was further extended by the Merchant Shipping Act 1854. The further development in English limitation of liability statutes took place after *Sutton* v. *Mitchell*.[48] A gang of robbers forcibly plundered the ship *Thames* which was moored in the river during the night. As the thieves acted on information supplied by one of the seamen on board, who later shared in the plunder, the shipowner's liability was limited. But if the thieves had acted in co-operation with somebody else rather than a member of the crew the owner's liability would have been unlimited. And in his judgment Judge Mansfield questioned whether such robbery was covered by the term "embezzlement". This judgment showed shipowners how easily their liability

43. 1733 H.C. Jour. 277.
44. *Ibid.*
45. *Ibid.*
46. 7 Geo. 2, c. 15 (1734).
47. The policy of Parliament was made clear by its preamble:
"WHEREAS it is of the greatest consequence and importance to this Kingdom, to promote the increase of the number of ships and vessels, and to prevent any discouragement to merchants and others from being interested and concerned therein;
And whereas it has been held, that in many cases, owners of ships or vessels are answerable for goods and merchandise shipped or put on board the same, although the said goods and merchandise, after the same have been so put on board, should be made away with by the masters or mariners of the said ships or vessels, without the knowledge or privity of the owner or owners, by means whereof merchants and others are greatly discouraged from adventuring their fortunes, as owners of ships or vessels, which will necessarily tend to prejudice of the trade and navigation of this Kingdom . . .".
48. 99 E.R. 948 (KB 1785).

could become unlimited and the dangers that they faced. As with *Boucher* v. *Lawson* a number of shipowners again petitioned Parliament after this case to expand the protection of the Act.

As a result of this petition another Responsibility of Shipowners Act was passed in 1786.[49] This Act was to explain and amend the 1734 Responsibility of Shipowners Act and give further relief to shipowners. It extended the relief to cases of robbery by persons other than the crew. It also provided that owners should not be liable for any loss or damage resulting from a fire[50] on board and also precluded recovery for loss to shipments of "gold, silver, diamonds, watches, jewels or precious stones" unless a bill of lading contained a description of the "nature, quality and value" of those particular goods.

Until the early nineteenth century the idea of limiting liability for damage resulting from collisions was not within the ambit of limitation statutes. In 1813 another Act concerning the liability of shipowners was passed. This was confined to registered British seagoing ships[51] and in its preamble it provided that the privilege of limitation was granted for reasons of public policy, not justice.[52] This Act provided for shipowners' limitation of liability for damages which arose as a result of negligence and for the first time extended limitation for loss or damage arising from collision. The Act's benefits were not to accrue to the owners of barges and vessels engaged in inland navigation. When the owner had paid into court the value of the vessel and its pending freight (or charter hire) the court would stay all other actions pending against such owner.

The principles stated under the 1734 Act was further extended by the Merchant Shipping Act 1854. This Act concerned cases of loss of life and personal injury. The calculation of the limit of liability was based on the previous principles, upon the value of the ship. In cases of loss of life and personal injury a minimum value of £15 was placed upon all seagoing vessels. But apart from this exception the owners of inferior ships had an advantage over the owners of well maintained ships as the value of their ship was much less. As a result of this disadvantage the method by which the limit of an owner's liability was to be calculated was altered by the Merchant Shipping Amendment Act 1862.[53] This Act extended the privilege of limitation to foreign ships, but the privilege of excluding liability was not extended. For all ships an average value of £15 or £8 per ton was accepted. The valuation would be £15 or £8 according to whether or not the incident was accompanied by loss of life or personal injury. In the case of sailing ships the tonnage taken was the

49. 26 Geo. 3, c. 86 (1786).

50. *Ibid.*, §2 states: "That no Owner or Owners of any Ship or Vessel shall be subject or liable to answer for or make good, to any or more Person or Persons, any Loss or Damage which may happen to Goods or Merchandize whatsoever which . . . shall be shipped, taken in, or put on board any such Ship or Vessel, by Reason or Means of any Fire happening to or on board the said Ship or Vessel."

51. Cited in Michael Thomas, p.1207, n.15.

52. The Responsibility of Shipowners Act 1813, 53 Geo. 3, c. 159 preamble provided that "WHEREAS it is of the utmost consequence and importance to promote the increase of the number of ships and vessels belonging to the United Kingdom, registered according to law, and to prevent any discouragement to merchants and others from being interested therein".

53. 25 & 26 Vict., c. 63 (1862).

registered tonnage, for steam ships it was the gross tonnage without the deduction of the engine room space. The provisions of the 1854 and 1862 Acts regarding limitation were, in substance, re-enacted in the Merchant Shipping Act 1894.[54] This general law regarding limitation of liability was later amended by the Merchant Shipping Acts 1906[55] and 1921[56] and the Merchant Shipping (Liability of Shipowners and Others) Acts 1900[57] and 1958.[58] The most important amendment was the Merchant Shipping (Liability of Shipowners and Others) Act 1958 which amended the 1894 Act to give effect to the 1957 Convention.

At present the law relating to the statutory right of shipowners and others to limitation and exclusion of liability is largely set out in the relevant provisions of the Merchant Shipping Acts 1894 and 1979, the latter implementing the Convention on the Limitation of Liability for Maritime Claims 1976, which came into effect on 1 December 1986. Section 185 of the Merchant Shipping Act 1995 restates sections 17 and 35(2) of and paragraph 3 of Schedule 5 to the 1979 Act and gives effect to the 1976 Limitation Convention which is set out in Schedule 7 to the Act.[59] Section 185(4) of the Merchant Shipping Act 1995 has the effect of preventing the Convention from having any application to claims for loss of life, personal injury or loss of or damage to property by a person who is aboard the vessel against which the claim is brought under a contract of service governed by the law of any part of the United Kingdom.[60]

It will be necessary here to consider the law under the 1894 Act, with its amendments under the 1958 Act, as the 1976 Convention was built on the practice established by the earlier convention.

THE 1957 LIMITATION CONVENTION: MERCHANT SHIPPING ACT 1894

(a) Persons entitled to protection

Originally the privilege provided by Part VIII of the 1894 Act was enjoyed only by the owner of the ship that caused the loss or damage. In this context the term

54. 57 & 58 Vict., c. 60 (1894).
55. 6 Edw. 7, c. 48 (1906).
56. 11 & 12 Geo. 5, c. 28 (1921).
57. 63 & 64 Vict., c. 32 (1900).
58. 6 & 7 Eliz. 2, c. 62 (1958).
59. Merchant Shipping Act 1995, s. 185 states:

"185.—(1) The provisions of the Convention on Limitation of Liability for Maritime Claims 1976 as set out in Part I of Schedule 7 shall have the force of law in the United Kingdom."

60. Merchant Shipping Act 1995, s. 185(4) states:

"185.—(4) The provisions having the force of law under this section shall not apply to any liability in respect of loss of life or personal injury caused to, or loss of or damage to any property of, a person who is on board the ship in question or employed in connection with that ship or with the salvage operations in question if—

(a) he is so on board or employed under a contract of service governed by the law of any part of the United Kingdom; and "owner" included equitable or unregistered owners.[61] Under the definition of the term if the crew of a barge negligently caused damage to a third party the hirer of the barge was fully liable. Similarly if a charterer appointed a master who negligently caused the ship to collide with another the charterer was liable in full, as the master was the charterer's servant. In order to avoid unfairness under the Act protection was extended by subsequent statutes. Under the 1958 Act the limitation right was extended to any charterer (demise time or voyage) and any person interested in a ship—in particular managers or operators.[62] For the purposes of Part VIII of the 1894 Act, the term "owner" now includes[63]:

(1) An owner, whether registered or unregistered;
(2) A charterer, whether by demise or otherwise;
(3) Any person interested in or in possession of a ship, and in particular, any manager or operator of a ship[64];
(4) The master and any member of the crew (with two exceptions);
(5) Some other servants of the above-mentioned persons in categories (1) to (3) are also protected if the claim arises from any act or omission of theirs during the course of their employment;
(6) If the master or crew are the servants of a person who is not included in categories (1) to (3) and the claim arises out of an act or omission of the master or member of the crew then that person is protected;
(7) The Crown. Under the Crown Proceedings Act 1947 the protection afforded by Part VIII of the 1894 Act was extended to Her Majesty in respect of HM ships. The extension also covers demise or subdemise charterers of ships from Her Majesty acting in the right of Her Government in the United Kingdom and to managers of HM ships.

In the rest of the Article the term "owner" covers all the persons who are entitled to seek protection of Part VIII of the 1894 Act.

(b) Types of ships entitled to limitation

In the 1894 Act a ship was defined to include "every description of vessel used in navigation not propelled by oars".[65] And by the same section, Part VIII of the Act, "vessel" was defined as including "any ship or boat, or any other description of vessel used in navigation". There are also many cases concerning what constitutes

(b) the liability arises from an occurrence which took place after the commencement of the Act. In this subsection, 'ship' and 'salvage operations' have the same meaning as in the Convention."

61. *The Spirit of the Ocean*, 167 Eng. Rep. 388, 390 (Adm. 1865).
62. Merchant Shipping Act 1894, s. 503, as amended by the Merchant Shipping (Liability of Shipowners and Others) Act 1958, s. 3.
63. See, Michael Thomas, pp.1209–1210.
64. In the relevant circumstances these words cover a shipbuilder or shiprepairer.
65. Merchant Shipping Act 1894, 57 & 58 Vict., c. 60, §742.

a ship or vessel in a particular case. But these cases make it clear that whether a particular object is or is not a ship or vessel depends on the facts of each case.

The definition of ship given in section 742 of the 1894 Act has been extended by a number of subsequent amendments. Under the Merchant Shipping Act 1921 it was extended to include "every description of lighter, barge, or like vessel used in navigation in Great Britain, however propelled: Provided that a lighter, barge, or like vessel used exclusively in non-tidal waters, other than harbours, shall not, for the purposes of this Act, be deemed to be used in navigation".[66] The extension under the 1958 Act states as follows: "any structure, whether completed or in the course of completion, launched and intended for use in navigation as a ship or part of a ship."[67] This statement shows that the 1958 Act has taken into account the modern methods of prefabrication in shipbuilding. The amendments in the 1958 Act have also extended the frame of the ships which are within the scope of the Act. Before the 1958 Act exclusion or limitation of liability could be claimed in respect of a British ship only if the ship was recognised as a British ship within the meaning of the 1894 Act. And the only way to recognise a ship as British was through registration.[68] So limitation could always be claimed in respect of British ships or foreign ships.[69] In *The Andalusian*[70] the vessel came into collision with another while being launched. The owners could not limit their liability because their vessel had not yet been registered. This decision has been criticised as in a similar case[71] the owners of the vessel being launched were held to be entitled to limit their liability. And in *The Andalusian* the issue of whether the vessel being launched was indeed a vessel or not was also not considered.[72]

By virtue of section 4(2) of the 1958 Act, Part VIII of the 1894 Act now applies to both unregistered and registered British ships. Now British ships can avail themselves of the right even if not yet registered. Under the Merchant Shipping Act 1921 the owners of docks, canals, harbour and conservancy authorities have also been given the right to limit their liability to ships using their installations.[73] In the case of foreign ships the right to limit liability depends on whether the owner is seeking to exclude or only to limit his liability. Based on the historical developments of these provisions the owner of a foreign ship may limit his liability under section 503, but he is not entitled to exclude it altogether under section 502.

66. Merchant Shipping Act 1921, 11 & 12 Geo. 5, c. 28, §1. This extension deals with Part VIII of the 1894 Act and also Part I which covers registration.
67. This extension only applies to Part VIII.
68. Merchant Shipping Act 1894, s. 508.
69. Merchant Shipping Act 1894, s. 503.
70. (1878) 3 P.D. 182.
71. See *St. Machar* (1939) 64 Ll.L.R. 27.
72. Cited in Michael Thomas, p.1212, n.50.
73. Merchant Shipping (Liability of Shipowners and Others) Act 1900, s. 2.

(c) Claims subject to limitation

Events for which limitation is sought are to be found in sections 502 and 503 of the 1894 Act. Section 502 sets out events for which liability may be excluded altogether. And section 503 sets out those for which liability may be limited.

There are two important points which have to be kept in mind in relation to the incidents with regard to which an owner may limit his liability. First, the limitation is only of the owner's liability in damages. Even if he can limit his liability he is not absolved from the consequences of his wrongful act. For instance in *The Stonedale* a dumb barge sank in the Manchester Ship Canal but the owners were held to be unable to limit their liability for the cost of raising her because the Canal Company made its claim under statutory powers which gave it the right to recover wreck-raising expenses, so the claim was not in damages.[74] So in cases where an owner is faced with a claim other than one for damages he cannot rely upon section 503. But under the 1958 Act an obligation or liability arising in respect of damage to harbour works, basins and navigable waterways is to be treated as a liability in damages. In theory the wreck-raising expenses of harbour authorities could be subject to limitation, but the United Kingdom has never brought into force the specific provisions of the 1957 Convention.

Secondly, the 1957 Convention did not allow a shipowner to limit his liability in relation to claims by his own crew if the law governing their contracts of service forbade it.[75] This protection was enacted into law in section 2(4) of the 1958 Act where none of the persons who were covered within the extended meaning of the "owner" were entitled to limit their liability in respect of any of the matters falling within section 503 if:

(a) the injured party was on board the ship or employed in connection with it under a contract of service with all or any of those persons; and
(b) that contract was governed by the law of any country outside the United Kingdom; and
(c) that law either did not set any limit on that liability or set a limit exceeding that set on it by section 503.

Under such protection as is offered by virtue of section 2(4) crews of British ships were at a disadvantage until the enactment of the Merchant Shipping Act 1979, section 35.[76]

74. (No.1) [1956] A.C. 1.
75. Art. I (4)(b).
76. S. 35 of the Merchant Shipping Act 1979 provides as follows:

"35.—(1) Nothing in Section 503 of the Merchant Shipping Act 1894 (which relates to the limitation of liability in certain cases of loss of life, injury or damage) shall apply to any liability in respect of loss of life or personal injury caused to, or loss of or damage to any property of, a person who is on board or employed in connection with the ship in question if—

(a) he is so on board or employed under a contract of service governed by the law of any part of the United Kingdom; and
(b) the liability arises from an occurrence which took place after the coming into force of this subsection and before the coming into force of the following subsection;

in this subsection 'ship' has the same meaning as in the said section 503.

In general liability may be limited in respect of claims for loss of life or personal injury suffered by any person on board a ship and any property on board the ship.[77] The owner may also limit for the faults of the persons not on board, provided that the wrongdoing occurred in the navigation or management of the ship or in relation to the carriage of passengers or cargo.

It is necessary to distinguish between incidents occurring on board the limiting ship and incidents occurring elsewhere. When an incident occurs on board, the owner may be entitled to limit his liability whatever the nature of the act or omission giving rise to liability and whether the person primarily at fault was on board the ship or elsewhere. On the other hand if the act or omission was otherwise than in the navigation or management of the ship, in the loading, carriage or discharge of its cargo, or in the embarkation, carriage or disembarkation of its passengers, the owner cannot limit his liability unless the person primarily at fault was on board his ship.

(i) Incidents on board

An owner may limit his liability for loss of life or personal injury to any person being carried in his ship[78] unless his ship is in fact a lighter or barge. For the purposes of limitation the definition of ship has been extended by the Merchant Shipping Act 1921, but under the extended definition[79] the owners of such vessels are not entitled to limit their liability in respect of loss of life or personal injury caused to any person carried therein.[80]

The words "carried in the ship" have been held to cover more than mere physical presence on board. In *Moore* v. *Metcalf Motor Coasters Ltd*[81] while returning to the ship from ashore the ship's chief engineer drowned as a result of falling from an allegedly defective gangway. In this case the court stressed the difference between the expression "being carried in the ship" and the expression "on board the ship" while discussing whether the seaman was "being carried in the ship" within the

---

(2) The provisions having the force of law under section 17 of this Act shall not apply to any liability in respect of loss of life or personal injury caused to, or loss of or damage to any property of, a person who is on board the ship in question or employed in connection with that ship or with the salvage operations in question if—

(a) he is on board or employed under a contract of service governed by the law of any part of the United Kingdom; and

(b) the liability arises from an occurrence which took place after the coming into force of this subsection;

and in this subsection 'ship' and 'salvage operations' have the same meaning as in those provisions."

77. Merchant Shipping Act 1894, s. 503(1)(a) and (b).

78. Merchant Shipping (Liability of Shipowners and Others) Act 1958, 6 & 7 Eliz. 2, c. 62, §503(1)(a).

79. Under the Merchant Shipping Act 1921, §1(1), the definition of "ship" includes "every description of lighter, barge or like vessel used in navigation" unless used exclusively in non-tidal waters.

80. Merchant Shipping Act 1921, 11 & 12 Geo. 5, c. 28, §3.

81. [1958] 2 Lloyd's Rep. 179, [1959] 1 Lloyd's Rep. 264 (CA).

meaning of section 503(1). It was held that the expression "on board the ship" means physically on board the ship. The former expression has a wider meaning and extends to persons who, although not actually on board, may be said to be carried by reason of some contractual arrangement. It includes the concept of the status of the person being carried under a contract of employment or as a result of some other contractual arrangement such as a passenger contract. Therefore the chief engineer who fell from a gangway whilst going on board was a person "being carried in the ship".

An owner also may limit his liability for any loss or damage to "any goods, merchandise, or other things whatsoever on board the ship".[82] In *The Stella* the argument that the meaning of the words "or other things whatsoever" ought to be restricted to things similar to goods or merchandise was rejected when a passenger had lost his luggage when the vessel sank.[83]

(ii) Incidents occurring elsewhere than on board the vessel

An owner may limit his liability for loss of life or personal injury to a person[84] and for loss of or damage to property not being carried in his ship where the death, injury, loss or damage arose from the act or omission of any person, whether on board the ship or not, in the navigation or management of the ship; in the loading, carriage or discharge of its cargo; in the embarkation, carriage or disembarkation of its passengers; or through any act or omission of any person on board the ship.

In general it is not very difficult to decide whether or not the act or omission took place during the loading, carriage or discharge of cargo or the embarkation, carriage or disembarkation of passengers. But the meaning of the words "navigation or management of the ship" has given rise to some difficulty as it is not easy to find cases concerning the meaning of the word "navigation" in the context of section 503 as amended by the 1958 Act.

In *The Warkworth* the ship collided with another vessel owing to a defect in its steam steering gear. This had been caused by the negligence of workmen and a superintendent on shore who were servants of the owner. The granting of a limitation decree was opposed on the ground that "improper navigation" referred only to something done in directing the progress of the vessel and did not apply to a mishap arising out of the negligence of the owner's servants ashore. Lord Brett rejected this argument and said: "... all damage wrongfully done by a ship to another whilst it is being navigated, where the wrongful action of the ship by which the damage is done due to the negligence of any person for whom the owner is responsible, is comprised within the statute".[85] It was held that the superintendent on shore could limit his liability. In *The Vigilant* it was stated that to cast off a tow

82. Merchant Shipping (Liability of Shipowners and Others) Act 1958, 6 & 7 Eliz 2, c. 62 §503(1)(b).
83. Cited in Michael Thomas, p.1218; [1900] P. 161, at pp.165–166.
84. Merchant Shipping (Liability of Shipowners and Others) Act 1958, 6 & 7 Eliz. 2. c. 62, §503(1)(b).
85. (1884) 9 P.D. 145 at p.147.

improperly during the course of towage is improper navigation rather than simply a breach of contract.[86] But in *The Alde* assisting the movement of another vessel by using a capstan to heave upon a rope fast to that vessel was held not to have been an action of navigation.[87]

The Athelvictor is of particular importance with regard to the meaning of the phrase "management of the ship". Until this case the phrase had been given a restricted meaning in cases under the Harter Act and the Carriage of Goods by Sea Act 1924. And in this case it was argued that the same narrow interpretation should be given to the phrase in section 503 of the 1894 Act. In *The Athelvictor* the owners sought to limit their liability for damage caused by an explosion and fire. The claimants disputed that the owners of the vessel were entitled to limit liability, and contended that the damage done was not due to the negligent navigation or management of the ship, but to negligence in the management of the cargo. The court reached the conclusion that in considering the meaning to be given to the words "improper management of the ship" in a limitation action, it was entitled to disregard the narrower construction put upon words in disputes arising under contracts evidenced by bills of lading. In giving the words their ordinary and natural meaning "improper management of the ship" included improper management of the ship's appliances; in this case, such appliances were fitted for both the ship's purposes and cargo purposes. The court was also satisfied that the narrower construction placed upon these words in related cases under the Harter Act or the Carriage of Goods by Sea Act was so placed because such a construction was the only one which enabled proper effect to be given to the shipowner's obligation under the overriding Act to take care of the cargo entrusted to him. In this case the court was not fettered by any such limiting considerations in construing the words in the sections of the Merchant Shipping Acts with which it was concerned. The court took the view that none of these sections regulated the position as between shipowner and cargo owner or in any way governed their relationship. The view of the court was that in the context of this case the words had a wider meaning than that which had been given to them in the bill of lading cases.[88] So the negligent failure of the ship's personnel to close the seavalves did not amount to improper navigation of the ship and therefore the shipowners were not entitled to limit their liability in respect of the loss of life or personal injury claims.

The comments of the court in *The Athelvictor* case were adopted by the court in *The Anonity*. In order to answer the question whether the negligence causing the damage constituted improper management of the vessel within the meaning of section 1 of the Act of 1900, the court adopted what had been said by the court in *The Athelvictor*. In this case causing or allowing sparks to escape from an oil-burning cooking stove, which should not have been burning at all while the vessel was at the jetty in question, was improper management of an appliance fitted solely for the ship's purposes and was therefore within the meaning of section 1 of

86. [1921] P. 312.
87. [1926] P. 211.
88. [1946] P. 42.

the Act of 1900. The court reached the conclusion that there was a misuse of this appliance and that this misuse, which caused the fire and the damage, was caused by improper management of the ship.[89]

In *The Tojo Maru* an attempt was made to extend this principle to the negligent act of salvors. On the basis of the judgments of Pilcher J. in *The Athelvictor* and *The Teal*[90] the salvors contended that the diver's act was in the "management" of the tug, because the bolt gun was an appliance of the tug, and "management of the ship" included management of its appliances. This argument was rejected on two bases: first, the judgments of Pilcher J. were wrong,[91] and, secondly, the bolt gun was not an "appliance of the ship" in the sense that it was not used in the running of the ship. Another argument that the diver's act fell within the category of "any other act or omission of any person on board the ship" also failed, as the act was not that of a person "on board the ship". The difficulties salvors faced in this case in limiting their liability and the fear of heavy claims led to the proposals for their benefit under the 1976 Limitation Convention.

(d) Duties of a shipowner with regard to limitation of liability

One of the prime objects of the limitation of liability provisions in the Merchant Shipping Act 1894 is to provide a measure of protection for the owner of a vessel from the results of negligence of his agents and servants but not of him himself. Therefore the shipowner is under certain duties with regard to the equipment of the ship and navigational matters in order to benefit from limitation of liability. In other words, to be able to benefit from the statutory right to limit liability the shipowner has to carry out certain duties.

(i) Duty to equip the ship

One of the prime considerations and duties of a shipowner wishing to avail himself of the statutory right to limit liability is to ensure that the vessel is properly manned. If there is any fault or privity on the part of an owner or shipowning company which leads to collision with another vessel or contributes to the cause of collision then it will not be possible to limit liability in respect of the damage done as a result of the collision as the owner or owners were not able to discharge the burden of proof upon them that the collision occurred without their actual fault or privity.

The duty of the shipowner with regard to equipping the ship covers providing effective gear, furnishing the master with proper loading instructions, providing the ship with adequate lights and supplying charts and navigational information.

89. [1961] 2 Lloyd's Rep. 117, affirming [1961] 1 Lloyd's Rep. 203.

90. In *The Teal* Pilcher J. said: "I am satisfied that any negligent act of the shipowner's servants which is carried out in furtherance of the cargo-carrying adventure, and which endangers the safety of the ship herself, constitutes improper management of the ship, for the purposes of this limitation action, even though the negligent act be one which is carried out solely or even primarily with reference to this cargo": (1949) 82 Ll.L.R. 414, 422 (Adm.).

91. In the CA Lord Denning expressly accepted that these two cases were correctly decided: [1969] 2 Lloyd's Rep. 193 (CA).

The employment of an inadequate and incompetent crew constitutes a breach of the duty to equip the ship. In *The Fanny*[92] the damage for which the shipowner sought limitation was partly attributable to the ship's incompetent crew, but the Court of Appeal found the crew competent and the limitation was allowed. In *The Empire Jamaica* the steamship came into collision with a motorship. At the time of collision the steamship was carrying a certificated master, a certificated first mate and two other officers who were not certificated. The court accepted that the vessel carried these three men to perform the duties of mates and that they were in fact employed and paid as such. It was held in the trial court that the failure to provide the requisite number of certificated officers was not itself enough to make an owner guilty of fault or privity in relation to a casualty occurring later. The court ruled that the owners of the *Empire Jamaica* were entitled to a decree of limitation of liability. In the House of Lords it was held that even if the owners of the *Empire Jamaica* were in breach of the relevant Ordinance in appointing an uncertificated man as second mate, that breach neither caused nor contributed to the collision.[93]

A similar question on the competence of the crew was also raised in *The Dayspring*. It was held that the crew's incompetence had not caused the loss.[94]

But the question whether employing an incompetent and unskilled crew would have amounted to a breach of duty to equip the ship adequately and in consequence caused the loss deprived the owner of the benefit of limitation was not answered in the above-discussed cases.

The shipowner is also under a duty to provide adequate lights in conformance with regulations, a safe system and also to maintain the steering machinery of the vessel properly. In *The Truculent* the Admiralty tried to limit liability for damage and loss of life when HM submarine *Truculent* collided with a ship and sank. And the Third Sea Lord admitted that he knew that the steering light did not conform with the regulations. It was held that there was fault in connection with the lights exhibited by the submarine and this fault also contributed to the loss and damage. At the same time as the fault belongs to a Third Sea Lord, a member of the Board of Admiralty, meaning the Crown, as owner of the submarine, was not able to limit liability.[95] In *Hook* v. *Consolidated Fisheries* a deckhand crushed his leg between the vessel and a dolphin when the master told him to jump on it to assist in securing the vessel. When the deckhand claimed damages the court recognised that this system was a common practice but it involved risks and the failure to adopt a safer system was indicative of the vessel's owner's actual fault or privity in regard to the deckhand's accident.[96] In *The Otterdal* the vessel came into collision with the *Tarbet* in the River Mersey and as a result the latter vessel sank. The court held that the collision was caused by a fault in the steering machinery of the *Otterdal*, a fault which had existed when the vessel sailed. In the limitation of liability proceedings

92. (1912) 28 T.L.R. 217.
93. (1957) A.C. 386, [1956] 3 W.L. R. 598, 100 S.J. 618, [1956] 3 All E.R. 144.
94. [1968] 2 Lloyd's Rep. 204.
95. [1951] 2 Lloyd's Rep. 308.
96. [1953] 2 Lloyd's Rep. 647 (CA).

the question arose whether the defect in the steering gear existed without the actual fault or privity of the owners of the vessel. The court said that the owner took a chance in allowing the ship to sail on that voyage without being assured that something thorough had been done to the faulty steering gear or that new gear had been installed. But the chance that he took turned out to be wrong and he was not entitled to a decree of limitation of liability. The Court of Appeal upheld this ruling.[97]

The shipowner is also under a duty to ensure that his vessels are navigated with the aid of up-to-date charts, corrected charts and sailing directions and provide her with a compass and radar. The owner or his *alter ego* also has to establish a system which will provide a managerial check to detect any failure in these duties. In *The England* a collision occurred on the Thames. The master did not have copies of the Port of London bye-laws on board. Such a failure was held to amount to actual fault or privity on the part of the owners. So they were unable to limit their liability.[98] With regard to shipowners' duties to supply charts, navigational and operational information *The Marion* has particular importance. Before this case there was a belief among many shipowners that the duty of providing the vessel with up-to-date charts, corrected charts, sailing directions and the like could properly be left to the master of the vessel unless the owner had reason to believe that the master was not carrying out his duties properly in this respect. But with *The Marion* it became clear that it is not sufficient for the shipowner merely to rely on the master of the vessel to obtain the necessary publications and information for navigational purposes. In *The Marion* a Liberian tanker left Hamburg on a voyage to Teesside. While waiting to berth at Teesside the master attempted to anchor the vessel and in doing so damaged an oil pipeline lying on the sea bed. The oil companies, having an interest in the pipeline, brought an action against the shipowners in Admiralty alleging that the damage to the pipeline had been caused by the negligence of the shipowner's servants or agents on board the vessel and claimed damages in excess of $US25 million. The owners admitted liability but claimed that the master had negligently used an out-of-date chart which did not have the pipeline marked on it and therefore sought to limit their liability. The oil companies contended that the immediate cause of the damage was the negligence of the master in navigating by reference to an obsolete chart on which the pipeline was not shown and that if he had had an up-to-date chart he would have been aware of the existence of the pipeline. The Admiralty judge granted limitation on the ground that the provision and maintenance of the charts was the sole responsibility of the master and it was his negligence in using an out-of-date chart which was the sole cause of the damage.

97. *Dampskibs Aktiessel* v. *Skabet Maritime ("Otterdal")* v. *"Tarbert" and Others* (1920) 4 Ll.L.R. 414 and (1920) 5 Ll.L.R. 368.

98. In the words of Wilmer L.J. the shipowner had failed "to take reasonable steps to see that the *Alletta* was properly equipped with the necessary publications. That . . . amounted to a fault on his part within the meaning of section 503 of the Merchant Shipping Act": *Rederij Erven H. Groen and Groen* v. *The "England" (Owners) and Others* [1973] 1 Lloyd's Rep. 373 at p.384.

The Court of Appeal reversed the decision on the ground that it was the ship manager's duty to ensure that there was an effective and properly supervised system of chart provision and maintenance and such a system was absent in this case. In the House of Lords it was held that this failing on the part of the ship's managers was directly causative of the oil companies' losses, and such a failure constituted in law actual fault of the owners of the vessel. Therefore they were denied the right to limit.[99]

Shipowners may also lose the right to limit their liability as an effect of failure to communicate information to their masters and ships. In *The Norman*[100] the trawler struck an uncharted rock off Cape Farewell with the loss of the trawler and 19 members of the crew. During its voyage her owners were given new information on dangers in the area in which the vessel was likely to be navigating but this information was not passed on to *The Norman* because its owners felt that this new wireless information to the vessel whilst at sea would have been an unnecessary departure from their normal practice. The owners also felt that they were entitled to rely on the judgement of their insurers who would if necessary have sent out a general warning to all trawlers fishing in that area. The dependants of those who had lost their lives made claims against the owners and the owners applied for a decree of limitation. In the House of Lords it was held that there was a duty on the part of the owners of the trawler to communicate the latest information that would assist navigation, and that their failure to do so was a fault.

The effect of the failure of owners to communicate information to their masters and ships upon their right to limitation of liability under the provisions of the Merchant Shipping Act is not limited to navigational matters. In *The Clan Gordon* the vessel capsized and sank together with her cargo during her voyage from New York to China. The owners of the vessel were unable to limit their liability as they had failed to transmit to the master certain information regarding water ballasting and instructions regarding the retention of water ballast in certain tanks.[101]

(ii) Navigational matters

In order to benefit from limitation the shipowner had to exercise certain duties with respect to navigation. When the vessel caused damage arising out of matters of negligent navigation then the shipowner had to prove that the accident involving his liability arose without, or was not contributed to by, his actual fault or privity.

The Norman sparked off a new approach to the extent of the duties of shipowners and ship managers, especially in regard to navigational matters. In the House of Lords it was held that there was a duty on the owners of a trawler to communicate the latest information that would assist navigation, and that the owners' failure to

99. *Grand Champion Tankers Ltd.* v. *Norpipe A/S and Others* [1984] 2 Lloyd's Rep. 1.
100. *Northern Fishing Company (Hull) Ltd.* v. *Eddom and Others (The Norman)* [1960] 1 Lloyd's Rep. 1.
101. (1923) 16 Ll.L.R. 367.

send on new information was a fault which they had failed to prove did not contribute to the loss of the vessel and had occurred without their fault or privity. This decision made it clear that it was no longer permissible for owners or managers to leave everything to the unassisted discretion of their masters. In other words it was not permissible for operators to wash their hands of all questions of navigation if they wished to benefit from a statutory right to limit liability. Under the law embodied in *The Norman* shipowners are under two main duties in order to benefit from limitation. When their vessel is at sea, first they must foresee the geographical scope of the navigation, and following that if there is a possibility of damage or loss within that area with regard to navigational hazards they must inform the vessel of new navigational hazards as a matter of routine prudence and good care. The shipowner has to show the degree of care which a person of ordinary prudence would exercise in the same or similar circumstances. Since this decision the courts have followed this approach.

As stated above in *The Marion* the House of Lords held that it was the duty of the managing director to ensure that an adequate degree of supervision of the navigation of the vessel and the master of the vessel, so far as the obtaining and keeping of up-to-date charts was concerned, was exercised by him or by his subordinate managerial staff, each of whom was fully qualified to exercise such supervision. And as the managing director had failed to perform his duty in this respect such failure in law constituted actual fault of the owners of the vessel.

As a consequence it is still the duty of the shipowners or their *alter ego* to supervise and to ensure that their masters are not negligent in the manner in which their ships are navigated.

The introduction of radar into merchant ships created new problems in relation to navigation in fog and compliance with the rules of the road at sea. In *The Lady Gwendolen*[102] the Court of Appeal stated that as high speed in fog was encouraged by radar, the installation of radar required particular vigilance from shipowners through the person (in this case the assisting managing director) who was responsible, in his capacity as owner, for the running of ships, and that the owners had failed in that vigilance and had failed to consider or appreciate the problems which were inherent in the gravity of the risks he was taking. It was held that such a failure was a contributory cause of the collision and that therefore the owners had failed to prove that the collision occurred without their actual fault or privity. The rule stated in this case was very clear. The shipowner had a duty to exercise some measure of supervision and control over the navigation of his ship. In *The Garden City* the court said that the top management of every shipowning corporation should institute a system for supervision of navigation and that where an individual owned one ship, it may be that he should take on that task himself. In this case criticism was made of the organisation regulations, which did not in terms provide that supervision of navigation was one of the main tasks within the responsibilities of the director for technical and investment affairs. They also did not provide in

102. [1965] 1 Lloyd's Rep. 335.

terms that it was one of the tasks of the chief navigator. As a consequence the court found that there was no fault on the part of the director general or director, in that they did what it was reasonable for the owner of a large number of ships to do—they appreciated the problems posed by the use of radar in fog, and impressed the urgency of such problems upon the masters of their vessels and took steps to ensure as far as they reasonably could that their vessels were safely navigated in fog. Therefore the owners were entitled to a decree of limitation.[103]

When *The Lady Gwendolen* and *The Garden City* are taken together they provide different results with regard to navigation in fog. In *The Lady Gwendolen* the nature of failures which made a shipowning company guilty and its liability unlimited have been illustrated. On the other hand in *The Garden City* it was shown that by the institution of a system for the supervision of navigation and detection of faults the shipowning company may be successful in proving it was not guilty of actual fault or privity and apply for a decree of limitation of liability.

In general it is possible to say that when an owner applies for a decree of limitation he must show not only that the casualty was caused without his actual fault or privity but also that he has not contributed to the casualty by any actual fault or privity on his part. It is not sufficient to employ a competent master[104] and leave him the responsibility to ensure that the vessel is navigated with the aid of corrected up-to-date charts. He must himself exercise supervision not only to ensure that the vessel properly carries up-to-date charts and other navigational aids but to make sure that the vessel is navigated with the aid of up-to-date charts, sailing directions, etc. Owners should also institute a proper system for the management and operation of their vessels, with particular reference to the supervision of navigation, the detection of faults and to giving adequate and proper notice to masters of certain dangers inherent in the operation of vessels at sea or in port.

(e) Exclusion of liability

Under section 502 of the Merchant Shipping Act 1894 which only applies to British ships an owner can exclude his liability for loss or damage arising from a fire on board or theft of undeclared valuables. In order to exclude liability such events should occur without actual fault or privity of the shipowner.

(i) Fire

As stated in *The Diamond* the owner may rely on the exception if the damage resulted from the water used to put out the fire or through smoke from the fire

103. [1982] 2 Lloyd's Rep. 382; *The Garden City (No. 2)* [1984] 2 Lloyd's Rep. 37.

104. As an exception to this general rule there are some cases which show that there are certain responsibilities which fall upon the master of the vessel. In *The Thordoc* the court found that there was no actual fault or privity of the owners of the vessel in regard to the condition of the compass which had been adjusted by a compass adjuster of repute before the casualty. The engagement of the helmsman and the putting of him at the wheel at the stage of the voyage in question were alike the responsibility

itself.[105] The owner does not have to show that the loss or damage was actually occasioned by flames or by heat. But the fire must be on board the ship. Hence, in *Morewood* v. *Pollok*[106] it was held that the owners were not entitled to exclude where the goods were destroyed by fire on board a lighter by which they were being conveyed to the ship. In *Tempus Shipping Co.* v. *Louis Dreyfus & Co.*[107] the goods heated by fire on board the ship were discharged into a lighter and then caught fire in the lighter. The shipowners relied upon the section as there was a causal connection between the loss and the initial fire on board the vessel. If the shipowner is not at fault or privy to the unseaworthiness and a fire comes about as a result of the unseaworthiness then the shipowner may rely upon the exclusion.[108] But if the unseaworthiness itself was not without the actual fault or privity of the owner then the owner is unable to rely on the exception as the loss happened with his actual fault or privity.[109]

(ii) Undeclared valuables

The provision relating to undeclared valuables is quite clear. Under it there are two possibilities: at the time of shipment the true nature and value of the valuables are either declared in writing or not declared. If the valuables are contained in a passenger's luggage they are within the exception. If they are lost in a robbery committed by a member of the crew, again the shipowner is protected. In *Smitton* v. *Orient Steam Navigation Company Ltd.* a person managed to get into the berth of a passenger through a fanlight that opened onto the deck and stole his watch, chain, gold cigar cutter and purse from the watchpocket. The shipowner relied upon the exclusion and avoided liability.[110]

(f) Actual fault or privity

(i) In general

A person may exclude his liability under section 502 or limit his liability under section 503 of the 1894 Act only if the incident giving rise to liability arose "without his actual fault or privity". Therefore the precise extent to which the shipowner's

---

and the work of the master of the vessel. And even if the master was wrong that was not the fault of the owners: *Paterson Steamship Ltd.* v. *Robinhood Mills, Ltd.* (1937) 58 Ll.L.R. 33.

105. [1906] P. 282.

106. Cited in Michael Thomas, p.1214, n.55; *Morewood* v. *Pollok* was decided under the Responsibility of Shipowners Act 1786, 26 Geo. 3, c. 86, §2. The relevant words of this Act were similar to those of s. 502 of the 1894 Act: *Morewood* v. *Pollok*, 118 E.R. 614 (QB 1853).

107. [1930] 1 K.B. 699, 709.

108. *Virginia Carolina Chem. Co.* v. *Norfolk & North American Steam Shipping Co.* [1912] 1 K.B. 229 (CA).

109. *Asiatic Petroleum Company Limited* v. *Lennard's Carrying Company Limited*, [1914] 1 K.B. 419 (CA).

110. (1907) 96 L.T. 848, 12 Com. Cas. 270, 10 Asp.M.L.C. 459, 23 T.L.R. 359.

conduct would deprive him of the statutory right to limitation or exclusion of liability depends on the true construction of the term "actual fault or privity". In considering whether the shipowner is guilty of actual fault his conduct has to be assessed. There have been some suggestions about the required standard of a shipowner's conduct.[111] The test suggested in *The Lady Gwendolen* was an objective one. The required standard for limitation purposes was the standard of the reasonable shipowner in the management and control of his vessel or fleet of vessels,[112] or the standard of a reasonably prudent and conscientious owner who has a vessel at sea.[113] In such a test the conduct of a shipowner is compared with that of the reasonable man. Under such a comparison the facts of the particular case have not been assessed. But when the wording of section 503 is examined in relation to the statement of Buckley J. in *Asiatic Petroleum Co. Ltd.* v. *Lennard's Carrying Ltd.* it becomes clear that the word "actual" points the way to the test intended to be applied. As stated by Buckley J. the words "actual fault or privity" "... infer something personal to the owner, something blameworthy in him as distinguished from constructive fault or privity of his servants or agents". The fault or privity have to be actually the shipowner's. If not he is not entitled to the benefit of the statute which states: "The owners of a ship ... where all or any of the following occurrences take place without their actual fault or privity ..." should not be liable above a certain amount. Under an objective test it is difficult to ascertain the shipowner's conduct clearly enough for limitation purposes and relate it to the facts of the given case.

On the other hand under the subjective test the connection between the shipowner and the act or omission giving rise to liability has been assessed. And with such an assessment the aim of the statute has also been achieved.

As stated above the words "actual fault or privity" infer something personal to the owner, something blameworthy in him. It is distinguished from constructive fault or privity like the fault or privity of his servants or agents. In order to benefit from the statutory defence the shipowner must show that he himself is not blameworthy for having either done or omitted to do something or has been privy to something. When someone is privy to something this generally means that he has a certain knowledge, confidential or otherwise, of some relationship or agreement or situation existing between two or more people.[114]

The meaning of "actual fault or privity" was explained in the leading case of *Asiatic Petroleum Co.* v. *Lennard's Carrying Co.* In this case the vessel was loaded with a cargo of benzine. During the course of her voyage she ran into heavy weather and because of the defective condition of her boilers was driven ashore. As a result of the resulting explosion the vessel and her cargo were lost. The question arose

111. See C.N. Cheka, "Conduct Barring Limitation" (1987) 18 *Journal of Maritime Law and Commerce*, 487–489.

112. In *The Lady Gwendolen* Wilmer, L.J. suggested that "... the test to be applied in judging whether shipowners have been guilty of actual fault must be an objective test", at p.386

113. See *The Norman*, *per* Lord Radcliffe at p.11; n.99 *supra*.

114. According to Hill one could virtually, but perhaps not entirely, describe the word "privity" as being equivalent to the word "knowledge": Christopher Hill, *Maritime Law* (4th ed. 1995), p.385.

whether the incident took place without the actual fault or privity of the owners of the ship. The owners were a limited company and the ship was managed by another limited company, namely John M. Lennard & Sons. Mr J.M. Lennard was the active director of the company and also a director of Lennard's Shipping Company Ltd. which claimed the protection of the Act. The court found that Mr J.M. Lennard was unable to prove that he did not know or could excuse himself for not having known of the defective condition of the boilers, which amounted to seaworthiness. It was also found that Mr Lennard was the person who was registered in the ship's register and was designated as the person to whom the management of the vessel had been entrusted. The court said that upon the true construction of section 502 of the Act the fault or privity is that of somebody who is not merely a servant or agent for whom the company is liable but somebody for whom the company is liable because his action is the very action of the company itself. It is not enough that the fault should be the fault of a servant. It must be one which is not the fault of the owner, or a fault to which the owner is privy.[115]

In *The Eurysthenes*[116] the vessel stranded and her cargo was substantially lost or damaged. The vessel was consequently alleged to be unseaworthy through lack of due diligence. The question was whether the shipowner was privy to the condition in which his ship had put to sea and following that whether he was entitled to limit his liability under the Merchant Shipping Act 1894. The Court of Appeal, referring to privity, commented that it did not mean that there was any wilful misconduct by the shipowner but only that he knew of the act beforehand and concurred in its being done. Privity did not mean that he himself personally did the act but only that someone else did it and he knowingly concurred. In other words "without his privity" is equivalent in meaning to "without his knowledge or concurrence".

It was also once thought that the owner was unlikely to be found at fault where the cause of the loss was a fault in the navigation of his vessel, a matter necessarily left in the hands of the master, officers and crew. But since the decision of the House of Lords in *The Norman*, where it was held that failure to send on the new information was fault on the part of the owners, the courts are less ready to accept the argument. Also following the decision in *The England*[117] the court said that an owner is not entitled simply to rely upon the master to obtain the necessary

115. In the words of Buckley L.J. "actual fault or privity ... infer something personal to the owner, something blameworthy in him, as distinguished from constructive fault or privity such as the fault or privity of his servants or agents. But the words 'actual fault' are not confined to affirmative or positive acts by way of fault. If the owner be guilty of an act or omission to do something which he ought to have done, he is no less guilty of an 'actual fault' than if the act had been one of commission. To avail himself of the statutory defence, he must shew that he himself is not blameworthy for having either done or omitted to do something or been privy to something. It is not necessary to shew knowledge. If he has means of knowledge which he ought to have used and does not avail himself of them, his omission so to do may be a fault, and, if so, it is an actual fault and he cannot claim the protection . . .": *Asiatic Petroleum Co.* v. *Lennard's Carrying Co.* at p.432.

116. *Compania Maritima San Basilio S.A.* v. *The Oceanus Mutual Underwriting Association (Bermuda) Ltd.* [1976] 2 Lloyd's Rep. 171 (CA).

117. See, n.98 *supra*.

regulations or otherwise instruct himself. In *The Marion* the vessel was owned by one limited company, Grand Champion Tankers Ltd., and the management of the vessel was delegated to another limited company, Fairwell-Maxwell Services Ltd. In this connection the court found that the person whose fault would constitute, as a matter of law, the actual fault of the owners of *The Marion* was the managing director of Fairwell-Maxwell Services. Amongst other things there were two actual faults of the owners: first, in the failure of the managing director of Fairfield-Maxwell Services to have a proper system of supervision in relation to charts and, secondly, in the failure, when he was away, to give his subordinate managerial staff instructions with regard to the matters about which he required to be kept informed. In *The Lady Gwendolen* the assistant managing director responsible for the running of the company's vessels was the *alter ego* of the company and the radar problem merited his personal attention. The contributory cause of the collision was the master's total lack of sense of urgency about the problem posed by radar navigation in fog, which should have been instilled into him from the highest level, but was not. In these circumstances the court ruled that the owners of the vessel were guilty of actual fault and therefore could not limit their liability. The Court of Appeal upheld the ruling.

In consideration of the abovementioned cases it became clear that certain acts or omissions constitute actual fault or privity on the part of owners where the loss was *prima facie* caused by a fault in navigation.[118] These acts are as follows: failure of the managing director to have a proper system of supervision in relation to charts—when the managing director fails to perform his duty it constitutes in law actual fault of the shipowners[119]; failure to insist upon the keeping of a proper logbook and to warn the master and mate of the dangers of allowing the vessel to be underway in the sole charge of a single deckhand[120]; allowing the ship to sail on a voyage without being assured that something thorough had been done to the defective steering gear or that new gear had been installed[121]; failure to pass on new information about dangers in the area of navigation[122]; failure to ensure that a copy of the Port of London Authority bye-laws containing prescribed manoeuvring signals was on board a vessel which regularly traded to London[123]; failure to consider or appreciate the problems arising out of the use of radar and to impress upon the master the gravity of the risks he was taking by proceeding at high speed in fog, relying merely upon his radar to warn him of impending danger.[124]

In order to deprive a shipowner of the protection of limitation or exclusion of liability the fault or privity on the owner's part must have been a cause of the loss or damage in question. In *The Empire Jamaica* the shipowners were admittedly in breach of a Hong Kong merchant shipping ordinance. But the evidence showed

118. See Michael Thomas, pp.1223–1224.
119. *The Marion.*
120. *The Dayspring.*
121. *The Otterdal.*
122. *The Norman* and *The Clan Gordon.*
123. *The England.*
124. *The Lady Gwendolen.*

that the uncertificated second mate was a fully competent seaman and that his lack of certificate had no causal connection with the fact of his negligent navigation. Thus the shipowners were held to be entitled to limit their liability even though they were at fault.[125]

(ii) The *alter ego* problem

In the old days when the owner of a ship was often the master of his ship it was easy to answer the question whether or not there had been actual fault or privity. Where the owner was an ordinary person it was not too difficult to determine whether or not there was actual fault or privity on his part. But the position is not so straightforward in cases where the owner is a company. In particular within today's large shipping organisations of subsidiary companies it is extremely difficult to find out whether or not the actual fault is that of the shipowner himself or somebody so closely identified with the company owning the ship that he can be said to be a person who spoke and acted as the shipowning company itself, or whether it was the neglect or default or act of a servant of the shipowner. Therefore the question is of which employee or agent can it be said that this act is the very act of the company itself? In cases where a limited company is concerned the *alter ego* of the company has to be determined. If a vessel is owned by one limited company and managed by another limited company the court will consider the managing company to find out whether the owners are guilty of actual fault. In such circumstances the *alter ego* of the managing company has to be determined.

In *Smitton* v. *Orient Steam Navigation Co.*, the argument was that the fault of the defendant company's marine superintendent was that of the company itself. But this argument was rejected by Channell J., who held that "in the case of a corporation the words mean the fault of the managing authority. In this country . . . that would be the board of directors."[126] This argument was rejected by Lord Dunedin in *Lennard's Carrying Company Ltd.* v. *Asiatic Petroleum Company Ltd.*, where he refused to assent to the proposition that the actual fault of the whole body of directors must be proved before a company could be found to have been actually at fault.[127] But interestingly the expression *alter ego* was first used in relation to an application for a decree of limitation of liability under the Merchant Shipping Act 1894 by counsel in this case. In this case the owners of the vessel sought the benefit of section 502 of the Act which provides immunity for the owner of the vessel for loss of cargo caused by fire, subject to the loss resulting without the actual fault or privity of the owners of the vessel. As stated above the managers of the shipowning company were another limited company. Viscount Haldane L.C. said of a company that it "is an abstraction. It has no mind of its own . . .; its active and directing will must consequently be sought in the person of somebody who for

125. *The Empire Jamaica*, at p.396.
126. [1907] 12 Com. Cas. 270.
127. [1915] A.C. 705, 715–716.

some purposes may be called an agent but who is really the directing mind and will of the corporation, the very ego and centre of the personality of the corporation."[128]

In the Act the words "*alter ego*" are not used. The Act says "owners". By subsequent legislation the Act was amended as follows: "owners of a ship . . . shall not, where . . . without their actual fault or privity . . . any loss or damage is caused to any property . . . through the act or omission of any person (whether on board the ship or not) in the navigation or management of the ship . . . be liable in damages beyond the following amounts . . .". These parts of the Act are primarily concerned with the protection of the owner against the negligent acts of his agents or servants.

Limitation of liability under the Act will be granted provided that the loss or damage was caused without the actual fault or privity of somebody who was not merely a servant or agent for whom the company owning the vessel was liable but somebody for whom the company was liable because that person's action is the very action of the company itself. In other words a company will lose the protection of the Act if the fault or privity "is the fault or privity of somebody who is not merely a servant or agent for whom the company is liable upon the footing respondent superior, but for whom the company is liable because his action is the very action of the company itself".[129] Courts subsequently faced with similar problems have followed this approach.

Thus, since this case whenever a limited company has been concerned it has been the practice to find out who is to be regarded as the *alter ego* of that company. This provides guidance in establishing whether or not the loss or damage caused by the vessel arose without the actual fault or privity of the owners of the vessel. It has been generally recognised that the words "actual fault or privity" infer something personal to the owners of the vessel, something blameworthy in them. This is distinguished from constructive fault or privity such as the fault or privity of their servants or agents.

The words "actual fault" are not confined to affirmative or positive acts by way of fault. Therefore if the owner is guilty of an act or omission to do something which he ought to have done, he is no less guilty of an "actual fault" than if the act had been one of commission.[130] If the shipowner wants to avail himself of the statutory defence he must show that he himself is not blameworthy for having either done or omitted to do something or been privy to something. If the owner has the means of knowledge which he ought to have used and does not avail himself of such knowledge his omission to do so may be a fault. Therefore it is not necessary to prove knowledge.

Under this approach, which was laid down by Viscount Haldane, L.C., in *Asiatic Petroleum Co.* v. *Lennard's Carrying Co.*, it is possible to think that the *alter ego* of a shipowning company must be found amongst the directors or at least senior

128. *Ibid.*, *per* Lord Chancellor, p.713.
129. *Ibid.*, pp.713–714.
130. See W.E. Astle, *Limitation of Liability. Merchant Shipping Act 1894; Section 503 and Subsequent Amendments* (1985), pp.7–14.

employees of that company. But under such a conclusion a company could not be guilty of actual fault if its directors had chosen a reputable management company and delegated the tasks of management. The managers would be negligent without the directors' knowledge.

In *The Lady Gwendolen* the Court of Appeal adhered to the approach laid down by Viscount Haldane, but also indicated that the court might be prepared to find the *alter ego* at a lower level in the management structure. In this case the shipowning side of the business of the owners of the *Lady Gwendolen* was a subsidiary to their main activity. The master of the ship habitually proceeded at excessive speed in fog, relying upon radar to warn him of impending danger. The owners of the vessel knew of his addiction to speed in fog and took no action to direct him in the proper use of radar. In the view of the court the radar problem was one of such great importance as to merit the personal attention of the owners and it was the duty of the *alter ego* to cure it. Then the question arose as to who was the *alter ego* of the company. The court said the duty of an *alter ego* rested on the assistant managing director of the company who was responsible for the company's vessels. It was held that the proper function and use of radar in fog was never considered by the assistant managing director or in fact anybody else. This sense of urgency and importance should have been instilled in the master from the highest level and this failure to do so rendered the owners of the vessel guilty of actual fault. The court said that the owners in this case or any other owners could not divest themselves of their obligations by delegating these to salaried officials and that there must be a limit on how far this or any other company might divest itself personally of such shipowning obligations as are of a fundamental or highly important nature.[131] In the Court of Appeal it was said that the owners of the *Lady Gwendolen*, in their capacity as owners of the vessel, were to be judged by the standard of conduct of the ordinary reasonable shipowner in the management and control of vessels. The court also said that the primary concern of a shipowner is the safety of life at sea, and properly manned and navigated ships. And as high speed in fog was encouraged by radar the installation of radar required the particular vigilance of shipowners through whoever was responsible for the running of the ships. In this case this duty was on the *alter ego*, namely on the assistant managing director who had failed in that duty.

It has been firmly established that the duty concerning the safe navigation of a vessel lies squarely with the owner of the vessel and/or the *alter ego* of the company. If the vessel is owned by one limited company and managed by another that duty will fall upon the managing company and/or the *alter ego* of that company.

The ruling of the House of Lords in *The Marion* dispelled all views to the contrary. In this case the ship was managed by another company. The court said that the managing director of the vessel's management company, the *alter ego*, had a duty to ensure that an adequate degree of supervision of the master in the keeping

131. See W.E. Astle, pp.9–10.

of up-to-date charts was exercised either by himself or by subordinate management staff. As the managing director had failed to perform his duty in that respect such a failure constituted actual fault of the shipowners.

Another question which arose on *alter ego* was whether the *alter ego* of a company can be held personally liable to an injured party and, if this is possible, whether such a person can limit his liability under the provisions of section 503 of the Merchant Shipping Act 1894. In *The Radiant*[132] a motor fishing vessel had gone aground in the Thames Estuary and when it was taken in tow by a sister ship the skipper's legs became caught up in the coil wire and were amputated. The skipper claimed against the owners of the vessels and also the managing director of the company, alleging that his injuries were caused by the unseaworthiness of both vessels due to their defective equipment and negligent navigation of the sister ship. With regard to the claim against the managing director, who was the *alter ego* of the company, the court said that he knew or had the means of knowing of the defects in the vessels which contributed to the accident. On his part there was a failure of management, as no adequate provision was made for periodical inspection. Accordingly the court ruled that the owners could not limit their liability.

It was also held that the skipper had discharged the burden upon him of proving that the managing director was a party to sending the vessels to sea in an unseaworthy condition and that his injury resulted from that condition. Therefore it was held that the managing director was personally liable and was not entitled to limit his liability under section 503 of the Act. So a personal action lay against the managing director of a company on the same facts as actual fault or privity was found against the company.

The judgment in *The Radiant* was referred to in *The Anonity*. But in this case Willmer L.J. stated that the question whether an injured claimant could successfully bring a personal action against a member of the company, whose conduct was held to amount to actual fault or privity of the company within the Merchant Shipping Acts, had to depend on the facts of the particular case.

Under the Merchant Shipping Act 1958 the injured party might have a right to recover from the negligent managers as this has been brought within the scope of the limitation provisions.[133] But as in *The Radiant* the burden of proof would lie upon the injured party to prove that the managers were guilty of contributory fault, whereas in a limitation action it is the shipowners who must prove that the occurrence took place without fault on their part.

(g) Limitation of liability for master or crew members

When the Merchant Shipping Act 1894 was passed it was intended that only the actual owner of a ship or the charterer by demise should be able to benefit from limitation of liability. The words "actual fault or privity" contained in section 503

132. [1958] 2 Lloyd's Rep. 596.
133. Merchant Shipping (Liability of Shipowners and Others) Act 1958, 6 & 7 Eliz. 2, c. 62, s. 3(1).

of the Merchant Shipping Act 1894 imply something personal to the owner and distinguished from the fault or privity of his servants. It is not enough that the owners of the vessel prove that somebody else, one of their servants for instance, was to blame for the collision or other circumstances resulting in the damage. The owner must go further than that and prove that such damage was caused without his own personal fault. Therefore even when navigation was in the sole charge of the owner's servants the owner had to make sure that there was no fault or omission on his part which might be said to have caused or contributed to the casualty. In *The Hans Hoth*[134] the vessel was involved in a collision with another vessel at the entrance to Dover Harbour. At the time of the collision, three persons owned the *Hans Hoth* and one of them was the master of the vessel at the time of the casualty. There was also a compulsory pilot on the bridge with the master. The collision was due to improper navigation of the *Hans Hoth* by her pilot who was solely in charge of her navigation for the purpose of taking the vessel into Dover Harbour. The owners of the vessel admitted liability in respect of the damage but sought to limit it under the Merchant Shipping Act. At the time of the collision and at all material times the master was on the bridge of the vessel, of which he was part owner, and was also controlling the engines. The question which had to be decided by the court was whether the master had to share responsibility for the fault of the pilot leading to the collision. Although there was a duty to assist in navigation by the master to a compulsory pilot, that duty did not extend to matters of purely local significance which would be within the particular knowledge of the pilot himself. It was therefore held that despite the failure of the master to acquaint himself with the significance of local signals he was entitled to a limitation decree. In *The Mint* the master and mate of the drifter were part-owners of the vessel and the collision arose out of their actual fault or privity. The owners brought a claim for a decree of limitation of liability and the court held that they were entitled to limit their liability except insofar as the master and mate were concerned.[135]

Obviously it is logical to allow a ship's master or crew member to benefit from limitation of liability as he is just as liable to be sued for damages resulting from his acts or neglect as the shipowner. In a case where the master or crew member is either the owner or at least a part owner of the vessel it is necessary to decide whether the fault or privity was that of the shipowner or the ship's master or crew member. Article 6(3) of the 1957 Convention also provides that where the master or member of the crew is at the same time the owner, charterer manager or operator of the vessel, such person will be entitled to limit his liability only if he commits the act, neglect or default in his capacity as master or as a member of the crew. The *Annie Hay* was concerned with the question whether the owner of the vessel, which collided with and sank a motor yacht, was entitled to limit his liability under the provisions of the Merchant Shipping Act 1894 as amended by the Merchant Shipping Act 1958. In this case the collision occurred due to negligent navigation of the ship's master who also happened to be the owner of the vessel and

134. [1952] 2 Lloyd's Rep. 341.
135. (1937) 59 Ll.L.R. 257, [1938] W.N. 17.

was in sole charge of the navigation. The vessel was acting as a patrol boat during a power-boat race. The motor yacht *Rosewarne* was returning to the River Fal. The *Annie Hay* was unaware of her existence and collided with her, causing her to sink. Claims were made against the owner of the *Annie Hay* under the provisions of section 3(2) of the 1958 Act. The owner of the *Annie Hay* alleged that the claims being made against him were in his capacity as master and in accordance with the natural meaning of the words in section 3(2) he was entitled to limit his liability. The court held that under the provisions of the Merchant Shipping Act the right to limit is permitted in respect of an act or omission of any person in his capacity as the Master or a member of the crew and section 3(2) of the Act of 1958[136] was broad enough in scope to cover a ship's master who happened also to be the owner of the ship. The words "any person in his capacity as master" were wide enough to include the owner of the *Annie Hay*. Therefore he was entitled to limit his liability. But if such a person commits a fault in his capacity as shipowner he loses his right to limit liability.[137]

In relation to the above-discussed cases it is possible to say that where there are part owners the fault or privity of one or more of them does not prevent the innocent part owners from excluding or limiting their liability. But if the part owner or owners are at fault or privy to the occurrence giving rise to the loss or damage then they remain liable to the full extent of their liability. As examined in *The Annie Hay* if the loss or damage occurred with the fault or privity of the master or a member of the crew and the master or the relevant crew member is also the owner or part owner of the ship then it is necessary to determine whether the fault or privity occurred in that person's capacity as owner or in the course of his duties about the ship. As stated in this case and in *The Hans Hoth* if a fault in the actual navigation of the vessel by an owner/master results in a collision then it is a fault or privity in his capacity as master and it is possible to limit his liability.

(h) Limitation actions

The right to limit liability does not arise automatically. The owner who wishes to limit his liability has to begin limitation proceedings in the Admiralty Court. Only if the court is satisfied that the incidents in respect of which limitation is sought arose without any fault or privity on the part of the owner will a decree of limitation

136. S. 3(2) of the 1958 Act provided that "... In relation to a claim arising from the act or omission of any person in his capacity as master or member of the crew or (otherwise than in that capacity) in the course of his employment as a servant of the owners or of any such person as is mentioned in subsection (1) of this section:—

(a) the persons whose liability is excluded or limited as aforesaid shall also include the master, the member of the crew or servant, and in a case where the master or member of the crew is a servant of a person whose liability would not be excluded or limited apart from this paragraph, the person whose servant he is; and
(b) the liability of the master, member of the crew or servant himself shall be excluded or limited as aforesaid notwithstanding his actual fault or privity in that capacity except in cases mentioned in paragraph (ii) of section five hundred and two of the said Act of 1894."

137. [1968] P. 341, [1968] 1 Lloyd's Rep. 141, [1968] 2 W.L.R. 353, [1968] 1 All E.R. 657.

be granted. Injured parties must show cause why the owner should not be allowed to limit his liability.

(i) Security and release from arrest

In order to obtain security in respect of a claim for damages against the owners of a vessel the right of arrest may be used. In the case of a plea of limitation of liability under the provisions of the Merchant Shipping Act 1894 some questions arise about the amount of security that may be demanded in order to secure the release of the vessel. When a vessel has been arrested the owner may seek to obtain the consent of the arresting party to the release of the vessel by offering bail in an amount satisfactory to that party. The arresting party may be satisfied with a guarantee given by a bank, insurance company or other satisfactory guarantor. In *The Charlotte* the court decided that if in an action in respect of damage done by a collision, the amount of the claim exceeds the statutory limitation under the provisions of the Merchant Shipping Act 1894, section 503, and the owners of the vessel raising the plea of limitation of liability on the ground that the collision occurred without their actual fault or privity are not faced with a denial of plea by the owners of the other vessel, then the security provided in order to release the vessel from arrest need be no more than the amount to cover the statutory limit plus costs. But, on the other hand, if the owners of the vessel claiming damages deny that the loss occurred without the actual fault or privity of the other vessel and deny the right of the owners of that vessel to limit their liability under the provisions of the Act, then security needs to be given to the full value of the loss.[138] In *The Coaster* the owners of the vessel were forced to provide security in a foreign, French, port. In order to avoid the arrest of their vessel the owners had to put up bail to the full value of their vessel at the time. In the subsequent proceedings the *Coaster* was found solely to blame and the owners met the claim of the other vessel under the judgment in the amount of security they had been forced to provide.[139]

In some cases an owner is faced by claimants in more than one country. When he puts up security or establishes the limitation fund this does not protect his vessel from arrest by another claimant in another port. Under section 5 of the 1958 Act an attempt has been made to preserve owners from multiple jeopardy. Under this section if a ship is involved in an incident in respect of which he is entitled to limit his liability then he should only be compelled to provide a limitation fund once and for all. If the owner makes the fund available in one country to meet all the limited claims, he should not be compelled to put up security for those claims in another country. Even if he is compelled to do so he should be able to get the security released. The meaning of security used in section 5(1) of the 1958 Act was discussed in *The Wladyslaw Lokietek*. In this case Brandon J. stated that in its ordinary and natural meaning the expression "security" was wide enough to cover

138. *The "Charlotte"* v. *"Theory" and Others* (1921) 9 Ll.L.R. 341.

139. *"T" Steam Coaster ("Coaster")* v. *Owners of Cargo Laden in "Dokka" and Others* (1921) 8 Ll.L.R. 368 and (1922) 10 Ll.L.R. 592.

all forms of security normally used, whether given inside or outside the court and that the High Court had the power to compel a plaintiff who has obtained security in the form of a written guarantee by a third party to release such security by ordering the guarantee to be cancelled or surrendered.[140]

If the owner has previously given security elsewhere and is compelled to put up further security in England, under section 5(1), (2)(a) and (b) of the Act, he is under certain obligations to have the security in England released. He must persuade the court that security in respect of the claims arising out of the incident has previously been given elsewhere. In other words the owner has to show that security had been given elsewhere before the vessel was arrested in England.[141] The form of the security must be satisfactory. The liability must be of a type to which a limit may be set under section 503. The security (the guarantee) previously given elsewhere should not be less than the limit set by section 503 or the amount of further security together with the guarantee should not be less than that limit. If the claim is finally established there should be no bar to the English claimant establishing his claim in the country in which the fund was set up and the claimant should be able to convert his share of the fund to sterling and withdraw it from that country. If any of the other claims brought against the fund are of a type for which a limit is not set by section 503 then the part of the fund actually available to the claimant in the English proceedings does not correspond to his claim. And as a last condition, based on subsection (4)(c), the part of the fund actually available to the claimant in the English proceedings should correspond to his claim.[142]

(ii) The limitation fund

Distinct and separate occasions

Under section 503 of the Merchant Shipping Act 1894, as amended by the Merchant Shipping Act 1958, one fund is established per incident. Section 503(3) provides that "the limits set by this section to the liabilities mentioned therein shall apply to the aggregate of such liabilities which are incurred on any *distinct occasion*, and shall also apply in respect of *each distinct* occasion without regard to any liability incurred on another occasion". In *The Harlow* there were two separate legal actions arising out of the casualty when a tug, having in tow five of the barges belonging to the tug owners, came into collision with the steamship *Dalton* in the River Thames, causing considerable damage to the *Dalton*. After the collision with the *Dalton* the tug made a turn in the river and hit another vessel. The question arose whether the latter casualty was a distinct and separate occasion from the first casualty, involving the *Dalton*. The court held that the damage to the vessel concerned in the second casualty happened on the same occasion as the damage to the *Dalton* and that the damages payable to the vessel concerned in the second

140. [1978] 2 Lloyd's Rep. 520 (QB) at p.533.
141. See *The Putbus* [1969] P. 136, [1969] 1 Lloyd's Rep. 253 (CA).
142. Cited in Michael Thomas, p.1237, n.168. S. 5(4)(c) of the 1958 Act was said by Lord Denning to be "unintelligible": *The Putbus* [1969] 1 Lloyd's Rep. 253, 256 (CA).

casualty had to come out of the one limitation fund.[143] In *The Lucullite* the vessel, *Lucullite*, was lying moored alongside and on the outside of another vessel. Due to a strong wind she bumped against the other vessel and caused considerable damage. So she cast off and proceeded ahead into harbour. As she was doing so her port bow struck the starboard quarter of another vessel and caused that vessel to sink. When the issue came before the court it was held that the second collision was not the necessary consequence of the first and that the two occasions of damage were separate and distinct. Therefore it was held that the owners of the *Lucullite* could not succeed in bringing the two separate claims within the same limitation fund.[144]

In deciding whether or not the various occasions of damage caused by the vessel are in fact separate and distinct the test is not lapse of time but whether both incidents are the result of the same act of negligence. Therefore, where the damage was caused to two vessels in succession by the improper putting to starboard of the vessel at fault, it was held that as there was time between the two collisions to correct the mistake, the second collision took place on a "distinct occasion" from the first one and two funds had to be established. The owners of the vessel who are seeking to limit liability have to prove that the damage did not arise on distinct occasions.[145]

Constitution of the limitation fund

Under English law it is not necessary to constitute a limitation fund prior to seeking a decree of limitation of liability. If the shipowners obtain a decree of limitation then the court will distribute the limitation fund amongst the various claimants. Until the 1958 Merchant Shipping Act the amount of the fund to be paid into court depended on the nature of the claim. If there were losses of life or personal injury claims either alone or together with loss of or damage to property, the figure was £15 per ton of the vessel's tonnage. If the only claims were for loss of or damage to property the figure was £8 per ton. Therefore before the limitation decree it was almost impossible to know the amount of the payment of the limitation fund into court. Under the 1958 Act a number of changes were made. Under section 1(1) of the 1958 Act figures of 3,100 and 1,000 gold francs respectively have been substituted for the figures of £15 and £8. Under section 1(2) the gold franc is defined by reference to a specific weight of gold of a specific millesimal fineness. Under an order made by statutory instrument the Secretary of State for Trade has the power to specify from time to time the amounts in pounds sterling that are to be taken as being equivalent to 3,100 and 1,000 gold francs respectively.[146]

Under such a system problems arose because of the delay between the incident giving rise to liability and the granting of a limitation decree. As a result of the difference in currencies at the date of the incident and at the date of the decree an

143. *Cory Lighterage, Ltd. ("Harlow") v. Dalton and Others* (1922) 10 Ll.L.R. 66, 169, 244, 488.
144. (1929) 33 Ll.L.R. 186.
145. *The Schwan–The Albano* [1892] P. 419 (CA) at p.429.
146. Merchant Shipping (Liability of Shipowners and Others) Act 1958, 6 & 7 Eliz. 2, c. 62, s. 1(3).

owner could find himself paying much more than he anticipated or a claimant could find himself receiving much less.

In England problems arose in 1968 following the devaluation of the pound in November 1967 and the resultant change in the sterling equivalents. *The Abadesa* was the first case under which such a problem arose. In this case it was held that conversion should take place in accordance with the sterling equivalents order in force at the date of the decree.[147] In the following case, *The Mecca*, the effect of section 1(4) of the 1958 Act was considered.[148] In conclusion it was said that "it may well be that, as the effect of the Act of 1958 comes to be fully appreciated, the practice in limitation actions will change and plaintiffs will seek to pay into court as early as possible and certainly before decree".[149] The term "an adequate payment into court" was also defined in this case as "a payment of an amount not less than the limit as ascertained in accordance with the order in force at the date when the payment was made".[150]

It has long been accepted that interest on a limitation fund ran from the date of the incident giving rise to liability. Until *The Abadesa (No.2)* the rate of interest remained constant at 4 per cent. In that case the rate was raised to 5½ per cent. But the following practice regarding the rate of interest payable upon a limitation fund was established in *The Funabashi*. In this case it was held that interest rates should be based on the principle that the owner of the offending vessel had kept the injured party out of his money and had had the use of it himself. And there was no particular merit in the stabilisation of interest rates. Therefore the rate of interest is now in the discretion of the court.[151]

Under section 7(2) of the 1958 Act all claimants are required to proceed against the fund once it is formed and barred from going against any other property of the shipowner. In general the court will advertise the fund and give potential claimants the opportunity to file a claim against it or take steps to set aside the decree of limitation. If the benefits of limitation were not granted all claims rank equally. And if the proceeds available from the sale of the ship were insufficient the unsatisfied claimant could only proceed against any other assets which the shipowner might have.

A limitation fund can be constituted in the shipowner's place of domicile of his choice. The place of the limitation fund may be relevant to the question of the appropriate jurisdiction to determine claims against the shipowner.

147. *The Abadesa (No.2)* (1968) P. 656, 663.

148. S. 1(4) of the 1958 Act provides as follows: "Where money has been paid into court ... in respect of any liability to which a limit is set aforesaid, the ascertainment of that limit shall not be affected by a subsequent variation of the amounts specified under subsection (3) of this section ... unless the amount paid ... was less than the limit as ascertained in accordance with the order then in force."

149. [1968] P. 665, at p.670.

150. *Ibid.*, p.669.

151. [1972] 1 Lloyd's Rep. 371.

Amount of limitation

When the privilege of limitation of liability was first introduced in 1734[152] the value of the ship and freight was taken as the measure. In some countries this rule has been retained, but in England the method of calculation based on tonnage was introduced under the Merchant Shipping Acts of 1854 and 1862. This method was substantially re-enacted in section 503 of the 1894 Act and amended by the Merchant Shipping Act 1958.

The tonnage to be taken as the basis of the owner's liability is that appearing on the ship's register in force at the time of the event giving rise to liability. The parties are not entitled to any subsequent change in the register. The tonnage appearing in the ship's register is only *prima facie* evidence and it is possible to prove that the registered tonnage was not properly calculated. "The tonnage of a steamship shall be her registered tonnage with the addition of any engine-room space deducted for the purpose of ascertaining that tonnage, and the tonnage of a sailing ship shall be her registered tonnage."[153]

Where a foreign ship was concerned in an action to limit, her tonnage for the purposes of limitation was measured according to British law. Special provisions made in section 503(2) with regard to the ascertainment of the tonnage of foreign ships are now applied to any ship structure that is unregistered or in the course of completion or construction. Measurement for foreign ships need only be made when the ship's tonnage, as stated in her national certificate or registry or other national papers, is not accepted as equivalent to tonnage denoted in the certificate of registry of a British ship. If it is impossible to ascertain the tonnage of the foreign ship under British law a certificate issued by the UK Surveyor General of Ships stating his opinion as to the relevant figure is sufficient for the satisfaction of the court. Crown ships are measured by the Chief Surveyor as if they were commercial vessels.

Under the 1958 Act there is a special rule for small ships in the case of personal injury and death claims. Under such claims if the registered tonnage of the offending vessel is less than 300 tons and the owner has obtained a decree limiting his liability the registered tonnage of the vessel will be taken to be 300 tons.[154] There is no minimum tonnage provision for claims for loss of or damage to property, whether alone or together with loss of life or personal injury claims. Therefore property claimants will be met by the actual limitation tonnage of the ship even if it is below 300 tons.

In a claim for which liability may be limited the court proceeds as it would in any other action when assessing damages. Thus the possibility of limitation is not taken into account. If the fund is adequate all claims are paid in full. But problems may arise if the fund is inadequate to meet the claims in full. In cases where there is only one claimant he receives the whole amount of the fund. The case becomes more

152. 7 Geo. 2, c. 15 (1734).
153. Merchant Shipping Act 1894, 57 & 58 Vict., c. 60, s. 503(2)(a).
154. Merchant Shipping (Liability of Shipowners and Others) Act 1958, 6 & 7 Eliz. 2, c. 62, s. 1(1).

problematic when there is not only more than one claimant but also more than one type of claimant. As a general rule if the fund is insufficient to meet all claims in full the court will ensure that each individual claimant will be paid a proportion of his claim. Where all the claimants are of the same type and all claims are in respect of loss of life or personal injury or in respect of loss of or damage to property, they are paid rateably, in proportion to the amount of damage that has been sustained. But if there are claims both for loss of life and personal injury there are certain priorities. In such a situation there is only one fund comprised of 3,100 gold francs multiplied by the ship's registered tonnage. Claimants in respect of loss of life or personal injury have priority over a proportion of this fund. Thus loss of life and personal injury claims are given priority against the fund to the extent of 2,100 gold francs multiplied by the ship's registered tonnage. If this sum is not sufficient to pay off all of these claims then the claimants must recover in proportion to their loss against the sum available. Depending on the extent to which the loss of life and personal injury claimants remain unpaid they rank equally with the damage claimants against the remainder of the fund which is comprised of 1,000 gold francs multiplied by the ship's registered tonnage.

If the incident involves a ship which is less than 300 tons and claims both for loss of life and personal injury then the calculations become complicated. The sum for loss of life and personal injury claims will be calculated on the basis of the difference between 300 tonnes multiplied by 3,100 gold francs and the result of multiplying the actual registered tonnage of the ship by 1,000 gold francs. If the claimants for loss of life and personal injury are not satisfied completely they will rank equally with the damage claimants against the sum obtained by multiplying the actual registered tonnage of the ship by 1,000 gold francs.

The principle of priority for life and injury claimants is only relevant when the shipowner succeeds in limiting his liability. If the owner of the offending vessel does not seek to limit his liability there is no priority. Loss of life or personal injury claimants rank equally with the other claimants. If the vessel is sold to satisfy claims and the proceeds are insufficient to meet all claims, then they are paid proportionately. Unsatisfied or partly unsatisfied claimants are left to proceed against whatever separate estate the owner may have and again rank equally against that estate. If there is only one claim then the shipowner may plead limitation of liability as a defence. In *The Mekhanik Evgrafov* the Admiralty judge commented that where there is only one claim section 503 ought to be pleaded as a defence.[155]

(i) Dock owners, harbour and pilotage authorities

Under section 2 of the Merchant Shipping Act 1900 the owners of any dock or canal, or a harbour authority or a conservancy authority, as defined by the Merchant Shipping Act 1894, are entitled to limit their liability for damage caused to vessels or their contents. Under this Act liability did not exceed £8 per ton. This

155. [1988] 1 Lloyd's Rep. 330.

figure is regularly amended by amendments to the Merchant Shipping Act. Obviously as it is not possible to measure the tonnage of a pier the Act fixes the limit according to the tonnage of the largest registered British ship which, at the time of such loss or damage occurring, is, or within the period of five years immediately prior to such loss or damage was, within the area over which such dock or canal owner, harbour authority, or conservancy authority, performs any duties or exercises any powers. A ship shall not be deemed to have been within the area over which a harbour authority or a conservancy authority performs any duty or exercises any powers by reason only that it was built or fitted out within such area, or that it has taken shelter within or passed through such area on a voyage between two places both situated outside that area or that it has loaded or unloaded mails or passengers within that area.

For the purposes of this section the term "dock" is defined widely to include docks and basins, tidal docks and basins, locks, cuts, entrances, dry docks, graving docks, gridirons, slips, quays, wharves, piers, stages, landing-places and jetties. The term "owners of a dock or canal" includes any person or authority having the control and management of any dock or canal. In *The Humorist*[156] the above-mentioned provisions came under review. In this case a barge while berthed and loading cargo alongside a wharf, took the ground on several tides and became waterlogged. Her bottom was set up owing to the presence of some iron weights in the berth. The barge owners sued the owners of the wharf for damages. The court ruled that the damage to the barge was sustained without the actual fault or privity of the wharf owners. It was held that the wharf owners' premises could properly be described as a "landing place", and therefore a "dock" for the purposes of the Act, and that the damage occurred within the area over which the wharf owners exercised control. Therefore it was held that the wharf owners were entitled to a limitation decree based upon the registered tonnage of the largest vessel using the berth within the previous five years. In *The Neapolis II*[157] the vessel suffered damage by grounding while she was tied up against a wharf leased and operated by the defendants. This case was based on the effect of section 2 of the Merchant Shipping Act 1900 on the liabilities of wharf owners and dock owners in respect of damage caused to a vessel while at its berth. In the lengthy proceedings which followed, the court found that it was negligent of the wharf owners not to put up clear warning notices and that had such notices been in position the *Neapolis II* would have seen that it was necessary to berth about 25 metres further upstream than she had done, so as to avoid the danger. As the wharf operators were aware of the absence of such notices and the existence of the danger, the court said, it must follow that such an absence of notices and the existence of the danger were well known to the managing director of the wharf operators. It followed that such absence was causally connected with the damage and that, therefore, the wharf operators could not rely on the limitation of liability under section 2 of the 1900 Act.

156. (1946) 79 Ll.L.R. 549, [1946] P. 198, 62 T.L.R. 602.
157. *Alberta Shipping Ltd.* v. *The Briton Ferry Stevedoring Co. Ltd.* [1980] 2 Lloyd's Rep. 369.

Another interesting point was discussed in *The Ruapehu*. In this case the relationship between the right of limitation of liability under section 2 and geographical area or capacity was discussed. Owing to the negligence of the ship repairers' servants a fire broke out in the *Ruapehu* while it was undergoing repairs in the ship-repairer's own drydock. The ship-repairers sought to limit their liability in respect of the damage caused. But the owners of the vessel contended that the ship repairers' liability arose out of their business as ship repairers and not out of their business as dock owners, and that therefore section 2 of the Act did not apply. The House of Lords held that the right to limit liability conferred upon dock-owners and others by section 2 was dependent on the area in which the dock-owners performed their duty and not in the capacity in which the dock-owners acted. As the ship-repairers were using the drydock in the ordinary course of business at the time of the incident they were entitled to limit their liability.[158] The meaning of the word "area" was discussed in another *Ruapehu* case. The court held in this case that upon the true construction of section 2(1) of the Merchant Shipping Act 1900, the word "area" meant that area over which the limiting dock-owner performed a duty or exercised a power which contained the dock in which the damage occurred.[159]

There may be practical problems in ascertaining the identity and tonnage of the biggest ship that had visited the area in the given period. As stated in section 2 ". . . where without their actual fault or privity any loss or damage is caused to any vessel or vessels, or to any goods, merchandise, or other things whatsoever on board any vessel or vessels . . ." the protection does not apply to personal injury or death claims, or to claims against the individual servants of the dock owner or harbour authority. A dock owner may claim the benefit of limitation even if in the particular special case he was doing work to a ship in his dock as a ship repairer.[160]

(i) Tug and tow situations

When one or more of the vessels concerned are involved in towage and a plea is made for a decree of limitation of liability then particular considerations have to be taken into account peculiar to damage caused during the performance of such operations. In the case of a collision between tug and tow the vessels involved may be in the same ownership or the one may be under different ownership from the other. When the tug and tow are in different ownership each owner is entitled to limit on the tonnage of his own vessel. However, where they are in the same ownership the position is more complex.

Tug and tow in the same ownership

In *The Ran and The Graygarth* the tug had a number of barges in tow. During the course of towage the barge *Ran* which was in the same ownership as the tug

158. (1929) 34 Ll.L.R. 402.
159. *Ibid.*
160. *Ibid.*

collided with and caused damage to another barge, the *Para*. In an action *in rem* by the owners of the barge *Para* against the owners of the barge *Ran* the court found that the collision was solely due to the negligent navigation of those on board the tug who had the control of the navigation of the *Ran*. It was held that the owners of the barge *Ran*, as the owners of the tug, were liable. During the course of application for a decree of limitation of liability the Court of Appeal held that as the action was brought against the barge *Ran* by the owners of the barge *Para* and as the *Ran* was improperly navigated by the servants of the owners of the *Ran* the owners of the barge *Para* were entitled to judgment against the defendants as owners of the *Ran*. The defendants were not entitled to limit their liability on the tonnage of the tug but had to bring their action to limit liability on the basis of the tonnage of the barge *Ran*.[161] *The Radiant* was based on an accident in the course of the towing of a fishing vessel by a sister ship. During the tow the tow rope parted and the vessel grounded for a second time. Wire was coiled on the deck of the vessel by the skipper and there made fast. The skipper's legs became caught up in the coil of wire and as the sister ship took up the slack both his feet were amputated. The skipper claimed against the company owning the vessels and the managing director of the company. The court ruled that as the company had failed to discharge the burden of proof upon it under section 503 it could not limit its liability under the Act. With regard to the managing director the claimant had discharged the onus of proving that this person was a party to the sending of the vessels to sea in an unseaworthy condition and that his injury resulted from that condition. It was held that the managing director was liable and that the claimant's right was a personal right against him in his personal capacity and therefore he was unable to limit his liability. As stated above the conclusion of the court in this case, that a personal action lay against the managing director on the same facts as actual fault or privity was found against the company, did not necessarily apply in all cases.[162]

In *The Harlow* the barges were all in the same ownership as the tug. The court ruled that where a collision with some other vessel is caused by the negligent navigation of those on board a tug and the tows, the owners may limit on the combined tonnage of the tug and such of her tows that were in contact with the other vessel or that, by their weight, contributed to the damage. This principle was later applied in a case where not only was the tug herself in collision but also there was no negligence on board the tow. In that case, *The Freden*, the collision was between tug and tow and a steamship where the tug and lighters were under the same ownership. The owners of the tug and the lighter astern were allowed to limit their liability to the tonnage of these two vessels.[163]

The Sir Joseph Rawlinson[164] established the principle that where a tug and a tow are under common ownership, their owners may limit their liability for collision damage to an amount calculated by reference to the tonnage of the tug alone,

161. *Rea, Ltd.* v. *Owners of Barge "Para" and Others* (1921) 8 Ll.L.R. 305.
162. See the comments of Willmer L.J. in *The Anonity*.
163. (1950) 83 Ll.L.R. 427.
164. *London Dredging Co. Ltd.* v. *Greater London Council* [1972] 2 Lloyd's Rep. 437.

where there is negligence for which they are liable on the part of the person in charge of the tug, but no negligence on the part of anyone on the tow. In other words, where the tug alone is negligent the owners of the tug and tow are entitled to limit their liability for negligence by taking into account the tug's tonnage only and disregarding the tonnage of the tow. This case was between the tug and her tow; a dumb barge collided with a sludge carrier in the Thames Estuary. The sludge carrier sank with loss of life. The ruling of the court was that the only causative negligence was that in the navigation of the tug and not negligence in the navigation of the tow or both tug and tow, and it was therefore held that the demise charterers were not liable in damages beyond an amount calculated by reference to the tug. Therefore where tug and tow are in the same ownership the limitation fund will be calculated by reference to their combined tonnage only where there is negligence, not only on the part of the person in charge of the tug, but also on the part of some person on board the tow, and their combined negligence results in the tow causing loss or damage to some other person. In any other case aggregation of tonnage is permitted and the limitation fund is calculated by reference to the tonnage of the tug alone.

In *The Smjeli* the owners of the tug and barge, which were both in the same ownership, admitted liability but sought to limit their liability under section 503 to the tonnage of the tug alone. This was denied on the ground that the council had a cause of action against the barge and also one against the tug, in that the master was negligent in his navigation of the tug in causing or allowing the wire to part and in failing to seek shelter toward North Foreland and instead attempting to hold the tow in the Dover Strait. The limit of liability was therefore to be calculated on both the tonnage of the tug and the barge in tow.[165]

Tug and tow in different ownership

As a general rule, where the tug and tow are in different ownership, each owner is entitled to limit on the tonnage of his own vessel. *The Bramley Moore* established that tug owners could limit their liability according to the tonnage of the tug alone. The casualty arose when the tug *Bramley Moore* was towing the barge *Millet*. The *Millet* collided with a motor vessel, the *Egret*, with the result that one of the crew of the *Millet* lost his life and two others were injured. The collision was partly the fault of those on board the *Bramley Moore*. The barges did not belong to the owners of the tug but were being towed under contract. In the liability proceedings the owners of the *Bramley Moore* claimed to limit their liability to a sum based on her tonnage alone. The question arose whether the decree of limitation of liability should be based on the combined tonnage of the tug and one or both of the barges, or on the tonnage of the *Bramley Moore* alone. The Court of Appeal ruled that under section 503 as amended by the 1958 Act the tug owners could limit their

165. [1982] 2 Lloyd's Rep. 74.

liability according to the tonnage of the tug alone and that the flotilla could not be taken as a unit.[166]

(j) Wreck raising expenses

In considering the right of a shipowner to include wreck raising expenses in a claim for limitation of liability, emphasis must be placed upon the words in section 503 of the Merchant Shipping Act 1894 "damage" and "property or rights". In *The Stonedale No. 1*[167] the question was whether the costs of the raising of a wreck constituted damages within the meaning of the Act of 1900. In this case the Manchester Ship Canal Company, having at its own expense raised the sunken vessel, then claimed the costs of the raising of the barge. The owners of the barge and the tug sought to have their liability limited in respect of this claim. In the House of Lords it was held that although the Act of 1900 extended the relief given to shipowners under section 503 of the Merchant Shipping Act 1894, it did not enable them to limit liability except where their liability lay in damage, and that the claim made by the Canal Company under the provisions of the Manchester Ship Canal Act 1936 was a claim to recover a debt for the expense of raising the sunken vessel and was not a claim for damages under the Act of 1900. Therefore the claim for limitation of liability failed.[168]

In a similar case, *The Urka*,[169] the court ruled that liability for the expense of removal was not liability in respect of any "loss or damage" to the owners of *Portugal* or their "rights". Nor was it liability in respect of any damage to the Harbour Commissioners' "property or rights". It was simply liability arising through the exercise by the Harbour Commissioners of their statutory powers, and accordingly the claim did not fall within the terms of the Acts. But this case was subsequently referred to by the court in *The Arabert*[170] as having been wrongly decided. In *The Arabert* the court came to a conclusion opposite to that found by the court in *The Urka* and decided that the question now to be borne in mind, in assessing whether or not a shipowner had the right to include wreck raising expenses in a claim for limitation of liability under the provisions of the Merchant Shipping Acts, was the nature of the claim. It was the nature of the claim which mattered.

166. [1963] 2 Lloyd's Rep. 429 (CA).

167. *The Stonedale No. 1 (Owners)* v. *Manchester Ship Canal Co.* [1956] A.C. 1, [1955] 2 Lloyd's Rep. 9 (HL), affirming [1954] P. 338, [1954] 1 Lloyd's Rep. 291 (CA), affirming [1953] 2 Lloyd's Rep. 319.

168. As in *Stonedale No. 1* in *The Millie* the court said the expenses incurred by the canal company in raising the vessel *Millie* were not "loss or damage" caused to their "property or rights": (1939) 64 Ll.L.R. 318.

169. [1953] 1 Lloyd's Rep. 478.

170. *The Arabert (No. 2)* [1961] 1 Lloyd's Rep. 363.

(k) Contracting out

The privilege of excluding or limiting liability, as provided for under sections 502 and 503 of the Act of 1894, may be waived by contract. The privilege of excluding or limiting liability is available to an owner only if he has not made some contrary provision in a contract with the party claiming against him or where the claim is not the type in which the owner can limit his liability.

Whether the owner has waived the benefit of sections 502 or 503 is a question of construction of the relevant contract. Thus if an owner wishes he may limit his liability for a lesser amount than provided for under section 503 or contract out of liability completely. But in such cases clear language must be used. The relevant limitation statutes can be excluded only by express contract. In *The Kirkness* a collision between a tug and its tow resulted in the loss of both tug and crew. It was held that the tow owners could limit their liability with regard to loss of life claims because the terms of the towage contract were not sufficiently clear to exclude the application of section 503. On the other hand the tow owners could not limit their liability for the loss of the tug because that liability arose under the contract and not merely as the result of damages action on the part of the tug.[171]

Apart from contracting out of liability altogether the owner also has an option to contract to be liable for the full amount of any loss or damage caused to another party or he may contract for a sum more than that provided for in section 503.

171. See Michael Thomas, p.1253; *The Kirkness* [1957] P. 51.

CHAPTER 20

THE CONVENTION ON LIMITATION OF LIABILITY FOR MARITIME CLAIMS 1976

IN GENERAL

(a) Introduction

The 1976 Convention is one of many international liability regimes which regulate legal relations between parties involved in international maritime adventures. The 1976 Convention is designed to replace the 1957 Limitation Convention internationally. On 1 December 1986 the Convention became effectively part of English law under section 17 of the Merchant Shipping Act 1979.[1] Following the suggestion of the CMI the title of the Convention is no longer "Limitation of the Liability of Owners of Sea-going Ships", as it was in the 1924 and in the 1957 Conventions, but "Limitation of Liability for Maritime Claims". The change was not simply in the title but in the basic concept underlying the construction of the system of limitation. Originally the benefit of limitation was granted only to owners of the ships as the obligations arising out of the operation of a vessel mainly rested on her owner and the limitation was realised by the surrender of the owner's interest in the ship. But gradually the benefit of limitation extended to bareboat charterers under section 186 of the US Limitation Act, to the operator and the "principal charterer" under Article 10 of the 1924 Brussels Convention and to any charterer, to the manager and operator of the ship under Article 6, paragraph 2, of the 1957 Brussels Convention.

Thus, before the 1976 Convention, as early as in the 1924 and the 1957 Conventions, limitation of liability was not granted merely to shipowners. The 1976 Convention regulates the amount of compensation which a shipowner must pay to various users of ships where the shipowner is liable to the user. Within the context of the Convention the term "user" means passengers, cargo owners and other parties who may have claims against the shipowner pursuant to the 1976 Convention. In general under the 1976 Convention the benefit of limitation is granted for claims arising in connection with the operation of the ship and it may be invoked by any person who is operating the ship, by salvors irrespective of whether or not they operate from a ship, by all persons for whose act, neglect or default the operator of the ship or salvor is liable, as well as by the insurers of liability for claims

1. As from 1 January 1996 the Merchant Shipping Act 1995.

subject to limitation. With the wider description of group of persons entitled to limit, the nature of the claims has become more important than the persons against whom the claims are made. Therefore the persons who can invoke limitation are of minor importance.

In some respects the 1976 Convention serves a similar function to other regimes pertaining to the carriage of goods by sea, such as the Hague Rules, the Hague-Visby Rules and the Hamburg Rules. But the Hague Rules Regime applies to contracts of carriage of goods by sea, while the 1976 Convention and its predecessors apply by force of international law and not by virtue of contractual arrangements.

The 1976 Convention applies wherever a shipowner or other person entitled under the 1976 Convention to limit liability seeks to limit his liability. The Convention does not deal with establishing liability which is a separate matter. In appropriate cases liability may be determined by the Hague Rules Regime or by some other contract law. In cases of oil pollution a separate legal liability regime applies to the liability of shipowners for pollution damage. The Convention entitles the owner to limit any liability established otherwise.

(b) The need for changes

In 1972 the CMI came to the conclusion that, although a relatively short time had elapsed since the adoption of the 1957 Brussels Convention on limitation of liability, the provisions of that Convention had already been overtaken by the changes which had occurred. Due to depreciation in monetary values the limitation amounts had become unrealistically low. The size of ships had increased, a number of difficulties of construction of the Convention had emerged, such as the question of its application to salvage. There was a need for the protection of other categories of operators. The need for revision became more urgent as a result of the 1969 Convention on Civil Liability for Oil Pollution Damage, which was in some respects inconsistent with the 1957 Convention.[2]

An international sub-committee was formed by the CMI in February 1972, which considered the problem whether the right of limitation should be accorded not only to ship operators and their servants but also to other persons rendering services in direct connection with the operation of a ship. The Committee chose an intermediate solution and extended the benefit of limitation to pilots who were not servants of the operator and to salvors who could not invoke limitation under the 1957 Convention. As regards claims excepted from limitation, the Committee decided that those subject to the provisions of the Civil Liability Convention of 1969 and those subject to any international convention or any national law governing liability for nuclear damage should be excepted. The conduct barring limitation was considered inadequate. The Committee recommended that the

2. See Francesco Berlingieri, and Kkaj Pineus, "The 1976 Convention", in United Nations (ed)., *Report on the Fourth Regional Seminar on Maritime Legislation, Rights and Responsibilities of Shippers and Carriers in the Transport of Goods by Sea*, 16–19 December 1991, Pattaya, pp.105–107.

right of limitation should be forfeited only because the person liable had acted with intent to cause damage or recklessly and with knowledge that damage would occur. The Committee agreed that the constitution of the limitation fund should not be compulsory, though specific rules were required to govern the fund when actually constituted.

In 1974 two alternative drafts, one in the form of a draft Protocol to amend the 1957 Convention and one in the form of a draft of a new Convention on Limitation of Liability for Maritime Claims were prepared for submission to the IMO. The Legal Committee of the IMO examined the two drafts and decided to use the draft of a new convention, as opposed to the draft of a Protocol to the 1957 Convention. At its eighth extraordinary session held on 3 November 1975 the Council of the IMO approved the convening in November 1976 of a conference to adopt a revised instrument to replace the 1957 Brussels Convention. In its ninth regular session this decision was endorsed by the Assembly of the IMO. Then an International Conference on Limitation of Liability for Maritime Claims was convened by the IMO from 1 to 19 November 1976 in London. The Convention entered into force on 1 December 1986.

The 1996 Protocol to the 1976 Limitation Convention was adopted on 3 May 1996. The Protocol increases the amount of compensation payable in the event of an incident substantially. It also introduces a "tacit acceptance" procedure for updating these amounts. In general the normal procedure for adopting amendments to an international treaty is by means of "explicit acceptance". Under this procedure the amendments enter into force so many months after being accepted by a specified number of parties to the original convention. The "tacit acceptance" procedure means that amendments which are nearly always adopted unanimously enter into force on a set date unless they are specifically rejected by a specified number of countries. Under "tacit acceptance" everyone involved knows exactly when an amendment will enter into force, which was impossible under the "explicit acceptance" system.

The 1996 Protocol will enter into force 90 days after being accepted by 10 states. By September 1997 there were nine signatory countries to the Protocol. These countries are the UK, France, Sweden, Finland, Germany, Canada, Norway, Denmark and the Netherlands. However, none of these countries have ratified the Protocol yet. Under the Protocol for ships not exceeding 2,000 g.t. liability is limited to 2 million SDRs for loss of life or personal injury and 1 million SDRs for other claims. Then liability increases with tonnage to a maximum above 70,000 g.t. of 2 million SDRs plus 400 SDRs per tonne for loss of life or personal injury and 1 million SDRs plus 200 SDRs per tonne for other claims.

ANALYSIS OF THE 1976 CONVENTION

(a) Persons entitled to limit their liability

Article 1(1) provides that "shipowners and salvors" may limit their liability in accordance with the rules of the Convention for claims set out in Article 2 thereof. In addition to persons entitled to limit under the 1957 Convention, Article 1 of the 1976 Convention extends the privilege to insurers and salvors. The term "shipowner" is, by Article 1(2), defined as meaning the owner, charterer, manager and operator of a seagoing ship. By paragraph 2 of Part II of Schedule 7 to the Merchant Shipping Act 1995, "ship" includes any structure (completed or in course of completion) launched and intended for use in navigation as a ship or part of a ship. And by paragraph 2 of Part II to Schedule 7 of the Act the right to limit is extended as part of English law to cover ships, whether sea-going or not. The shipowner must be construed accordingly. The 1979 Act, by Schedule 5, paragraph 4, extends the right to limit to hovercraft as if they were ships.

Probably the most significant of the newly introduced groups of people in Article 1 is salvors, who may limit their liability under the Convention. Benefits of limitation are extended to salvors and to any person for whom a salvor is responsible. Under the 1957 Convention salvors were merged with shipowners and they were not recognised as a separate category of individuals pursuing a specialised type of operation. If a salvor's salvage vessel or tug is involved in a collision then he is in the same position under section 503 of the 1894 Act with regard to claims in tort as any other shipowner in relation to limitation of liability. But in some cases liability may arise out of the act or omission of a person not on board the tug and not concerned with the navigation or management of the tug as in *The Tojo Maru*, and the salvor may face unlimited liability which will put him out of business. Therefore the salvor's entitlement to limit his liability was introduced into the 1976 Convention. To a large measure it was inserted to overcome the effect of the House of Lords decision in *The Tojo Maru*. Now salvors are permitted to limit their liability even when not operating from a tug or other salvage vessel.

Before the 1976 Convention it had already been incorporated by reference into LOF 1980 that a salvor in respect of any claim by a person with whom he had contracted under LOF 1980 (but not otherwise) was entitled to limit his liability.

Under the Convention a salvor is broadly defined as a person rendering services in direct connection with salvage operations, including the removal, destruction or the rendering harmless of the cargo of the ship and measures taken in order to avert or minimise loss. Salvage operations are not as such defined in the Convention but it is submitted that they include not only claims on a no cure, no pay basis but claims where the operations are being carried out for a lump sum or at a daily rate. Therefore the salvor is entitled to limit his liability whether he is engaged under LOF or under a lump sum or daily hire contract. But a shipowner himself cannot limit his liability as against the salvor in relation to remuneration under a contract of salvage.

Under Article 1(4) it is possible for any person for whose act, neglect or default the shipowner or salvor is responsible to be entitled to avail himself of the right of limitation provided for in the Convention. Therefore it is not possible to evade limitation of liability by claiming against the employees of shipowners and salvors. According to Meeson[3] in order to define the persons who are covered by this provision it is necessary to start with the shipowner or salvor.

"If a claim can be brought against the shipowner or salvor for the act, neglect or default of the person concerned, then that person will be entitled to limit his liability if sued directly. If it is not possible to bring a claim against the shipowner or salvor, a claim brought against the person concerned will not be one against which he will be entitled to limit his liability. His right to limit liability is parasitic on the shipowner's or salvor's right."

Under this paragraph it is possible for a wider circle of persons to limit their liability than under the 1957 Convention.

(b) Claims subject to limitation

There are six categories of claim listed in Article 2 of the 1976 Convention as being claims subject to limitation of liability. Article 2 further provides that liability may be limited "whatever the basis of liability may be". Therefore it is no longer the case that only a claim for damages may be the subject of limitation. Two principal changes were introduced by the Convention: first, claims now qualify for limitation whatever the basis of liability may be, and, secondly, claims now qualify for limitation if they arise on board or in direct connection with the operation of the ship or with salvage operations.

Under the previous limitation regime limitation was restricted to acts or omissions done by a person on board or in the navigation or management of a ship, or in the loading, carriage or discharge of its cargo, or in the embarkation, carriage or disembarkation of its passengers. Sometimes these restrictions have resulted in unfortunate decisions as in *The Tojo Maru*[4] where it was held that salvors were not entitled to limit their liability since the negligent act of a diver was not an act done either in the management of or on board the tug. In *The Stonedale (No. 1)*[5] it was held that an owner could not limit his liability for wreck removal expenses payable under statute since such expenses were in the nature of a debt rather than damages.

The 1976 Convention deals with this problem by inserting a wider definition of claims which are subject to limitation and it removes the requirement that the claim must sound in damages. The first category of claims contained in Article 2(1)(a) expressly covers both loss of life and personal injury claims and also property claims including consequential loss. Where loss of life or personal injury or loss or damage to property occurs otherwise than "on board the ship", then it is required that it should occur in direct connection with the operation of the ship or

3. Nigel Meeson, *Admiralty Jurisdiction and Practice* (1993), p.226.
4. [1971] 1 Lloyd's Rep. 341, [1972] A.C. 242, [1971] 2 W.L.R 970, [1971] 1 All E.R. 1110.
5. [1956] A.C. 1.

with a salvage operation. Under this Article the courts need to decide upon what incidents are "directly connected" as required.

Limitation may be invoked not only in respect of direct physical loss or damage, but also in respect of consequential loss. This is made clear by the express reference to consequential loss resulting from loss of life or personal injury or loss of or damage to property.

Article 2(1)(b) gives a right to limit for "claims in respect of loss resulting from delay in the carriage by sea of cargo, passengers or their luggage". As the opening sentence provides "whatever the basis of liability may be" the right to limit includes liability in contract as well as in tort.

Under Article 1(1) of the 1957 Convention and section 2(1) of the 1958 Act limitation was possible where rights had been infringed. This right has been preserved under the 1976 Convention but restricted in scope of limitation to non-contractual rights. Article 2(1)(c) gives a right to limit liability in respect of "claims in respect of other loss resulting from the infringement of rights other than contractual rights, occurring in direct connection with the operation of the ship or salvage operations". The words "other loss" probably refer to losses other than those defined in paragraphs (a) and (b). The rights referred to in this paragraph must be non-contractual and exclude rights referred to in sub-paragraphs (a) and (b).

Article 2(1)(d) deals specifically with claims in respect of removal of wreck expenses except where the claim relates to remuneration under a contract with the person liable under Article 2(2). For a claim in respect of wreck removal expenses or other statutory liability to be subject to limitation it would be necessary to show that it fell within Article 1(d). Quite often oil cargoes have to be removed as part of an operation to minimise pollution damage or the risk of oil pollution damage. As claims for oil pollution damage and nuclear damage are excluded, by Article 3, from the ambit of limitation of liability under the Convention the scope of this paragraph is very restricted. In the UK the policy of unlimited liability for wreck removal expenses has been kept. By paragraph 3 of Schedule 7 to Part II of the Merchant Shipping Act 1995 a reservation is made in respect of Article 2(1)(d) of the 1976 Convention. Therefore liability for claims for the cost of wreck removal remains unlimited insofar as the operation is performed pursuant to statutory powers.

By Article 2(1)(e) claims in respect of "the removal, destruction or the rendering harmless of the cargo of the ship" are subject to limitation of liability. There is no definition or restriction on the ordinary meaning of cargo. There is no restriction on who may make the claim for removal of the cargo for that claim to fall within paragraph (e). In any case by Article 2(2), claims under paragraph (e) are not subject to limitation of liability if they relate to remuneration under a contract with the person liable. The final sub-paragraph (f) of Article 2(1) makes subject to limitation of liability "claims of a person other than the person liable in respect of measures taken in order to avert or minimise the loss for which the person liable may limit his liability in accordance with this Convention, and further loss caused by such measures". In order to limit liability for claims of third parties in respect of measures taken to avert or minimise loss, the loss should be that in respect of which the right to limit liability has accrued to the person liable. The expenses must have been incurred by persons other than the person who is liable and potentially able to limit. By Article 2(2), which aims to prevent the shipowner seeking to limit his liability to his own contractor, such claims are not subject to limitation of liability if they are claims for remuneration under a contract with the person liable. In other words, if there is a contractual relationship between the person liable and the third party who has taken the measures to avert or minimise the loss and thus incurred expenses, such a claim may not be subject to limitation. As limitation is not available in the UK under Article 2(1)(d), claims in respect of cargo removal qualify for limitation before the ship is sunk, wrecked, stranded or abandoned but there is no limitation after any such event has occurred.

(c) Claims excepted from limitation

Article 3 sets out those claims which are excepted from limitation of which there are five classes. A shipowner is not entitled to limit his liability in respect of a claim for salvage made against him by a salvor, nor against a claim for a general average contribution. But if cargo owners claim an indemnity in respect of salvage and general average contributions the shipowner will be able to limit his liability against such claims. In *The Breydon Merchant* the vessel suffered a serious fire in her engine room. The shipowners applied for limitation under the 1976 Convention. Cargo owners argued that the vessel was unseaworthy and the shipowners' claim for damages, including a salvage contribution, was not subject to limitation because salvage claims were excluded under Article 3 of the 1976 Convention. The court held that under Article 2 of the Convention shipowners may limit their liability for various claims, including loss of or damage to property occurring on board, and consequential loss "whatever the basis of liability may be". It did not matter whether the shipowners' liability arose in contract, tort or by statute. They could limit their liability for the claim by cargo-owners arising out of their breach of contract.[6]

Exclusions cover claims for oil pollution damage within the meaning of the Civil Liability Convention 1969 and those for nuclear damage. Claims by servants or agents of a shipowner whose duties are connected with the ship or the salvage operation, in relation to carrying out their duties, are excluded from the scope of the Convention. Such an exclusion prevents a shipowner from limiting liability in relation to a claim made by his servants, such as the master or crew. In the United Kingdom by section 185(4) of the Merchant Shipping Act 1995 the Convention does not apply to claims for loss of life, personal injury or loss or damage to property by a person aboard the vessel against which the claim is brought who is under a contract of service governed by the law of any part of the United Kingdom.

6. [1992] 1 Lloyd's Rep. 373.

(d) Loss of right to limit liability

(i) Conduct barring limitation

One of the major changes introduced by the 1976 Convention was in the provision applying to barring the capacity to limit liability. Under the 1957 Convention an owner was entitled to limit his liability "... unless the occurrence giving rise to the claim resulted from the actual fault or privity of the owner". An owner lost the right to limit liability if there was "fault or privity" on his part leading to the accident. The owner had to establish the absence of actual fault or privity. It was also established under the 1957 Convention that the relevant fault or privity had to be found in the management of the owner. In related cases the scope of relevant management responsibility has been expanded by the courts to include situations where the management of the shipowning company failed to inform the master of navigational dangers, or required the master and crew to work extraordinary hours in dangerous situations.

Under the 1976 Convention this "actual fault or privity" rule was replaced by Article 4 which provides as follows: "a person liable shall not be entitled to limit his liability if it is proved that the loss resulted from his personal act or omission committed with the intent to cause such loss or recklessly and with knowledge that such loss would probably result". The test under the 1976 Convention is similar to that in the Warsaw Convention, the Hague-Visby Rules, the Hamburg Rules and to the CMR and CMI Conventions pertaining to road and rail transportation. The new formulation is clearly protective of defendant shipowners. By imposing the "fault or privity" test with the reckless knowledge test Article 4 imposes a subjective test which in most cases will make the limitations unbreakable.

Under Article 4 the shipowner or salvor will be barred from limiting liability only where he is personally at fault in the manner specified in the Article. If the loss has resulted from the intentional or reckless conduct of his servants or agents for whom he is only vicariously responsible, he will be entitled to limit his liability.[7] But if the conduct is that of the shipowner or salvor personally he will not be entitled to do so. Obviously in the case of individuals it is not difficult to apply this provision. But in the case of a corporate body it is more difficult to find out when that body has been personally at fault. The answer to this question is related to the concept of the *alter ego*. The matter received attention in several cases such as *The Lady Gwendolen*, *The Garden City* and *The Marion*.

In *Asiatic Petroleum Co. Ltd.* v. *Lennard's Carrying Company Ltd.*, Hamilton L.J. said that "in the case of a company, the 'owners' within the meaning of the section must be the person or persons with whom the chief management of the company's business resides ...".[8] In the House of Lords the question was dealt with in more detail by Viscount Haldane who said of a company "its active and directing will must consequently be sought in the person of somebody who for some purposes may be called an agent, but who is really the directing mind and will of the

7. The person concerned would not be able to limit if sued directly.
8. *Per* Hamilton L.J. at p.437, *supra*, Chapter 19, n. 109.

corporation, the very ego and centre of the personality of the corporation . . .".[9] If this approach had been strictly adopted it would have greatly limited the possibility of breaking limitation, because in normal circumstances the "directing mind" of a company is not aware of the actual management and operation of the ships owned or operated by his company. In *The Lady Gwendolen* some flexibility appears to have been applied.[10] In *The Marion* the test applied comes closer to the strict test applied in *Asiatic Petroleum Co. Ltd.* v. *Lennard's Carrying Company Ltd.* In this case although the fault which caused the accident was attributable to the operations manager of the company and to his assistant, Lord Brandon of Oakbrook found in fact that if the managing director of the company had not been informed of a report of the Liberian authorities on the deficiencies found in the maintenance of the charts that was partly his own fault.[11]

Therefore if a person who is identified as an *alter ego* commits a personal act, this act will prevent the principal shipowner from limiting his liability on that particular occasion. On the other hand if the person who commits the act is not defined as an *alter ego* then his act or omission prevents him from limiting his liability but the shipowner is allowed to limit liability.

The intentional act barring limitation has to be an act intended to cause the loss which occurred. The term "recklessness" under the Convention has also caused difficulties of definition. In the context of criminal law recklessness was considered in *R.* v. *Caldwell*.[12] In this case Lord Diplock stated that ". . . in ordinary speech it [recklessness] includes not only deciding to ignore a risk of harmful consequences resulting from one's acts that one recognised as existing, but also failing to give any thought whether or not there is any such risk in circumstances where, if any thought were given to the matter . . .". Under Article 4 recklessness has to be coupled with knowledge that such loss would probably result. In *Goldman* v. *Thai Airways*[13] the Court of Appeal considered the meaning of the phrase in the Warsaw Convention as amended by the Hague Protocol "recklessly and with knowledge that damage would probably result". It was held that "recklessly" was not to be construed in isolation, but in its context and with the qualification that the act or omission had to have been done both "recklessly" and "with knowledge that damage would probably result". It is suggested that Article 4 is to be construed in a similar way. Under such a consideration proper emphasis will be given to each case. The question which has to be answered is not only whether the act or omission was "reckless" but also whether it was done with knowledge that such damage would probably result. In other words it has to be proved that there has

9. *Ibid.*, *per* Viscount Haldane at p.713, *supra*, Chapter 19, n. 127.

10. In *The Lady Gwendolen*, Willmer L.J. at p.346 said that "On the principles stated in these cases I should be disposed to say that actual fault on the part of Mr Boucher, as registered ship's manager and head of the traffic department, would be sufficient in the particular circumstances of the present case to constitute actual fault or privity of the company. But I do not find it necessary to reach any final conclusion upon this point . . .".

11. See the statement of Lord Brandon of Oakbrook in *The Marion* at p.8.

12. [1982] A.C. 341 (HL).

13. [1983] 1 W.L.R. 1186 (CA).

been reckless behaviour and with knowledge that loss would probably result. This may be proved by an objective test but the burden of proving the intent is based on a more subjective test. But whatever the nature of the test the instances of recklessness or intention to cause loss are clearly narrower than "actual fault or privity". It becomes much harder to break the limit by replacing the fault or privity test with the reckless knowledge test. Therefore the formulation of intent and recklessness contained in Article 4 seems to provide a result which is more favourable to limitation. It is more difficult under the 1976 Convention to break limitation. Obviously it is also possible to lose the right to limit liability by contract.[14]

(ii) Burden of proof

Previously the burden of proof lay upon the person seeking to establish the right to limit. That person had to show that he personally had no fault or privity directly resulting in or contributing to the incident giving rise to the liability. Under the 1976 Convention the onus of proof shifted from what it was under the 1957 Convention. The party seeking to deny limitation has the burden of proof in showing that the conduct displayed by the person liable fell within the definition laid down in Article 4.[15]

(e) Counterclaims

In collision cases one incident may give rise to cross-claims. Where both ships are to blame and both are damaged the question arises as to whether one of the parties applies limitation of liability and each party claims the other's limitation fund or whether limitation is applied only to the balance after setting off one claim against the other. Under English Admiralty practice where, as a result of collision, there were counterclaims, the established procedure was to set them off. The shipowner was entitled to limit his liability on the balance. Since the decision of the House of Lords in *The Khedive*[16] this has been the law in England in the case of a collision. The 1957 Limitation Convention, Article 1(5), followed this practice, and similarly shipowners entitled to limit their liability under the 1976 Convention are entitled to set off their limited liability against any claim which the claimant may have against them. This is dealt with by Article 5 of the Convention which provides as follows: "where a person entitled to limitation of liability under the rules of this Convention has a claim against the claimant arising out of the same occurrence,

14. See *The Satanita* where the House of Lords held that upon the true construction of the rules the words "all damages" excluded the operation of limitation of liability under the statute: *Clarke* v. *The Earl of Dunraven* [1897] A.C. 59 (HL).

15. In *The Bowbelle* it was stated that with the effect of Articles 2 and 4 specified claims were subject to limitation of liability unless the person making the claim proved that the loss resulted from a personal act or omission of the shipowner committed with intent to cause such loss or recklessly and with knowledge that such loss would probably result. Obviously such a statement imposes upon the claimant a very heavy burden: [1990] 1 Lloyd's Rep. 532.

16. (1882) 7 App. Cas. 795 (HL).

their respective claims shall be set off against each other and the provisions of this Convention shall only apply to the balance, if any." Thus the provisions of the Convention apply only to the balance. So the claims, unaffected by any entitlement to limit, have to be decided in the first place and then, if there is a balance, the Convention applies only to that. This practice will be relevant only if the shipowner is the party liable to pay the balance after the set-off. As stated above, Article 5 of the 1976 Convention provides that claims and counterclaims arising out of the same occurrence must be set off against each other and limitation is to be applied only to the balance, if any, payable. All persons entitled to limitation of liability are entitled to set off if they have a counterclaim.

The only problem in respect of counterclaims may arise if there is a counterclaim by a shipowner in respect of negligence in the performance of salvage services.[17] As Article 5 applies to claims which are both the subject of limitation of liability it is suggested that this Article does not apply to cases where there is a requirement to set off the salvage award and the claim for damages before applying limitation of liability. In order to apply limitation to the balance of the claims it should be possible to claim limitation for both claims if the balance is a claim against which limitation of liability is not applicable, in which case the practice under Article 5 does not make sense. Therefore it is said that under the true construction of the 1976 Convention Article 5 applies only claims within the ambit of Article 2, because only in such circumstances does a party need to strike any balance in order to determine which party may be entitled to limit his liability.[18]

(f) The amount of limitation

(i) In general

The 1976 Convention follows the tonnage system of limitation as opposed to the older methods calculated according to the extent of the shipowner's interest in the adventure, namely the value of his vessel. Under the value-based system if the vessel has a low value the limit of liability is similarly low. This creates a great disadvantage after serious marine casualties like the *Torrey Canyon*, which caused about $6 million worth of pollution damage but the owners claimed a limit of $50 for a single salvaged lifeboat. Therefore the tonnage system prevailed internationally in 1957 and was modified in the 1976 Convention. Under the 1976 Convention the value of the vessel is not relevant.

The amount of limitation available to the shipowner and salvor depends on the nature of the claim and on the tonnage of the vessel. As far as the nature of the claim is concerned Article 6 of the 1976 Convention draws a distinction between claims for loss of life or personal injury on the one hand and "other" claims on the other hand. Property claims, including cargo claims, fall into the "other" category.

With the exception of limits applicable to passenger claims for loss of life or

17. See Nigel Meeson, pp.229–230 for a detailed discussion.
18. *Ibid.*

personal injury, the limits of liability contained in the Convention are calculated in special drawing rights by reference to the tonnage of the relevant ship.

(ii) The limitation unit

The original British tonnage-based limitation statutes fixed the limits at £15 or £8 per ton. The limits, which were expressed in francs under the 1957 Convention, were defined in Article 3(6) as being units consisting of 65.5 milligrams of gold millesimal fineness 900. There was a flat rate of Poincaré gold francs (3,100) multiplied by the ship's tonnage. When this system was created it was thought to be stable and uniform. But the differences in practice caused problems, as in some countries gold value was translated into national currencies at official rates and in others at market rates. To avoid the problems with gold the international community has turned to the special drawing rights (SDR) of the International Monetary Fund (IMF). The SDR is a unit of account whose value is determined daily by the IMF on the basis of a basket of currencies. The IMF publishes its rates daily but they may also be found as against the various national currencies in the daily editions of the *Financial Times* or *Lloyd's List*. The important point about the SDR is that it is issued daily and based on market exchange rates, by reference to which limitation amounts may be assessed and calculated.

Under Article 6 of the 1976 Convention the gold franc has disappeared in favour of the unit of account, which is defined by Article 8 as the SDR. The appropriate value date is when the limitation fund is constituted, payment is made or security is given. Article 8(1) of the Convention does not refer to the date on which the decree is pronounced. The only relevant date is that of payment. Therefore the fluctuations of national currency as respects the SDR normally affect the limit only after the person liable has made payment either by constituting the fund or by paying the claimant. The only exception is where the fund is constituted by producing a guarantee under the law of the state party in whose territory the fund is constituted.

The value of a national currency in terms of special drawing rights of a state party which is not a member of the International Monetary Fund is calculated in a manner determined by that state party. Article 8(4) provides that such calculation must be made so as to express in the national currency of the state party in question as far as possible the same value of SDRs and further provides that states parties shall communicate to the depositary the manner of calculation at the time of ratification. If a state party is not a member of the IMF it has to comply with this provision. Thus it may provide that the value of a national currency in terms of SDRs is calculated on the basis of the rate of exchange at the date either when the limitation fund is constituted or when payment is made or security is given (whichever is relevant) which under the law of that state is equivalent to such payment.

Article 8 also provides that if the law of those states which are not members of the

IMF does not permit the calculation of the value of their national currency in terms of the SDRs they may declare that the limits of liability provided for in the Convention shall be fixed in monetary units of 65½ milligrams of gold of millesimal fineness 900, i.e. in Poincaré francs. Such limits are indicated in Article 8 and are calculated by converting in round figures the limits in SDRs on the basis of the ratio 1:15.

(iii) The extent of the limit

The limits provided for in the 1957 Convention were not sufficient. Personal claims were treated differently from property claims. The Convention provided a uniform limit whatever the size of the vessel. With the 1976 Convention the method of calculating limits has been changed. Instead of a uniform figure multiplied by the tonnage of the ship, now under Article 6 there is a sliding scale, with the amount of currency reducing in proportion to the greater amount of tonnage. The substantial minimum tonnage and the graded sliding scale are established to ensure that the limitation fund of smaller ships is relatively higher than that for larger ships. Article 6 provides five slices for personal claims and four for property claims. Consequently the limitation limits provided in the 1957 Convention increased significantly under the 1976 Convention. The 1976 Convention maintains the special position of personal claimants under the 1957 Convention.

There is a significant change in the treatment of small ships under the 1976 Convention. Under Article 6 there is a minimum limitation tonnage of 500 tons, which applies to ships whose actual tonnage is less than 500 tons. Article 15(2) also allows a state to regulate by specific provisions of national law the system of limitation of liability to be applied to ships of less than 300 tons. It is possible to find such a provision in the Merchant Shipping Act 1995, Schedule 7, Part II, paragraph 5, which leaves the position in the Convention unaltered for ships with a tonnage of 300 to 500 inclusive. For ships of less that 300 tons, Article 6 is to have effect as if the figure in Article 6(1)(a)(i) referred to 166,667 SDRs and that in Article 6(1)(b)(i) referred to 83,333 SDRs. Under UK law the minimum limits also apply to both personal and property claims.

The limit is calculated by multiplying the appropriate SDR amount for each slice by the tonnage of the ship. The Merchant Shipping Act 1854 was the first to use tonnage for limitation purposes. Tonnage for this purpose was "the registered tonnage with the addition of any engine room space deducted for the purpose of ascertaining that tonnage". Under the 1976 Convention this method changes in favour of the limitation tonnage being the gross tonnage calculated in accordance with the tonnage measurement rules contained in the International Convention on Tonnage Measurement of Ships 1969. This Convention entered into force on 18 July 1982.[19] From this date the 1969 Tonnage Convention applies to all new ships,

19. Enacted in the UK by the Merchant Shipping (Tonnage) Regulations 1982 (SI 1982 No. 841).

those which undergo major alterations and those which are voluntarily remeasured. Ships which have their tonnages calculated under the old systems until 18 July 1994 may retain their present tonnages. But from 18 July 1994 all UK registered vessels, irrespective of their date of build, have to be remeasured under the 1969 Tonnage Convention. Under the new regulations every ship is given a gross tonnage and a net tonnage. The net tonnage represents the actual size of the ship in physical terms. If at the time when limitation is claimed, the tonnage has not been and cannot be ascertained in accordance with regulations 4 and 6 of the Merchant Shipping (Tonnage) Regulations 1982,[20] as amended by the Merchant Shipping (Tonnage) (Amendment) Regulations 1988,[21] the best evidence available of the measurements of the ship shall be used in calculating the tonnage of the ship according to those regulations. Under the 1969 Tonnage Convention some ships will have greater tonnage for limitation purposes. This is because the Convention removed some anomalies under the old measuring systems whereby certain ships had abnormally low tonnages. Some ships, such as roros, container ships, vehicle carriers or supply boats, may be particularly affected and have their tonnage doubled or trebled.

The 1976 Convention sets out five groups of vessels: those up to 500 tons, those from 501 to 3,000 tons, those from 3,001 to 30,000 tons, those from 30,001 to 70,000 tons, those in excess of 70,000 tons. The overall limits have been increased, but instead of the same rate per ton being payable whatever the size of the vessel, there is a sliding scale as the size increases. So the general effect is to increase the limitation fund for smaller vessels more substantially than for those of greater tonnage.

The sliding scale for the calculation of limitation funds is as follows.

Claims for loss of life or personal injury of passengers

- —a fund is constituted by multiplying the number of passengers which the ship is certificated to carry by 46,666 SDR;
- —the fund is subject to a maximum of 25,000,000 SDR.

Claims for loss of life or personal injury of persons other than passengers

—tonnage not exceeding 500 tons	333,000 SDR
—if a ship exceeds 500 tons, the following sums must be added:	333,000 SDR:
—tonnage between 501 and 3,000 tons	extra 500 SDR per ton
—tonnage between 3,001 and 30,000 tons	extra 333 SDR per ton
—tonnage between 30,001 and 70,000 tons	extra 250 SDR per ton

20. SI 1982 No. 841. It implements the rules for measuring gross tonnage set out in Annex 1 of the International Convention on Tonnage Measurement of Ships, 1969 (Cmnd. 4332).
21. SI 1988 No. 1910.

—tonnage in excess of 70,000 tons	extra 167 SDR per ton
—for a salvor not operating from any ship or operating solely on the ship being salved	833,000 SDR

Other claims

—tonnage not exceeding 500 tons	167,000 SDR
—if a ship exceeds 500 tons, the following sums must be added:	
—tonnage between 501 and 30,000 tons	extra 167 SDR per ton
—tonnage between 30,001 and 70,000 tons	extra 125 SDR per ton
—tonnage in excess of 70,000 tons	extra 83 SDR per ton
—for a salvor not operating from any ship or operating solely on the ship being salved	334,000 SDR

A separate limitation fund is established for each category of claim. If the event gives rise to claims both for loss of life and personal injury to persons other than passengers and other claims and the fund available for personal injury or death claims is not sufficient to pay all these claims then the fund for other claims will be available to meet the unsatisfied personal injury or death claims. In other words if the amount available for a payment to personal injury claimants under Article 6(1)(a) is insufficient to pay the claimants, those claimants can then have access to the property funds set up under Article 6(1)(b). In such circumstances the unsatisfied personal injury and death claims rank rateably and not in priority to other claims. So the personal injury claims which need to go against the cargo fund due to insufficiency of the primary personal fund rank only rateably with the cargo claims and do not have a preference over them.

Under Article 6(3) a state party may provide in its national law that claims in respect of damage to harbour works, basins and waterways and aids to navigation shall have priority over other property claims but not over loss of life or personal injury claims. Therefore when a ship collides with such an installation and causes extensive property damage, and if there was also extensive death and personal injury resulting in significant claims against the shipowner then it is possible that there may be very little or nothing left over to pay the cargo owner. As the UK has not yet implemented this provision harbour works claims remain, for the time being, with other property claims.

But in cases where there is a collision between vessels and/or between vessels and shore installations problems may become complicated. The best precaution to be taken in such circumstances might be to have the adequate marine insurance cover and leave the problem to the insurer.

Under the 1976 Convention salvors have been allowed to limit in their own right as salvors. Under Article 6(4) where a salvor is not at the time of the incident operating from a tug or salvage vessel, he may limit his liability by reference to a tonnage of 1,500 tons. The 1,500 ton limit applies where a salvage team is helicoptered onto the casualty direct from land.[22] If the negligence is committed by the salvage crew operating from the salvage vessel the limit is calculated according to the tonnage of the vessel.[23] As the limitation is applicable for any salvor not operating from any ship or for any salvor operating solely on the ship to, or in respect of, which he is rendering salvage services the burden of proof that one of these circumstances is true and correct falls upon the salvor seeking to limit.

(iv) The limit for passenger claims

Article 7 of the 1976 Convention introduced a new and separate limit of liability for claims made by passengers. This Article deals in particular with claims in respect of loss of life or personal injury to passengers. The limit which applies to claims made by passengers against the carrying ship is 46,666 SDRs multiplied by the number of passengers that the ship is authorised to carry, up to a maximum of 25 million SDRs. In passenger claims the provisions of the International Convention on the Carriage of Passengers and Their Luggage 1974 (the Athens Convention) as enacted by section 14 of the Merchant Shipping Act 1979 and Schedule 3 to the Act may also apply. And a similar Article, Article 7 of the 1976 Convention, can be seen in the Athens Convention. The actual limitation figure of 46,666 SDRs or units of account is the same in each Convention.[24] Under the Athens Convention the limit is calculated by multiplying the figure (46,666) by the number of passengers who are actually on board and are travelling and are injured/killed. But under the 1976 Convention the limit is fixed at an amount reached by multiplying the same figure by the number of passengers which the ship is allowed by its certificate to carry. According to Hill[25] Article 7 of the 1976 Convention is in direct conflict with Article 7(1) of the Athens Convention. The only explanation of this coincidence is that, with regard to Article 19 of the Athens Convention,[26] where the provisions of the 1976 Convention permit the carrier to limit liability to a lower figure than he could in a similar situation limit under the Athens Convention, the lower 1976 figure will be the one to apply. But according to Gaskell[27] Article 7 of the 1976 Convention does not create any liability for passenger carriage. That will be decided according to the existing law. The Athens Convention is the one which creates a regime of liability and limitation for

22. Art. 9(1)(c).

23. Art. 9(1)(b).

24. The Athens Convention: the limit is 46,666 SDRs, to passengers claiming. The limit established applies *per* "carriage", which is interpreted as being the whole period over which the passenger is being carried. The 1976 Convention: the limit shall be 46,666 units of account multiplied by the number of passengers which the ship is authorised to carry etc., with an overall limit of 25 million units of account. This fund applies to any distinct occasion.

25. See Christopher Hill, pp.412–413.

26. Article 19 of the Athens Convention provides that "this Convention shall not modify the rights or duties of the carrier, the performing carrier and their servants or agents provided for in international conventions relating to the limitation of liability of owners of seagoing ships."

27. Nicholas Gaskell, "The Amount of Limitation", in Institute of Maritime Law, The University of Southampton (comp.), *The Limitation of Shipowners' Liability. The New Law* (1986), pp.53–57.

passenger carriage. And with regard to similarities between the two Conventions, although at first sight the Athens Convention limit may seem identical to the 1976 Limitation Convention limit, there are two major differences. First, the 1976 Convention limit is not applied in relation to each passenger claim, as under the Athens Convention, but is multiplied by the theoretical number of passengers that the ship is authorised to carry according to her certificate. The importance of this point is that in cases where the ship is half full the whole Article 7 limit will be available to passengers.[28] Secondly, the 1976 Convention has the overall maximum limit of 25 million SDRs, but under the Athens Convention there is no equivalent figure. Another difference between the two Conventions is that under the Athens Convention the limit of liability of 46,666 SDRs applies to each passenger, as opposed to the global fund under the 1976 Convention, and applies "per carriage" as opposed to on "any distinct occasion" as under the 1976 Convention. In conclusion it is possible to say that the 1976 Convention will prevail as far as the shipowner is concerned, but on the other hand the owner may also benefit from the limits provided for in the Athens Convention as Article 14 of the Athens Convention provides that "no action for damages for the death of or personal injury to a passenger, or for loss of or damage to luggage, shall be brought against a carrier or performing carrier otherwise than in accordance with this Convention".

The separate limits under the 1976 Convention apply only where there are claims for loss of life or personal injury to passengers. Under Article 7(2) a passenger is any person carried in that ship "under a contract of passenger carriage or who, with the consent of the carrier, is accompanying a vehicle or live animals which are covered by a contract for the carriage of goods". Thus a passenger includes one who has a contract of carriage and is being carried under it. This definition excludes stowaways, visitors, guests of the owners or crew and those working on the ship under a contract of employment with somebody else, like dockers. The claim under Article 7 has to be made by or on behalf of a person carried in that ship. Article 7(2)(b) includes in the definition of passenger certain persons accompanying vehicles or live animals, but in both cases the consent of the carrier is required. Carrier will include a charterer as well as the shipowner. As a result of this separate treatment of passengers they will not have a share in the general limitation fund.

(g) Distinct occasion and aggregation of claims

The 1976 Convention applies once and for all global aggregation of claims. Once a shipowner seeks to limit his liability under the 1976 Convention any claimant who is not a party to the limitation proceedings misses out altogether. This principle is provided under Article 9 of the 1976 Convention as the limits of liability applying

28. At this point Hill reaches the same conclusion as Gaskell by stating that under the different method of calculation the limitation fund under the 1976 Convention is not influenced by the actual number of passengers travelling on board.

"... to the aggregate of all claims which arise on any distinct occasion ..." against the shipowner or other person entitled to limit liability under Article 1. Under Article 9(2) there is separate aggregation for passenger claims subject to the limits of Article 7. Therefore once a person is defined as a passenger it is not possible for that person to have a share in that ship's Article 6 funds to the extent that he has claims unsatisfied because of the limits under Article 7.

Article 9 states that the limits of liability determined under Article 6 apply to claims which arise on any distinct occasion. The expression "on any distinct occasion" is a matter for determination by the relevant court. In cases where one collision occurs at the beginning of the voyage and another at the end each collision constitutes a distinct occasion. And there is a limitation fund for each occasion. But if a ship is successively involved in a number of incidents problems may arise. In *The Rajah* a ship struck a tug and also the tow. It was held that there was one act of negligence, one distinct occasion and therefore one limitation fund.[29] In *The Schwan* it was stated that the important point was not the passage of time between the two incidents but whether the second loss resulted from the same act of negligence.[30] Obviously the approach set out in *The Schwan* is easy to follow if the court can draw a line between the two acts of negligence, but in some cases like *The Calliope* the court was unable or not prepared to draw such a line. In this case it was again stated that the 12 hours difference between the two incidents alone was not decisive as the wording required an examination of the causation issue.[31] As a general principle it was stated that a second incident may not be wholly distinct from the first. The onus of proving that the damage did not arise on distinct occasions rests with the ship that is seeking to limit its liability to a single occasion. In other words the owner has to prove that there is only one distinct occasion giving rise to a claim. The courts may allow more than one limitation fund where the occasions are separate but the causative acts of negligence are not completely distinct.

(h) The limitation fund

Under the 1957 Limitation Convention the constitution and distribution of the limitation fund and rules of procedure were left to the national law of the state in which the fund was constituted. This situation has led to wide variations of practice between the states in which the 1957 Convention has been ratified. On the other hand the 1976 Convention has detailed provisions on the administration and distribution of the limitation fund.

(i) Limitation of liability without constitution of a limitation fund

Under English law it has always been possible to plead limitation of liability as a defence to a claim brought in the English courts. When a person pleads limitation

29. (1872) L.R. 3 A. & E. 539.
30. [1892] P. 419.
31. [1970] 1 All E.R. 624.

as a form of defence in an action against him there is no obligation to constitute a limitation fund prior to judgment being given. This procedure is appropriate only where an incident gives rise to one single claim, as, for instance, in the case of damage to a bulk cargo on board the limiting vessel. Because when limitation is used as a defence to an ordinary liability action it does not concentrate all claims against that same person arising out of that same incident into one jurisdiction. Therefore if another claimant brings another action against the same person in respect of loss or damage arising out of the same incident the defendant may be found liable to pay the limitation figure over again, assuming that the loss or damage was in excess of the limitation figure. Another method under English law of pleading limitation of liability is to institute a limitation action. Under this method the person who takes this action is entitled to limit his liability against all persons who may claim in respect of the loss or damage which has arisen out of that particular incident.

Article 10 of the 1976 Convention provides for a situation where liability is limited without a limitation fund being constituted. Under this Article limitation of liability may be invoked even though no limitation fund has been constituted. If limitation of liability is invoked in these circumstances the provisions for the distribution of that fund apply correspondingly. Questions of procedure must be decided in accordance with the national law of the state party in which action is brought.

(ii) Constitution of the fund

The 1957 Limitation Convention did not provide much guidance on how and where the fund was to be constituted. This was left to the domestic law of each country. The person liable may not take the initiative to set up the fund unless legal proceedings are brought against him in respect of claims subject to limitation in a state party and may then set up the fund with the court or other competent authority of that state. Therefore the person liable has a choice of place in which to set up the fund only if legal proceedings are brought against him in respect of claims subject to limitation in two or more states parties.

Article 11 of the 1976 Convention is concerned with the constitution of the limitation fund. Under paragraph 1 it is provided that any person alleged to be liable may set up a fund in the amounts specified in Articles 6 and 7 with the court or other competent authority in any state party in which legal proceedings are instituted in respect of claims subject to limitation. By paragraph 11 of Part II of Schedule 7 to the Merchant Shipping Act 1995 "the court" is specified as the High Court in England and Wales and the Court of Session in Scotland. Under UK law interest is to be included in the calculation of the fund running from the date of the incident which gave rise to the liability to the date on which the fund was set up. The fund is constituted in the sum of the relevant limits under Articles 6 and 7. These are, under paragraph 1 of Article 11, the limits which apply to claims for which the person constituting the fund "may be liable". Therefore although the legal proceedings have been brought in respect of one particular type of claim only, such as a property claim, the person intending to constitute the fund must, when calculating the amount for which the fund must be constituted, take into account all types of claims in respect of which he may be liable.

The rate of the interest is left to the law of the country in which the fund is constituted. Any fund set up in this way is to be available only for the payment of claims in respect of which limitation of liability can be invoked. If there are other claims against the defendant in respect of which he is unable to limit his liability then these claims will not rank against the fund, and if these claims need enforcing they will not be enforced against the fund but against any other assets which the person who is liable may have. Under paragraph 2 a fund may be constituted either by depositing the sum or by producing a guarantee acceptable under the legislation of the state party in which the fund is set up and considered to be adequate by the court or other competent authority. Under English law it has always been preferable to establish a fund only by cash deposit. As there is nothing under the 1979 Merchant Shipping Act referring in particular to this issue it is possible to state that UK law will also come to the same point as the 1976 Convention by accepting the idea that the deposit of a limitation fund by way of a guarantee acceptable to the court or other competent authority of the state party is the equivalent of a deposit by cash.

In practice, payments of cash into court are rare, and usually the fund will be constituted by the shipowner, salvor or other entitled person or his insurer by providing a guarantee or letter of undertaking acceptable to the claimant and to the court pursuant to the relevant court rules.

Under paragraph 3 the term "or his insurer" has been referred to in the constitution of the limitation fund. But the insurer is not a person mentioned in the relevant sub-paragraph of Article 9 as "any person alleged to be liable". Therefore it is not clear whether a fund set up by an insurer on behalf of his assured will satisfy the provisions of the 1976 Convention with regard to the constituting of the fund.[32]

(iii) Distribution of the fund

Under Article 12 the limitation fund must be distributed among the claimants in proportion to their established claims against it. Established claims means those recognised by the court. The fund referred to must be that which is properly constituted and defined under Article 11 of the Convention. No lien or privilege to which a claimant may be entitled will enable that claimant to claim priority over the other claims against the fund.[33]

Article 12 preserves rights of subrogation in favour of persons who have made payments to settle claims against the fund. Therefore if, before the fund is distributed, the person liable or his insurer has settled a claim against the fund, he acquires by subrogation, up to the amount he has paid, the rights which the person

32. See Christopher Hill, pp.415–416.
33. Para. 9, Part II, Schedule 7 to the Merchant Shipping Act 1995.

so compensated would have enjoyed under the provisions for the limitation of liability for maritime claims. Under paragraph 3 the right of subrogation may also be exercised by other persons in respect of any amount of compensation which they may have paid, but only to the extent that such subrogation is permitted under the applicable national law.

Under Article 4 a person liable is protected by preserving his subrogated rights for the future. If the person liable or his insurer knows that he will have a liability arising out of the same occurrence but which at the time of the distribution of the fund has not been presented or proved against him. That person may set aside an appropriate portion of the fund for use at a future date to cope with this expected claim.

(i) Bar to other actions

Under Article 13 there is further protection for the shipowner or other person entitled to limit a claim by providing that where a limitation fund has been constituted in accordance with Article 11, any person having made a claim against a fund is barred from exercising any right in respect of such claim against any other assets of a person by or on whose behalf the fund is being constituted. Therefore if the fund is not constituted in the sum required by the Convention it is not treated as a limitation fund under the Convention and does not prevent the claimant from exercising his rights against other assets. Basically no other assets of the person liable are to be exposed to legal action and the claimant's rights may be exercised solely against the constituted fund.

This Article is a successor to Article 5 of the 1957 Convention where an English court would not permit the release of a ship arrested in the UK, even if a foreign limitation fund had been established, unless such a reliable person was cleared of any serious question of actual fault or privity. This has been incorporated into English law by section 5 of the Merchant Shipping Act 1958 and discussed in *The Wladyslaw Lokietek*. In this case following a collision the plaintiffs arrested a sister ship of the defendants. The defendants brought proceedings in the Polish court and constituted a fund. They applied for a release of the guarantee under section 5 of the 1958 Act. The Admiralty Court held that even if a limitation fund was established in another court, in this case Poland, the vessel would not be released from arrest in England unless the defendant could prove that there was no serious question to be tried in relation to absence of actual fault or privity on his part under section 503 of the 1894 Act. The defendant failed to discharge this burden of proof.[34] The courts have had considerable difficulty in applying section 5 of the 1958 Act.[35]

Under the 1976 Act the requirements of section 5 of the 1958 Act seem to have been disposed of. By Article 13 where a limitation fund has been constituted, and providing the claimant may bring a claim against the fund before the court

34. [1978] 2 Lloyd's Rep. 520.
35. See also *The Putbus* [1969] 1 Lloyd's Rep. 253.

administering it and the fund is actually available and freely transferable in respect of that claim against thc fund, the claimant is barred from exercising any rights against other assets of the party seeking to limit, and if any ship or other property of that party has already been arrested or attached within the jurisdiction of a state party it may be released or must be released if the limitation fund has been constituted in any of the following places:

(i) at the port where the occurrence took place, or, if it took place out of port at the first port of call thereafter; or
(ii) at the port of disembarkation in respect of claims for loss of life or personal injury; or
(iii) at the port of discharge in respect of damage to cargoes; or
(iv) in the state where the arrest is made.

The protection provided for by Article 13 is particularly important for shipowners. It offers a legal advantage by providing for the release of the ship or the property which has already been arrested, because most shipowners organise their affairs by operating their vessels in one-ship companies, where the ship is the only asset of the company. And once the shipowner has applied to constitute a limitation fund it is entitled to a stay in any other proceedings brought by a claimant.

Exclusions and reservations

The 1976 Convention applies to seagoing ships. Non-seagoing ships are excluded. Under Article 15(5) the Convention does not apply to air-cushion vehicles or floating platforms constructed for the purpose of exploring the sources of the seabed. Under Article 15(4) there are also prohibitions on states against applying the 1976 Convention to ships constructed or adapted for drilling purposes where the particular state has established its own system of legislation dealing with liability.

Article 15(1) of the Convention accords states parties the right to exclude wholly or in part from the application of the Convention any person referred to in Article 1 who does not, at the time when the rules of the Convention are invoked, habitually reside in or have his principal place of business in a state party and any ship which does not fly the flag of a state party.

Countries ratifying the Convention are also given certain rights to carry out the terms of the Convention when adopted. Under Article 15(2) a state party may regulate specific provisions of national law for a system of limitation of liability to be applied to vessels which are intended for navigation in inland waterways and are of less than 300 tonnes. The Convention does not automatically apply to ships of less than 300 tonnes and any country may exclude them from the right to limit.

CHAPTER 21

CONFLICTS OF LAW

In tort cases involving international maritime claims there is always a problem about the most appropriate jurisdiction in which the matter may be resolved. There are two reasons for such a problem. First, in the case of maritime torts, of which the collision at sea is the classic example, the choice of jurisdiction lies with the parties.[1] Secondly, by its nature maritime law is quite likely to have conflicts of law problems as it embraces civil law in its origin, common law in its modern evolution and is also found in national statutes and international conventions.[2]

Limitation of liability is frequently a significant factor in determining the most appropriate jurisdiction. Obviously the final decision will change depending on the paying party or recovering party. For instance, a person is the recovering party if he will seek a jurisdiction where limits of liability are larger, or where there is a reasonable prospect of defeating the opponents' right to limitation. The law which provides for the right to invoke limitation and the law which would govern the size of the limitation fund itself has to be determined. Thus the limitation regimes of countries which apply different measures of limitation of liability under different international limitation conventions have to be compared. The first question which has to be answered under the international limitation conventions is whether the Convention applies. The answer is given in Article 7 of the 1957 Convention and Article 15(1) of the 1976 Convention where the system under both Conventions is substantially identical.[3] Under these Articles a state may avail itself of the "right to exclude" the application of the Convention as a detail of its implementation of the Convention into its national law. Therefore it will be

1. On the contrary, in the case of most maritime contracts such as charterparties or contracts for sale and purchase of ships the parties make an express choice of law and jurisdiction.

2. For detailed information on shipowners' limitation of liability and conflicts of law see William Tetley, "Shipowners' Limitation of Liability and Conflicts of Law: The Properly Applicable Law" (1992) 23 *Journal of Maritime Law and Commerce* 585–606.

3. Art. 7 of the 1956 Convention states: "... Nevertheless, each Contracting State shall have the right to exclude, wholly or partially, from the benefits of this Convention any non-Contracting State, or any person who, at the time when he seeks to limit his liability or to secure the release of a ship or other property arrested or the bail or other security in accordance with the provisions of Article 5 hereof, is not ordinarily resident in a Contracting State, or does not have his principal place of business in a Contracting State or any ship in respect of which limitation of liability or release is sought which does not at the time specified above fly the flag of a Contracting State". Art. 15(1) of the 1976 Convention substantially restates the position as it was under this Art. of the 1957 Convention.

necessary to examine the law of each contracting state to establish whether it in actual fact availed itself of the right wholly or partially to exclude non-Convention persons or ships from the application of the Convention.

In cases where a 1957 Convention country has not availed itself of the right to limit the scope of application of the Convention the question is whether any shipowner or other person entitled to limit its liability from any country may go to a contracting state and constitute a fund there. If he can he will be able to profit from the benefits of the Convention even if the collision took place in the territorial waters of a non-contracting state between the two non-Convention ships. If such a practice is accepted it will encourage a debtor to go to a Convention state and assume the benefits of the Convention. In other words, the debtor would be encouraged to forum shop. Obviously it cannot be accepted that as soon as a collision has occurred somewhere any debtor may go to a contracting state and benefit from the rules of the Convention. Thus there is a need to find a way to prevent excessive use of the opportunities provided for by the Convention. Under the 1924 Convention, by Article 8, national laws determined procedure and time limits in applying the rules on arrest of ships and force of security. Article 14 provides that "Nothing in the foregoing provisions shall be deemed to affect in any way the competence of tribunals, modes of procedure, or methods of execution authorized by the national laws". If, under the national law, the definition of "procedure" was broad enough there was always a possibility that national laws could replace the basic terms of the 1924 Convention on the calculation of the fund and even the right to limit. Under the 1957 Convention a couple of Articles affected the choice of law under the Convention. Under Article 1(6) the *lex fori* determines who has the burden of proof of whether or not there was actual fault or privity on the part of the shipowner. Article 4 provides that rules relating to the constitution and distribution of the limitation fund as well as of procedure shall be governed by the national law of the state in which the fund is constituted. And by Article 5(5) the law of the contracting state in which an action is brought shall govern the questions of procedure and time limits.[4]

The 1976 Convention provides a uniform structure for limitation of liability in respect of persons entitled to limit claims subject to limitation and rules concerning the constitution and distribution of the limitation fund. Although it does not contain any express jurisdictional provisions its choice of law provisions are confined to the constitution and distribution of the limitation fund. The problem of preventing the excessive use of the opportunities provided by the Conventions has been solved under the 1976 Convention by providing in Article 11 that a fund may *only* be constituted in a state where *legal proceedings are instituted* in respect of claims subject to limitation. Articles 14 and 10(3) of the Convention also deal with some aspect of choice of law. Under Article 14 a directive provides that the constitution and distribution of a limitation fund, as well as any relevant procedure, will be decided by the law of the state in which the fund is constituted.

4. See William Tetley, pp.588–590.

Article 10(3) provides that procedure relating to limitation where no fund has been constituted shall be decided by the law of the contracting state in which the action is brought.

With regard to jurisdictional agreements the approach will depend on the force given to jurisdictional clauses by the national law. For instance in cases where England is the forum and there is a jurisdiction clause, bearing in mind that the 1976 Convention is mandatory, it is possible to argue that this clause will not be upheld as it would lead to the application of a different limitation statute. Thus the validity of a jurisdiction clause will be affected by the Convention only to the extent that a national law sees the Convention as mandatory or not. Therefore the question whether a court of a contracting state would apply a choice of law which avoids the Convention is left open.[5]

In *Caltex Singapore Pte. Ltd., Caltex Trading Pte. Ltd. and Caltex Services Pte. Ltd.* v. *BP Shipping Ltd.*[6] the defendant owners of the ship *British Skill* were domiciled in England. On 18 July 1994 their vessel struck a jetty in Singapore. The plaintiffs were domiciled in Singapore. The plaintiffs sued the defendants in England and Wales for damages in the sum of $10.5m. The defendants had already begun limitation proceedings arising out of the incident in Singapore in which they admitted liability for the collision. But they applied to stay the English action on *forum non conveniens* grounds. The plaintiffs contended that if the English action were stayed, with the result that the action had to be tried in Singapore, they would be deprived of a legitimate juridical advantage in that their prospects of recovery would be very much less in Singapore because of the lower limit of liability. The limit would be about $5.8m. The judge first considered whether the defendants had shown that there was another available forum of competent jurisdiction in which the case might be tried more suitably for the interests of all parties and the ends of justice. The alleged tort had been committed in Singapore and that was a significant connecting factor. But the judge attached little weight to the mere fact that the plaintiffs were based in Singapore. In fact he considered it more important to examine where the trial would most appropriately be held. It was stated that the dispute could more conveniently be tried in Singapore where the bulk of the oral and documentary evidence on quantum was situated. Coupled with the place of commission of the tort and the plaintiffs being based in Singapore, the judge held that the defendants had established that Singapore was the natural forum for the determination of the issues in the action. Then the judge considered whether the plaintiffs had shown that the stay would deprive them of a legitimate juridical advantage by leading to the application of the Singapore law of limitation of liability. By discussing this point the judge adopted the principle in *Dicey and Morris*[7] that questions of quantification of damage are procedural and governed by

5. For detailed information on jurisdiction and the 1976 Convention see David Jackson, "The 1976 Convention and International Uniformity of Rules", in Institute of Maritime Law, University of Southampton (comp.), pp.126–143.

6. [1996] 1 Lloyd's Rep. 286.

7. A.V. Dicey and J.H. Carlile, *Dicey and Morris on the Conflict of Laws* (ed. Lawrance Collins) (12th ed., 1993), p.183.

the *lex fori*. The limitation provisions did not qualify or attach to the plaintiff's right of recovery. In a multi-claim case the plaintiff would recover the judgment in full against the defendant who would then begin a limitation action. The right to limit was similar to a partial immunity from arrest or execution, and attached to a shipowner rather than to a claim. If it were substantive, there could be a situation where many different limitation regimes applied to the same incident by reason of differing proper laws. That would be impractical. The judge held that it was English public policy as expressed by Parliament in the 1979 Act that the 1976 Convention limit should be applied. He held that if the matter had to be litigated in Singapore, the plaintiffs would, on the basis of their claim as formulated, be deprived of a legitimate juridical advantage. Accordingly it provided that, as the plaintiff's claim exceeded the Singapore limit, there would not be a stay of the English action. The test was whether the plaintiffs had shown an arguable case for recovery of damages in excess of the limit. The judge adjourned the hearing of the application to allow the plaintiffs to adduce further evidence on this aspect.

CHAPTER 22

THE INSURANCE VIEWPOINT

LIMITATION OF LIABILITY AND INSURANCE RELATION

The original rationale for enabling shipowners to limit their liability was the risky nature of shipping and the need to encourage people to go into the business of owning and trading ships. Until the twentieth century the limitation was usually based on the value of the ship. The Merchant Shipping Act of 1854 provided for a limitation fund based on a sterling multiple of the ship's tonnage and made it impossible to provide for liability as part of normal hull insurances. Thus, the present-day marine insurance practice of shipowners' liability insurance flows from those concepts of limitation. In other words it is possible to say that today's shipowners' liability insurance is literally built on concepts of limitation. Obviously limitation provides a mechanism for sharing costs between all those involved in and benefiting from the maritime adventure. The other important function of limitation is that it brings a degree of comparability between different ships trading in different areas and exposed to different risks. Under the right to limit a shipowner obtains cover for his total exposure, not just for some proportion. Shipowners are able to have some certainty in assessing their operating costs.

It is sometimes argued by the critics of the limitation concepts embraced in the 1957 and 1976 Conventions that they put the shipowner and other parties benefiting from the limitation concepts in a favoured position. In other words they argue that shipping is not exceptional and why should shipowners have a special form of protection? It is difficult to deny that these Conventions put the shipowner in a protected position, but most of the transporters have some form of limit to their liability either by contract or by statute. For instance air carriers are able substantially to limit their liability pursuant to the Warsaw Convention 1929 as amended by the Hague Protocol 1955. These Conventions deal with liability in the same way as does the Hague Rules Regime and impose a limitation system which is, like the 1976 Convention system very difficult to break. Limitations are also available to international rail and road carriers under various Conventions.

Under hull insurance the hull underwriter always knows the maximum amount of his exposure to a risk, namely the insured value. And as a result of this certainty he will not be paying more than this amount and any legal fees associated with the claim. But P&I insurers are in a different position as their cover is best described as "unlimited coverage of a shipowner's legal liabilities". Sometimes it is suggested that the substantial reinsurance cover which is currently purchased by the International Group of P&I clubs means that limitation is no longer necessary. But the real position is that the existence of limitation makes the extent of the cover possible. And club cover is not in fact "unlimited" but rather limited to an owner's legal liability. In other words, P&I clubs can provide unlimited cover as the laws of maritime nations stipulate a reasonable amount to which shipowners can usually limit their liabilities. Therefore, in particular for P&I club cover, limitation of shipowner's liability has a particular importance. Furthermore clubs are dependent in turn on their reinsurance costs and the concept of unlimited liability ignores the problem of realistic insurable limits. The insurable limit is the maximum amount of overall coverage available at a realistic cost in respect of any one catastrophe. As a consequence that amount determines the cost to the assured and the cost to the consumer of goods being carried by the shipowner. Therefore whenever the shipowner's liability exposure is increased underwriters must revise their figures for the amount of coverage available and for the cost of such coverage.[1]

THE CURRENT LAW

In most jurisdictions there are insurance provisions which allow for a direct right of recovery by a third party claimant from the liability insurer in cases where the assured shipowner goes bankrupt. The basic principle under these provisions is that the third party may recover from the insurer the same amount as the assured himself would have received. Therefore it is possible for the insurer to avail himself of all the terms of the policy by including deductibles and policy conditions.

Under the English Third Parties (Rights Against Insurers) Act there is no question of true direct action against an insurer. This Act applies only in the event of the insolvency of the insured against whom the third party has his claim. There is sometimes confusion between the right of a shipowner to limit his liability to third parties and the right of the shipowner's liability insurer to restrict his cover for the shipowner. Under this confusion it is also possible to think that if the shipowner loses his right to limit his liability, he will also lose his insurance cover or at least any cover which is in excess of the amount of his relevant limitation fund. But in fact the insurance cover is different and quite separate from the shipowner's limitation. The only link between the insurance cover and the shipowner's right to limit is that the insurer covers only a shipowner's legal liabilities. Thus if a shipowner is legally entitled to limit his liability he is supposed to limit his liability because he has no cover in excess of the amount of the relevant limitation fund. It is possible to find this requirement in P&I cover where it is stated that where the member or a co-assured is entitled to limit his liability pursuant to any rule of law, the maximum

1. See Leslie J. Buglass, "Limitation of Liability From a Marine Insurance Viewpoint" (1979) 53 *Tulane Law Review* 1364–1365.

recovery under a P&I entry is the amount to which the member or the co-assured may limit his liability. This rule states that if a member or co-assured is entitled to limit liability, the maximum amount recoverable from the Association cannot exceed the limitation sum.[2]

The general law which enables an insurer to deny cover to a shipowner on the basis of behaviour can be found in the English Marine Insurance Act 1906. Section 39(4) states that "A ship is deemed to be unseaworthy when she is reasonably fit in all respects to encounter the ordinary perils of the seas of the adventure assured" and Article 39(5) reads as follows: "In a time policy there is no implied warranty that the ship shall be seaworthy at any stage of the adventure, but, where, with the privity of the assured, the ship is sent to sea in an unseaworthy state, thc insurer is not liable for any loss attributable to unseaworthiness". English courts have interpreted these provisions very restrictively and under these provisions the fault must be proved, it must cause the loss, and its existence must be known or ought to have been known. Most importantly the knowledge by the assured shipowner that the defect made the ship unseaworthy in the sense of its being unlikely to complete its voyage must be proved. The element of recklessness on the part of the shipowner is also required. The insurer is also absolved from his liability to pay if a ship is deliberately sunk or for other deliberate acts of the assured.[3]

LIABILITY INSURER UNDER THE 1976 CONVENTION

Article 1(6) of the 1976 Convention introduced another class of person hitherto unable to limit, namely an insurer of liability for claims subject to limitation. Under Article 1(6), where a person entitled to limit liability is insured, the insurer is entitled to the benefits of the Convention "to the same extent" as the assured himself. There is no such provision in either the 1957 Convention or the 1958 Act. Thus it is not possible to evade limitation of liability by bringing an action against a liability insurer and a liability insurer should not be able to limit his liability if the assured cannot.

As stated above, in the United Kingdom a direct action may be brought in certain circumstances by a third party against the insurer under the Third Parties

2. It is possible to find the same Rule in different P&I Club Rules. For instance in the UK Club Rules, Rule 5B(i), states that "Subject to these Rules and to any special terms and conditions upon which a ship may be entered, the Association insures the liability of the Owner in respect of an entered ship as this liability may be determined and fixed by law including any laws pertaining to limitation of liability. The Association shall in no circumstances be liable for any sum in excess of such legal liability . . .". Similarly in the Britannia Club Rules, Rule 27(1) provides that "Subject to these Rules and to any special terms and conditions upon which a Ship may be entered, the Association insures the liability of a Member in respect of an Entered Ship as this liability may ultimately be determined and fixed by law, including any laws pertaining to limitation of shipowners' liability. The Association shall in no circumstances be liable for any sum in excess of such legal liability . . .".

3. S. 55(2)(a) of the 1906 Act provides as follows: "The insurer is not liable for any loss attributable to the willful misconduct of the assured, but unless the policy otherwise provides, he is liable for any loss proximately caused by a peril insured against, even though the loss would not have happened but for the misconduct or negligence of the master or crew."

Act 1930. But as stated in *The Fanti and The Padre Island*[4] the third party claimant's rights under the Act are restricted by the express provisions of the insurance cover. In an action brought under this Act, the third party is in no better position than the insured. Again it was stated in *The Fanti and The Padre Island* that "in a case where the insurer would have had a defence to a claim by the insured before the statutory transfer of his rights to the third party, the insurer will have precisely the same defence to a claim made by the third party after such transfer".[5] Therefore when insurers find themselves in the shoes of their assured they should have the same limitations and defences as the assured has available to him. Where the assured was not entitled to limit his liability by reason of his personal act or omission, committed with the intent to cause such loss, or recklessly and with knowledge that such loss would probably result, very often an insurer has a defence to the claim by the assured under the terms of the insurance policy or under section 55(2) of the Marine Insurance Act 1906. If the statement in Article 1(6) on the insurer's right to limit "to the same extent" as the assured is interpreted narrowly the assured may not be entitled to rely upon the provisions of the policy or the Marine Insurance Act and may have to meet in full the claim of the third party claimant if the assured has forfeited his right to limit liability. But as the aim of the new limitation legislation is to provide stable affordable insurance cover such a result cannot have been intended by the drafters of the Convention. Therefore the effect of Article 6(1) of the Convention in this context appears to be that the insurer being sued under the 1930 Act can plead limitation of liability where this is applicable and had not been raised by the assured.[6]

4. It was held by the House of Lords in *The Fanti/The Padre Island* that a third party claimant stepping into the shoes of the assured under the 1930 Act and thereby taking over the assured's rights under the P&I policy could not recover from the P&I insurer under the Act unless and until the assured had complied with the Club's Rules and first paid the claim: [1990] 2 Lloyd's Rep. 191.

5. [1991] 2 A.C. 1 (HL), *per* Lord Brandon at p.29H.

6. See Christopher Hill, p.403; Nigel Meeson, pp.226–227; Patrick Griggs and Richard Williams, *Limitation of Liability for Maritime Claims* (2nd ed., 1991), pp.14–16.

CHAPTER 23

THE EXCLUSION CLAUSES

Bills of lading, both before and since the adoption of the Hague Rules and the Hague-Visby Rules, have contained limitation of responsibility clauses or stipulations. In these clauses the carrier has sought to limit or exclude his responsibility for loss of or damage to cargo. In general these clauses are called "limitation clauses", "non-responsibility clauses", "exculpatory clauses", "exception clauses", "exclusion clauses" and "disclaimer clauses". Under these limitation clauses the carrier is either absolved from his responsibility under the contract or his responsibility is partly lessened.

Article 3(8) of the Hague and Hague-Visby Rules invalidates any clause which purports to alleviate or lessen the responsibilities of the carrier and the ship as they are set out in the Rules. Before the introduction of the Hague Rules it was customary for shipowners to include in the bill of lading a clause to the effect that the owner was to have the benefit of any insurance that might have been effected on the goods carried. Under such a clause the subrogation rights of the insurer against the owner were worthless. Shipowners also included wide exemption clauses in the bills of lading to exempt themselves from liability in respect of cargo loss or damage, however caused. By the end of the nineteenth century the cargo-owning nations were becoming dissatisfied with the developments in the cargo-owning countries as they allowed carriers to exclude virtually all responsibility for anything that had happened to the goods during the voyage. The first step to find a fairer regime for cargo interests was taken in America under the Harter Act 1893. Following that the Hague Rules became a draft Convention, signed at Brussels in 1924 and incorporated into English law by the Carriage of Goods by Sea Act 1924 with minor amendments. The current Carriage of Goods by Sea Act came into force in June 1977 and adopts the Hague Rules as amended in 1968 and now known as the Hague-Visby Rules.

With the introduction of the Hague Rules the practice on limitation clauses changed. The Rules were adopted to provide a fair balance of rights and responsibilities between carriers and shippers and they provided for the minimum responsibilities and liabilities of the owners. Permitting the carrier to limit his liability by limitation clauses would destroy the fair balance the Rules were intended to strike.

After the introduction of the Rules it became evident that owners could not limit their liability to the invoice value of the goods, in other words to any value other than the sound market value of the goods on the day on which they were delivered by the carrier, or the day upon which they should have been delivered. The authority for this is to be found in *The Arpad.*[1] In this case it was said that if there is a market at the time the goods are to be delivered in which goods of a kind fit to implement the contract made by the owner of the bill of lading at the time of the breach of contract can be bought or sold, the measure of damage is the value ascertained by the market price of similar goods. When there is no such market the value must be ascertained in some other way. It was established in this case that any clause, covenant or agreement in the bill of lading that might have the effect of lessening the liability of the carrier below this figure was null and void. As the invoice value of the goods is normally less than the sound market value, a clause in the bill of lading purporting to agree that claims in respect of loss or damage to cargo shall be adjusted on the basis of the invoice value are of no value to the shipowner.[2]

In general a limitation clause is invalid if it contravenes either the Hague Rules or Hague-Visby Rules or reduces the carrier's liabilities under those Rules. Article 3(8) of the Hague and Hague-Visby Rules provides that:

> "Any clause, covenant or agreement in a contract of carriage relieving the carrier or the ship from liability for loss or damage to or in connection with, goods arising from negligence, fault or failure in the duties and obligations provided in this Article or lessening such liability otherwise than as provided in these Rules, shall be null and void and of no effect. A benefit of insurance in favour of the carrier or similar clause shall be deemed to be a clause relieving the carrier from liability."

With regard to the term "any clause" used in the Article it is possible to state that any clause in a bill of lading which purports to lessen the liability of the carrier, otherwise than as provided for in the exceptions contained in the Rules, will be rendered null and void. Under this Article if a valuation clause sets the limit of liability at a specific amount per package which is lower than the amount specified by Article 4(5) and when the clause purports to base the limit on the invoice value of the goods then the clauses are invalid. If an invoice value clause increases the recovery to more than the actual loss such a clause will be held valid. The benefit of insurance clauses is specifically declared invalid in the last sentence of Article 3(8).

There are cases in which limitation clauses have been held valid. If the clause does not contravene Article 3(8) of the Rules it will be deemed valid. The limitation clause also will be upheld where the Hague and Hague-Visby Rules do not apply and the clauses do not contravene any applicable law having public policy or public statute.[3]

1. (1934) 49 Ll.L.R. 313.
2. See W.E. Astle, *Shipping and the Law* (1980), pp.61–63.
3. See William Tetley, pp.220–221.

PART 5

JURISDICTION OVER THE SEA AND SHIPS

CHAPTER 24

THE LAW OF THE SEA

HISTORICAL DEVELOPMENT OF THE LAW OF THE SEA

Ships operating on international voyages visit many different states. While they are in the territorial waters of other states they must obey the laws of those states. This takes us into the area of international law which regulates relations between states. The development of the law of the sea is inseparable from the development of international law in general.

The law of the sea[1] has rested on two traditional concepts. These are the high seas, where the freedom of the seas prevails, and the territorial sea, which is under the sovereignty of the coastal state. In the early concepts of the law of the sea navigation on the high seas was open to everybody, but during the fifteenth and sixteenth centuries powerful maritime states claimed sovereignty over the high seas. In opposition to the principle of maritime sovereignty, the principle of the freedom of the high seas began to develop. Britain was one of the leading powers which pursuaded and consolidated such a policy. From the beginning of the nineteenth century until the end of the Second World War the freedom of the high seas was a firmly established principle of international law. Freedom of fishing and navigation were the most important freedoms. However since 1945 there have been international political changes and they have also affected the freedom of the seas. Many states including coastal states gained independence and began to rethink their own interests. Obviously the freedom of the high seas and the delimitation of maritime zones were within their priorities. As a result of these developments new concepts, such as the continental shelf, exclusive fishery zones and exclusive economic zones, were introduced, which in turn led to the problems of jurisdiction over these areas.

In retrospect Brown[2] divides the post-war history of the law of the sea into two distinct periods. The first period was between 1945 and 1960, during which new law of the sea concepts were consolidated, clarified and codified, largely in response to technological changes. These are reflected in the four Geneva

1. For detailed information about the law of the sea see I.A. Shearer, *Starke's International Law* (11th ed. 1994), pp.218–228; E.D. Brown, *The International Law of the Sea*, Vol. I. (1994), pp.5–13; R.R. Churchill and A.V. Lowe, *The Law of the Sea* (2nd ed., 1988), pp.4–24.

2. E.D. Brown, Vol. I, pp.8–11.

Conventions on the Law of the Sea produced by the United Nations International Law Commission in 1958. The second period began in the 1960s and is still continuing.

SOURCES OF THE LAW OF THE SEA

The law of the sea encompasses all aspects of the uses and resources of the oceans. Maritime law constitutes that specialised branch of the law which governs maritime transport and international sea-borne trade.

The general framework for the majority of the public law conventions in the maritime field is contained in the United Nations Convention on the Law of the Sea 1982 (UNCLOS 1982). The Convention establishes a comprehensive framework for the regulation of the whole ocean area. The provisions of the Convention govern, *inter alia*, the spatial limits and regimes of national sovereignty or jurisdiction over the zones of ocean space adjacent to the coasts, access to the seas, navigation, protection and preservation from pollution of the maritime environment, exploitation of living resources and their conservation, exploitation of non-living resources, marine scientific research, mining of the sea-bed and the settlement of disputes.

(a) The 1958 Geneva Conventions

Since the First United Nations Conference of the Law of the Sea in 1958 (UNCLOS I) the international law of the sea has been one of the most dynamic areas of international law. As a result of this Conference four multilateral conventions were concluded. These Conventions covered various aspects of the law of the sea. These are the 1958 Geneva Conventions on the Territorial Sea and Contiguous Zone, The Continental Shelf, The High Seas and The Fishing and Conservation of Living Resources of the High Seas.[3] All these Conventions are in force. The first three of these have been ratified by substantial numbers of states. The Conventions codify customary law, so it is possible to say that although they are binding only on states which are parties to them, many of their provisions can be used as evidence of customary law even against states which are not parties to them. The 1958 Conference failed to reach agreement on a number of issues, in particular on the width of the territorial sea.

A second conference, UNCLOS II, was convened in 1960 to discuss the width

3. Convention on the High Seas, 13 U.S.T. 2312, T.I.A.S. 5200, 450 U.N.T.S. 82, 52 A.J.I.L. 842 (1958); Convention of the Continental Shelf, 15 U.S.T. 471, T.I.A.S. 5578, 499 U.N.T.S. 311, 52 A.J.I.L. 858 (1958); Convention on the Territorial and Contiguous Zone, 15 U.S.T. 1606, T.I.A.S. 5639, 516 U.N.T.S. 205, 52 A.J.I.L. 834 (1958); Convention on Fishing and Conservation of the Living Resources of the High Seas, 17 U.S.T. 138, T.I.A.S. 5969, 559 U.N.T.S. 285, 52 A.J.I.L. 851 (1958).

of the territorial sea and fishery limits. But it failed to adopt the formula which provided a six-mile territorial sea plus a six-mile fishery zone.

Since 1960 some states have become dissatisfied with some rules which were laid down in the 1958 Conventions and technological developments created a need for new rules. In 1967 Dr Arvid Pardo, the Maltese ambassador, called for the recognition of the deep-sea bed lying beyond the limits of national jurisdiction over the continental shelf. Under the 1958 Conference there was no provision for the legal regime of the deep-sea bed, as during the 1958 Conference mineral wealth of the deep-sea bed was not appreciated and the technology was not sufficiently developed. Consequently, in 1973 a third conference, UNCLOS III, was convened. In UNCLOS III there were two groups of states. The majority was in favour of having the deep-sea bed and its resources declared as the common heritage of mankind. Obviously such a claim involved the revision of parts of the 1958 Conventions on the Continental Shelf and on the High Seas. Other states were not in favour of such a change. They were reluctant to review the law of the sea. But, as stated above, the newly independent states provided the substantial majority and the earlier law was reviewed. After meeting for nine years the Conference finally adopted the text of the United Nations Conference on the Law of the Sea in 1982.

(b) The United Nations Convention on the Law of the Sea 1982

The Law of the Sea Convention 1982 (UNCLOS 1982) was the result of nine years of work by the Third United Nations Conference on the Law of the Sea. With its entry into force there was created an overlap between the treaty regimes as many Articles of the 1982 Convention repeat verbatim, or with minor changes, the text of the corresponding Articles of the 1958 Conventions. In other words many of its Articles restate the provisions of the 1958 Geneva Conventions and restate conventional rights and duties in a more accessible form. Under such an overlap how would the relations between the parties be governed? If states are both parties to the 1982 Convention the relations will be governed by the 1982 Convention irrespective of whether either was a party to any of the 1958 Conventions. In other words the 1958 Conventions are superseded by the 1982 Convention between parties to the 1982 Convention (Article 311 of UNCLOS 1982).[4] Relations between states which are both parties to a 1958 Convention but of which only one is a party to the 1982 Convention will be governed by the 1958 Convention. If a state is not a party to any of these Conventions its relations with every other state will be governed by customary law. If states parties to the UN Convention are also parties to another agreement compatible with it, the UN Convention will not alter the rights and obligations arising under such other agreement.

4. UNCLOS 1982, Art. 311(1) states: "1. This convention shall prevail, as between States Parties, over the Geneva Conventions on the Law of the Sea of 29 April 1958."

(c) Customary law

The activities of states in the international arena may give rise to binding law. For instance if all maritime nations declare the width of territorial sea at 12 miles that may become customary law. If states in general permit innocent passage for foreign merchant ships through their territorial sea a customary law comes into existence. Obviously all acts cannot give rise to customary law, there are two elements which must be present[5]: (a) the generality of states must habitually observe a particular usage; (b) states should believe themselves to be legally bound to do so.

Customary rules may also crystallise from usages or practices which have evolved in circumstances such as diplomatic relations between states, the practice of international organs, state laws, decisions of courts and state military or administrative practices. For instance, as a basic rule of international law conventions are binding upon states which have become parties to them once the Convention has come into force. However, the rules contained in a Convention may receive such wide acceptance and adoption in practice that they may be considered to have become rules of international customary law binding on all states, whether parties or not.

CURRENT STATUS OF THE LAW OF THE SEA

From the description of the sources of the law of the sea it becomes clear that the whole law of the sea is not contained in one single text. The 1958 Geneva Conventions provided a basic framework for most of it but the 1982 Law of the Sea Convention replaces them. Obviously rules of customary law will continue to be important in determining the detailed rights and duties of states, in particular when they are not parties to any of the Conventions. Because of its importance to its importance in the law of thc sea the 1982 Convention will be discussed in detail below.

(a) Current status of the United Nations Convention on the Law of the Sea 1982

After very long-term, continuous, intense and difficult negotiations between the nations of the world the UN Convention was adopted by UNCLOS III on 30 April 1982. Negotiations were protracted because it was decided that all decisions should be based on consensus rather than on a majority vote. But the interests and political attitudes of the participants were too diverse, so in the end the Convention was adopted by majority vote: 130 votes in favour to four against, with 17 abstentions. It was opened for signature at Montego Bay, Jamaica, on 10

5. The elements of customary law mainly derive from a series of decisions of the Permanent Court of Justice and the International Court of Justice. The well known decisions are: the *North Sea Continental Shelf Cases*, *The Lotus Case* [1927] P.C.I.J. Ser.A. No.10; *the Anglo-Norwegian Fisheries Case* [1951] I.C.J. Rep. 116; *Nicaragua* v. *USA* [1986] I.C.J. Rep. 14.

December 1982 and remained open for signature until 9 December 1984 (Article 305(2)). At that date 155 states and four other entities had signed it. Under Article 308 the Convention came into force 12 months after the date of deposit of the 60th instrument by ratification or accession. The Convention received its 60th ratification or accession on 16 November 1993 and therefore entered into force on 16 November 1994.

(b) Status of the states under the Law of the Sea Convention

(i) States that ratify the Convention

States which have ratified or acceded to the Convention are obliged to refrain from any act which would undermine the purpose or object of the Convention. With the entry into force of the Convention states which are parties become bound by its terms.

Under Article 311(3)[6] two or more states parties to the Convention may conclude agreements modifying or suspending the operation of the Law of the Sea Convention as it applies to the relations between them. But such agreements are restricted. They must not relate to a provision derogation from which is incompatible with the effective carrying out of the object and purpose of UNCLOS 1982. It is also provided that any such agreement must not affect the application of the basic principles embodied in UNCLOS 1982 and must not affect the enjoyment by other states parties of their rights or the performance of their obligations under the Convention. It is possible that Article 311 contemplates the likelihood that parties will want to modify their legal position by entering into treaties.[7] For instance two or more states may enter into an agreement to delimit their shared continental shelf or their exclusive economic zones. These agreements would not derogate from the overall purpose of UNCLOS 1982. But in this context it would be difficult to determine the "object and purpose" of the Convention or "its basic principles". Under Article 311(6) there is another clear undertaking. That Article states that the parties will not enter into any agreement in derogation of the basic principle relating to the common heritage of mankind set forth in Article 136.[8]

6. UNCLOS 1982, Art. 311(3) states:

"3. Two or more States Parties may conclude agreements modifying or suspending the operation or provisions of this Convention, applicable solely to the relations between them, provided that such agreements do not relate to a provision derogation from which is incompatible with the effective execution of the object and purpose of this Convention, and provided further that such agreements shall not affect the application of the basic principles embodied herein, and that the provisions of such agreements do not affect the enjoyment by other States Parties of their rights or the performance of their obligations under this Convention."

7. See T.W. Bennett, "The Status of the Law of the Sea Convention", T.W. Bennett and W.H.B. and Dean *et al.*, *The Law of the Sea* (1986), p.8.

8. UNCLOS 1982, Art. 311(6) states:

"6. States Parties agree that there shall be no amendments to the basic principle relating to the common heritage of mankind set forth in Article 136 and they shall not be party to any agreement in derogation thereof."

(ii) States that accede to only part of the Convention

The package deal principle underlying the Convention is expressed in Article 309. This Article specifically provides that no reservations or exceptions may be made unless expressly permitted by other Articles of the Convention.[9] One of the reasons for the establishment of such an article was the United States' dissatisfaction with the regime established for deep-sea-bed mining.[10] As is widely known, in 1982 President Reagan declared that the United States would not sign the Convention because of objections to Part XI, the proposed regime for deep-sea-bed mining. But when it became apparent that the Convention would receive the number of ratifications necessary for entry into force before the end of 1994 informal consultations began to meet the objections of the industrialised states. The consultations resulted in a draft "Agreement relating to the implementation of Part XI of the United Nations Convention on the Law of the Sea of 10 December 1982" and a draft resolution by which the UN Assembly would adopt the Agreement and urge states to adhere to it and to the Convention.[11] The resolution was adopted on 28 July 1994.[12] Under Article 2 of the 1994 Agreement it is to be interpreted and applied together with Part XI of the Convention as a single instrument. In the event of any inconsistency between the Agreement and Part XI of the UN Convention, the provisions of the Agreement prevail.[13]

(iii) The Law of the Sea Convention and third states

As stated in the Vienna Convention on the Law of Treaties, "a treaty does not create either obligations or rights for a third State without its consent". The International Court of Justice reaffirmed this principle in the *North Sea Continental Shelf* cases.[14] There are, however, exceptions to this principle. For instance under

9. UNCLOS 1982, Art. 309 states:

"**Reservations and exceptions**

No reservations or exceptions may be made to this Convention unless expressly permitted by other articles of this Convention."

10. For detailed information about relations between the USA and UNCLOS 1982 see Bernard H. Oxman (Rapporteur), "United States Interest in the Law of the Sea Convention, Panel on the Law of the Ocean Uses" (1994) 88, *American Journal of International Law* 168–178; Louis B. Sohn, "Law of the Sea Forum: The 1994 Agreement on Implementation of the Seabed Provision of the Convention on the Law of the Sea, The 1994 Agreement and the Convention" (1994) 88 *American Journal of International Law* 687–704.

11. UN Doc. A/48/950 (1994).

12. GA Res. 49/263 (28 July 1994).

13. New York Agreement of 28 July 1994 Relating to the Implementation of Part XI of the United Nations Convention on the Law of the Sea 10 December 1982, Art. 2 states:

"1. The Provisions of this Agreement and Part XI shall be interpreted and applied together as a single instrument. In the event of any inconsistency between this Agreement and Part XI, the provisions of this Agreement shall prevail."

14. *North Sea Continental Shelf Cases Federal Republic of Germany/Denmark; Federal Republic of Germany/Netherlands* [1969] I.C.J. Rep. 3, 25–26 (Judgment of 20 February). It is stated in these cases that ". . . it would simply be told that, not having become a party to the convention it could not claim any rights under it until the professed willingness and acceptance had been manifested in the prescribed form".

UNCLOS 1982 terms such as "all states", "any state" and "every state" are used. It is not clear from these words whether the third states enjoy the rights under the Convention. The only way to ascertain this is by examining the official statements made by the representatives of states participating in the Third United Nations Conference on the Law of the Sea. When the delegations' speeches on the subject of third states' rights are examined[15] it is possible to argue that the reference to "all states" or "every state" reflects the assumption that all states would be parties to the Convention.[16] Another expectation would be that some of the rules of the Convention would become customary law in the future.

Another possible expectation is related to the obligations of third states. Under Article 35 of the Vienna Convention on the Law of the Treaties,[17] third states do not need to be specifically named before an obligation can be imposed on them. In this case the obligations of third states may be inferred from expressions like "all states" and "every state". Thus, if third states accept the obligations in writing many provisions of the Law of the Sea Convention may have the effect of imposing obligations on them. For instance under Article 3 of the Convention: "Every State has the right to establish the breadth of the territorial sea up to a limit not exceeding 12 nautical miles, measured from baselines determined in accordance with this Convention." According to this Article the corresponding duty of every state is to recognise the breadth of the territorial sea established by other states up to 12 miles. Although a number of countries now claim 12 miles as their territorial sea some of them retain three mile and six mile limits. Under such practice it becomes clear that the 12 mile limit may not yet have assumed the status of customary law. The continued refusal of some states to accept the 12 mile limit as applicable to them means that the 12 mile limit cannot be regarded as a universal customary norm. Therefore if third states assert rights under the Convention they either have

15. Ambassador Arias-Schreiber of Peru stated that "No State can claim that the new rules and rights established by the Convention apply to that State if it is not a party to the Convention": 185th meeting. See also the speeches of other delegations which were against third states' rights: Canada (185th mtg. of 6 December 1982, UN Doc. A/CONF.62/PV. 185, at 31(1983); Cameroon (*ibid.* at 37); Tanzania (187th mtg. of 7 December 1982, UN Doc. A/CONF.62/PV. 187, at 83 (1983); Byelorussian SSR (189th mtg. of 8 December 1982, UN Doc. A/CONF.62/PV. 189, at 46 (1983); Indonesia (186th mtg. of 6 December 1982, UN Doc. A/CONF.62/PV.186, at 16 (1983); Iran (191st mtg. of 9 December 1982, UN Doc. A/CONF.62/PV. 191, at 32 (1983); and the USSR (*ibid.* at 41).

16. According to Luke T. Lee, "The Law of the Sea Convention and Third States" (1993) 7 *American Journal of International Law* 549, the terms "all states" and "every state", as used in the Convention, may be held to have any of the following four meanings:

(1) the plain and natural meaning to include all states, whether or not parties to the Convention, as beneficiaries of rights in the Convention;
(2) as synonyms for "states parties" where the intent of the parties is to that effect, as indicated in some statements made in Montego Bay;
(3) as a denotation of customary rules, and hence binding upon all states; or
(4) as an intent to generate customary rules.

17. Vienna Convention on the Law of Treaties, 1969, Art. 35 states:

"An obligation arises for a third State from a provision of a treaty if the parties to the treaty intend the provision to be the means of establishing the obligation and the third State expressly accepts that obligation in writing."

to establish an agreement with those state parties that accept such rights or grant those rights under customary international law.

(c) The jurisdictional zones

(i) In general

Until the end of the Second World War the seas were divided into four jurisdictional zones. These were internal waters, the territorial sea, the contiguous zone and the high seas. During the 1958 Geneva Conventions attempts were made to introduce new maritime zones. The most important of these developments was the assertion of jurisdiction and control over the natural resources of the continental shelf, which was referred to by President Truman in his Presidential Proclamation No. 2667 on 28 September 1945. The principle of coastal state rights over the continental shelf was established by the International Court of Justice 24 years later.[18] Another important development was the introduction of the differentiation between the concept of the territorial sea and that of the exclusive fishing zone. In 1972 the exclusive economic zone concept was introduced into the law of the sea. The new concept of archipelagic waters was accepted widely under UNCLOS 1982. As a consequence of all these developments, five more zones are now generally recognised. These are the continental shelf, the exclusive fishing zone, the exclusive economic zone, archipelagic waters and the area beyond the limits of national jurisdiction.

UNCLOS 1982 sets out different legal regimes for the freedom of navigation on the high seas, which also applies to the exclusive economic zone, and for passage in the territorial sea, through international straits and in archipelagic waters.

(ii) Internal waters

Internal waters were defined as "waters on the landward side of the baseline of the territorial sea" in Article 5(1) of the Geneva Convention on the Territorial Sea and the Contiguous Zone 1958. Internal waters consist of ports, harbours, rivers, lakes and canals and also waters on the landward side of the baselines used for measuring the width of the territorial sea. Article 8(1) of UNCLOS adopts the same definition. The definition is subject to one variation which is about the territorial sea baselines of archipelagic states. These states are the subject of special provisions in Part IV of UNCLOS.[19]

18. *North Sea Continental Shelf*, Judgment [1969] I.C.J. Rep. 3.
19. UNCLOS 1982, Art. 8 states:
"**Internal Waters**:
1. Except as provided in Part IV, waters on the landward side of the baseline of the territorial sea form part of the internal waters of the State.
2. Where the establishment of straight baseline in accordance with the method set forth in article 7 has the effect of enclosing as internal waters areas which had not previously been considered as such, a right of innocent passage as provided in this Convention shall exist in those waters."

Baselines

The baseline[20] serves the purpose of defining the limits of a state's territory. Normal baselines have been considered under Article 5 of UNCLOS which is the same as Article 3 of the Geneva Convention except for a minor drafting change.[21] Under Article 5 the difficulty concerns determining the low-water line along the coast. There are many low-water lines to choose from and international law leaves the choice to the discretion of the coastal state.

If a coastline has different characteristics such as the coast of Norway or of north-west Scotland the establishment of a normal baseline causes practical difficulties. Therefore Article 4(1) of the Geneva Convention and Article 7(1) of UNCLOS declared in identical terms[22] that in certain geographical circumstances it is permissible to draw straight lines across the sea on a map.[23] These lines could be from headland to headland or from island to island and the territorial sea is measured from those straight lines.[24] Obviously the fact that states may use a system of straight baselines does not mean that every application of it is necessarily lawful. As it was stated in the *Anglo Norwegian Fisheries case*, while actual delimitation will be left to the state itself, "the validity of the delimitation with regard to other states depends upon international law".[25]

Exercise of sovereignty and the right of access to ports and internal waters

Internal waters are parts of the sea which are not either the high seas or relevant zones or the territorial sea. They are classified as an integral part of the coastal state, which enjoys full territorial sovereignty over them. In relation to the sovereignty of the coastal state there is no right of innocent passage through internal waters as exists through the territorial sea. In other words sovereignty of a coastal state over its internal waters is not limited by an obligation to grant a right of innocent passage to foreign shipping. Limitations on sovereignty may only arise under international customary law or under treaties entered into by the coastal

20. See R.R. Churchill and A.V. Lowe, Chapter 2.
21. UNCLOS 1982, Art. 5 states:
"Normal Baseline
Except where otherwise provided in this Convention, the normal baseline for measuring the breadth of the territorial sea is the low-water line along the coast as marked on large-scale charts officially recognised by the coastal State."
22. Under Art. 7 of UNCLOS 1982 there is only a minor modification in respect of "low tide elevations" (rocks visible only at low tide).
23. The Geneva Convention on the Territorial Sea 1958, Art. 4 states:
"1. In localities where the coastline is deeply indented and cut into, or if there is a fringe of islands along the coast in its immediate vicinity, the method of straight baselines joining appropriate points may be employed in drawing the baseline from which the breadth of the territorial sea is measured." See also Art. 7 of UNCLOS 1982.
24. The straight baseline system was legally recognised for the first time by the International Court of Justice in its judgment in the *Anglo-Norwegian Fisheries case* in 1951 ([1951] I.C.J. Rep. 116). At that time the Court's decision was regarded as an innovation but today many states use the straight baseline system and it is employed on the west coast of Scotland.
25. See Martin Dixon, *Textbook on International Law* (3rd ed., 1993), p.192.

state. As a general rule there is no right of navigation for foreign ships through internal waters. When the existence of sovereignty over internal waters and the absence of any general right of innocent passage through them are considered together it is possible to reach the conclusion that in customary international law there is no right for foreign ships to enter a state's ports and other internal waters. The only exception to this rule is contained in Article 5(2) of the Geneva Convention and Article 8(2) of UNCLOS 1982. Under these provisions where a straight baseline is established in accordance with the method set out in Article 7 it has the effect of enclosing as internal waters areas which were not previously considered as such, and a right of innocent passage is preserved in those waters.

It is clear that a coastal state is entitled to prohibit entry into its ports by foreign warships but it is arguable whether it has a similar right to forbid foreign merchant ships from entering its ports. According to Brown[26] a right of access to ports exists under international customary law and it exists for the parties of the Geneva Convention and Statute on the International Regime of Maritime Ports 1923, as well as under various bilateral commerce and navigation treaties. In the absence of any clear right of access to ports in customary law most states have such rights under a treaty. Obviously developments in international trade and commerce also affect customary international law, so states which are keen to improve their trade welcome foreign ships to their ports.

Freedom of access granted to foreign vessels is not an absolute freedom. States have the right to nominate those ports of theirs which are open to international trade. A state may also close its international ports to protect its vital interests without violating customary international law. States also have a right to prescribe conditions for access to their ports.[27] In cases where freedom of access continued to exist the coastal state would be under an obligation to publicise changes in the nature of access channels and ensure that they were kept safe.

Jurisdiction in internal waters

When a ship enters a foreign port or other internal waters it is under the territorial sovereignty of the coastal state. A coastal state may exercise its jurisdiction over foreign ships within its internal waters to enforce its laws. Although the coastal state may apply and enforce its laws in full against foreign merchant ships in its internal waters this principle is subject to a number of exceptions. For instance the courts of the flag state may also try people for crimes committed on board a ship. In such cases there will be concurrent jurisdiction. For instance in *R.* v. *Anderson* an American national committed manslaughter on board a British vessel in French internal waters. It was held by the Court of Criminal Appeal in the UK that the American national was subject to the jurisdiction of the British courts although he

26. E.D. Brown, Vol. I, pp.37–40.

27. In the *Nicaragua* case the International Court of Justice stated that internal waters are subject to the sovereignty of the state and that it is "by virtue of its sovereignty that the coastal State may regulate access to its port": [1986] I.C.J. Rep. 14, at 111. See R.R. Churchill and A.V. Lowe, pp.52–54.

was within the sovereignty of French justice.[28] Another exception to local jurisdiction is where merely disciplinarian issues related to the ship's crew are involved. The coastal state will not interfere with the exercise of disciplinary powers by the captain over his crew. If the crime committed by a member of the crew does not concern the maintenance of peace within the territory of the coastal state it will usually allow the matter to be dealt with by the authorities of the flag state. Another important point about the coastal state's jurisdiction over foreign ships arises in relation to the right of exit of foreign ships from their ports. Obviously the right of access to ports and internal waters also implies a right to leave. But this right of exit is subject to limitations. Ships in port are liable to arrest as security in civil action such as for customs offences or in actions *in rem* against the ship. In cases of pollution offences port state jurisdiction is considerably extended. Under the International Convention Relating to Intervention on the High Seas in Cases of Oil Pollution 1969 the right of intervention of coastal states[29] on the high seas in cases of oil pollution casualties is expressly recognised.[30] It may be exercised anywhere on the high seas.

The jurisdiction of the coastal state as regards foreign warships is quite limited. A foreign warship observes the coastal state's laws about navigation and health but the authorisation of the captain or of the flag state comes before the coastal state's jurisdiction. Members of the crew are immune from prosecution by the coastal state for crimes committed on board the ship. The immunity will also cover crimes committed on shore if the crew were in uniform and on official business at the time of the crime.

(iii) Territorial sea

The width of the territorial sea

The width of the territorial sea is one of the most controversial questions in international law. Throughout the entire story of the territorial sea there has always been disagreement on how far it may extend from the baselines. In the sixteenth and seventeeth centuries limits of visibility determined the extent of waters over which control was claimed. In the eighteenth century it became generally accepted that the width of the territorial sea should be the same as the range of a cannon (the cannon shot rule). But in the nineteenth century this was transmuted into the three mile rule, which was supported in particular by the United Kingdom and the United States, but on the other hand the Scandinavian states claimed four miles of territorial sea and Spain and Portugal claimed six miles. The 1930 Hague

28. 1 Cox C.C. 198. See also *Wildenhus*, where a crew member of a Belgian vessel murdered another Belgian national when the ship was docked in the port of Jersey City in New York. It was held by the US Supreme Court that the American court had jurisdiction to try the crew member (120 U.S. 1 (1887)); M.N. Shawn, *International Law* (2nd ed., 1986), pp.295–296.

29. The right of intervention is limited to a maritime casualty resulting in a case of pollution involving a privately owned ship of one of the contracting parties.

30. See Art. 1 of the 1969 Intervention Convention.

Conference attempted to reach agreement upon the width of the territorial sea and failed. The 1958 Geneva Convention on the Territorial Sea did not include an Article on the width of the territorial sea owing to disagreement among the states.[31] The second UN Conference in 1960 failed narrowly to agree on a compromise solution. When the diversity of state practice and the failure of the 1958 and 1960 Conferences are examined together it is difficult to say what the customary law is about the width of the territorial sea.

Finally Article 3 of UNCLOS 1982[32] says that all states have the right to establish the breadth of the territorial sea up to a limit not exceeding 12 nautical miles from the baselines. This can be accepted as the internationally agreed maximum limit. When the territorial sea claims were examined as at June 1993[33] it became clear that many states including those which had ratified or acceded to UNCLOS still claimed more than 12 miles. Obviously now states parties to the Convention are bound by the Convention with its entry into force and cannot claim more than 12 miles.

By customary international law foreign merchant ships have the right of innocent passage through the territorial sea of a coastal state. Under Article 14 of the Convention on the Territorial Sea 1958, the coastal state must not hamper innocent passage and must publicise any dangers to navigation in the territorial sea of which it is aware. Article 17[34] of UNCLOS followed closely the language of Article 14(1) of the Geneva Convention. Article 19(2) of UNCLOS has developed the notion of innocent passage.[35] Under Article 21(4) foreign ships exercising the right of innocent passage are required to comply with all such laws and regulations and all generally accepted international regulations relating to the prevention of collisions at sea.[36]

31. At the Geneva Conference of 1958 the UK suggested 6 miles as the width of the territorial sea. Later, the USA brought another proposal which was for a 6-mile territorial sea and a further 6-mile zone in which the coastal state would have exclusive fishing rights. The UK's proposal was withdrawn in favour of the USA's proposal. Other states also suggested that the width of the territorial sea should be fixed at 12 miles. At the end of the Conference no agreement was reached.

32. UNCLOS 1982, Art. 3 states:

"**Breadth of the territorial sea**

Every State has the right to establish the breadth of its territorial sea up to a limit not exceeding 12 nautical miles, measured from baselines determined in accordance with this Convention."

33. See E.D. Brown, Vol. I, pp.46–49.

34. UNCLOS 1982, Art. 17 says:

"**Right of innocent passage**

Subject to this Convention, ships of all States, whether coastal or land-locked, enjoy the right of innocent passage through the territorial sea."

35. Art. 14(4) of the Geneva Convention stated that "passage is innocent so long as it is not prejudicial to the peace, good order or security of the coastal state . . .". But the Convention does not provide any guidance on what acts or circumstances may be prejudicial to the peace, good order or security of the coastal state. Under Art. 19(2) of UNCLOS 1982 this term has been clarified. The Art. says what is meant by "prejudicial".

36. The reference to generally accepted international regulations relating to the prevention of collisions at sea will be read principally as a reference to the 1972 Convention on the International Regulations for Preventing Collisions at Sea, as amended in 1981, 1987 and 1989.

The significance of the outer limit of the territorial sea

As explained above the width of the territorial sea has always been a controversial problem in the law of the sea. The main reason for such controversy was the interests of states. A wide territorial sea was in the interests of some states but against the interests of others. The most obvious conflicts of interest were about fishing. Nowadays states are entitled to claim exclusive fishery zones beyond their territorial sea but until 1960 the only way to extend fishing limits was by extending the territorial sea. Another problem was related to security. The controversy between maritime states, which wished to keep the coastal states' jurisdiction to a minimum, and coastal states, which wished to have exclusive control over the sea adjacent to their coast for fishing and mineral resources activities, caused the development of the continental shelf and the EEZ. Prior to the emergence of the concept of the EEZ the outer limit of the territorial sea was of fundamental importance, because the outer limit divided the more landward area from the more seaward area. In the more landward area the principle of sovereignty predominated and in the more seaward area that of the freedom of the high seas prevailed.

The position has changed under UNCLOS 1982. Obviously the coastal state still has sovereignty in the territorial sea, and in case of doubt there are strong legal presumptions in favour of the sovereignty of the coastal state. Beyond the outer limit of the territorial sea the concept of the EEZ was introduced. The international community has freedom in this area but it is not a part of the high seas. Under UNCLOS 1982 the EEZ became an area where the coastal states and other states enjoyed rights.

Juridical nature of the territorial sea

A state's power to control the activities of foreign ships in territorial waters and adjacent zones is called coastal state jurisdiction. In the territorial sea the coastal state enjoys sovereignty but has to respect the rights of foreign ships. As was mentioned above the fundamental restriction upon the sovereignty of the coastal state is the right of other nations to innocent passage through its territorial sea. Right of innocent passage distinguishes the territorial sea from the internal waters of the state where the coastal state has complete jurisdiction. Articles 17 and 18 of UNCLOS 1982 contain the right of innocent passage of foreign ships through the territorial sea. Under Article 24 the coastal state may not hamper innocent passage. Article 21(2) provides for the legislative jurisdiction of a coastal state in the territorial sea, and indicates the matters on which a state may adopt laws and regulations. Ships exercising the right of innocent passage shall comply with all such laws and regulations. Under Article 21(2) such laws and regulations shall not apply to the design, construction, manning or equipment of foreign ships unless they give effect to generally accepted international rules and standards. Apart from these limitations the coastal state enjoys full sovereignty in its territorial sea.

In the territorial sea a coastal state may take steps to prevent passage that is not innocent. If a foreign ship is engaged in any of the activities set out in Article 19(2) its passage shall be considered prejudicial to the peace, good order and security of the coastal state. Stopping and inspection of a ship may take place for any violation of the applicable laws in the territorial sea, as provided for in relation to pollution offences.

Criminal and civil jurisdiction over foreign ships

Since a state has sovereignty over its territory, this also pertains to foreign ships in ports on its territory or at offshore terminals in its territorial sea. The specific port state powers include the inspection of a ship's certificate, physical inspection of the ship, if warranted by the evidence, and detention of the ship. The exercise of criminal and civil jurisdiction on board foreign ships, whether merchant or government ships operated commercially, in passage through the territorial sea is dealt with in Articles 27 and 28 of UNCLOS 1982.[37]

Under Article 27(1) the criminal jurisdiction of a coastal state is not to be exercised on board a foreign ship passing through the territorial sea in order to arrest any person or to conduct any investigation in connection with any crime committed on board the ship during its passage, save only in the following cases:

(a) if the consequences of the crime extend to the coastal state;
(b) if the crime is of a kind to disturb the peace of the country or the good order of the territorial sea;
(c) if the assistance of the local authorities has been requested by the master of the ship or by a diplomatic agent or consular officer of the flag state; or
(d) if such measures are necessary for the suppression of illicit traffic in narcotic drugs or psychotropic substances.

If a ship passes through the territorial sea after having left the internal waters of a coastal state then the coastal state may act in any manner prescribed by its law for the purpose of arrest and investigation on board (Article 27(1)).[38] The coastal state's right is not restricted by the terms of Article 27(1). Except as provided in parts of the Convention dealing with the EEZ or the protection of the marine environment the coastal state may not exercise criminal jurisdiction or undertake any criminal investigation on board a foreign ship in the territorial sea concerning a crime committed before its entry therein. In other words the coastal state cannot act where the crime was committed before the ship entered the territorial sea, providing the ship is not entering or has not entered internal waters (Article

37. Jurisdiction over foreign ships is dealt with in Art. 19(1) of the 1958 Convention of the Territorial Sea. The provisions are reaffirmed in Art. 27 of the LOSC 1982. LOSC 1982 Art. 27(1) only added "or psychotropic substances" to Art. 19(1).

38. UNCLOS 1982, Art. 27(2) states:

"**Criminal jurisdiction on board a foreign ship**

2. The above provisions do not affect the right of the coastal State to take any steps authorised by its laws for the purpose of an arrest or investigation on board a foreign ship passing through the territorial sea after leaving internal waters."

27(5)).[39] In determining whether or how an arrest is to be made, the local authorities are to have regard to the interests of navigation (Article 27(4)).[40]

Civil jurisdiction is provided for in a different manner. The coastal state may not exercise its civil jurisdiction over a foreign ship passing through its territorial sea under Article 28(1) of UNCLOS 1982. It is not to stop or divert a ship for the purpose of exercising civil jurisdiction in relation to a person on board.[41] The coastal state may not levy execution against or arrest a ship unless obligations are involved which were assumed by the ship itself in the course of or for the purpose of its voyage through the waters of the coastal state or unless the ship is passing through the territorial sea on its way from internal waters. These prohibitions are, however, without prejudice to the coastal state's right to take civil measures against a foreign ship in the territorial sea after leaving internal waters (Article 28(2)(3)).[42]

Warships are defined and dealt with under Articles 29–32 of UNCLOS 1982.[43] The coastal state may not exercise its jurisdiction over them and other government ships operated for non-commercial purposes. But if a warship is not complying with the coastal state's laws and regulations on passage in the territorial sea and disregarding a request to conform thereto, it may be required to leave the territorial sea immediately (Article 30).[44]

Where damage is done to a coastal state owing to non-compliance by a warship

39. UNCLOS 1982, Art. 27(5) states:
"**Criminal jurisdiction on board a foreign ship**
5. Except as provided in Part XII or with respect to violations of laws and regulations adopted in accordance with Part V, the coastal State may not take any steps on board a foreign ship passing through the territorial sea to arrest any person or to conduct any investigation in connection with any crime committed before the ship entered the territorial sea, if the ship, proceeding from a foreign port, is only passing through the territorial sea without entering internal waters."

40. UNCLOS 1982, Art. 27(4) states:
"**Criminal jurisdiction on board a foreign ship**
4. In considering whether or in what manner an arrest should be made, the local authorities shall have due regard to the interests of navigation."

41. UNCLOS 1982, Art. 28(1) states:
"**Civil jurisdiction in relation to foreign ships**
1. The coastal State should not stop or divert a foreign ship passing through the territorial sea for the purpose of exercising civil jurisdiction in relation to a person on board the ship."

42. UNCLOS 1982, Art. 28(2) and (3) states:
"**Civil jurisdiction in relation to foreign ships**
2. The coastal State may not levy execution against or arrest the ship for the purpose of any civil proceedings, save only in respect of obligations or liabilities assumed or incurred by the ship itself in the course or for the purpose of its voyage through the waters of the coastal State.
3. Paragraph 2 is without prejudice to the right of the coastal State, in accordance with its laws, to levy execution against or to arrest, for the purpose of any civil proceedings, a foreign ship lying in the territorial sea, or passing through the territorial sea after leaving internal waters."

43. See also Art. 21–23 of the Geneva Convention on the Territorial Sea and the Contiguous Zone 1958.

44. UNCLOS 1982, Art. 30 states:
"**Non-compliance by warships with the laws and regulations of the coastal State**
"If any warship does not comply with the laws and regulations of the coastal State concerning passage through the territorial sea and disregards any request for compliance therewith which is made to it, the coastal State may require it to leave the territorial sea immediately."

or other government ship operated for non-commercial purposes, or owing to the breach by such vessels of the Convention or other rules of international law the flag state bears international responsibility.[45] Apart from exceptions as are contained in subsection A (Articles 17–26) and Articles 30–31, nothing in the Convention is to affect the immunities of warships and other government ships operated for non-commercial purposes (Article 32).

(iv) The contiguous zone

In general

The idea of a contiguous zone was virtually formulated in the 1930s by the French writer Gidel. The Geneva Convention on the Territorial Sea and the Contiguous Zone 1958 gave general conventional recognition to the legality of this practice for the first time. Under Article 24 of the Geneva Convention states were not permitted to extend the contiguous zone beyond 12 miles from the territorial sea baseline. The potential breadth of contiguous zone which would be claimed depended upon the breadth of territorial sea claimed. But under UNCLOS 1982 the territorial sea was extended to 12 miles and Article 24 of the Geneva Convention no longer had much effect. Therefore by Article 33(2) of UNCLOS states are permitted to extend the contiguous zone to 24 miles from the territorial sea baselines. Under this provision if a state claims 12 miles' territorial sea its contiguous zone is limited for up to a further 12 miles. Unlike the territorial sea contiguous zones have to be specifically claimed.

Juridical nature of the contiguous zone

Article 24 of the Geneva Convention describes the contiguous zone as a "zone of the high seas contiguous to its territorial sea". Under Article 33 of UNCLOS 1982 it is referred to as "a zone contiguous to its territorial sea". Under Article 55 of UNCLOS 1982 exclusive economic zone is defined as "an area beyond and adjacent to the territorial sea" and it is open to states to claim one, although it is not obligatory.

The legal status of the contiguous zone became more complicated under UNCLOS 1982. The juridical nature of the contiguous zone may change depending on the exclusive economic zone. For instance if an EEZ were claimed

45. UNCLOS 1982, Art. 31 states:
"**Responsibility of the flag State for damage caused by a warship or other government ship operated for non-commercial purposes.**

The flag State shall bear international responsibility for any loss or damage to the coastal State resulting from the non-compliance by a warship or other government ship operated for non-commercial purposes with the laws and regulations of the coastal State concerning passage through the territorial sea or with the provisions of this Convention or other rules of international law."

the contiguous zone and EEZ areas would overlap. If an EEZ were not claimed the contiguous zone would be a zone of the high seas because the high seas part of the UN Convention applies, *inter alia*, to all parts of the sea that are not included in the EEZ, territorial sea or internal waters.[46] Another point which affects the juridical nature of the contiguous zone is its association with the high seas or the EEZ. If the contiguous zone is accepted as a zone of the high seas, which was the case under the Geneva Conventions, it will be governed under the principle of the freedom of the high seas and the coastal state will have exceptional rights or powers in the zone. If the contiguous zone is accepted as a part of the EEZ there is no legal presumption in favour of the freedom of the high seas or the coastal state's sovereignty, as it is accepted as a zone which is *sui generis*. Under UNCLOS the contiguous zone falls within the EEZ. As a consequence the presumption against a coastal state's jurisdiction is removed. If a coastal state claims jurisdictional rights which are not expressly granted under the UNCLOS the dispute will be resolved on the basis of equity and taking into account the respective importance of the parties concerned.

Under Article 33(1) of UNCLOS the coastal state may exercise the control necessary to prevent infringement of its customs, fiscal, immigration or sanitary laws and regulations within its territory or territorial sea. The coastal state may also exercise the control necessary to punish an infringement of the above laws and regulations committed within its territory or territorial sea. The preventive control authorised under Article 33(1) is exercisable only in relation to incoming vessels. The coastal state may exercise whatever control is necessary to prevent infringement of its laws and regulations. Punitive control under Article 33(1)(b) will be directed against foreign vessels after they have left the territorial sea. Under such control the coastal state may use all necessary powers in order to punish those who have violated the laws and regulations of the coastal state in its territory or territorial sea. The coastal state may also use its right of arrest upon the ship.

(v) The exclusive economic zone (EEZ)

In general

One of the major reasons for the establishment of the EEZ was the controversy over fishing zones. Since about 1960 there has been a tendency for states to claim exclusive fishery zones beyond their territorial sea. The EEZ is not to be found in the 1958 Geneva Conventions. For instance the 1958 Geneva Convention on the Territorial Sea did not reach agreement on the creation of fishing zones and Article 24 of the Convention does not give exclusive fishing rights in the contiguous zone. The EEZ has its origins in the exclusive fisheries zones but it is recognised in the *Fisheries Jurisdiction* cases by the International Court of Justice.[47] Until the entry into force of UNCLOS the EEZ was a concept of customary law alone.

46. E.D. Brown, Vol. I, p.130.
47. *Fisheries Jurisdiction cases* (*UK* v. *Iceland*) Merits, Judgment [1974] I.C.J. Rep. 3.

Article 57 of UNCLOS provides that the zone shall not extend beyond 200 nautical miles from the baselines from which the breadth of the territorial sea is measured. Article 58 lays down the rights and duties of other states in the EEZ. These are basically the high seas freedoms of navigation, overflight and laying submarine cables and pipelines. In exercising these rights and performing duties states must have due regard to the rights, duties and laws of the coastal state. If there were any conflict over the attribution of rights and jurisdiction in the EEZ the problems would be solved on the basis of equity and in the light of all the relevant circumstances.[48] Under the old practice in conflict between a coastal state's rights and other states' rights it would have been solved by reference to the freedom of the high seas. If particular rights had not been given by the Convention to the coastal state it would have been presumed that the coastal state did not have those rights. Under the new approach adopted by UNCLOS, Article 59, there are no longer any legal presumptions concerning rights and jurisdiction in the EEZ.

Sui generis nature of the EEZ

Before the establishment of the EEZ concept the contiguous zone, the continental shelf and the exclusive fishery zone were recognised beyond the outer limit of the territorial sea. The coastal state has limited rights and powers over these maritime zones and such rights and powers constitute exceptions to the principle of the freedom of the high seas. But the EEZ departs from this common pattern and is regarded as *sui generis*. The EEZ is a belt of sea, adjacent to the coast, extending up to 200 miles from the baselines of the territorial sea, which means that the EEZ is 188 miles in width. But although the territorial sea ends at 12 miles, the 188 miles of the EEZ are not high seas. Under Article 55 of UNCLOS the EEZ area is no longer regarded as part of the high seas.[49] There is no reference in Article 55 to the high seas. And Article 86 also states that the provisions of Part VII (high seas) apply to all parts of the sea that are not included in the exclusive economic zone. In the EEZ a coastal state exercises rights and powers to various degrees. First, within the EEZ, the coastal state is given sovereign rights for the purpose of exploiting, conserving and managing the natural resources, whether living or non-living, of the waters super-adjacent to the sea-bed and of the sea-bed and its subsoil, and

48. UNCLOS 1982, Art. 59 states:
"**Basis for the resolution of conflicts regarding the attribution of rights and jurisdiction in the exclusive economic zone**
In cases where this Convention does not attribute rights or jurisdiction to the coastal State or to other States within the exclusive economic zone, and a conflict arises between the interests of the coastal State and any other State or States, the conflict should be resolved on the basis of equity and in the light of all the relevant circumstances, taking into account the respective importance of the interests involved to the parties as well as to the international community as a whole."

49. UNCLOS 1982, Art. 55 states:
"**Specific legal regime of the exclusive economic zone**
The exclusive economic zone is an area beyond and adjacent to the territorial sea, subject to the specific legal regime established in this Part, under which the rights and jurisdiction of the coastal State and the rights and freedoms of other States are governed by the relevant provisions of this Convention."

with regard to other activities for the economic exploitation and exploration of the zone (Article 56(1)(a)). The coastal state also enjoys jurisdiction with regard to the establishment of artificial islands and other installations, marine scientific research and the protection and preservation of the marine environment (Article 56(1)(b)). UNCLOS has also recognised the rights and freedoms that states other than the coastal states in question enjoy in the EEZ. However, the freedoms listed in Article 87 (in Part VII on the high seas) will not wholly apply to the EEZ. Article 58 only refers to freedom of transport and communication. By specifying certain freedoms, Article 58(1) underlines that there is no presumption in favour of other freedoms. The specifically recognised freedoms are to be enjoyed subject to the relevant provisions of UNCLOS. In any event, states shall have due regard to the rights and duties of the coastal states and shall comply with the laws and regulations adopted by the coastal states in respect of this area (Article 58).

Article 73 sets out the limits subject to which a coastal state may enforce the laws and regulations governing the exercise of its sovereign rights and its jurisdiction in the EEZ. The Article instances such measures as boarding, inspection, arrest and judicial proceedings. The penalty of imprisonment is excluded. In the exercise of its powers a broad and general discretion has been left to a coastal state.[50]

(vi) The continental shelf

The origin of the continental shelf as a juridical concept lies in customary law. Traditionally the first move in this direction was the Truman Proclamation of 1945.[51] In the Proclamation with Respect to the Natural Resources of the Subsoil and Seabed of the Continental Shelf US President Truman put forward the idea that a continental shelf belongs to the coastal state. The United States asserted its right to explore and exploit the natural resources of the continental shelf "beneath the high seas but contiguous to" its coast. After this declaration many states claimed rights over the oil, gas and mineral deposits believed to exist in the shelf. The first Conference on the Law of the Sea adopted the 1958 Convention on the

50. UNCLOS 1982, Art. 73 states:
"**Enforcement of laws and regulations of the coastal State**
1. The coastal State may, in the exercise of its sovereign rights to explore, exploit, conserve and manage the living resources in the exclusive economic zone, take such measures, including boarding, inspection, arrest and judicial proceedings, as may be necessary to ensure compliance with laws and regulations adopted by it in conformity with this Convention.
2. Arrested vessels and their crews shall be promptly released upon the posting of reasonable bond or other security.
3. Coastal State penalties for violations of fisheries laws and regulations in the exclusive economic zone may not include imprisonment, in the absence of agreements to the contrary by the States concerned, or any other form of corporal punishment.
4. In cases of arrest or detention of foreign vessels the coastal State shall promptly notify the flag State, through appropriate channels, of the action taken and of any penalties subsequently imposed."

51. Under the Truman Proclamation the US government regards "the natural resources of the subsoil and seabed of the continental shelf beneath the high seas but contiguous to the coasts of the United States as appertaining to the United States, subject to its jurisdiction and control."

Continental Shelf, which defined the shelf in terms of its exploitability as the submarine area adjacent to the coast but outside the territorial sea to a depth of 200 metres or, beyond that limit, to where the depth of the super-adjacent waters admits of the exploitation of the natural resources. Under this definition the inner limit of the shelf is the outer edge of the territorial sea but the seaward limit is a depth of 200 metres or any area capable of exploitation. Under such a provision the extension of coastal state jurisdiction is directly related to its ability to extract natural resources. The outer limits of the shelf which is subject to the jurisdiction of the coastal state are unclear. The definition in Article 1 of the 1958 Convention was criticised and in Article 76 of UNCLOS 1982 the seaward limit of the shelf is determined by the following rules:

1. All states, regardless of the physical shape of the shelf, have a continental shelf in law up to 200 miles from the baselines of the territorial sea. This is so even if there is no physical shelf or if it stops short of 200 miles (Article 76(1)).
2. If the physical extent of the shelf goes beyond 200 miles, the coastal state has a shelf in law "throughout the natural prolongation of its land territory to the outer edge of the continental margin" (Article 76(1)). In cases where the shelf goes beyond 200 miles as a physical feature, the coastal state has shelf rights to the outer edge of the continental rise except as provided in rule (3).
3. The continental shelf in law cannot extend beyond 350 nautical miles from the baselines of the territorial sea or 100 nautical miles from the 2,500 metre isobath (depth line) (Article 76(5)).

Under the above-stated conditions all states have a shelf up to 200 miles irrespective of physical conditions and there is a seaward limit even if the shelf continues further.

The coastal state may exercise sovereign rights over the continental shelf for the purposes of exploring it and its natural resources under Article 77(1) of UNCLOS.[52] The rights referred to in Article 77(1) are exclusive, which means that no other state may undertake such activities without the express consent of the coastal state (Article 77(2)). The rights of the coastal state over the continental shelf do not depend on occupation, effective or national, or on any express proclamation (Article 77(3)).

The rights of the coastal state over the continental shelf do not affect the legal status of the super-adjacent waters as high seas, or that of the airspace above the waters (Article 78).

When the relationship between the continental shelf and the EEZ is examined it becomes clear that both give the coastal state sovereign rights to explore and exploit the natural resources of the sea which is adjacent to its coast. Under the UNCLOS the continental shelf and the EEZ are legally autonomous. There is a geographical overlap within the 200 mile limit with the EEZ in respect of

52. See Art. 2 of the Geneva Convention on the Continental Shelf 1958.

non-living resources, such as oil and gas. However, despite this overlap there are also some differences between the two regimes.[53] But within the 200 mile limit they operate with respect to the same area and similar non-living resources.

(vii) The high seas

Definition

The term "high seas" was defined in Article 1 of the Geneva Convention on the High Seas 1958 as "all parts of the sea that are not included in the territorial sea or in the internal waters of a state". In relation to the developments in the law of the sea this definition needs modifying. The definition in Article 86 of UNCLOS covers "all parts of the sea that are not included in the exclusive economic zone, in the territorial sea or in the internal waters of a State, or in the archipelagic waters of an archipelagic State".

The general principle of the high seas is that they are *res communis*. All states, whether coastal or not, have the right to exercise high-seas freedoms.[54] The freedom of the high seas is exercised under the conditions laid down by UNCLOS and other rules of international law. The freedoms are quite wide, such as the freedom of navigation and overflight, the freedom to fish, the freedom to lay submarine pipelines, the freedom to conduct scientific research and the freedom to construct artificial islands (Article 87). All these freedoms must be exercised with due regard to the rights of other states and to the special regime established for the deep-sea bed. The high seas may not be subject to the sovereignty of any state.[55]

Jurisdiction on the high seas

In public international law, the organising principle on the national territory is that of exclusive jurisdiction of the sovereign state. Each state has full jurisdiction to legislate and enforce its legislation on its own territory. In exercising this jurisdiction a state shall respect the sovereignty of other states and shall observe the rules of international law. In public international law, according to the principle of territoriality, the scope of legislative and enforcement jurisdiction is restricted to the territory of the state. In public international maritime law the scope of jurisdiction is not restricted to a state's territory. As a general rule a ship on the high seas is subject only to international law and to the laws of the flag state. The flag state will enforce the rules and regulations of its own municipal law and international law. The flag state's jurisdiction is based on the assumption that the

53. For the differences between the continental shelf and the EEZ see Martin Dixon, pp.203–204.
54. UNCLOS 1982, Art. 90 says:
"**Right of navigation**
Every State, whether coastal or land-locked, has the right to sail ships flying its flag on the high seas."
55. UNCLOS 1982, Art. 89 states:
"**Invalidity of claims of sovereignty over the high seas**
No State may validly purport to subject any part of the high seas to its sovereignty."

ship is a floating part of the flag state's territory. The flag state's jurisdiction follows the ship wherever it goes and it is certainly not restricted to its territory. Therefore it becomes important to know which state is the flag state. As a general rule only the flag state may exercise jurisdiction over a ship on the high seas.[56] It is an accepted principle of international law that the state whose flag the vessel flies has jurisdiction over that vessel. Ships may sail under the flag of one state only and, apart from exceptional cases provided for in international treaties or in the Convention, they are subject to the exclusive jurisdiction of that state on the high seas.[57] Article 94 sets out the duties of the flag state with respect to its ships on the high seas. According to Article 94(1) "every state shall effectively exercise its jurisdiction and control in administrative, technical and social matters" over its ships.

Warships and ships owned or operated by a state and used only on government non-commercial service on the high seas have complete immunity from the jurisdiction of any state other than the flag state.[58]

Piracy apart, the ordinary rules of international law concerning criminal jurisdiction apply to crimes committed on the high seas. For this purpose a ship is treated as if it were the territory of the flag state. This principle was criticised in *R.* v. *Gordon-Finlayson, ex p. An Officer*,[59] where it was stated that a ship is not a part of the territory of the flag state but that jurisdiction is exercisable over the ship by that state in the same way as over its own territory. The concept of the national territoriality of vessels on the high seas continued to be adhered to. But controversies have arisen in connection with criminal liability for collisions at sea. In *The Lotus* a French ship, the *Lotus*, collided with a Turkish ship on the high seas. People on the Turkish ship were drowned. The *Lotus* reached a Turkish port and there, Lieutenant Demons, who had been steering the *Lotus* at the time of the collision, was arrested and prosecuted for manslaughter. France complained that Turkey's jurisdiction was contrary to international law. But the Permanent Court of International Justice held that[60] the collision had taken place on the Turkish ship

56. There are certain exceptions to the exclusivity of flag state jurisdiction.
57. UNCLOS 1982, Art. 92(1) states:
"**Status of ships**
1. Ships sail under the flag of one State only and, save in exceptional cases expressly provided for in international treaties or in this Convention, shall be subject to its exclusive jurisdiction on the high seas. A ship may not change its flag during a voyage or while in port of call, save in the case of a real transfer of ownership or change of registry."
58. UNCLOS 1982, Art. 95 says:
"**Immunity of warships on the high seas**
Warships on the high seas have complete immunity from the jurisdiction of any State other than the flag State."
Art. 96 states:
"**Immunity of ships used only on government non-commercial service**
Ships owned or operated by a State and used only on government non-commercial service shall, on the high seas, have complete immunity from the jurisdiction of any State other than the flag State."
59. [1941] 1 K.B. 171, 110 L.J.K.B. 289, 57 T.L.R. 270, 85 Sol.Jo. 214. See I.A. Shearer, p.246.
60. The Court laid down the following rule:
"Jurisdiction cannot be exercised by a State outside its territory except by virtue of a permissive rule derived from international custom or convention. It does not, however, follow that international

and therefore the officer responsible could be prosecuted by Turkey as well as by France. This decision was based on an objective territorial principle.

The *Lotus* decision met with severe criticism. The rule which it adopted was reversed by the 1952 Brussels Convention for the Unification of Certain Rules relating to Penal Jurisdiction.[61] The principle set out in Article 1 of the Brussels Convention received approval in Article 11 of the Geneva Convention on the High Seas 1958 and Article 97 of UNCLOS 1982, which states:

> "1. In the event of a collision or any other incident of navigation concerning a ship on the high seas, involving the penal or disciplinary responsibility of the master or of any other person in the service of the ship, no penal or disciplinary proceedings may be instituted against such person except before the judicial or administrative authorities either of the flag State or of the State of which such person is a national.
>
> 2. In disciplinary matters, the State which has issued a master's certificate or a certificate of competence or licence shall alone be competent, after due legal process, to pronounce the withdrawal of such certificates, even if the holder is not a national of the State which issued them.
>
> 3. No arrest or detention of the ship, even as a measure of investigation, shall be ordered by any authorities other than those of the flag State."

Both the 1958 Geneva Convention on the High Seas and UNCLOS 1982 reserve penal or disciplinary proceedings in cases of collision or other navigational incidents to the authorities of the state in whose ship the defendant served or, if different, the state of which he is a national. Under these provisions a state retains jurisdiction over its nationals wherever they may be.

Under Article 98 of UNCLOS states are under a duty to require masters of ships flying their flags to render assistance in cases of danger, distress and collisions.

(viii) Miscellaneous matters

International straits

The rights of passage through straits and whether the waters of the strait are high seas or territorial sea primarily depend on the 1958 Geneva Conventions. If the navigable channel in an international strait is comprised of high seas then foreign ships enjoy the freedom of navigation. While they are passing through the strait they will be free from coastal jurisdiction or control. But if the strait is comprised of the territorial waters of one or more states is there a right of navigation and may it

---

law prohibits a State from exercising jurisdiction in its own territory, in respect of any case which relates to acts which have taken place abroad and in which it cannot rely on some permissive rule of international law": Permanent Court of International Justice (P.C.I.J.) Ser. A No.10 (1927).

61. Art. 1 of the Convention states: "In the event of a collision or any other incident of navigation concerning a seagoing ship and involving the penal or disciplinary responsibility of the Master or of any other person in the service of the ship, criminal or disciplinary proceedings may be instituted only before the judicial or administrative authorities of the State of which the ship was flying the flag at the time of the collision or other incident of navigation."

be restricted by the coastal state? Foreign ships enjoy only the right of innocent passage. This right of passage can be temporarily suspended by the coastal state if it is necessary for security. But in straits used for international navigation between one part of the high seas and another innocent passage cannot be suspended. The rule adopted in the 1958 Territorial Sea Convention reflected the decision in the *Corfu Channel case.*[62] But Article 16(4) included not only straits connecting two areas of the high seas but also those connecting the high seas with the territorial sea of another state.[63]

After 1958 the coastal states' jurisdiction was extended, and the extension of the territorial sea and the emergence of the EEZ in particular provoked the need to re-examine the right of ships to pass through straits used for international navigation. Consequently, under UNCLOS 1982 a new right called "transit passage" was defined. The right of transit passage exists through straits used for international navigation between one part of the high seas or an EEZ and another part of the high seas or an EEZ. It applies to those straits which pass through the territorial sea and does not apply to straits where the waters are actually high seas or EEZ. The right also does not apply to territorial sea straits formed by a coast and an island where a suitable navigable route exists around the island in the high seas or EEZ (UNCLOS Articles 36 and 38). Another exception concerns long-standing International Conventions. Transit passage will not apply to those straits which are already governed by long-standing International Conventions in force such as the Montreux Convention 1936 governing the Bosphorus and Dardenelles.

During transit passage ships must comply with generally accepted international regulations, procedures and practices for safety at sea and for the prevention of pollution from ships.[64] Thus under Article 39(2), Conventions such as SOLAS and the IMO pollution Conventions apply to ships during transit passage even if their flag states are not parties to those Conventions. Obviously the duty to comply with international safety and pollution standards is independent of coastal legislation. But if they are implemented in coastal legislation they will become directly enforceable by the coastal state authorities.

Under the condition of being adopted by the competent international organisation, the IMO, the coastal state may prescribe sea lanes and traffic separation schemes in the strait (UNCLOS, Article 41). Under Article 42(1) a coastal state may legislate for passing vessels only in respect of fishing and the taking on board or

62. The *Corfu Channel case* established that, as a matter of customary law, warships (and, *a fortiori*, merchant ships) had a right of innocent passage through international straits which could not be suspended by the coastal state: [1949] I.C.J. Rep. 4.

63. Geneva Convention on the Territorial Sea and the Contiguous Zone 1958, Art. 16(4) states: "There shall be no suspension of the innocent passage of foreign ships through straits which are used for international navigation between one part of the high seas and another part of the high seas or the territorial sea of a foreign State."

64. UNCLOS 1982, Art. 39(2) states:

"**Duties of ships and aircraft during transit passage**

2. Ships in transit passage shall:

(a) comply with generally accepted international regulations, procedures and practices for safety at sea, including the International Regulations for Preventing Collisions at Sea;

putting overboard of any commodity, currency or person in violation of local customs, fiscal, immigration or sanitary regulations. Under Article 42(1) international standards are not imposed, but these laws like others must not discriminate among foreign ships or have the practical effect of impairing the right of transit passage.[65] Although the coastal states have the right to adopt laws and regulations relating to transit passage through straits it is not clear how these laws would be enforced against such ships. According to Churchill and Lowe[66] the general territorial sea rules will apply (UNCLOS, Article 34). Under these rules enforcement may only be exercised where the good order of the territorial sea or coastal state is disturbed or the flag state requests assistance. As a general rule under Article 42(4) foreign ships exercising the right of passage shall comply with all laws made within legislative competence allowed under the Convention.

Archipelagos

An archipelago means a group of islands, including parts of islands, interconnecting waters and other natural features which are so closely interrelated that they form an intrinsic geographical, economic and political entity, or which historically have been regarded as such (UNCLOS, Article 46). UNCLOS 1982 lays down a specific regime for archipelagic states, that is states constituted wholly of one or more archipelagos. Archipelagic states may use strait baselines to join the outer islands of the archipelago. These baselines then will serve as baselines from which the territorial sea is measured. The waters inside the baselines are known as "archipelagic waters" and the state has full sovereignty over them. The concept of archipelagic waters is new in international law. They are neither internal waters nor territorial sea. The archipelagic state's sovereignty is subject to a number of rights enjoyed by third states. An archipelagic state must respect rights enjoyed by third states deriving from existing agreements (UNCLOS, Article 51(1)). The right of innocent passage exists in archipelagic waters. But the archipelagic state may without, discrimination in form or in fact among foreign ships, suspend temporarily in specified areas of its archipelagic waters the innocent passage of foreign ships if such suspension is essential for its security (UNCLOS, Article 52). A right of archipelagic sea lane passage also exists through archipelagic waters in respect of routes normally used for international navigation. Archipelagic sea lane passage is not suspendable.

(b) comply with generally accepted international regulations, procedures and practices for the prevention, reduction and control of pollution from ships."

65. UNCLOS 1982, Art. 42(2) states:

"Laws and regulations of states bordering straits relating to transit passage

2. Such laws and regulations shall not discriminate in form or in fact among foreign ships or in their application have the practical effect of denying, hampering or impairing the right of transit passage as defined in this section."

66. See R.R. Churchill and A.V. Lowe, pp.90–93.

Islands

An island is defined in the Conventions as "a naturally formed area of land, surrounded by water, which is above water at high tide".[67] The territorial sea of an island is measured in accordance with the general rules on baselines (UNCLOS, Article 121(2)). This means that every island has a territorial sea. If islands are close to the coast of a state this will involve a seaward extension of the main territorial sea.

67. UNCLOS 1982, Art. 121(1) states:
"**Regime of islands**
1. An island is a naturally formed area of land, surrounded by water, which is above water at high tide".
See also the Geneva Convention on the Territorial Sea and the Contiguous Zone 1958, Art. 10(1).

CHAPTER 25

ADMIRALTY JURISDICTION

IN GENERAL

The Admiralty Court is a part of the High Court of Justice, Queen's Bench Division. It has special powers in Admiralty matters and exercises jurisdiction *in rem* and *in personam*. Its jurisdiction is statutory and governed by sections 20 to 24 of the Supreme Court Act 1981. Since 1 January 1982, the statutory framework of Admiralty jurisdiction in England and Wales has been found in the Supreme Court Act 1981 (sections 20–24) and in the County Courts Act 1959 as amended by the Supreme Court Act 1981, section 149(1), Art. 3, paragraphs 4 and 5. Along with most of the Act, sections 20–24 came into effect on 1 January 1982.

As will be seen below the relevant provisions of the Supreme Court Act encompasses statutory development since 1956. Before the Supreme Court Act the Administration of Justice Act 1956[1] (the 1956 Act) was the statute which governed the Admiralty jurisdiction of the High Court. The 1956 Act was founded upon two international conventions, namely the International Convention Relating to the Arrest of Seagoing Ships and the International Convention on Certain Rules concerning Civil Jurisdiction in Matters of Collision. Both Conventions were signed at Brussels on 10 May 1952. The principal purpose behind the passing of the 1956 Act was to give effect to the international legal obligations incurred by the United Kingdom government under the terms of these Conventions. The 1956 Act regulated the invocation of the court's jurisdiction *in rem* in relation to the vast majority of maritime claims within the cognizance of the court. The Act was also the source of the so-called sister ship action *in rem* which will be discussed later.

The Supreme Court Act 1981 broadly made two important changes in Admiralty jurisdiction. One of them was to make ships under demise charter liable to actions *in rem*. The other was to put an end to arguments that the sister ship remedy was limited to ships in the same ownership.

1. The other principal statutes were the Admiralty Court Act 1861 and the Judicature (Consolidation) Act 1925.

THE FRAMEWORK OF THE SUPREME COURT ACT 1981

The geographical jurisdiction of the Admiralty Court extends over the world-wide maritime environment. Generally the court has jurisdiction over ships and aircraft whatever the nationality, domicile or residence of their owners. It also entertains claims wherever they arise. But the jurisdiction of the Admiralty Court is limited by subject-matter. By section 24(1) of the Supreme Court Act 1981 "ship" includes every description of vessel used in navigation and a reference to a ship includes a reference to a hovercraft. The list of claims within Admiralty jurisdiction is set out in section 20 of the Supreme Court Act 1981.

Section 20(1)(a) of the Supreme Court Act[2]

"*(a) Any claim to the possession or ownership of a ship or to the ownership of any share therein*"

This head of jurisdiction applies only to ships. It does not cover aircraft. Until 1840 the High Court of Admiralty had no jurisdiction over questions of ownership.[3] Today it has jurisdiction over all problems relating to the ownership of a ship whatever its nationality, or the domicile or residence of the owner. As stated in *The Bineta*[4] in an ownership action the court has power to grant a declaration that the plaintiff is the legal owner of a ship and is entitled to be registered as the legal owner. It was also stated in *The Rose*[5] that the court has the power to rectify the register of a British ship where some incorrect entry has been made. If a person is unable to obtain registration of a ship, as he does not have a bill of sale from the last registered owner he will benefit from such a statement of the High Court of Admiralty.[6]

"*(b) Any question arising between the co-owners of a ship as to possession, employment or earnings of that ship*

(c) Any claim in respect of a mortgage of or a charge on a ship or any share therein

(d) Any claim for damage received by a ship"

Under the 1956 Act an action *in rem* lay on the basis of a claim for "damage received by a ship". The 1981 Act follows the view expressed in *The Jade*[7] and limits this ground of claim to the foundation for an action *in personam*. In this case Lord Diplock pointed out that this ground of claim does not appear in the 1952 Arrest Convention as a maritime claim for which a ship may be arrested. He also

2. Under this Part the subject-matter over which the Admiralty Court has jurisdiction has been explained in detail with regard to collision and oil pollution. For detailed information about Admiralty Court jurisdiction see Nigel Meeson, *Admiralty Jurisdiction and Practice* (1993) and D.C. Jackson, *Enforcement of Maritime Claims* (2nd ed., 1996).

3. Under the Admiralty Court Act 1840 jurisdiction was conferred by s. 4 "to decide all questions as to the title to or ownership of any ship or vessel, or the proceeds thereof remaining in the registry, arising from any cause of possession".

4. [1966] 2 Lloyd's Rep. 419.

5. (1873) L.R. 4 A. & E. 6.

6. See Nigel Meeson, pp.22–24.

7. *The Jade* [1976] 2 Lloyd's Rep. 1.

added that in the draft of the 1956 Act an action *in rem* may only be permitted in connection with the ship in respect of which the claim arose. This means that where the claim was based on damage received by a ship a claimant may arrest only his own ship. This view has been changed by the Supreme Court Act 1981 which provides that such a ground can be the basis only of a claim *in personam*. Under the 1981 Act the claim must be in connection with a ship and based on damage received by a ship and the ship is that of claimants.

This head of jurisdiction covers damage done to a ship by something other than a ship, like a pierhead or a buoy. It was said in *Hindustan Steam Shipping Co. Ltd.* v. *Siemens Bros. & Co. Ltd.*[8] that a products liability claim against a manufacturer whose defective equipment has caused damage to a ship is covered under this head of jurisdiction. In cases where damage was done to a ship in collision with another ship it is not possible to bring an action against the cargo laden on board the ship doing wrong even if it belongs to the owner of that ship.[9]

As the paragraph is confined to damage done to a ship damage done to any vessel which is not a ship is not within the scope of the paragraph. Similarly damage done to maritime property such as cargo, passengers' baggage or structures such as docks, harbours and oil rigs or personal injury suffered by individuals is not within the scope of the paragraph.

"*(e) Any claim for damage done by a ship*"

Lord Diplock in *The Eschersheim* made the principal statement about the phrase "damage done by a ship". He said the ship must be the actual instrument by which the damage was done. In the case of a collision physical contact between the ship and whatever object sustains damage is not essential. Even if a ship negligently causes a wash and damages some other vessel or some property on shore the damage will be covered by Admiralty Court jurisdiction.[10] If damage is done by the wrongful act of the master and crew other than in navigating the ship the damage is not done by the ship. As stated in *The Dagmara and Ama Antixine* the damage done need not be physical. If a vessel deliberately prevents another vessel from fishing and causes financial loss the claim will be covered under this paragraph as the plaintiffs have a cause of action in tort for such financial damage.[11] The following kinds of damage are examined under this head of jurisdiction[12]: damage by

8. [1955] 1 Lloyd's Rep. 167.
9. See Nigel Meeson, p.28.
10. In *The Eschersheim* Lord Diplock said that "to fall within the phrase (*damage done by a ship*) not only must the damage be the direct result or natural consequence of something done by those engaged in the navigation of the ship, but the ship itself must be the actual instrument by which the damage was done. The commonest case is that of collision, . . . but physical contact between the ship and whatever object sustains the damage is not essential—a ship may negligently cause a wash by which some other vessel or some property on shore is damaged": [1976] 1 W.L.R 430 (HL), emphasis added.
11. [1988] 1 Lloyd's Rep.431. See Nigel Meeson, pp.28–29.
12. Nigel Meeson, *ibid.*

collision due to faulty steering gear,[13] damage to a telegraph cable by the mate cutting it,[14] damage by collision with an anchor,[15] damage by a falling derrick,[16] damage to cargo caused by leaving a seacock open,[17] damage done by a wreck,[18] damage caused by a ship's wash.[19] The damage done relates not only to damage or loss done directly but also to that which is consequential.

In cases of collision there may be a substantial overlap between paragraphs (d) and (e). If one ship alone is at fault "damage would be done" by the delinquent ship and "damage received" by the innocent ship. In cases where both ships are negligent each ship would be both the doer and receiver of damage. According to Thomas[20] notwithstanding the overlap, in cases of collision damage jurisdiction primarily arises under paragraph (e), damage done by a ship.

Oil pollution

Subsection 2(e) extends to any claim for damage done by a ship and covers claims in respect of liability incurred under the Merchant Shipping (Oil Pollution) Act 1971 and any claim in respect of liability falling within the International Oil Pollution Compensation Fund under Part I of the Merchant Shipping Act 1974.[21]

"*(f) Any claim for loss of life or personal injury sustained in consequence of any defect in a ship, her apparel or equipment, or in consequence of the wrongful act, neglect or default of*

(i) The owners, charterers or persons in possession or control of a ship; or

(ii) The master or crew of a ship, or any other person for whose wrongful acts, neglects or defaults the owners, charterers or persons in possession or control of a ship are responsible;

being an act, neglect or default in the navigation or management of the ship, in the loading, carriage or discharge of goods on, in or from a ship, or in the embarkation, carriage or disembarkation of persons, in or from the ship."

Under this paragraph the loss of life or personal injury should be linked to a defect in a ship, neglect or default in navigation, the management of the ship, loading, carriage or discharge of goods, embarkation, carriage or disembarkation of passengers. An action may be brought in respect of personal injury or loss of life

13. *The Warkworth* (1884) 9 P.D. 145.
14. *The Clara Killam* (1870) L.R. 3 A. & E. 161.
15. *Re St Margetts and Ocean Accident and Guarantee Corporation* [1901] 2 K.B. 792.
16. *The Minerva* [1933] P. 224.
17. *Good* v. *London Steamship Owners Association* (1871) L.R. 6 C.P. 593.
18. *The Zelo* [1922] P.9.
19. *The Batavier* (1854) 9 Moo. P.C. 286, *Luxford* v. *Large* (1832) 5 C. & P. 421, *The Royal Eagle* (1950) 84 Ll.L.R. 543, *The Royal Sovereign* (1950) 84 Ll.L.R. 549.
20. D.R. Thomas, *Maritime Liens, British Shipping Laws*, Vol. 14 (1980), pp.107–108.
21. Section 20(5) is amended by the Merchant Shipping Act 1995, Sch. 13, para. 59. Supreme Court Act 1981, s.20, Admiralty Jurisdiction of the High Court (as amended) states:

"(5) Subsection (2)(e) extends to—

(a) any claim in respect of a liability incurred Chapter III of Part IV of the Merchant Shipping Act 1995;

(b) any claim in respect of a liability falling on the International Oil Pollution Compensation Fund or on the International Oil Pollution Compensation Fund 1984 under Chapter VI of Part VI of the Merchant Shipping Act 1995."

suffered by a person on board the defective ship, but also in respect of personal injury or loss of life suffered by a person on board another ship or ashore. The claims are linked to the individual liability and responsibility of the owner, charterer or person in possession or control or any person for whose acts those persons are responsible or the master or crew.

"*(g) Any claim for loss of or damage to goods carried in a ship*"

The claim under this paragraph clearly covers a claim based on physical damage. For the purposes of the paragraph a ship is construed as meaning the ship on which the goods are carried. The question under this paragraph is whether loss of or damage to goods is restricted only to such loss or claim or includes any claim based on such loss or damage. As the paragraph does not include any restriction it is suggested that consequential loss should be recoverable. So damages for delay or loss of profit may be recoverable. But when the cause and consequence are examined indemnity for pollution liability may not be covered because indemnity liability, it may be argued, is hardly the result of damage to goods unless caused by the goods.[22]

"*(h) Any claim arising out of any agreement relating to the carriage of goods in a ship or to the use or hire of a ship*

(j) Any claim (i) under the Salvage Convention 1989; (ii) under any contract for or in relation to salvage services; or (iii) in the nature of salvage not falling within (i) or (ii) above.

(k) Any claim in the nature of towage in respect of a ship or an aircraft

(l) Any claim in the nature of pilotage in respect of a ship or an aircraft

(m) Any claim in respect of goods or materials supplied to a ship for her operation or maintenance

(n) Any claim in respect of the construction, repair or equipment of a ship or dock charges or dues

(o) Any claim by a master or member of the crew of a ship for wages (including any sum allotted out of wages or adjudged by a superintendent to be due by way of wages)

(p) Any claim by a master, shipper, charterer or agent in respect of disbursements made on account of a ship

(q) Any claim arising out of an act which is or is claimed to be a general average act

(r) Any claim arising out of bottomry

(s) Any claim for the forfeiture or condemnation of a ship or of goods which are being or have been carried or have attempted to be carried, in a ship, or for the restoration of a ship or any such goods after seizure, or for droits of admiralty".

22. See D.C. Jackson, p.58. According to Thomas the terms do not extend to including a claim for indemnity. D.R. Thomas, pp.109–110.

EXERCISE OF JURISDICTION

(a) In general

Claims falling within Admiralty jurisdiction may be exercised *in personam* or *in rem*. All cases within the jurisdiction of the Admiralty Court may be brought by an Admiralty action *in personam*. But an action *in rem* may be pursued in certain cases and in certain circumstances. An action *in rem* is an action against a thing, or more precisely relates to interests of a person in the thing.[23] If the defendant does not appear in an action *in rem* judgment will only be enforceable against the *res*. The defendant will not be personally liable. But once the defendant submits to the jurisdiction of the court personally, the action continues as an action *in rem* and *in personam*. In *The Gemma*[24] the Court of Appeal held that, by appearing, the owners had rendered themselves personally liable and that payment of the unsatisfied balance could accordingly be enforced by a writ of *fieri facias*.[25] In *The Tatry*[26] the Court of Appeal held that after acknowledgment of service in an Admiralty action *in rem* the action does not lose its *in rem* character, but proceeds as hybrid, being both *in rem* and *in personam* even though the *res* may have been released by the court. In order to respond to arguments by counsel that actions *in rem* and actions *in personam* are in the end merely different ways of proceeding against a defendant and that an action *in rem* in no way differs in its nature from an action *in personam*, Fletcher Moulton L.J. made the following statement and explained the nature of an action *in rem* in English law:

"... action *in rem* is an action against the ship itself. It is an action in which the owners may take part, if they think proper, in defence of their property, but whether or not they will do so is a matter for them to decide, and if they do not decide to make themselves parties to the suit in order to defend their property, no personal liability can be established against them in that action. It is perfectly true that the action indirectly affects them. So it would if it were an action against a person whom they had indemnified."[27]

If the judgment is not fully satisfied by enforcement against the *res* execution proceedings may be initiated against the defendant personally. In other words, as long as the proceeds of sale are insufficient to cover damages a party may bring a subsequent action *in personam* in respect of the same claim. Hence, judgment in an action *in rem* is not a bar to a subsequent action *in personam*. Proceeding *in rem* has

23. In *The Henrich Bjorn* Lord Watson said that "The action *in rem*, that being, as I understand the term, a proceeding directed against a ship or other chattel in which the plaintiff seeks either to have the *res* adjudged to him in property or possession, or to have it sold, under the authority of the Court, and the proceeds, or part thereof, adjudged to him in satisfaction of his pecuniary claims. The remedy is obviously an appropriate one in the case of a plaintiff who has a right of property or other real interest in the ship, or a claim of debt secured by a lien which the law recognizes": (1866) L.R. 11 P.C. 270.

24. [1899] P. 285.

25. "... that the persons so appearing as the defendants have done in the present case, become parties to the action, and thereby become personally liable to pay whatever in the result may be decreed against them; and the action, though originally commenced *in rem*, becomes a personal action against the defendants upon appearance." *Ibid.*, p. 291.

26. [1992] 2 Lloyd's Rep. 552 (CA).

27. *The Burns* [1907] P.137.

some advantages. By proceeding against the *res* a plaintiff obtains security for his claim. It is possible to enforce the judgment against the *res* by arresting the ship and selling it. As long as the property proceeded against is within the jurisdiction it is possible to bring an action *in rem* against that property. But the same claim could not be brought *in personam* against foreign owners of that property unless the conditions were fulfilled to obtain a leave to serve a writ *in personam* out of the jurisdiction. It is not possible to bring an action *in rem* for every claim within the Admiralty jurisdiction. Action *in rem* is not available for limitation actions, applications under the Merchant Shipping Acts and any claim for damage received by a ship.

(b) Availability of the action in rem

The action *in rem* lies only for claims within Admiralty jurisdiction. Under the 1956 Act claims listed in the Act, as within Admiralty jurisdiction, could all be enforced by an action *in personam* or *in rem*. As to the action *in rem*, claims were divided into three categories. These were (a) maritime liens; (b) claims such as ownership, including co-ownership, mortgage and forfeiture; and (c) all claims save those in (b) not qualifying for or not enforced through a maritime lien. Apart from a change in relation to damage sustained by a ship the Supreme Court Act 1981 adopts identical categories. In relation to (a) and (b) there is no substantive change, but as to (c) claims in relation to the ship in respect of which a claim arises and claims in relation to "sister" ships were added.

Under the 1956 Act if an action *in rem* is brought against a ship and the beneficial owner of the ship is sued at the same time the beneficial owner must be liable *in personam*. The action *in rem* linked to the liability *in personam* of the user who was or by the time of the issue of the writ had become the owner. An action *in rem* against the ship in respect of which the claim arose would lie only if the person liable on the claim *in personam* was the beneficial owner "as respects all the shares" in the ship when the "jurisdiction was invoked". Reliance on "invoking of jurisdiction" led to problems of interpretation. In *The Monica S*[28] the invoking of Admiralty jurisdiction linked to the bringing of an action *in rem* meant the issue of a writ *in rem*. Consequently, it was held that the sale of a ship after issue of the writ did not defeat the claim. In *The Berny*[29] the argument was that if Admiralty jurisdiction was not invoked by the issue of the writ a plaintiff could issue only one writ. It was held that the jurisdiction was not invoked until the service of the writ. Under the 1981 Act the use of the phrase "invoking jurisdiction" was replaced by the "bringing of an action".

The requirement under the 1956 Act that the person liable *in personam* be the beneficial owner at the time the action was brought was clearly contrary to the 1952 Arrest Convention as far as the ability to arrest was concerned. In the Arrest Convention liability *in personam* was not mentioned and it provided that the ship

28. [1968] P. 741.
29. [1977] 2 Lloyd's Rep. 533, [1979] Q.B. 80.

might be arrested whoever was liable on the claim. There was no reference to ownership or any other link at the time the action was brought.

In order to bring an action *in rem* under the 1981 Act two conditions have to be fulfilled. First, the claim must arise in connection with a ship. Secondly, the person who would be liable on the claim in an action *in personam* was, when the cause of action arose, the owner or charterer or in possession or control of the ship, and when the action was brought was either the beneficial owner of that ship as respects all the shares in it or the charterer of it under a charter by demise. The phrase "when the cause of action arose" refers to the happening of the event on which the claim is based. In relation to "when the action is brought" the person who would be liable *in personam* must be either the beneficial owner as respects all the shares in the ship or the charterer by demise when the action is brought. Under the 1981 Act the time when the action is brought is related to the issue of the writ *in rem*. Therefore any change in ownership after the issue of the writ is not relevant to the availability of the action *in rem*. In practice, the writ must be issued as soon as possible.

(i) The sister ship issue[30]

When an Admiralty claim arises in connection with a ship, it is often possible to enforce that claim by bringing an action *in rem* against that ship. In most cases, instead of bringing an action against the ship in connection with which the claim arises, it is also possible to bring an action against another ship owned by the person liable for the claim.

There is no problem if both ships are wholly owned by the person liable. It is for the claimant to choose which ship to arrest. But if the person liable is not the owner of the ship (ship X) in connection with which the claim arises but owner of another ship (ship Y), is it possible to bring an action against ship Y? There are different views on the issue. One is quite restrictive. Under this view, for either ship to be arrested both must be wholly owned by the person liable. Hence, it is not possible to bring an action against ship Y. According to the opposite view any ship owned by the person liable may be arrested. The only requirement is that at the time of the casualty the person liable was the owner or charterer or in possession or control of the ship X. Therefore it is possible to bring an action against Y. Under the middle view there is no need for ownership, but at the time of the casualty the person must have been either the owner or demise charterer of ship X.

Until the 1956 Act there was no question about the ability to arrest any ship other than the offending ship. The possibility of arresting another vessel arises out of the enactment of section 3(4) which reads:

30. For detailed information on sister-ship action *in rem* see, Rhidian Thomas, "The Sister Ship Action *in rem*" [1979] L.M.C.L.Q. 158–168; Steven Hazelwood, "Gaps in the Action *in rem*" [1982] L.M.C.L.Q. 210–217, 422–426; A.M. Tettenborn, "The Time Charterer, One Ship Company and the Sister Ship Action *in rem*" [1981] L.M.C.L.Q. 507–509; S.J. Tabbush, "Arrest of Ships Owned by Charterers" [1982] L.M.C.L.Q. 585–591; D.C. Jackson, "Admiralty Jurisdiction—The Supreme Court Act 1981" [1982] L.M.C.L.Q. 236–248.

"In the case of any such claim as is mentioned in paragraphs (d) to (r) of subsection (1) of section one of this Act, being a claim arising in connection with a ship, where the person who would be liable on the claim in an action *in personam* was, when the cause of action arose, the owner or charterer of, or in possession or in control of the ship, the Admiralty jurisdiction of the High Court may be invoked by an action *in rem* against:

(a) that ship, if at the time when the action is brought it is beneficially owned as respects all shares therein by that person, or
(b) any other ship which, at the time when the action is brought, is beneficially owned as aforesaid."

Under the 1956 Act it was clear that an action could be brought against a ship other than that in respect of which the claim arose if owned by the owner of the ship in respect of which the claim arose. By its wording the Act created uncertainty in two areas. First, it was not clear whether an action could be brought against ships owned by the charterers liable in respect of a claim. Secondly, if such a remedy was available against ships owned by demise charterers it was not clear whether it extended to ships owned by time and voyage charterers. The interpretation of section 3(4) has presented a dilemma. The more liberal construction is the above-mentioned wider view: the operation of the two paragraphs of section 3(4) independently. Any other ship beneficially owned by the wrongdoer may be arrested, notwithstanding that the wrongdoer was not the owner of the particular ship. The restrictive interpretation is that any other ship in paragraph (b) may be arrested only if it is in the same ownership as the offending ship. The two ships must be sister ships.

The Court of Appeal did not give much guidance on reaching any conclusion on this issue, considered by the High Court of Hong Kong in *The Ledesco Uno*.[31] In this case the court held that a ship beneficially owned by charterers could not be arrested in respect of a claim which had arisen in respect of a vessel on a voyage charter to them because the vessels were not sister ships. In this case it appears to be suggested that the term "charterer" is to be confined to meaning demise charterer. However, there was uncertainty in this case about the exact context in which the term "charterer" was being discussed. But in *The Permina 108* the Court of Appeal of Singapore rejected the conclusion reached in *The Ledesco Uno* and held that there was no requirement for the ships to be sister ships. In this case the court took into account the terms of the 1952 Arrest Convention where it limits the effect of the sister ship provision to vessels owned by owners or by demise charterers. In *The Permina 108* the court refused to refer to the language of the 1952 Arrest Convention. The *Permina 108* decision enabled a plaintiff to invoke Admiralty jurisdiction against any other ship, if at the time when that action was brought the other ship was beneficially owned by the person liable on a claim *in personam*, who was, when the cause of action arose, the owner or charterer or in possession or control of the ship in connection with which the claim arose.[32]

31. *Antaios Compania Naviera S.A.* v. *Ledesma Overseas Shipping Corporation* [1978] 2 Lloyd's Rep. 99.
32. [1978] 1 Lloyd's Rep. 311.

In the UK the issue was considered in *The Eschersheim*.[33] In the House of Lords Lord Diplock said: "It is clear that to be liable to arrest, a ship must not only be the property of the defendant to the action, but must also be identifiable as the ship in connection with which the claim made in the action arose, or a sister ship of that ship."

In *The Maritime Trader*[34] a company called MTS owned the ship. The company MTO owned all the shares in MTS. Consequently, MTS was a wholly owned subsidiary of MTO. In 1976 MTS took delivery direct from the builders of the vessel *Maritime Trader*. In 1980 MTO chartered the vessel *Antaios* from the plaintiffs on a demise charter. The charterers allegedly failed to pay the hire. The plaintiffs sought to arrest the *Maritime Trader* in an English port. The plaintiffs alleged that they had the right to arrest under section 3(4). The charterparty provided for disputes to be referred to arbitration in London according to English law. The plaintiffs based their claims on the provisions of section 1(1)(h) of the 1956 Act[35] and issued a writ *in rem* against the *Maritime Trader* pursuant to the provisions of section 3(4) of the 1956 Act. In seeking to set aside the writ MTO contended that under section 3(4) the "arrestable ship", the *Maritime Trader*, should be in the same beneficial ownership as the "offending ship", the *Antaios*. In other words the vessels should be sister ships. That was not so in this case. Sheen J. followed the statement of Lord Diplock in *The Eschersheim* and held that the *Maritime Trader* was not a ship against which the jurisdiction of the court could be invoked *in rem* because she was not a sister ship of the *Antaios*. He also ruled out the possibility of arresting a ship owned by a demise charterer on the ground that Article 3(4) of the Arrest Convention[36] had not been implemented by Parliament.

Obviously the restrictive approach in *The Maritime Trader* was in line with the way in which the 1952 Arrest Convention has been interpreted in the UK. The other signatories to the 1952 Arrest Convention have not followed the restrictive construction and do not require the vessels to be in the same ownership. The decision in this case has been criticised and regarded as an unfortunate conclusion. The restrictive approach allows shipowners to charter extra tonnage and commit

33. [1976] 2 Lloyd's Rep. 1.
34. [1981] 2 Lloyd's Rep. 153.
35. The Administration of Justice Act 1956, s.1 states:
"The Admiralty jurisdiction of the High Court shall be as follows, that is to say, jurisdiction to hear and determine any of the following questions or claims . . .
(h) any claim arising out of any agreement relating to the carriage of goods in a ship or to the use or hire of a ship."
See Supreme Court Act 1981, s.20(2)(h).
36. International Convention For the Unification of Certain Rules Relating to the Arrest of Sea-Going Ships 1952, Art. 3(4) says
"When in the case of a charter by demise of a ship the charterer and not the registered owner is liable in respect of a maritime claim relating to that ship, the claimant may arrest such ship or any other ship in the same ownership of the charterer by demise, subject to the provisions of this Convention, but no other ship in the ownership of the registered owner shall be liable to arrest in respect of such maritime claims.
The provisions of this paragraph shall apply to any case in which a person other than the registered owner of a ship is liable in respect of a maritime claim relating to that ship."

the type of acts within section 1(1), paragraphs (d)–(r) of the 1956 Act as they know that their own vessels will sail free of arrest.[37] It is also clear that Article 3(4) of the 1952 Arrest Convention does not require offending and arrestable vessels to be in the same ownership.

In *The Span Terza*[38] the problem of the construction of section 3(4) was again raised. The question in this case was whether an action by shipowners against time charterers of their ship would make the arrest of the ship possible. The Admiralty Registrar and Sheen J. both followed Lord Diplock's observations in *The Eschersheim* and refused to order the arrest of a vessel owned by time charterers. But the Court of Appeal, contrary to the decision in *The Maritime Trader*, allowed the appeal and held that the proper interpretation of the 1956 Act was the widest one which allowed the arrest of vessels owned by charterers generally.[39] The conclusion in *The Span Terza* was strengthened in the Supreme Court Act 1981. Under section 21(4) of the Supreme Court Act 1981 an action *in rem* will lie against ships owned by charterers of or those in possession or control of the ship in respect of which the claim arose. Charter by demise was specified in section 21(4) and as a result it became clear what particular type of charter was meant.

The Supreme Court Act 1981, section 21(4) now reads:

(i) that ship, if at the time when the action is brought the relevant person is either the beneficial owner of that ship as respects all the shares in it or the charterer of it under a charter by demise; or

(ii) any other ship of which, at the time when the action is brought, the relevant person is the beneficial owner as respects all the shares in it.

The purpose of allowing the arrest of sister ships *in rem* can only be to found jurisdiction against the person liable and give the plaintiff security for his claim. In order to serve this purpose and avoid doing injustice to third parties the only requirement is that the ship must be wholly owned by the person liable.

(ii) Beneficial ownership and the corporate veil

It is necessary for the court to identify the beneficial owner of all the shares in a ship which is sought to be proceeded against *in rem* in order to determine whether the claim may be brought against that ship. The meaning of the phrase "beneficial owner" has given rise to controversy. In 1970 in *The Andrea Ursula*[40] Brandon J. held that demise charterers were beneficial owners within the meaning of the 1956 Act. But in 1977 in *I Congreso del Partido* Robert Goff J. held that the phrase

37. See Steven J. Hazelwood, Part Two, May 1982, pp.211–214; Part Three, August 1982, pp.423–424. S.J. Tabbush, at pp.590–591, also said that Sheen J.'s decision could be justified on the ground that the vessel which it was sought to arrest was owned not by the charterer but by a subsidiary company of the charterer.

38. [1982] 1 Lloyd's Rep. 225 (CA).

39. In this case Donaldson L.J., dissenting, favoured the middle interpretation, allowing the arrest of vessels owned by an owner or demise charterer only. The decision in the *Span Terza* was accepted as the correct decision but as the appeal had to be heard *ex parte* and in a great hurry it cannot be regarded as settling the whole problem.

40. [1973] Q.B. 265.

"beneficial owner" referred only to cases of equitable ownership, whether or not accompanied by legal ownership, and it was not wide enough to include cases of possession or control without ownership, however full and complete such possession might be and however much control over the ship the owner might have.[41] In *The Father Thames* Sheen J. agreed that demise charterers did not qualify as beneficial owners and held that the phrase "beneficially owned" in the 1956 Act did not apply to a demise charterer.[42] Obviously all these cases were decided under the 1956 Act. The condition under the Act which requires that the person liable *in personam* should be the beneficial owner at the time the action was brought was clearly contrary to the Arrest Convention. The Convention provides that the ship in respect of which the claim arose may be arrested, whoever is liable, and there is no reference to ownership or any other link at the time the action was brought.

In the 1981 Act an important change was made in the wording of the relevant provisions and specific reference to a demise charterer was introduced. Under the 1981 Act if a ship sought to be arrested is the ship in connection with which the claim arose the ship may be arrested providing that the person liable *in personam* is the demise charterer. But if the issue is related to the arrest of another ship it is clear that the ship cannot be arrested unless the person liable *in personam* is its owner and not merely the demise charterer.

Another controversial issue concerning beneficial ownership is the one-ship company. In the shipping business it has long been the practice to register several ships in the names of separate companies. These companies' only assets are the particular ships registered in their names but they are financed by a common source and managed and operated as a common fleet. These one-ship companies are generally registered in countries where the identification of shareholders in companies is not a matter of public record. As stated by Meeson[43] the courts have recognised that the "one ship" company is a legitimate business arrangement. In law the beneficial owner of the ship is the company, which is a separate and distinct legal entity or person from the beneficial owner of the company. But in the absence of fraud it is not permissible to lift the corporate veil in order to look behind the "one ship" company structure to identify the beneficial owner of the company in order to say that the beneficial owner of the company is the beneficial owner of the ship, because in strict English legal theory the corporate veil provides substantive protection.

In *The Maritime Trader* the second ground upon which the defendants sought to have the writ set aside was that it had been issued in respect of a ship which was not beneficially owned by the person who would be liable in an action *in personam*. In

41. [1978] Q.B. 500. In this case Robert Goff J. also stated that "the intention of Parliament in adding the word 'beneficially' before the word 'owned' . . . was simply to take account of the institution of the trust, thus ensuring that if the ship were operating under the cloak of a trust, those interested in the ship should not thereby be able to avoid the arrest of the ship". As stated by Meeson in that case the relevant person was the operator and manager and the ship was held not to be beneficially owned by them: see Nigel Meeson, pp.76–81.

42. [1979] 2 Lloyd's Rep. 364.

43. Nigel Meeson pp.78–81.

this case the defendants claimed that at the time the action was brought the *Maritime Trader* was not beneficially owned as respects all the shares therein by the person who would be liable on the claim in an action *in personam*. Sheen J. held that although all the shares in MTS were owned by MTO, a shareholder had no property in the assets of the company. Therefore MTO had no property in the *Maritime Trader*, which formed the assets of MTS. It could not be said that the *Maritime Trader* was beneficially owned as respects all the shares therein by MTO unless the corporate veil could be lifted. Sheen J. stated that he would not hesitate to lift that veil if the evidence suggested that it obscured from view a mask of fraud rather than the true face of the corporation. In divining the ownership of a vessel Sheen J. cited *The Aventicum*.[44] In that case there was a dispute about the beneficial ownership of a vessel and Slynn J. said[45]: "where damages are claimed by cargo owners and there is a dispute as to the beneficial ownership of the ship, the Court in all cases can and in some cases should look behind the registered ownership to determine the true beneficial ownership". He also added that[46] "it is plain that . . . the Act intends that the Court shall not be limited to a consideration of who is the registered owner or who is the person having the legal ownership of the shares in the ship; the directions are to look at the beneficial ownership. Certainly where there is a suggestion of a trusteeship or nominee holding, there is no doubt that the Court can investigate it . . .".[47]

But in *The Maritime Trader* there was no evidence that the ship had been purchased by MTS in order that it would not be available as security for a judgment against MTO. As a result Sheen J. refused to lift the corporate veil. He indicated that, in his view, it was only where the relationship between the corporate owner and the ship was clearly a sham or amounted to fraud that any other legal person could be said to be the beneficial owner. Under such an approach there was a heavy onus on the plaintiff. But in *The Saudi Prince*[48] Sheen J. held that the plaintiff had satisfied the onus and that on the facts of the case he would have been prepared to recognise that a transfer was a sham. In Hong Kong the legitimacy of the one-ship company structure was also endorsed in *The Neptune*. In that case it was held that the use of one-ship companies as a means of limiting liability did not raise an inference of fraud so as to justify the lifting of the corporate veil.[49]

44. [1978] 1 Lloyd's Rep. 184 (QB).
45. *Ibid.*, p.187, col. 1.
46. *Ibid.*, p.187, col. 2.
47. These two passages from the judgment of Slynn J. are uncontroversial. But some other passages in the judgment which appear to suggest that the court is empowered to embark upon a wide-ranging investigation of the corporate veil simply for the purpose of the investigation of beneficial ownership, are *obiter dicta* and, it is submitted, are also wrong. "At the time when *The Aventicum* was decided the divergence of view between Mr Justice Brandon in the *Andrea Ursula* and Mr Justice Robert Goff in *The I Congreso del Partido* was a live issue and had not been finally resolved. Mr Justice Slynn expressly declined to resolve it in the case before him as he considered it unnecessary on the facts of the case. Therefore many of his remarks in this case must be taken in the context of considering a possible wider meaning of beneficial ownership which cannot be applicable to cases decided under the 1981 Act": cited in Nigel Meeson, pp.77–78.
48. [1982] 2 Lloyd's Rep. 255.
49. *The Neptune* [1986] H.K.L.R. 345 (High Ct.).

When the decision in *The Maritime Trader* is examined in relation to the restrictive attitude of English law towards lifting the corporate veil, Sheen J.'s decision is probably defensible, but such an approach encourages growth of "one ship" companies. Under such a practice when a shipowner chooses to own each of its ships through a separate "one ship" company it will be immune from the sister ship jurisdiction because none of its ships will be in the same "beneficial ownership" as another of "its" ships in respect of which a claim arises.

As was the *Maritime Trader*, the *Evpo Agnic*[50] was proceeded against *in rem*. It was arrested in respect of a cargo claim arising out of the sinking of another ship, the *Skipper I*. Both ships were owned by separate Panamanian companies. The companies' officers and shareholders were the same individuals. The same Panamanian company also managed the ships. The cargo interests contended that lifting the corporate veil would establish that the two ships were in the same beneficial ownership and under such a discovery the question of beneficial ownership would be investigated. But the Court of Appeal rejected this argument and held that the right of arrest under section 21(4)(ii) did not extend to a ship owned by a sister company of the company owning the ship in connection with which the claim arose.

The *Maritime Trader* and *The Evpo Agnic* were distinct from *The Aventicum* and *The Saudi Prince*. In the latter cases the defendant shipowners alleged a change in beneficial ownership after the claim arose but before the issue of the writ. And such a change in the beneficial ownership required investigation. But in the former cases there was no change of ownership, so investigation was not needed. As a result it is for the court to choose between the restrictive and liberal approaches to what is now section 21(4) of the Supreme Court Act 1981.

(iii) The Mareva injunction

Until the beginning of the nineteenth century the Admiralty Court was possessed of three distinct ways of proceeding which were: the writ *in rem*, the writ *in personam* and a decree of attachment of the goods and chattels of the defendant, which could be added to the writ *in personam* if the defendant was not found in the jurisdiction.

The Mareva injunction was granted for the first time in 1975 to restrain defendant charterers from disposing of their assets in England outside the jurisdiction on the basis of section 45(1) of the Supreme Court of Judicature (Consolidation) Act 1925. In *Nippon Yusen Kaisha* v. *Karageorgis*[51] the charterers failed to pay the hire, which reached a substantial sum. They could not be traced but there was evidence that they had funds at a bank in London. The Court of

50. [1988] 1 W.L.R. 1090 (CA), [1989] 1 W.L.R. 127 (HL).

51. [1975] 2 Lloyd's Rep. 137, [1975] 1 W.L.R. 1093. s.45 of the Supreme Court of Judicature (Consolidation) Act 1925 stated as follows: "If it appears that the debt is due and owing—and there is a danger that the debtor may dispose of his assets so as to defeat it before judgment—the court has jurisdiction in a proper case to grant an interlocutory judgment so as to prevent him disposing of those assets."

Appeal granted an injunction restraining the charterers from disposing of or removing from the jurisdiction any of the assets. This case was followed by *Mareva Compania Naviera S.A.* v. *International Bulk Carriers S.A.*, from which the injunction took its name.[52] The test for a Mareva injunction was developed in *Rasu Maritima S.A.* v. *Pertamina*. In this case the Court of Appeal affirmed the practice. Lord Denning held that "an order restraining removal of assets can be made whenever the plaintiff can show that he has a 'good arguable case' ".[53] The Court of Appeal judgment in *Third Chandris Shipping Corporation* v. *Unimarine S.A.* was the first case in which the general principles for the grant of a Mareva injunction and how those should operate in practice were stated.[54] In addition to these principles another condition has emerged in relation to the position of London as a central forum for foreigners to litigate or arbitrate disputes. So the courts are now also asking whether foreigners who have agreed to arbitrate or litigate in London under a jurisdiction clause should be subject to a Mareva injunction.[55] In *Nippon Yusen Kaisha*[56] or in *The Mareva*[57] there had been no distinction expressly drawn between foreign and English-based defendants. But in *The Pertamina*[58] it was clear that the remedy was applicable only to foreign-based defendants. But in *Rahman (Prince Abdul) bin Turki al Sudairy* v. *Abu-Taha and another* the Court of Appeal firmly established that a Mareva injunction could be granted against a defendant even if based in England, as long as the risk of dissipation of assets existed.[59] Under this practice the provisions of section 45 of the 1925 Act were amended and the common law Mareva injunction was recognised by statute in section 37 of the Supreme Court Act 1981, which reads as follows:

"(1) The High Court may by order (whether interlocutory or final) grant an injunction or appoint a receiver in all cases in which it appears to the court to be just and convenient to do so.

(2) Any such order may be made either unconditionally or on such terms and conditions as the court thinks just.

52. [1975] 2 Lloyd's Rep. 509 (CA). *Per* Lord Denning MR: "If it appears that the debt is due and owing—and there is a danger that the debtor may dispose of his assets so as to defeat it before judgment—the court has jurisdiction in a proper case to grant an interlocutory judgment to prevent him disposing of those assets."

53. *Rasu Maritima S.A.* v. *Perusahaan Pertambangan Minyak Dan Gas Bumi Negara (Pertamina) and Govt. of Indonesia (Intervener)* [1978] 1 Q.B. 644, at p.661, [1977] 2 Lloyd's Rep. 397.

54. These principles are stated as follows: the plaintiff should make a full and frank disclosure of all matters in his knowledge which are material for the judge to know. He should give particulars of his claim against the defendant. He should give some grounds for believing that the defendant has assets here. He should give some grounds for believing that there is a risk of the assets being removed before the judgment or award is satisfied. He must give an undertaking in damages, in case he fails in his claim or the injunction turns out to be unjustified: *Third Chandris Shipping Corporation* v. *Unimarine S.A.* [1979] Q.B. 645 (CA).

55. See William Tetley, "Attachment, the Mareva Injunction and *Saisie Conservatoire*" [1985] L.M.C.L.Q. 58–80.

56. See n.51 *supra*.

57. See n.52 *supra*.

58. [1978] Q.B. 644 (CA).

59. [1980] 1 W.L.R. 1268 (CA).

(3) The power of the High Court under subsection (1) to grant an interlocutory injunction restraining a party to any proceedings from removing from the jurisdiction of the High Court, or otherwise dealing with, assets located within that jurisdiction shall be exercisable in cases where that party is, as well as in cases where he is not, domiciled resident or present within that jurisdiction."

Apart from the physical presence of a defendant in England in some cases he may be within the extended or assumed jurisdiction of the courts with regard to rules for issue and service of a writ overseas. Under R.S.C. Order 11 the court has discretion to give leave for the writ first to be issued and then served. These are distinct procedures and the first can be involved without the second. Before the Civil Jurisdiction and Judgments Act 1982, Order 11 rule 1(1)(b) did not operate to give jurisdiction over a main action merely because a Mareva injunction was sought.[60] But under section 25 of the Civil Jurisdiction and Judgments Act 1982 an English court is allowed to grant "interim relief" such as a Mareva injunction in respect of proceedings which have been or are to be commenced in another Contracting State to the Brussels Convention on Jurisdiction and the Enforcement of Judgments in Civil and Commercial Matters or in another part of the United Kingdom.

Mareva injunctions are applicable to ships even though this is an area where the Admiralty jurisdiction is paramount. In cases which are not covered under the heads of a claim in section 20 of the Supreme Court Act a Mareva injunction can be applied for. In England the arrest of a ship under the Supreme Court Act 1981 perfects the security of a plaintiff's claim and the ensuing maritime lien makes the plaintiff a secured creditor, subject only to the claims of other secured creditors. Similarly the Mareva injunction induces a defendant to put up security in the same way as the threat of arrest does in Admiralty actions. But the Mareva does not provide any security in the property frozen. Therefore although it puts practical pressure on a defendant it is less effective than the action *in rem*. It operates against the assets of the defendant but does not create any rights *in rem*. Whether or not covered by the Mareva injunction a plaintiff with a judgment against the defendant can enforce it against his assets provided that it is subject to the rights of other people in the property.

In some cases proceedings are possible in more than one jurisdiction including England. A defendant may claim that there is a foreign jurisdiction clause or the principles of *lis alibi pendens* apply, so the foreign jurisdiction is more convenient. If the order staying the English proceedings is expressed in terms which do not preclude the grant of a Mareva injunction and if jurisdiction itself is not challenged successfully there will be no inconsistency between a Mareva injunction and the foreign litigation. If the English courts decide that a foreign court has exclusive jurisdiction apart from the 1968 Convention there can be no Mareva.

60. See *Siskina and Others* v. *Distos Compania Naviera S.A. (The Siskina)* [1979] A.C. 210, [1978] 1 Lloyd's Rep.1.

CHAPTER 26

RESTRICTIONS ON THE EXERCISE OF JURISDICTION

JURISDICTIONAL AGREEMENTS

Contracts of carriage by sea frequently involve an international dimension. In the event of any dispute problems may arise as to the court in which proceedings may be instituted and as to the appropriate law which is applicable to the transaction. In maritime transactions the parties involved are frequently of different nationalities and the places at which the contract is made and is to be performed are found in states where different laws are in existence. As a result, the problem of the choice of the governing law arises. Many of the standard bill of lading and charterparty forms make express provision by including clauses which specify where any dispute is to be brought. Such a term is known as a choice of jurisdiction clause and provides that all disputes shall be litigated in the courts of a particular system.

Subject to statutory provisions under international Conventions to which the United Kingdom is a party, the English courts recognise the general principle of freedom to contract. In general they are prepared to accept jurisdiction even though the contract has no connection with the chosen forum. The courts also recognise a foreign jurisdiction clause, and if proceedings are brought in an English court in contravention of such clause they are prepared to grant a stay. The principles applicable in such situations were formulated in *The Eleftheria*.[1] Subject to limits the court will give effect to the parties' choice of law. The common law conflict rules centred on the proper law of the contract which was the law chosen by the parties or, in default of choice, the legal system with which the contract had the closest and most real connection.

There is no international Convention regulating conflict of law problems on a worldwide basis. The common law rules on choice of law have been superseded by the Rome Convention on the Law Applicable to Contractual Obligations. The Convention was done at Rome on 19 June 1980 and came into force on 1 April 1991. It was incorporated into English law by the Contracts (Applicable Law) Act 1990.[2] The Convention puts the English conflict of law rules on contracts onto a

1. [1970] P. 94, [1969] 1 Lloyd's Rep. 237.
2. For detailed information about the Rome Convention 1980 see Peter Kaye, *The New International Law of Contract of the European Community: implementation of the EEC's contractual obligations convention in England and Wales under the Contracts (Application Law) Act 1990* (1993).

statutory basis for the first time and replaces the common law rules. Article 17 provides that in individual contracting states the Convention applies only to contracts made after the date of its entry into force in those states. Therefore the pre-existing English contract conflicts law continues to operate in relation to contracts falling within the Convention's ambit if they were concluded on or before 1 April 1991. The Act will also be interpreted in relation to the earlier common law rules.

The Convention supersedes all pre-existing national rules of private international law of contract of the contracting states. But after the Convention has entered into force for a contracting state the state may adopt a new international law rule and replace the corresponding rule in the Convention. Obviously such a replacement could be made only under certain conditions. Under Article 23 the new national rule in question should be restricted to a particular category of contract, rather than being generalised. The contracting state should communicate its intentions to the other signatory states as set out in Article 23. And if requested by another signatory state consultations must take place over the proposed derogation from the Convention's provisions.

(a) Scope of the Rome Convention

The Convention's objective is to unify the private international law of the contract among contracting states. By standardising the choice of law rules the Convention seeks to minimise the advantages of forum shopping. It applies to contractual obligations involving a choice between the laws of two different legal systems. Its application is obligatory, not optional or discretionary, and is not restricted to international cases. In contrast to the Brussels Convention, the countries concerned under the Rome Convention are not limited to member states. As stated in the Report on the Convention[3] some cases involve one or more elements foreign to the internal social system of a country and thereby give the legal systems of several countries claims to apply. These are precisely the situations in which the uniform rules are intended to apply. Article 2 expressly provides that any law specified by the Convention shall be applied whether or not it is the law of a contracting state. Under Article 2(3) the Convention applies to conflicts involving different parts of the United Kingdom. It will also apply to a dispute which is solely between the laws of different territorial units within the same state, such as the USA. In such a case the dispute should come before the courts of a member state of the European Community where the Convention is in force. Therefore whenever there is a choice of forum clause and a foreign law is chosen the Convention will apply even if the case is a purely domestic one. In the context of some contracts, in particular contracts of carriage, Article 21 provides that the Convention "shall not prejudice the application of international Conventions to which a contracting state is, or becomes, a party".

3. The history of the preparatory work and negotiations is set out in the Report on the Convention by Professors Giuliano and Lagarde, [1980] OJ C282/1 (hereafter cited as "Report").

(b) Agreement on choice of law

Article 3 of the Convention confirms the position which had been reached by the English common law. The parties are free to choose the law applicable to their contract. The leading pre-Convention authority for English law is the decision of the Privy Council in *Vita Food Products Inc.* v. *Unus Shipping Co. Ltd.*[4] This decision provides the authority for the proposition that the parties are free to submit the validity of their contract to any law of their own choosing, irrespective of any connection with the contract. But the choice must be *bona fide* and not contrary to English public policy. Again as with the English common law provisions, any reference to an applicable law under the Convention is a reference to the domestic law of the country in question. The doctrine of *renvoi* is specifically excluded (Article 15). Also in this respect the Convention shows similarity to the position reached under common law.[5]

By virtue of Article 3(1) of the Convention, a contract is to be governed by the law chosen by the parties, which will govern the substance of their contractual relationship. According to the last sentence of Article 3(1) parties may choose different laws to govern different parts of the contract. Therefore they may choose different legal systems to govern different clauses in the contract. For instance they may choose a law to govern the interpretation of the contract and a different law to govern its discharge.[6] Obviously a partial or multiple choice should be logically consistent and not give rise to contradictions.

The parties may expressly choose a law to govern their contract either by a term in their contract or by way of a separate agreement. The statement in the contract must clearly indicate a territorial legal system or a standard form of contract must contain such a provision. It is also possible to have an express choice in a charterparty or carriage contract by stating that "this contract shall be governed by the law of the flag". But in practice such a statement may cause some difficulties.[7] The parties' choice of law can be made at any time after the conclusion of the contract.

In the absence of an express choice the choice may be inferred from the terms of the contract or the circumstances of the case, provided that this is demonstrated with reasonable certainty. Therefore if the parties have not set out their choice the court should not proceed to determine the most closely connected country's law under Article 4, but instead should try to ascertain whether the parties had a

4. [1939] A.C. 277, [1939] 1 All E.R. 513.

5. See *Amin Rasheed Shipping Corpn.* v. *Kuwait Insurance Co.* [1984] A.C. 50, [1983] 2 All E.R. 884, [1983] 3 W.L.R. 241 (HL).

6. The Rome Convention, Art. 3(1) states:

"A contract shall be governed by the law chosen by the parties. The choice must be expressed or demonstrated with reasonable certainty by the terms of the contract or the circumstances of the case. By their choice the parties can select the law applicable to the whole or a part only of the contract."

7. See *Compagnie Tunisienne de Navigation S.A.* v. *Compagnie d'Armement Maritime S.A. [1971]* A.C. 572, [1970] 3 All E.R. 71 (HL). In this case the express selection of "laws of the flag of the vessel carrying the goods" was held insufficient to give effect to the contract where it contemplated performance in a number of different vessels flying different flags.

particular but unexpressed choice of applicable law. The Report gives examples of factors which may enable an inference to be made. These include: the use of a standard form, for example Lloyd's policy of marine insurance; a previous course of dealing under contracts containing a choice of law clause, the omission of which from the instant contract was not the consequence of a decision to change the parties' policy; a choice of court or arbitration clause; a reference to rules of a particular system of law.[8] It is worth noting that although the Report instances a Lloyd's policy of marine insurance as being known to be governed by a particular system of law, even though there is no express choice of law, the English judiciary has not applied it in order to ascertain the proper law. For instance in *Amin Rasheed Shipping Corp.* v. *Kuwait Insurance Co.* the policy was a standard Lloyd's policy, as set out in the Marine Insurance Act 1906, Schedule 1. The House of Lords inferred that English law was the governing law *but not* from the fact that the policy was in an English form, but because the parties' rights and obligations could only be determined by reference to that Act and its judicial interpretation.

The parties may vary any choice of law that was made previously. But the variation should not prejudice the formal validity of the contract under the rules set out in Article 9, or adversely affect the rights of third parties (Article 3(2)). The new or revised choice of law may apply to the whole contract or to any severable part of it.

In relation to freedom of choice one might consider whether the freedom is so wide that the parties are free to choose the application of choice of law rules other than those in the Convention. For instance could the parties before an English court choose to have applied to their contract the common law choice of law rules replaced by the Convention under the Contracts Act 1990? If such freedom of choice were allowed it would undermine the objective of the Convention as it aims to provide uniform choice of law rules within the European Community.[9]

(c) Limitations on freedom of choice

The apparent freedom of the parties to choose any law to govern a contract under the Convention is subject to mandatory rules and public policy.

Under the Convention in three situations the mandatory rules of a domestic legal system will apply irrespective of a choice of law by the parties. These are situations where the contract is wholly connected with a single country and the parties have chosen a different law as the applicable law, where there is a consumer contract within the provisions of Article 5 and in individual contracts under Article

8. Under the common law rules the strongest indication of an implied choice of law is a choice of forum clause or a clause providing that any dispute arising out of the contract shall be decided by arbitration in a particular country. But such presumptions are not conclusive and they are taken into account as one of the factors to be taken into consideration when reaching a decision as to the proper law. See *Tzortzis* v. *Monark Line* [1968] 1 W.L.R. 406, *Compagnie Tunisienne de Navigation S.A.* v. *Compagnie d'Armement Maritime S.A.* and *The Parouth* [1982] 2 Lloyd's Rep. 351 (CA).

9. See The Report C282/8 and Peter North, *Private International Law Problems in Common Law Jurisdictions* (1993) pp.126–127.

6. Under Article 3(3) the parties will not be able to avoid a mandatory rule of a national state by the choice of a foreign law in circumstances where all other relevant elements of the contract at the time of that choice are connected exclusively with that particular national state. Mandatory rules are defined in Article 3(3) as rules "which cannot be derogated from by contract". Thus if a contract is for all purposes English it is not possible for the parties to evade the provisions of the Unfair Contract Terms Act 1977 by merely stipulating that the contract is to be governed by another system of law. If such a contract was litigated in France the French courts would be directed to apply the 1977 Act. An English statute, such as the Unfair Contract Terms Act 1977 or COGSA 1971, which expressly or impliedly provides that its provisions cannot be excluded in a contract is a mandatory rule for the purpose of this Convention.

The mandatory application of COGSA 1971 became the practice after *Vita Food Products Inc.* v. *Unus Shipping Co. Ltd.* As a result of this case, that the parties could choose as the proper law the law of any country, there opened up the possibility of deliberate evasion of the Hague Rules. Under COGSA 1924 it was possible for shipowners to avoid the effect of the Hague Rules by providing that disputes under their bills of lading were to be resolved either in jurisdictions which did not apply the Hague Rules regime or under the laws of such jurisdiction.[10] Because of the *Vita Food* case the Hague-Visby Rules adopted a different drafting technique.[11] In the UK the Rules apply to all bills of lading issued on outward voyages from the UK, and wherever the bill of lading either incorporates the Rules themselves or expressly provides that UK law is to govern the contract. Under the COGSA 1971 the Rules "have the force of law".[12] The effect of mandatory application of the Hague-Visby Rules became clear in *The Morviken.* In this case the bill of lading provided in the same clause (clause 2)[13] for exclusive jurisdiction

10. Under COGSA 1924 it was possible for the shipowners to avoid the effect of Article III Rule 8: "Any clause, covenant or agreement in a contract of carriage relieving the carrier or the ship from liability for loss of or damage to or in connection with goods arising from negligence, fault or failure in the duties and obligations provided in this Article or lessening such liability otherwise than as provided in these Rules, shall be null and void and of no effect.

A benefit of insurance or similar clause shall be deemed to be a clause relieving the carrier from liability."

11. Article X of the Hague-Visby Rules which states when the Rules apply is a new provision. It was drafted to avoid the consequences of the *Vita Food* decision. Art. X(c) provides that "The provisions of these Rules shall apply to every bill of lading . . . if the contract contained in or evidenced by the bill of lading provides that . . . the legislation of any state giving effect to them are to govern the contract, whatever may be the nationality of the ship, the carrier, the shipper, the consignee, or any other interested person."

12. COGSA 1971, s.1(3) states: "Without prejudice to subsection (2) above, the said provisions shall have effect (and the force of law) in relation to and in connection with the carriage of goods by sea in ships where the port of shipment is a port in the United Kingdom, whether or not the carriage is between ports in two different states within the meaning of Article X of the Rules."

13. Bill of lading condition 2 provided as follows: "Law of application and jurisdiction: The law of the Netherlands in which the Hague Rules as adopted by the Brussels Convention of 25th August, 1924 are incorporated . . . shall apply to this contract. The maximum liability package is D.fl. 1250 . . . All actions under the present contract of carriage shall be brought before the Courts of Amsterdam and no other Court shall have jurisdiction with regard to such action." [1983] 1 Lloyd's Rep. 1.

in the Court of Amsterdam and choice of law and a maximum carrier's liability according to the Hague Rules. At that time Dutch law gave effect to the Hague Rules under which the carrier's liability would have been limited to about £250. But the package limitation under the Hague-Visby Rules, as applied in the UK, was £11,490. And Article IV, Rule 5(g), of the Rules provided that the package limit could by agreement be increased but not reduced. As the voyage was from a port in Scotland to Bonaire there was no dispute that the bill of lading was one to which COGSA 71 applied. The Court of Appeal considered that unlike the Hague Rules the Hague-Visby Rules have the coercive force of law. Therefore anything in the bill of lading, including a proper law or jurisdiction clause, which was inconsistent with the Rules or which derogated from their effect was to be rejected. The House of Lords accepted that the application of clause 2 would subject the dispute to a jurisdiction the substantive law of which offered a lower package limit than that prescribed by the Hague-Visby Rules. Therefore an English court was bound to treat such a clause as null and void.[14] After *The Morviken* was decided the 1968 Brussels Convention on Jurisdiction and the Enforcement of Judgments in Civil and Commercial Matters was incorporated into English law by the Civil Jurisdiction and Judgments Act 1982 (1982 Act) and the Contracts Act 1990. If the 1982 Act had been in force at the time *The Morviken* came before the courts how might they have dealt with the conflict between section 2(1) of the 1982 Act[15] and section 1(2) of COGSA 1971?[16] Under these sections both the 1968 Brussels Convention and the Hague-Visby Rules have the force of law in the UK and under Article 17 of the Brussels Convention 1968 exclusive jurisdiction is provided for the courts of a contracting state.[17]

In relation to Article 57 of the 1982 Act[18] it is possible to argue that the English court would be entitled to assume jurisdiction, notwithstanding the exclusive jurisdiction provision in the bill of lading, because the UK is a party to the Hague-Visby Rules. But if the view were to be taken that the Hague-Visby Rules were not a Convention which governs jurisdiction it would be difficult for the court to avoid exclusive jurisdiction in the bill of lading. As a result the court would face the problem of determining whether the Hague-Visby or the Brussels Convention 1969 should prevail. But when Article 7(2), which provides that nothing in the

14. See also *The Benarty* [1984] 2 Lloyd's Rep. 244 (CA).

15. S. 2(1) of the Civil Jurisdiction and Judgments Act 1982 provides: "The Brussels Convention shall have the force of law in the United Kingdom, and judicial notice shall be taken of them."

16. S. 1(2) of COGSA 1971 provides: "The provisions of the Rules, as set out in this Schedule to this Act, shall have the force of law."

17. Art. 17 of the Brussels Convention 1968 provides that: "If the parties, one or more of whom is domiciled in a Contracting State, have agreed that a court or the courts of a Contracting State are to have jurisdiction to settle any disputes which have arisen or which may arise in connection with a particular legal relationship, that court or those courts shall have exclusive jurisdiction."

18. S. 57 of the Civil Jurisdiction and Judgments Act 1982 states that: "This Convention shall not affect any Conventions to which the Contracting States are or will be parties and which, in relation to particular matters, govern jurisdiction or the recognition or enforcement of judgments."

Convention shall restrict the application of the rules of the forum in a situation where they are mandatory irrespective of the law otherwise applicable to the contract, and Article 3(3) are considered together the Hague-Visby Rules would be regarded as "mandatory rules" of English law and the effect of the Rules would be identical. It is also worth noting that a choice of foreign law is admissible as long as the applicable law does not decrease but increases the carrier's liability under the Rules.

The application of COGSA 1971 was also considered in *The Komninos S*. In this case the bill of lading covered a shipment from Greece to Italy. It contained a clause which conferred exclusive jurisdiction on the British courts. And the action was brought in London. As there was no explicit choice of law in the contract at first instance the closest and most real connection rule applied to determine the governing law, and Greek law was held to be the applicable law. But the Court of Appeal reversed the decision and held that the British courts referred to in the bill of lading were to be interpreted as meaning the English courts, and such a choice of forum clause was enough to infer an implied choice of English law by the parties. As the parties had impliedly chosen English law the Court of Appeal had to decide whether the parties had also impliedly chosen the application of COGSA 1971. Bingham L.J. held that COGSA 71 did not apply, on the basis of the choice of forum clause in which the parties intended that "the legislation of the United Kingdom giving effect to the rules should govern the contract".[19]

The Unfair Contract Terms Act 1977 places restrictions on the validity of exclusion or exemption clauses in various types of contracts. This Act is an important example of mandatory rules in English law. Section 27(2) of the Act[20] is aimed at preventing the parties to a contract most closely connected with the UK from contracting out of the restrictions laid down by the 1977 Act, by choosing the law of a country outside the UK. If the requirements of section 27(2) are met the courts of the UK will apply the parties' choice of law to the contract which is subject to the restrictions of the 1977 Act.[21]

In relation to the mandatory rules of other countries Article 7(1) provides:

"... When applying under this Convention the law of a country, effect may be given to the mandatory rules of the law of the other country with which the situation has a close connection, if and in so far as, under the law of the latter country, those rules must be applied whatever the law applicable to the contract. In considering whether to give effect to these mandatory rules, regard shall be had to their nature and purpose and to the consequences of their application or non-application."

19. [1990] 1 Lloyd's Rep. 541; *Financial Times*, 16 January 1990 (CA); [1991] 1 Lloyd's Rep. 370.
20. S. 27(2) of the Unfair Contract Terms Act 1977 provides that:
"This Act has effect notwithstanding any contract term which applies or purports to apply the law of some country outside the United Kingdom, where (either or both):
(a) the term appears to the court, or arbitrator or arbiter to have been imposed wholly or mainly for the purpose of enabling the party imposing it to evade the operation of this Act; or
(b) in the making of the contract one of the parties dealt as consumer, and he was then habitually resident in the United Kingdom, and the essential steps necessary for the making of the contract were taken there, whether by him or by others on his behalf."
21. See Abla J. Mayss, *Conflict of Laws* (1994), pp.80–82.

Under this Article, in limited circumstances, a court may apply a mandatory rule which is not that of the forum or the applicable law or the law imported into the contract by Article 3, 5 or 6. The application of Article 7(1) in the UK is specifically excluded by section 2(2) of the Contracts Act 1990, in accordance with Article 22 of the Convention. As this provision has no application to UK courts the rules of a foreign legal system which is not the applicable law could only be taken into account by the UK courts if they were required to do so under the Convention or in the exercise of their own public policy.

Article 7(2) is related to the mandatory rules of the forum. The Article provides that "nothing in this Convention shall restrict the application of the rules of the law of the forum in a situation where they are mandatory irrespective of the law otherwise applicable to the contract". Under Article 7(1) and (2) mandatory rules may be applied whether the law is chosen by the parties or determined by the rules of the Convention in default of choice. When this Article is examined literally the forum may apply its own exclusions, choice rules, and any other provisions as long as they are mandatory, which means ignoring the whole regime under the Convention.[22] "The application of a rule of the law of any country specified by the Convention may be refused only if such application is manifestly incompatible with the public policy (*ordre public*) of the forum" (Article 16). Under Article 16 the application of the law of any country may be refused if to apply it would be manifestly incompatible with the public policy of the forum. But the public policy exception to the application of a rule of the applicable law is intended to be of very limited scope. So the objection is not to the foreign law itself but only to its application by the forum in that particular case.

(d) Applicable law in the absence of choice

In the absence of choice by the parties the applicable law will be determined in accordance with Article 4. Paragraph (1) of this Article provides: "To the extent that the law applicable to the contract has not been chosen in accordance with Article 3, the contract shall be governed by the law of the country with which it is most closely connected." The basic principle in Article 4(1) is exactly that of the common law approach which involved a test for the system of law with which the contract was most closely connected.[23]

The test in Article 4(1) is an objective one under which the applicable law is ascertained by looking at the connecting factors rather than trying to discover what the parties would have chosen if they had thought about the matter. As with Article 3(1), in the last sentence of Article 4(1) there is room for dépeçage as well for it provides that "nevertheless, a severable part of the contract which has a closer connection with another country may, by way of exception, be governed by the law

22. See Raymond Smith, *Conflict of Laws* (1993) pp.127–133.

23. In *Compagnie Tunisienne de Navigation S.A.* v. *Compagnie d'Armement Maritime S.A.*, the general rule was stated to be that "where the parties have not exercised their choice, the proper law of the contract is the system of law with which it has closest and most real connection": at p.609.

of that country". As dépeçage is intended to operate in only very limited circumstances this provision applies only in exceptional cases.

Under the general rule set out in Article 4(1) regard must be had to the law of the country with which there is closest connection, rather than, as in the common law jurisdictions, with the legal system with which there is the closest connection.[24] In determining the country of closest connection it is also possible to take account of factors which supervened after the conclusion of the contract.[25]

The common law systems had not developed presumptions to identify the proper law and in recent years the aim was to reject the use of presumptions in trying to determine with which country or legal system a contract was most closely connected. In contrast, the Convention has three rebuttable presumptions, which are not to be applied if it appears from the circumstances that the contract is more closely connected with another country.[26]

The first presumption in Article 4(2) is the concept of "characteristic performance". Under the Article it is to be presumed that the contract is most closely connected with the country where the party who is to effect the characteristic performance of the contract has his habitual residence at the time of conclusion of the contract. Although characteristic performance is not defined under the Convention, Article 4(2) places emphasis on characteristic performance. Under the provision for all contracts, other than those relating to immovables or carriage of goods, there is a rebuttable presumption that the contract is most closely connected with the country where the party who is to effect the characteristic performance of the contract has his habitual residence or, in the case of a body corporate or incorporate, where it has its central administration. If the contract is entered into in the course of the trade or profession of the party who is to effect the characteristic performance, then the applicable law is presumed to be that of the country of that party's principal place of business or of the place of business through which, under the terms of the contract, performance is to be effected. When the party carrying out the characteristic performance has been identified, the Convention examines the law of his habitual residence, central administration or place of business. The Report provides further assistance in identifying the country or place of characteristic performance of a contract. In an ordinary bilateral contract under which one party is to pay money in return for the provision of the goods or service the characteristic performance is the provision of the goods or service, rather that the payment of the price. In an international contract the presumption under Article 4(2) is in favour of the application of the

24. In the *Tunisienne* case it was not very clear whether the reference should be made to the law of the country or the system of law with which the connection was to be shown. In the later cases the more common formulation has been to refer to the system of law. And in some cases like *James Miller & Partners Ltd.* v. *Whitworth Street Estates (Manchester) Ltd.* [1970] A.C. 583 it has been said that the tests of system of law and the law of the country should be combined. And if the connections with each of them are different they should be weighed in balance. See also *Coast Lines* v. *Hudig & Veder Chartering NV* [1972] 2 Q.B. 34.

25 See the Report C282/20 (referred to in n.3).

26. Art. 4(5).

law of the seller's place of business but in a major international contract it will be more difficult to identify the characteristic performance.

If the characteristic performance cannot be found the applicable law will be that of the country with which the contract is most closely connected.

The second presumption in Article 4(3) provides that if the subject matter of the contract is a right in immovable property, then it is to be presumed that the contract is most closely connected with the country in which the immovable property is situated.

The Convention contains special rules in relation to consumer contracts and individual employment contracts, because the consumer and the employee are considered to be weak contracting parties.

(e) Contracts for the carriage of goods by sea

The Convention excludes from its scope certain contractual obligations.[27] In relation to carriage of goods by sea the provisions clearly apply to charterparties and contracts where non-negotiable transport documents are issued. But the Convention does not apply to obligations arising under bills of exchange and other negotiable instruments "to the extent that the obligations . . . arise out of their negotiable character". The Report states that bills of lading and similar documents issued in connection with transport contracts are only excluded if they are regarded as negotiable instruments by the law of the forum. Therefore if a document is not regarded as a negotiable instrument it falls outside the exclusion. Whether a document is to be characterised as a negotiable instrument or not, it is not governed by the Convention, but is a matter for the forum's law including its rules of private international law. Under English private international law such negotiability is to be conferred by the law of the country in which the instrument is transferred.

It was deemed inappropriate to submit contracts for the carriage of goods to the presumption relating to characteristic performance having regard to the peculiarities of this type of contract. Therefore special presumptions were provided for transport contracts in Article 4(4). Carriage of passengers remains subject to the general presumption in Article 4(2).

In default of choice in contracts for the carriage of goods including single voyage charterparties the presumed closest connection of the contract will be with the country in which at the time of contracting[28]:

—the carrier has his principal place of business and

—either the place of loading or discharge or the consignor's principal place of business is situated.

27. For a complete list, see Art. 1.2.

28. For detailed discussion of the application of the Contracts Act 1990 to contracts for the carriage of goods by sea see Regina Asariotis, "Contracts for the Carriage of Goods by Sea and Conflict of Laws: Some Questions Regarding the Contracts (Applicable Law) Act 1990" (1995) 26 *Journal of Maritime Law and Commerce* 293–314.

The terms "carrier", "consignor" and "principal place of business" are not clearly defined in the Convention. The Report states that "carrier" relates to the contractual, not the actual, carrier. The carrier for this purpose is the party who undertakes the carriage, whether he does any carrying himself or arranges for someone else to do it.

In cases in which the carrier's principal place of business coincides with neither of the places listed in Article 4(4) there will be no presumption. It is also not possible to apply the general presumption in Article 4(2) to contracts for the carriage of goods. If the contract also appears to be more closely connected with another country the presumption will be rebutted. Another problem may arise in cases of flags of convenience where a company is incorporated in one of the flag-of-convenience countries but the vessel is managed by a different company with offices in another country. In such a case where is the relevant place of business? Obviously such circumstances will restrict the application of the presumption in practice.

In the Report the "consignor" refers to any person who consigns goods to the carrier, and the places of loading and unloading are those agreed at the time of conclusion of the contract, in other words they are only the contractual and not the actual places of loading and discharge.

If the required contract elements are lacking under Article 4(4) then the court falls back on the closest connection under Article 4(1). Even if the combination is satisfied the presumption is rebuttable under Article 4(5) if in all the circumstances the contract is more closely connected with another country.

(f) Floating choice of law clauses

In recent years the courts were faced with clauses in maritime contracts under which the law governing the contract is not identified and left to the subsequent choice of one of the contracting parties. When the attitude of the courts to the proper law of contracts is considered it becomes clear that only one legal system should govern the contract as a whole.[29] It is not possible to make a contract without a governing law during its currency. If there is no choice of law clause in the contract the proper law cannot be determined by an event occurring after the making of the contract.

There is not much authority on the validity of floating choice of law clauses. In *The Armar* a bill of lading contained a clause which gave the carrier/shipowner the right to select the place of general average adjustment and its laws. The shipowner exercised his right to select London as the place of adjustment. At the same time the consignees' cargo insurers, who were not party to the bill of lading, signed a Lloyd's Average Bond. Thus, the question before the court was about the proper

29. It was stated by Lord Diplock that: "... contracts are incapable of existing in a legal vacuum. They are mere pieces of paper devoid of all legal effect unless they were made by reference to some system of private law which defines the obligations assumed by the parties ...": *Amin Rasheed Corporation* v. *Kuwait Insurance Co.*, at p.65.

law of the Lloyd's Average Bond. It was held that by the act of selection English law became the law governing the adjustment and, as a consequence, the law governing the adjustment bond was also English. But the Court of Appeal strongly disagreed with this view and held that the proper law of the contract has to be ascertainable at the time of the making of a contract and cannot fall to be determined by reference to a subsequent event.[30] Megaw L.J also added that he did not think it "... an attractive, or a possible, concept of English private international law that the governing law, initially being, say, the law of Algeria, should thereafter change into the law of England".

In relation to the Contracts (Applicable Law) Act 1990 one may consider other possibilities, for instance whether it is possible to change the governing law or to exercise a choice of law after the conclusion of the contract. Under Article 3(2) parties may exercise or vary their choice at any time. Thus, in principle it would be possible for the parties to exercise or vary a choice of law after the conclusion of a contract. This principle seems to be clearly contradictory to the common law position as expressed by Megaw L.J. in *The Armar*. As long as freedom of choice under the Act is exercised by the agreement of the parties nothing shall restrict the parties' freedom to agree or change their agreement at any stage. Therefore when the choice of law in *The Armar* is considered, under the Act it would be invalid. As the cargo insurers were not party to the bill of lading, as a matter of English law, it is not possible to say that they had consented to the unilateral choice of law made by the shipowner on the basis of the bill of lading clause. In other words a third party cannot be bound by a unilateral choice of law by a carrier on the basis of a clause in the bill of lading.

What would be the situation if a third party indorsee had a bill of lading containing such a clause? In a situation where the carrier chooses the applicable law on the basis of a clause in the bill of lading will the consignee, who becomes a party to the contract on thc tcrms of the bill of lading, be found to have consented to the unilateral choice of law of the carrier. As stated in the Report a third party who has acquired rights under a contract concluded by two other parties will not lose these rights as a consequence of a subsequent change of the applicable law. As stated above Article 3(2) allows the parties to exercise or vary their choice at any time, but any variation by the parties of the law to be applied made after the conclusion of the contract shall not adversely affect the rights of third parties. Thus a consignee who has title to sue under the Carriage of Goods by Sea Act 1992 should not lose his rights by a subsequent choice of law if that chosen law does not recognise such a right.

In general it is possible to think that if a consignee is bound by the terms of the

30. The statement made by Megaw L.J. in *The Armar* was that "there must be a proper law of any contract—a governing law—at the time of making that contract ... As a matter of legal logic, I find insuperable difficulty in seeing by what system of law you are to decide what, if any, is the legal effect of an event which occurs when a contract is already in existence with no proper law; but instead with a 'floating' non-law": *The Armar Shipping Co. Ltd.* v. *Caisse Algérienne d'Assurance et de Réassurance (The Armar)* [1980] 2 Lloyd's Rep. 450 at p.455.

bill of lading he will also be bound by a choice of law made by the carrier, as the right of choice is embodied in the bill. But when one considers the requirement under the Act for the agreement of the parties, i.e. consent by the consignee, this general principle does not seem to apply. In common law a similar problem was considered in *The Iran Vojdan*. In that case the bill of lading in dispute covered a cargo shipped in Hamburg on an Iranian vessel. The clause in the bill of lading gave the carrier the right of choice between Iranian, German or English law as the governing law of the contract. In this case the *Armar* decision was followed and the trial judge refused to give effect to the clause in the bill of lading. In the opinion of Bingham J., "the proper law is something so fundamental to questions relating to the formation, validity, interpretation and performance of the contract that it must . . . be built into the fabric of the contract and cannot float in an indeterminate way until finally determined at the option of one party".[31]

Bingham J. also held that as the part of the clause relating to jurisdiction was not severable from the invalid choice of law provision there was no valid jurisdiction clause. He also added that if the clause had confined itself to conferring three options for the choice of jurisdiction on the carrier alone the effect would be given properly and without difficulty.[32] Based on this view and the decision of the European Court of Justice in Case 71/83[33] the consignee of the bill of lading would have been contractually bound to the choice of forum exercised by the carrier unilaterally. In principle the consignee would be bound by the agreement between the original parties to the contract.

In *The Star Texas* the validity of a floating-choice curial law was discussed. In this case the arbitration clause did not expressly confer on either party the right to choose a particular system of law. It conferred a right to select a jurisdiction in which the arbitration was to be heard. When an international contract contains an arbitration clause different systems of law may be relevant to disputes arising under it. These are the laws governing the substantive obligations of the parties, the arbitration clause and the "curial law", and that governing the procedure under which the arbitration is to be conducted. In this case the appellant argued that, even if there was no implied choice of substantive law, there was still an implied choice of the curial law. As English law does not countenance a floating choice of law the arbitration clause should be void. The court held that there was no objection to a floating choice of curial law. In fact Lloyd L.J. said that it was common to have an arbitration clause which provided for a choice of

31. *Dubai Electricity Co. and Others* v. *Islamic Rebuplic of Iran Shipping Lines* [1984] 2 Lloyd's Rep. 380, at p.385.

32. *Ibid.*

33. Regina Asariotis, p.308, n. 78. In *Partenreederei MS Tilly Russ and Ernest Russ* v. *N. V. Haven en Vervoerbedrijf Nova and NV Goeminne Hout* (Case 71/83) [1984] E.C.R. 2417, [1985] Q.B. 931, the European Court of Justice held that if a jurisdiction clause in a bill of lading was effective as between the carrier and the shipper, it was also effective as between the carrier and a holder of the bill of lading, provided that, under the applicable law, the holder of the bill of lading succeeded to the shipper's rights and obligations.

venues.[34] Therefore the dictum of Lord Diplock in the *Amin Rasheed* case that every contract had to have a proper law did not apply to curial law. Based on the view in *The Star Texas* floating choice of law clauses are effective only as regards the crucial law. They are not effective in relation to substantial obligations of the parties or the arbitration agreement. The law governing a contract to arbitrate cannot be determined by the election of one of the parties at some time after the date on which the agreement is concluded. But it is possible to have a clause in the contract which provides that one law shall apply in one event but that another shall apply in another event. Thus in *The Mariannina* the clause which provided that if a provision for arbitration in London were to be ruled unenforceable the contract should be governed by Greek law was held valid by the Court of Appeal.[35]

With regard to the validity of floating choice of law clauses the current view of English law can be summarised as follows[36]: the validity of a floating choice of forum clause is determined by the proper law of the contract, but it is not clear whether the law of the forum or the proper law of the contract determines the validity of such a clause. If the parties have chosen a law to govern the contract that law will decide whether a floating choice of forum clause is valid or not. But if a contract contains a floating choice of law clause and its validity is not governed by English law as the *lex fori* the validity of that clause is determined by the objectively assessed proper law. If English domestic law applies, a floating choice of law clause will be accepted as invalid. If English law is the proper law of a contract a floating choice of forum clause will be held valid.

ARBITRATION AGREEMENTS

(a) Incorporation of arbitration clauses into charterparty bills of lading

As parties are submitting to the common law to select the governing law for their contract they are also free to decide whether disputes will go to court or to arbitration. Most contracts of affreightment include a clause which provides that any dispute arising thereunder shall be referred to arbitration. Most charterparties contain an arbitration clause and most disputes involving them are also referred to arbitration. Arbitration clauses are less frequently found in bills of lading except

34. Lloyd L.J. said that "Nobody has suggested ... that that makes the arbitration clause void for uncertainty or otherwise unworkable. It makes good commercial sense that the law governing the arbitration procedure should be the law of the country where the arbitration takes place. The objections which apply to a floating proper law do not apply to a floating curial law. A contract without proper law cannot exist. It is, it has been said, no more than an abstraction or piece of paper. But an arbitration agreement can exist perfectly well without it being known at the time the agreement is entered into what law will govern the arbitration procedure": *Star Shipping A.S.* v. *China National Foreign Trade Transportation Corp.* [1993] 2 Lloyd's Rep. 445.

35. *Astro Venturoso Compania Naviera* v. *Hellenic Shipyards S.A.* [1983] 1 Lloyd's Rep. 12 (CA).

36. See Adrian Briggs, "The Validity of 'floating' Choice of Law and Jurisdiction Clauses", [1986] L.M.C.L.Q. 508–518 and also M.N. Howard, "Floating Choice of Law Clauses, Case and Comment" [1995] L.M.C.L.Q. 1–8; Andrew Beck, "Floating Choice of Law Clauses" [1987] L.M.C.L.Q. 523–533.

where the bill incorporates an arbitration clause in the charterparty under which it is issued.

In relation to incorporation of arbitration clauses into charterparty bills of lading the main issues are whether the language used in the bill of lading has the effect of incorporating the arbitration clause in the charterparty and also whether the arbitration clause itself was apt to be incorporated in the bill of lading. Bills of lading play a specific role in international trade. They are negotiable commercial instruments and often used by parties without any knowledge of the terms of the corresponding charterparties. So the terms contained in bills of lading should be clear, certain and unambiguous. For the incorporation of an arbitration clause in a bill of lading in particular, the language used in a bill of lading is the most important factor.

In general the English law of contract has taken a benevolent view of the use of words to incorporate by reference standard terms to be found elsewhere. But for the incorporation of charterparty arbitration clauses into bills of lading the courts have developed a stricter test. The development of this attitude can be traced back to *Hamilton & Co.* v. *Mackie & Sons* where the bill of lading had the words "all terms and conditions as per charterparty". The charterparty contained the clause "all disputes under this charter shall be referred to arbitration". The court held that the arbitration clause referred only to disputes "under this charter" and did not include disputes under the bill of lading.[37] Following this case, in *T. W. Thomas & Co. Ltd.* v. *Portsea Steamship Co. Ltd.* it was stated that ". . . if it is desired to put upon the holders of a bill of lading an obligation to arbitrate because that obligation is stated in the charterparty, it must be done explicitly".[38] Therefore the general words "all terms and conditions" were not enough for incorporation. Similarly in *The Njegos*[39] the bill of lading was stated to incorporate "all the terms, conditions and exceptions" and it was held that the arbitration clause was not incorporated into it. In these cases the language used was too general to incorporate a term such as an arbitration clause.

Controversy often arises about the true interpretation and effect of the words "terms and conditions" or "all terms" or "conditions". These terms were considered in the abovementioned cases. *The Merak* was authority for the proposition that reference to "clauses" is enough to permit incorporation of an aptly drafted arbitration clause. In this case clause 32 of the charterparty expressly stipulated for arbitration of "any dispute arising out of this charterparty or any bill of lading issued thereunder". The Court of Appeal reached the conclusion that the words in bills of lading "all the terms, conditions and clauses . . . contained in the said charterparty" were sufficient to incorporate the charterparty arbitration clause into the bills of lading.[40] In *The Annefield*[41] this new line of authority was

37. (1889) 5 T.L.R. 677, at p.677.
38. *T.W. Thomas & Co. Ltd.* v. *Portsea Steamship Co. Ltd.* [1912] A.C. 1.
39. (1935) 53 Ll.L.R. 286, [1936] P. 90.
40. [1964] 2 Lloyd's Rep. 527, [1965] P. 223.
41. [1970] 2 Lloyd's Rep. 252.

commented on and it was stated that general words in the bill of lading would suffice if the arbitration clause in the charterparty was expressly applicable to bills of lading. But in *The Federal Bulker*[42] a strict test was applied and the Court of Appeal held in this case that an arbitration clause is not a term, and in *The Varenna* that an arbitration clause is not a condition.[43] The approach of the Court of Appeal in *The Federal Bulker* was not followed in *The Rena K* and it was held that an arbitration clause could be incorporated from a charterparty into a bill of lading, provided that this was done clearly and explicitly in either document.[44]

In order to incorporate an arbitration clause in the charterparty into a bill of lading there should be specific words of incorporation in the bill of lading. If the arbitration clause refers to "disputes arising under this charterparty" and general words of incorporation have been used the clause will not be regarded as incorporated into the bill of lading. Therefore if there are general words of incorporation in the bill of lading the arbitration clause or any other provision in the bill of lading should make it clear that the arbitration clause under the bill of lading governs disputes under both the bill of lading and the charterparty. If the arbitration clause is specifically incorporated in the bill of lading it should make sense in the bill of lading and should not conflict with the express terms of the bill of lading.

At common law the existence of a jurisdiction or arbitration clause in a contract was taken to be a strong indicator of a tacit choice of the law of the court or arbitration. The Convention does not cover arbitration and choice of law agreements and their form, validity and consent thereto. But contracts in which arbitration and jurisdiction clauses may be contained are not excluded. Therefore in cases where an arbitration clause is an integral part of a larger contract and not an independent agreement the exclusion relates only to the arbitration clause and not to the remainder of the contract. Although jurisdiction and arbitration agreements are excluded they may still be considered in order to decide upon an implied choice of law to govern the main contract under Article 3(1) or the country of closest connection under Article 4(1). The validity of such clauses will not be decided by the law which would govern the contract under the Convention, but will be decided according to the national rules of the forum.

In cases where a bill of lading incorporates the terms of the charterparty the Convention will determine whether the incorporation indicates a contractual choice of law or not. But the validity of the incorporation will be determined by the common law conflict of law rules. Thus, if a charterparty contains a London jurisdiction or arbitration clause the court will apply English law to determine the validity of the bill of lading incorporating the clause.

42. *Federal Bulk Carriers Inc.* v. *C. Itoh & Co. Ltd. and Others* [1989] 1 Lloyd's Rep. 103 (CA).
43. [1983] 2 Lloyd's Rep. 592 (CA).
44. [1978] 1 Lloyd's Rep. 545; [1979] 1 All E.R. 397; [1979] QB 377.

(b) Stay of proceedings under arbitration

At common law but not under the Convention English courts used to have a general discretion to order a stay of proceedings.[45] Under the 1996 Arbitration Act the abolition of the distinction between domestic and non-domestic agreements with the court's discretion to refuse a stay was removed in all cases. Section 9 of the Arbitration Act 1996, which applies to all arbitration agreements, removes the discretion of the court to refuse to stay its own proceedings. Section 9 permits judicial action to go ahead only in certain closely defined circumstances where the arbitration agreement itself is for some reason unenforceable. But section 86 of the Arbitration Act 1996 modifies section 9 in its application to domestic agreements and retains the court's discretion to refuse to stay its proceedings. The reason for such an arrangement is that under section 88 of the Arbitration Act 1996 the Secretary of State has the power to amend or repeal section 86. Therefore if section 86 were repealed the distinction between domestic and non-domestic agreements would be removed and the general provisions of section 9 of the 1996 Act would govern all agreements. The repeal of section 86 did not take place on the implementation of the 1996 Act on 31 January 1997 but the government also did not bring Section 86 into effect along with the rest of the Act. Therefore until section 86 is either repealed or brought into effect all arbitration agreements, domestic and non-domestic, are governed by section 9 of the 1996 Act and the court has no discretion to refuse to stay its proceedings.

THE APPROPRIATE FORUM (*FORUM CONVENIENS*)

An English court may refuse to exercise its jurisdiction on the basis that it is an inappropriate forum. Until relatively recent times the courts denied that English law contained any general doctrine of *forum non conveniens*. The doctrine was unanimously rejected by the House of Lords in *The Atlantic Star*. But again this case represented a development of law in this area. Before it, under the traditional English rules, in the absence of *lis alibi pendens* a defendant seeking a stay of proceedings had to prove that the continuance of proceedings would be "vexatious and oppressive", which meant the plaintiff's harassment of the defendant. In *The Atlantic Star* it was accepted that this traditional English rule should be interpreted more liberally.[46] The balance had to be found between the advantages to the plaintiff of trial in England and the disadvantages to the plaintiff. This liberalisation process continued in *MacShannon* v. *Rockware Glass Ltd.*, where Lord Keith said that it would be necessary to weigh the advantages to the plaintiff and the disadvantages to the defendant. Since this case the starting point has been to ask whether there is a more appropriate forum for trial abroad. In this case Lord Keith

45. The issues to be taken into account in ordering a stay of proceedings are analysed in *The Eleftheria* (1969] 2 All E.R. 641.

46. [1974] A.C. 436.

referred to the country with which the action had the most real and substantial connection.[47]

The House of Lords expressly embraced the doctrine of *forum non conveniens* in *The Abidin Daver*.[48] In this case a Cuban vessel was in collision with a Turkish ship in the Bosphorus within Turkish waters. An action was started by the Turkish owners in the Turkish court in Istanbul. The Cuban owners began an action *in rem* in the English Admiralty Court. The Turkish owners asked for a stay of this action, which was granted. The Court of Appeal lifted the stay and the Turkish owners appealed to the House of Lords. Before *The Atlantic Star* and *McShannon* v. *Rockware Glass Ltd.* if a case involved *lis alibi pendens* a defendant was granted a stay only if he could bring his case under the vexation and oppression criteria. But in *The Abidin Daver* the court was directly confronted with a case of *lis alibi pendens*. The House of Lords allowed the appeal and restored the stay of the English proceedings. It looked at the advantage to the plaintiff of trial in England. The Turkish court was the natural and more appropriate forum for the trial. It was also convincing that the plaintiffs would not obtain any advantage by coming to England for trial. This case introduced a new principle to deal with cases of *lis alibi pendens*. In order to obtain a stay in *lis alibi pendens* cases it has only to be shown that the alternative forum abroad is a natural forum.

In order to find the appropriate forum the court will consider whether, by staying proceedings, the plaintiff will be deprived of some advantages that he would have obtained from trial in England. The advantage to the plaintiff can be personal or juridical. The burden of establishing a juridical or personal advantage is on the plaintiff. In *Spiliada Maritime Corp.* v. *Cansulex Ltd.* the House of Lords greatly reduced the significance of the advantage to the plaintiff. The mere fact that the plaintiff has a legitimate personal or juridical advantage in proceedings in England cannot be decisive. So appropriateness is to be the main criterion for the exercise of the discretion to stay.[49]

47. [1978] A.C. 795, [1978] 1 All E.R. 625.
48. [1984] 1 Lloyd's Rep. 339, [1983] 3 All E.R. 46 (CA) [1984] A.C. 398, [1984] 1 All E.R. 470.
49. *The Spiliada* [1987] A.C. 460.

CHAPTER 27

THE CIVIL JURISDICTION AND JUDGMENTS ACT 1982

APPLICATION TO ADMIRALTY ACTIONS *IN REM*

For practical purposes all Member States of the EU may be treated as parties to the Brussels Convention on Jurisdiction and the Enforcement of Judgments in Civil and Commercial Matters 1968. The Brussels Convention was incorporated into English law by the Civil Jurisdiction and Judgments Act 1982 (the 1982 Act).[1] The Act substantially came into force on 1 January 1987. It applies to proceedings commenced on or after that date. The 1982 Act has the same effect upon Admiralty actions *in personam* as on any other High Court action. But the relationships between the Convention and the established rules concerning the jurisdiction of the Admiralty Court in an action *in rem* have raised questions.

Under Article 1 the Convention "shall apply in civil and commercial matters whatever the nature of the court or tribunal". In relation to Article 1 the Convention *prima facie* applies to many maritime claims which are the subject of proceedings in the UK courts. Article 2 provides the basic rule: "... persons domiciled in a Contracting State shall, whatever their nationality, be sued in the courts of that State". Therefore if a defendant is domiciled in a contracting state in the case of a maritime claim or other shipping dispute he may be sued in that state. Under this principle of conferring jurisdiction on the courts of one state and denying it to the courts of others the risk of reaching inconsistent decisions by different courts in the same dispute is removed or minimised. But under English Admiralty law the service of the writ in an action *in rem* gives the court jurisdiction to decide the case whoever is the owner of the *res*. So do the jurisdictional rules of the Convention prevail over this principle of English law?

Under Article 2 the Convention does not define "domicile", but lays down choice of law rules for the court seised of the matter to apply in order to determine

1. The 1968 Convention is amended by accession treaties following the accession of Denmark, the United Kingdom, Ireland, Greece, Portugal and Spain. In order to allow Austria, Finland and Sweden to become parties a further accession treaty will emerge. The 1982 Act was amended by the Civil Jurisdiction and Judgments Act 1991 which came into force on 1 May 1992.

The 1991 Act brings into force the Lugano Convention on Jurisdiction and Enforcement of Judgments in Civil and Commercial Matters 1988. This almost identical convention of 1988 was made between the Member States of the European Union and the states of the European Free Trade Association—Austria, Finland, Iceland, Norway, Sweden and Switzerland.

"domicile". The domicile of an individual depends upon the coincidence of two factors: residence in the state and having a substantial connection with that state.[2] In order to ascertain the domicile of a corporation the Convention lays down a much clearer rule. Article 53 provides that the domicile of a corporation may be either the state in which it is incorporated and has its registered office or official address or the state where its central management and control are exercised.[3] Section 42 of the 1982 Act states that if the corporation or association is domiciled in the UK it must have been incorporated or formed under the law of part of the UK and have its registered office or some other official address in the UK or have its central management and control exercised in the UK. To be domiciled in England a corporation or association must have its seat in the UK and have its registered office or some other official address or its central management and control or a place of business in England.

If a shipowning or chartering company is incorporated in Liberia, Panama or Cyprus which are favoured by shipowners but not contracting states, but at the same time its central management and control are exercised in the United Kingdom, it will have its seat in the UK and will be domiciled in England and the provisions of the Act will therefore apply. In an Admiralty action *in rem* the action is brought against a ship. It is debatable whether a ship has domicile, so Article 2 could not apply to an Admiralty action *in rem*.

In *The Deichland*[4] the plaintiffs were the owners of a cargo of steel which was shipped on board the vessel and which was alleged to be damaged on discharge. The plaintiffs served a writ *in rem* on the vessel. The shipowner's P&I Club gave a guarantee but did not make any express submission to the jurisdiction of the English courts so that the ship was not arrested. The demise charterers disputed the English court's jurisdiction, arguing that they were domiciled in Germany and that, under Article 2 of the Convention, they were entitled to be sued there. The plaintiffs contended that there was no person being sued, so the Convention was inapplicable to *in rem* proceedings. The Court of Appeal concluded that, in reality, the demise charterers were being sued and therefore the Convention did apply. Consequently, as the defendants were domiciled in Germany the English courts were obliged to decline jurisdiction. Then the plaintiffs advanced an alternative argument. They contended that they could invoke the benefit of Article 7 of the Arrest Convention, by virtue of Article 57 of the 1968 Convention. But then they were faced with the problem that Article 7 appeared to require an actual arrest. Although the writ *in rem* had been served on *The Deichland* she had not been arrested. The plaintiffs argued that it was sufficient that a usual form of guarantee had been given by a P&I club specifically in order to prevent the arrest, but this argument was rejected. Therefore the jurisdiction of the Admiralty Court was not

2. See s.41 of the 1982 Act.

3. Art. 53 of the Convention provides: "For the purposes of the Convention, the seat of a company or other legal person or association of natural or legal persons shall be treated as its domicile. However in order to determine that seat, the court shall apply its rules of private international law."

4. [1990] 1 Q.B. 361 (CA).

derived from the Arrest Convention and, as a consequence, Article 57 did not prevent the Brussels Convention from applying. In this case the plaintiffs' failure to arrest the vessel was fatal, and as a result the English court did not have jurisdiction. To make sure that the English courts do have jurisdiction where the defendant is domiciled in a contracting state the plaintiff in Admiralty proceedings must actually arrest the ship or should obtain an express agreement by the defendant to submit to the jurisdiction of the English court to avoid arrest at the same time as obtaining security.

This decision was criticised by Meeson.[5] His argument was that under an Admiralty action *in rem* the person whose appearance is being compelled is the owner of the property proceeded against. In other words in an Admiralty action *in rem* the person "being sued" is the shipowner. In *The Deichland*, at the time when the writ was served the ship was in new ownership under the name the *Barracuda*. The only interest Deich had in the proceedings was the security provided by a P&I club for the claim on its behalf. The case does not appear to have drawn the distinction between the time when the writ is issued and that when the writ is served, because when the writ *in rem* is issued the person liable on the claim in an action *in personam* is the same person as the person against whose ship the action is brought. Therefore the person being sued is the shipowner, just as if the action were *in personam*. When the writ was served the result was not the same in this case.

The next relevant decision was *The Po*. In this case two vessels, one Italian-owned and the other American-owned, were in collision in Rio de Janeiro harbour. In 1988 the owners of the *Bowditch* served a writ *in rem* on the *Po* when she was in Southampton. The P&I club put up security to prevent the arrest of the *Po* but the letter of undertaking failed to provide expressly for English jurisdiction. The *Po* was owned and controlled by an Italian corporate body. The defendants again disputed the jurisdiction of the English court on the basis that the action should have been brought in Italy, the country of their domicile, and applied for a stay under Article 2 of the Convention. With regard to the arrest of the *Po* the court reached a different conclusion. It held that the English court had jurisdiction under the Collision Convention, and that therefore Article 57 of the Brussels Convention applied, which meant that the Convention did not deprive the English courts of jurisdiction.[6] Article 1 of the Collision Convention expressly allowed the English court to assume jurisdiction, despite the fact that there had been no arrest, since "arrest could have been effected" and "other security had been furnished".

In *The Prinsengracht*[7] the plaintiffs issued a writ *in rem* against a vessel owned by the defendants who were domiciled in England. A bail bond had been issued by the owners' P&I club. In general a bail bond is given to the court by a third party, which is usually the P&I club. If the defendant fails to pay any judgment then the judgment may be enforced against the guarantor. The bail bond did not contain an

5. Nigel Meeson, pp.94–98. See also Nigel Meeson, "The Effect of the Brussels Convention 1968 on the English Action *in rem*", *P&I International*, April 1992, pp.5–7.
6. [1990] 1 Lloyd's Rep. 418; [1991] 2 Lloyd's Rep. 206 (CA).
7. [1993] 1 Lloyd's Rep. 41.

express submission of the substantive dispute by the defendant to the jurisdiction of the English court. Despite the existence of the bond the plaintiffs proceeded to arrest the ship but then released her almost immediately. The owners disputed jurisdiction and argued that Article 2 of the 1968 Brussels Convention gave jurisdiction to the courts of the Netherlands, their domicile. They also argued that by providing the bail voluntarily they had deprived the plaintiffs of the right to arrest the vessel and therefore to establish jurisdiction under Article 7 of the Arrest Convention. The other argument brought by the owners was that the only way to establish jurisdiction was to arrest the vessel before the bail bond had been lodged. As this was not the case the court's jurisdiction could not be invoked by an arrest. The owners' arguments were rejected on different grounds. First, the owners had voluntarily acknowledged the issue of the writ even before it was served. Secondly, by giving bail they had submitted to the jurisdiction. Thirdly, in any event, and notwithstanding the fact that bail had been lodged, the arrest was not wrongful.

Following these cases one may conclude that if the ship is within the jurisdiction and if the claim falls within the Collision Convention the provision of security to prevent arrest should be sufficient to establish *in rem* jurisdiction. For maritime claims other than collision actions, provided the vessel is arrested, the Arrest Convention should enable the court to assume *in rem* jurisdiction regardless of the domicile of the defendant.

SPECIAL JURISDICTION

Article 17 sets out provisions regarding certain written jurisdiction clauses. They have the effect of granting the chosen court exclusive jurisdiction in relation to the dispute even though the relevant clause itself does not purport to confer exclusivity. Provided the conditions imposed by Article 17 are satisfied a written agreement to English jurisdiction which may be contained in a charterparty or bill of lading or other contract will be sufficient to establish jurisdiction under the Convention. But with regard to a bill of lading particular problems may arise because the bill of lading containing a jurisdiction clause is often issued after the contract between the shipper and the carrier has been concluded without mentioning jurisdiction. And even if the jurisdiction clause is agreed between the shipper and the carrier how does it affect a third party indorsee of the bill? In *Partenreederei MS Tilly Russ* v. *Haven en Vervoebedrijf Nova NV*[8] the original wording of Article 17 was considered by the European Court of Justice. The Court held that a jurisdiction clause printed in a standard form bill of lading was an agreement in writing or an oral agreement evidenced in writing for the purposes of Article 17. In order to reach this conclusion the agreement of the parties to the bill should have been evidenced in writing. If the clause had been orally agreed by the parties before its incorporation in the bill it should be signed by the carrier to confirm it in writing. The bill must otherwise be part of a continuing business

8. [1985] 1 Q.B. 931.

relationship between the parties, which itself was governed by general conditions including the jurisdiction clause. If under the relevant national law the holder of the bill of lading succeeds to the shipper's rights and obligations then the clause which was effective between the carrier and the shipper would also be effective between the carrier and a holder of the bill of lading. The jurisdiction agreement may also be "in a form which accords with practices which the parties have established between themselves, or in international trade or commerce, in a form which accords with a usage of which the parties are or ought to have been aware and which in such trade or commerce is widely known to, and regularly observed by, parties to contracts of the type involved in the particular trade and commerce concerned". If Article 17 applies there is no question of *forum non conveniens.*

Under Article 5(5) of the Convention "a person domiciled in a Contracting State may, in another Contracting State, be sued . . . as regards a dispute arising out of the operations of a branch, agency or other establishment, in the courts for the place in which the branch, agency or other establishment is situated". Under this Article "branch, agency or other establishment" is subject to the direction and control of the parent body. Therefore in shipping cases if the defendant shipowner or charterer was domiciled in France and merely had an agent or broker acting *ad hoc* in England it is not possible to rely on Article 5(5).

In matters related to tort Article 5(3) provides that "a person domiciled in a Contracting State may, in another Contracting State, be sued . . . in matters relating to tort, delict or quasi-delict, in the courts for the place where the harmful event occurred". "The place where the harmful event occurred" includes both the place where the defendant acted and where the harm occurred.[9] But this concept may give rise to difficulties in the case of accidents on board a ship or involving a ship, such as a collision. For instance if the accident takes place on board a ship on the high seas the courts of the state whose flag she flies will be the relevant place for the purposes of Article 5(3). If the accident takes place on board a ship in the territorial waters of a state that state's courts will have jurisdiction irrespective of the flag of the ship. If the accident is between two or more ships on the high seas, then English law will apply irrespective of the flags of the ships.[10] In cases where the accident is between two or more ships in the territorial waters of a state again that state in which the accident occurred will govern jurisdiction.

Article 5(1) of the Convention provides that "a person domiciled in a Contracting State may, in another Contracting State, be sued . . . in matters relating to a contract, in the courts for the place of performance of the obligation in question". This jurisdiction applies only in respect of defendants who are "domiciled in a Contracting State". Reliance on this jurisdiction is not obligatory. In shipping cases it is difficult to ascertain the "obligation in question" which forms the basis of the plaintiff's claim. In an example given by Brice the characteristic

9. See Geoffrey Brice, "Maritime Claims: The European Judgments Convention" [1987] L.M.C.L.Q. 286–287.

10. But if the wrongful act took place on board one or more ships the courts might hold that the place of the tort was related to the flag of both or either of the ships.

obligation of the owner under a charterparty is presumed to be providing the ship in accordance with the charter terms.[11] With regard to the "place of performance" there are difficulties in charterparty disputes as it may not be easy to decide the place of performance. In some charterparty cases the place of performance may be a place other than the territory of any contracting state where the plaintiff must rely on Article 2 as Article 5(1) cannot apply.

The Convention applies in civil and commercial matters, but it does not apply to arbitration. Therefore parties are free to arbitrate their disputes. The Convention does not apply to proceedings to enforce or set aside an award or decision incorporating an award or to any matter concerned with the conduct of arbitration proceedings. In cases where a charterparty dispute is based on a contract containing an arbitration clause or parties have agreed to submit their disputes to arbitration the courts of contracting states will give effect to those agreements to arbitrate although they may have jurisdiction under the Convention. The parties' right to arbitrate will be limited to the extent which their national laws permit.

JURISDICTION UNDER INTERNATIONAL CONVENTIONS

Contracting states to the Convention may be parties also to other Conventions. Article 55 specifies a list of bilateral Conventions between Member States concerning jurisdiction and the recognition of judgments which were superseded by the 1968 Convention. These Conventions continue in force in respect of matters outside the 1968 Convention (Article 56). Under Article 57 the Convention shall not affect any other Convention which in relation to a particular matter governs jurisdiction.

(a) The 1952 Arrest Convention

The Convention specifies the jurisdictions in which proceedings against defendants domiciled in a contracting state may be brought in civil and commercial matters. But in some cases a claimant may wish to bring proceedings in a state other than the one specified, to be able to arrest the defendant's ship. Such right is preserved by Article 57 if the case is the one covered by the Arrest Convention, which provides for jurisdiction over the merits of a maritime claim as defined in Article 1 through arrest. Jurisdiction may be founded under Article 7 of the 1952 Arrest Convention and that will be within the terms of the Brussels Convention by virtue of its Article 57.[12]

11. Geoffrey Brice, pp.285–286.

12. The applicability of the Arrest Convention under Art. 57 was clearly recognised in *The Deichland*.

(b) The 1952 Collision Convention

Article 1 of the 1952 Convention provides that in a collision action it is not necessary to effect an arrest if bail or other security has been provided in lieu of arrest. But the ship should have entered the jurisdiction so that it could have been arrested.[13]

(c) The 1992 Oil Pollution Conventions

Under Article IX of the International Convention for Civil Liability for Oil Pollution Damage 1992 jurisdiction is limited to the state or states in which the damage occurs or preventive measures have been taken to prevent or minimise pollution damage. Under the Fund Convention 1992 an action against the fund for compensation shall be brought only before a court competent under Article IX of the 1992 Liability Convention.

STAY OF PROCEEDINGS

The Convention aims to provide for the determination of disputes by one court only on the basis of rules provided in the Convention. Thus in cases where it is possible to establish jurisdiction under the Convention the court may be required to decline jurisdiction under Article 21 or Article 22. These Articles provide as follows:

> "21. When proceedings involving the same cause of action and between the same parties are brought in the courts of different Contracting States, any court other than the court first seised shall of its own motion stay its proceedings until such time as the jurisdiction of the court first seised is established. Where the jurisdiction of the court first seised is established, any court other than the court first seised shall decline jurisdiction in favour of that court.
>
> 22. Where related actions are brought in the courts of different Contracting States, any court other than the court first seised may, while the actions are pending at first instance, stay its proceedings."

Article 21 applies only in proceedings involving the same cause of action. The proceedings must be brought in the courts of different contracting states and must be between the same parties. Under these conditions a court must decline jurisdiction. The matter must be considered, whether or not raised by the parties. The court has to decline jurisdiction "of its own motion". Under the Article the choice given by the power is between declining jurisdiction and staying the proceedings pending a decision on the jurisdiction of the earlier court. The choice is not between staying and continuing the proceedings.

In *The Kherson* the question was whether an action *in rem* and one *in personam* can constitute proceedings involving "the same cause of action and between the

13. In *The Bowditch* v. *The Po* it was confirmed by the CA that, under Art. 57 of the Brussels Convention, jurisdiction may properly be established under the 1952 Collision Convention.

same parties" under Article 21. The plaintiff cargo owners had commenced proceedings *in personam* in Rotterdam but were unable to arrest the vessel. Then they commenced proceedings in England *in rem* and arrested the vessel. The court held that although the English action was *in rem*, the cause of action was clearly the same in both Holland and England. It was also held that as the shipowners had acknowledged service in England the action then took effect as one *in personam* and not *in rem* so that the same parties were involved. Consequently, the court decided that it had to decline jurisdiction in favour of the Rotterdam proceedings.[14]

In addition to the proceedings brought in England if there are proceedings on the merits in another contracting state the requirement of having proceedings in different contracting states will be fulfilled. Difficulties have arisen in relation to the requirements that the proceedings must involve the same cause of action and be between the same parties.

Article 22 confers power on a court other than the court first seised to stay proceedings. For the purposes of this Article, actions are deemed to be related where they are so closely connected that it is expedient to hear and determine them together to avoid the risk of irreconcilable judgments resulting from separate proceedings. The power may be exercised only when both proceedings "are pending at first instance". A power is given to decline jurisdiction if one of the parties so applies, the court other than the first seised has the power to consolidate the actions and the court first seised has jurisdiction over both actions. When is a court first seised of proceedings? Is it when the writ has been issued or when it is served? In *The Freccia del Nord*[15] it was held that the court is seised of *in rem* proceedings when the writ is served. Similarly in *The Duke of Yare* the Court of Appeal held that in an ordinary and straightforward case the English court does not become seised of an action for the purposes of Article 22 on the issue of the writ, but only when the writ is served on the defendant.[16] For the purposes of section 21(4) of the Supreme Court Act 1981, an *in rem* action is brought when the writ is issued. The jurisdiction of the court is only fully invoked in an *in rem* action when the writ is served on the *res*.

COLLISION ACTIONS AND THE 1982 ACT

Prior to the coming into force of the 1982 Act the English courts had a discretion to stay an action on the basis that there was a clearly more approprite forum for the resolution of the dispute, that that other forum had competent jurisdiction and that the matter could be tried more suitably in the interests of all the parties and the ends of justice.[17] When the relationship between the 1952 Arrest Convention and Articles 21 and 22 of the Jurisdiction Convention considered together is examined,

14. [1992] 2 Lloyd's Rep. 261.
15. [1989] 1 Lloyd's Rep. 388.
16. *Dresser UK Ltd.* v. *Falcongate Freight Management Ltd.* [1992] 2 All E.R. 450.
17. See *The Abidin Daver*, n.48, Chapter 26 *supra*.

a confusion has arisen as a result of *The Netty*. In that case[18] it was held that Article 57 of the Jurisdiction Convention provided that the Convention did not apply if a special international agreement provided otherwise and that the Arrest Convention of 1952 was such an agreement. Article 2 of the Arrest Convention provided that a ship flying the flag of one of its contracting states could be arrested in the jurisdiction of any of the contracting states and Article 7 provided that the courts of the country in which the arrest was made would have jurisdiction to determine the merits of the case. This case was decided before the Jurisdiction Convention became part of English law. Therefore the court was not concerned with the provisions of the Jurisdiction Convention relating to *lis alibi pendens*. Despite the confusion following this case it was clear that *in rem* jurisdiction, allowing service on and arrest of a vessel, was not prohibited by the Convention. *The Nordglimt* shows the problems encountered where the defendant in an *in rem* Admiralty action is domiciled in a contracting state. This case involved a Danish ship owned by a company incorporated in Copenhagen. The receivers of a cargo and their insurers had brought an action in Holland against the owners of the carrying cargo and claimed damages for damage and shortage. *In rem* proceedings were then begun by the receivers in England against a sister ship.

The judge applied Article 22 of the Jurisdiction Convention, since the 1982 Act was then in force, and held that as the 1952 Arrest Convention does not deal with *lis alibi pendens* the 1952 Convention and the 1968 Convention should be read together and, in so far as any dispute over the case proceeding in another court or a problem with a related action arose, the solution would be found in the 1968 Convention. So the judge directly applied Article 22 and on the basis of the ordinary principles of *forum conveniens*, which he considered was still within the inherent jurisdiction of the court. Hobhouse J. held that in order for the parties to be the same in both jurisdictions for the purposes of Article 21, the actions in both jurisdictions had to be either *in rem* or *in personam*, since an action *in rem* is not, at the time when it is commenced, an action between the same parties as an action *in personam*.[19] The problem was further considered in *The Linda*. In March 1987, a collision occurred in international waters off the coast of Belgium. On 14 April, the owners of the *Linda* arrested the other ship, the *Arco Humber*. Security was provided and she was released from arrest. On 16 April, the owners of the *Linda* began proceedings in Holland claiming damages from the owners of the *Arco Humber*. On 18 April, the owners of the *Arco Humber* issued a writ *in rem* out of the Admiralty Registry in England against the *Linda* and the writ was served on the *Linda* in England. She was arrested, then released on 22 May after providing security. So the English action which had started *in rem* had become an action *in personam* by reason of the release of the *Linda* and the acknowledgment of service of the writ by her owners. The owners asked the Admiralty Court to decline jurisdiction in favour of the Dutch court pursuant to Articles 21 and 22 as the English and Dutch proceedings involved the same cause of action, both

18. [1981] 2 Lloyd's Rep. 57.
19. [1987] 2 Lloyd's Rep. 470.

proceedings were between the same parties and the Dutch court was the first seised. In this case it was held that the 1952 Arrest Convention contained no provisions which prescribed what should happen if two actions arising out of the same incident were commenced in different countries. Accordingly, Articles 21 and 22 did not affect the Arrest Convention and the English court had to give effect to one or other of those Articles. Sheen J. applied the Jurisdiction Convention and considered that the decision in *The Norglimt* could be properly distinguished so that Article 21 applied, on the basis that the English action *in rem* against the *Linda* had become an action *in personam* by reason of the release of the *Linda*.[20]

In a collision case if a ship is owned by a person or company domiciled in a contracting state, an action *in personam* may be brought against that person in the contracting state in which he is domiciled. If the collision occurs in the territorial waters of a contracting state and the defendant in the collision action is domiciled in a contracting state, then under Article 5(3) it is possible to sue the defendant in the courts of the country in which the collision occurred. If two actions *in personam* are brought in different contracting states by the two shipowners involved in a collision, then Article 21, rather than Article 22, ought to apply to the situation. If two actions *in personam* are brought in different contracting states by the owners of two vessels involved in a collision then the second court seised must decline jurisdiction in favour of the court seised originally, unless the jurisdiction of the first court is contested, in which case the second court may stay its proceedings.

20. [1988] 1 Lloyd's Rep. 175.

APPENDIX I

TOTAL LOSSES BY REGISTRATION

Registration	1991			1992			1993		
	No.	GT	Age	No.	GT	Age	No.	GT	Age
Australia	1	1 211	17	...	...	...	...	...	...
Bahamas	6	79 607	17	1	13 176	27	2	55 183	15
Brazil	...	...	...	2	8 429	18	1	18 318	20
Canada	5	952	22	4	4 449	37	2	625	40
China	1	4 791	21	1	9 705	18	3	23 489	23
China (Taiwan)	2	781	16	1	1 051	23	1	1 720	1
Cyprus	17	238 342	20	3	24 171	22	5	115 019	21
Denmark	4	826	18	1	149	24	1	398	29
Denmark (Dis)	1	1 167	4	1	1 599	8	1	1 354	3
France	2	2 805	17	1	791	35	...	...	...
Germany	2	625	32	...	...	...	2	1 209	42
Greece	12	175 897	24	9	111 605	23	6	58 009	23
Honduras	15	13 977	32	17	15 789	21	20	39 593	26
Hong Kong	1	75 330	9	...	...	...	1	3 738	20
India	2	93 426	20	2	34 518	16	1	10 092	15
Indonesia	7	14 078	17	8	14 928	19	13	19 249	20
Italy	11	126 900	22	4	4 367	26	5	2 729	28
Japan	20	17 721	13	13	3 807	12	14	11 963	13
Korea (South)	19	108 754	20	15	82 264	21	13	40 744	23
Liberia	6	150 291	16	3	168 021	19	3	68 881	20
Malaysia	1	892	23	2	2 183	25	3	8 244	32
Malta	16	134 403	26	13	176 263	22	8	52 348	21
Mexico	1	377	9	5	2 032	16	...	...	...
Netherlands	3	1 592	17	...	...	...	1	5 613	25
Norway	...	...	...	12	4 052	30	7	3 358	25
Norway (Nis)	3	17 830	15	5	7 132	23	2	6 996	23
Panama	34	169 140	21	40	403 949	21	33	176 480	21
Peru	...	...	...	1	266	...	1	7 902	23
Philippines	9	77 575	21	4	8 749	19	6	18 944	23
Poland	...	...	...	1	107	13	1	3 015	16
Russia	...	...	...	1	10 977	19	8	23 495	17
(USSR)	3	6 580	19	...	...	...	...	...	...
Saint Vincent	12	8 089	31	9	13 529	28	16	68 537	23
Singaporc	3	46 900	22	2	2 594	16	2	18 779	21
Spain	11	4 433	23	6	3 617	21	11	6 158	19
Sweden	2	5 313	31	...	...	...	1	499	17
Turkey	9	128 668	14	7	75 928	22	3	1 478	14
United Kingdom	4	1 684	24	3	854	24	1	120	19
United States of America	12	16 495	20	19	4 333	32	19	25 657	24
Other Flag States	64	209 674	22	44	214 497	20	45	62 241	22
Totals	321	1 937 126	21	260	1 429 881	22	262	962 177	22

SOURCE: Lloyd's Register of Shipping

1994			1995			1996		
No.	GT	Age	No.	GT	Age	No.	GT	Age
...	...	...	2	22 090	17	...	...	...
3	34 240	21	5	36 295	19	2	15 333	22
...	...	...	...	...	...	...	...	...
2	254	23	2	24 298	39	1	105	7
1	2 161	19	2	12 602	18	7	45 209	17
...	...	...	1	3 518	21	2	7 484	26
14	440 239	21	4	161 358	16	7	137 673	20
4	1 884	18	...	...	...	...	...	...
1	1 510	11	...	...	...	...	...	...
3	685	21	1	625	19	3	519	42
...	...	...	3	2 233	36	2	3 443	29
4	35 639	31	3	5 265	32	5	25 872	26
10	15 255	27	12	16 293	26	9	6 109	33
...	...	...	1	15 865	3	...	...	...
2	19 241	19	4	2 783	20	1	10 011	20
6	5 773	24	9	19 144	18	2	598	20
3	29 317	32	2	2 188	30	2	400	11
6	1 266	12	24	9 413	14	11	5 303	9
11	11 286	21	7	6 530	24	1	1 574	25
2	96 434	22	4	211 832	16	2	55 828	6
...	...	...	2	8 238	16	2	11 779	23
10	229 206	21	4	106 254	22	5	77 803	23
1	16 087	14	...	...	...	...	...	...
...	...	...	2	1 375	20	1	344	16
3	1 794	17	5	2 648	22	...	...	...
1	499	16	1	419	28	...	...	...
29	235 910	22	22	134 851	24	27	185 964	21
1	355	1	3	1 044	23	1	274	7
4	14 151	20	3	1 920	21	2	5 253	22
3	8 797	20	...	...	...	...	...	...
13	56 493	25	12	21 955	24	10	29 145	20
...	...	...	...	...	...	...	...	...
10	29 181	24	6	15 768	24	6	30 493	25
...	...	...	3	1 892	8	...	...	...
4	1 648	13	9	3 110	21	6	3 904	19
...	...	...	1	252	35	...	...	...
4	69 889	19	5	3 926	34	6	35 639	23
7	1 688	18	2	534	26	2	661	38
12	11 375	28	16	4 090	18	14	2 781	24
40	373 501	25	43	89 367	30	40	191 850	26
214	1 745 758	23	225	949 975	23	179	891 351	23

APPENDIX II

TOTAL LOSSES BY REGISTRATION AND SHIPTYPE CATEGORY

Registration	*Shiptype Category*	*Total Losses* No.	GT	Age	**ATL** No.	GT	Age	**CTL** No.	GT	Age
Antigua & Barbuda	General Cargo	2	648	43	1	350	40	1	298	46
Argentina	Fish Catching	1	396	23	1	396	23	...	...	...
Bahamas	General Cargo	1	1 034	18	...	...	...	1	1 034	18
	Container	1	14 299	25	...	...	...	1	14 299	25
Belgium	Fish Catching	1	224	11	1	224	11	...	...	...
Belize	Other Bulk Dry	1	762	22	1	762	22	...	...	...
	General Cargo	8	6 462	26	8	6 462	26	...	...	...
	Fish Catching	1	495	21	1	495	21	...	...	...
Brunei	Offshore Supply	1	832	7	1	832	7	...	...	...
Canada	Fish Catching	1	105	7	1	105	7	...	...	...
Chile	General Cargo	1	197	36	1	197	36	...	...	...
China	Bulk Dry	1	16 473	15	1	16 473	15	...	...	...
	General Cargo	5	19 053	19	5	19 053	19	...	...	...
	Container	1	9 683	11	1	9 683	11	...	...	...
China (Taiwan)	Other Bulk Dry	1	6 698	21	1	6 698	21	...	...	...
	General Cargo	1	786	31	1	786	31	...	...	...
Cyprus	Bulk Dry	3	107 404	19	3	107 404	19	...	...	...
	General Cargo	3	28 848	19	2	15 539	19	1	13 309	21
	Towing/Pushing	1	1 421	26	1	1 421	26	...	...	...
Ecuador	Passenger	1	2 204	34	1	2 204	34	...	...	...
Estonia	General Cargo	1	196	23	1	196	23	...	...	...
Faeroes	General Cargo	1	134	41	1	134	41	...	...	...
France	Other Bulk Dry	1	196	57	1	196	57	...	...	...
	Fish Catching	2	323	35	2	323	35	...	...	...
Germany	General Cargo	1	372	34	...	...	...	1	372	34
	Fish Catching	1	3 071	23	...	...	...	1	3 071	23
Ghana	Fish Catching	1	604	52	1	604	52	...	...	...
Greece	Oil	1	827	24	...	...	...	1	827	24
	Bulk Dry	1	16 565	18	...	...	...	1	16 565	18
	Other Bulk Dry	1	4 045	24	1	4 045	24	...	...	...
	General Cargo	1	1 640	26	1	1 640	26	...	...	...
	Passenger/Ro-Ro Cargo	1	2 795	38	1	2 795	38	...	...	...
Guyana	General Cargo	1	200	31	1	200	31	...	...	...
Honduras	General Cargo	7	5 556	32	7	5 556	32	...	...	...
	Offshore Supply	1	199	40	1	199	40	...	...	...
	Other Activities	1	354	34	1	354	34	...	...	...
India	General Cargo	1	10 011	20	...	...	...	1	10 011	20
Indonesia	General Cargo	1	402	14	1	402	14	...	...	...
	Passenger/Ro-Ro Cargo	1	196	26	1	196	26	...	...	...
Iran	General Cargo	1	281	58	1	281	58	...	...	...
Italy	Passenger	1	262	6	1	262	6	...	...	...
	Fish Catching	1	138	16	1	138	16	...	...	...
Japan	Other Bulk Dry	2	1 107	11	1	610	12	1	497	9
	General Cargo	5	1 893	5	5	1 893	5	...	...	...
	Refrigerated Cargo	1	499	14	...	...	...	1	499	14
	Passenger/Ro-Ro Cargo	1	1 327	16	1	1 327	16	...	...	...
	Fish Catching	1	124	16	1	124	16	...	...	...
	Dredging	1	353	11	1	353	11	...	...	...
Korea (North)	General Cargo	1	6 676	16	1	6 676	16	...	...	...
Korea (South)	General Cargo	1	1 574	25	1	1 574	25	...	...	...
Liberia	Container	2	55 828	6	1	16 915	2	1	38 913	10
Malaysia	General Cargo	2	11 779	23	2	11 779	23	...	...	...
Malta	Liquefied Gas	1	10 909	27	...	...	...	1	10 909	27
	Bulk Dry	2	51 114	23	1	35 427	23	1	15 687	23
	General Cargo	2	15 780	22	1	10 964	15	1	4 816	29
Mauritania	Fish Catching	1	359	6	1	359	6	...	...	...
Morocco	Fish Catching	1	160	25	1	160	25	...	...	...
Namibia	Fish Catching	1	173	32	1	173	32	...	...	...
Netherlands	Fish Catching	1	344	16	1	344	16	...	...	...
Nigeria	General Cargo	1	10 983	17	...	...	...	1	10 983	17

Registration	*Shiptype Category*	*Total Losses*			**ATL**			**CTL**		
		No.	GT	Age	No.	GT	Age	No.	GT	Age
Panama	Liquefied Gas	1	2 102	22	1	2 102	22	...	...	...
	Chemical	2	5 677	19	2	5 677	19	...	...	...
	Oil	5	8 922	24	5	8 922	24	...	...	...
	Bulk Dry	2	73 645	21	1	9 996	18	1	63 649	23
	General Cargo	10	57 276	18	8	43 698	19	2	13 578	17
	Container	2	14 052	25	...	...	...	2	14 052	25
	Refrigerated Cargo	1	1 232	28	1	1 232	28	...	...	...
	Other Dry Cargo	1	20 884	26	1	20 884	26	...	...	...
	Fish Catching	2	1 967	25	1	149	16	1	1 818	34
	Research	1	207	18	1	207	18	...	...	...
Peru	Fish Catching	1	274	7	1	274	7	...	...	...
Philippines	General Cargo	1	998	15	1	998	15	...	...	...
	Other Dry Cargo	1	4 255	29	1	4 255	29	...	...	...
Portugal	Other Bulk Dry	1	2 917	25	...	...	...	1	2 917	25
	Dredging	1	1 287	32	1	1 287	32	...	...	...
Portugal (Mar)	Oil	1	140 277	19	...	...	...	1	140 277	19
Romania	General Cargo	1	3 374	15	1	3 374	15	...	...	...
Russia	Oil	1	4 198	23	...	...	...	1	4 198	23
	General Cargo	2	2 735	29	2	2 735	29	...	...	...
	Fish Catching	4	8 264	19	2	2 940	16	2	5 324	22
	Other Fishing	3	13 948	13	3	13 948	13	...	...	...
Saint Vincent	Bulk Dry	2	21 703	26	2	21 703	26	...	...	...
	General Cargo	3	6 846	27	2	3 493	25	1	3 353	33
	Refrigerated Cargo	1	1 944	16	1	1 944	16	...	...	...
Somali Republic	Research	1	445	31	1	445	31	...	...	...
Spain	General Cargo	1	1 389	30	1	1 389	30	...	...	...
	Ro-Ro Cargo	1	1 254	13	1	1 254	13	...	...	...
	Fish Catching	4	1 261	18	4	1 261	18	...	...	...
Syria	General Cargo	1	1 928	27	1	1 928	27	...	...	...
Thailand	General Cargo	1	1 597	28	1	1 597	28	...	...	...
	Passenger	1	2 210	27	...	...	...	1	2 210	27
Togo	Offshore Supply	1	196	34	1	196	34	...	...	...
Turkey	General Cargo	5	33 739	24	3	9 807	23	2	23 932	26
	Passenger/Ro-Ro Cargo	1	1 900	20	1	1 900	20	...	...	...
United Arab Emirates	General Cargo	1	3 330	24	...	...	...	1	3 330	24
United Kingdom	General Cargo	1	382	40	1	382	40	...	...	...
	Fish Catching	1	279	35	1	279	35	...	...	...
United States of America	Fish Catching	11	1 828	23	8	1 225	27	3	603	15
	Offshore Supply	1	633	25	1	633	25	...	...	...
	Towing/Pushing	2	320	24	1	148	30	1	172	17
Uruguay	Fish Catching	1	294	17	1	294	17	...	...	...
Vanuata	General Cargo	1	1 491	5	1	1 491	5	...	...	...
Vietnam	General Cargo	1	518	25	1	518	25	...	...	...

APPENDIX III

CALENDAR OF MAJOR OIL SPILL INCIDENTS AND INTERNATIONAL RESPONSE

Year	Summary of events and ramifications
1967	*Torrey Canyon* ran aground near the south coast of England—salvage attempts were thwarted by an explosion in the engine room and precipitous government action—120,000 tonnes of oil was spilt and grave environmental damage resulted with clean-up costs exceeding $16m. This was the first major pollution incident to come to public attention.
1969	International conference relating to intervention on the high seas in cases of oil pollution gave right to coastal states to intervene to protect their interests.
1969	International Conference on Civil Liability for Oil Pollution Damage (CLC) provided uniform rules for compensating victims of oil pollution.
1971	*Wafra* became immobilised off Cape Agulhas, South Africa. The engine room flooded and the ship drifted ashore after an attempted tow. It was later refloated and towed 200 miles offshore where it was sunk by high-explosive missiles—40,000 tonnes of oil burnt or spilt—resulting in serious environmental damage to coastal habitats.
1971	Conference on the Establishment of an International Compensation Fund for Oil Pollution Damage (FUND)—provided compensation to supplement CLC.
1973	International Conference on Oil Pollution (MARPOL)—covering tanker design, equipment and operation.
1973	*Andros Patris* was abandoned off the northwest coast of Spain when a 50 foot crack in her hull developed, an explosion and fire followed; about 50,000 tonnes of oil was spilt and 30 people died. The ship was towed south and the remaining oil transferred to other tankers.
1974	*Metula* of 210,035 dwt grounded and suffered serious bottom damage in Magellan Straits, Chile, 53,000 tonnes of oil leaked from the ship covering 1,000 square miles of water and coastline with heavy environmental loss—the remaining cargo, some 140,000 tonnes, was transferred to other tankers and the ship towed to Brazil and sold for scrapping in Spain.
1975	*Jacob Mærsk* struck bottom in Leixos Harbour, Portugal, causing an explosion in her engine room and setting the whole ship on fire—six people died—smoke from burning oil badly affected local inhabitants and oil polluted 20 miles of beaches and fishing beds—the ship eventually broke up—spillage totalled 80,000 tons of oil.
1976	*Urquiola* struck bottom while entering La Coruña, Spain, causing oil leakage. Explosions and a massive fire followed. Widespread pollution damage to shellfish beds and the marine environment resulted. Further explosion caused damage in the town and smoke affected local inhabitants. 108,000 tonnes of oil spilt and total pollution damage and clean-up expenses were put at $62m, the ship was declared a CTL with a settlement of £10.4m.
1977	*Hawaiian Patriot* developed a 100ft crack in her hull 300 miles off Honolulu. Oil leaked into the sea resulting in a large explosion with fire engulfing the ship and oil on the surrounding sea. The ship later broke in two and sank. One man died and 99,000 tons of oil were spilt which drifted away from the islands. The total insured value of ship and cargo amounted to $32m.

Year	*Summary of events and ramifications*
1978	Protocol to MARPOL providing for incidents involving harmful substances and arbitration.
1978	*Amoco Cadiz* was wrecked on north coast of France after steering gear failure in bad weather. Despite attempts to tow the ship offshore after grounding 232,000 tonnes of oil was spilt, causing grave environmental damage. France later introduced measures to distance tankers from her coasts and IMO took measures to improve the performance standards of steering gear on large ships. This was the biggest oil spill ever recorded and litigation on behalf of the French government and other interested parties continued for over 10 years. Final settlement amounted to $32m for ship and cargo and over £253m for claims by French interests.
1979	*Betelgeuse* blew up whilst discharging oil in Bantry Bay, Ireland. There was extensive damage to the terminal, heavy oil pollution and 50 people died; final liability claims amounted to about $120m.
1979	*Atlantic Empress* in collision with another VLCC, the *Aegean Captain*, off Tobago in West Indies. The ships were locked together and on fire after the collision. 29 deaths resulted. The *Atlantic Empress* eventually sank with her cargo of 270,000 tonnes of crude oil. There was also heavy leakage from the *Aegean Captain*. The casualty resulted in a record pay-out by the London marine insurance market.
1979	*Burmah Agate* in collision with bulk carrier *Mimosa* off Texas. Fire and explosions ensued which continued for weeks. About 35,000 tonnes of oil was burnt or spilled. There were 32 deaths. The insured value of both ships was about $13m.
1979	*Independenta* with a cargo of 93,000 tonnes of oil collided with the dry cargo ship *Evrialy* just south of the Bosphorus, Turkey. The *Independenta* caught fire, which spread to the *Evrialy*, 42 crew members of the *Independenta* died and, still burning, she drifted ashore. The ship continued to burn for weeks causing great ecological damage. Eventually the fire burnt itself out. The insured value of the one year old *Independenta* was over $40m and the value of the *Evrialy*, £2.5m.
1983	*Assimi*, laden with 53,000 tonnes of oil, experienced a major fire in her engine room when 55 miles off Muscat, Oman. Fire spread to the rest of the ship and to oil on the surrounding seas: the ship was abandoned and later towed offshore where she sank.
1983	*Castillo de Bellver*, laden with 250,000 tonnes of light crude oil from the Gulf bound for Spain, burst into flames off Saldanha Bay, South Africa. The crew abandoned ship; three men died and the ship sank after a huge explosion. An offshore wind pushed the oil away from beaches but black oily rain from the fire caused damage to crops and freshly shorn sheep. The ship was insured for $18m and the cargo for $54m.
1985	*Nova* sank 20 miles south of Iran in the Gulf with the loss of 70,000 tonnes of oil.

Year	*Summary of events and ramifications*
1988	*Odyssey* sank 700 miles off Nova Scotia with a loss of 132,000 tonnes of oil.
1989	*Exxon Valdez* ran aground off coast of Alaska spilling 37,000 tonnes of oil and causing grave environmental damage. The total cost of the clean-up, fines and compensation exceeded $1 billion and up to $10bn, if current court awards are upheld. This led to OPA 90 and other measures.
1989	*World Prodigy* grounded south of Newport, RI, USA, spilling 1,400 tonnes of oil which threatened rich sea life in Narragansett Bay.
1989	*Kharg V*, 248,000 dwt, exploded off the north west coast of Africa, spilling 70,000 tonnes of crude oil. She had to be towed 1,500 miles south (leaking oil) for transferral of remaining cargo, as no nearby nation would allow her near their coast.
1989	International Conference on Salvage—providing uniform rules on salvage and making provision for limited compensation to salvors for preventing oil pollution.
1990	The US Oil Pollution Act (OPA 90) calls for double hull tankers and other stringent anti-pollution requirements.
1990	International Conference on Oil Pollution Preparedness, Response and Co-operation calls for worldwide organisation for oil spills: effective 1995.
1991	*ABT Summer* sank 700 miles off Angola with loss of 260,000 tonnes of oil.
1991	*Agip Abruzzo*, carrying 80,000 tonnes of light crude was struck at night while lying at anchor off Livorno, Italy, by the ro-ro ferry *Moby Prince*. The ferry caught fire and 143 persons died. The fire lasted seven days and moderate oil pollution affected a wide area of coast. The collision resulted in legal action in the Italian courts and declaration of limitation of liability by the owner of the *Moby Prince* on the grounds of crew negligence. Claims totalling about $50m have been presented against the owners of the *Moby Prince*, whereas the limit of liability will amount to about $4.4m.
1991	*Haven*, laden with 144,000 tonnes of crude oil, caught fire and suffered a series of explosions whilst at anchor off Genoa. The ship broke into three main parts and was towed into shallower water where she sank. Much of the oil was consumed by fire but over 10,000 tonnes of oil and burnt oil residue escaped. Claims made on IOPCF amounted to £485m ($728m) in 1993, greatly exceeding total compensation available under the CLC and Fund Conventions—put at £41m by the IOPCF.
1992	*Aegean Sea* grounded at La Coruña, Spain, in heavy weather. The ship broke in two and caught fire—72,000 tonnes of oil was spilt and wide environmental damage resulted. Subsequent restrictions on fishing affected 3,000 fishermen; claims for clean-up costs exceeded $23m in 1993; and claims for compensation from fishermen and others totalling $152m had been lodged in the Spanish courts. Claims resulting from the *Aegean Sea* and *Haven* casualties include a large number for "pure economic loss": that is economic loss suffered by persons whose property has not been contaminated as a result of the incident. Where claims were seen to be based upon a loss directly resulting from the

Year	Summary of events and ramifications
	casualty, IOPF met the claim, but it is obviously a difficult area where disputes are likely.
1990–95	Extensive amendments to MARPOL 73/78 including requirements under Annex 1 for double hulls on new tankers and enhanced surveys for existing tankers.
1990–95	Extensive amendments to SOLAS 1974 including the mandatory introduction of the International Safety Management (ISM) Code effective from 1988, but earlier for some ships in European waters.
1993	*Braer*, laden with 84,000 tonnes of North Sea crude, experienced engine failure south of the Shetland Islands, UK. The ship was driven ashore in severe weather and broke up on coastal rocks. All cargo and bunker oil escaped, amounting to about 85,000 tonnes, and caused widespread damage to fishing grounds and the environment; oily spray blown ashore contaminated farm-land. This led to extensive claims from the Shetland Island Council and others for the cost of cleaning buildings and damage to amenities, and from the UK government to recompense them for clean-up activities. In January 1996, although claims amounting to $75m had been met, claimants sought a further $100m in compensation. As a result of the *Braer* casualty, the Donaldson Inquiry was set up by the UK.
1994	Conference to agree major amendments to the International Convention on Standards of Training Certification and Watchkeeping. Effective 1997.
1996	*Sea Empress* stranded off entrance to Milford Haven, UK, when entering port. Bad weather made refloating difficult and 65,000 tonnes of oil leaked into the sea and caused grave damage to highly sensitive coastal areas and to fishing. The aggregate value of claims is likely to exceed those connected with the *Braer*. The ship was eventually floated and taken into harbour where remaining oil was transferred to other ships.
1996	International Conference on Carriage of Hazardous and Noxious Substances by Sea.
Notes	1. War casualties are not included. 2. The trend towards claims based on "pure economic loss" is likely to continue in the *Braer* and *Sea Empress* settlements.

Sources: ITOPF, LLP

APPENDIX IV

LARGEST TANKER SPILLS

	Flag	*Vessel Name*	*Year Built*	*DWT*	*Date*	*Approximate Amount Lost/ Spilled: Tons*
1	Greek	Atlantic Empress	1974	297,361	19.7.79	257,000
2	Spanish	Castillo De Bellver	1978	271,540	6.8.83	239,000
3	Liberian	Amoco Cadiz	1974	237,439	16.3.78	221,000
4	Liberian	Odyssey	1974	138,392	10.11.88	132,000
5	South Korean	Sea Star	1968	122,230	19.12.72	121,000
6	Liberian	Torrey Canyon	1959	118,285	18.3.67	118,000
7	Liberian	Hawaiian Patriot	1965	99,447	23.2.77	101,000
8	Romanian	Independenta	1978	152,408	15.11.79	95,000
9	Spanish	Urquiola	1973	111,225	12.5.76	91,000
10	Liberian	Braer	1975	89,730	5.1.93	86,825
11	Greek	Irenes Serenade	1965	105,460	23.2.80	82,000
12	Iranian	Khark 5	1975	284,632	19.12.89	76,000
13	Greek	Aegean Sea	1973	114,036	3.12.92	72,500
14	Norwegian	Nova	1975	239,435	6.12.85	70,000
15	Greek	Epic Colocotronis	1965	65,673	13.5.75	58,000
16	Liberian	Sinclair Petrolore	1955	56,089	6.12.60	56,000
17	Japanese	Yuyo Maru No. 10	1966	53,684	9.11.74	52,000
18	Greek	Assimi	1964	59,032	7.1.83	50,000
19	Liberian	Abt Summer	1974	267,801	28.5.91	48,750
20	Greek	Andros Patria	1970	222,173	31.12.78	48,000
21	Liberian	World Glory	1959	47,179	14.6.68	46,000
22	British	British Ambassador	1958	45,650	10.1.75	45,000
23	Netherlands Antilles	Metula	1968	210,719	9.8.74	45,000
24	Greek	Pericles G.C.	1967	60,044	9.12.83	44,000
25	Liberian	Mandolin II	1958	42,074	28.2.68	41,000
26	Danish	Jakob Mærsk	1966	89,412	29.1.75	41,000
27	Liberian	Burmah Agate	1963	62,663	1.11.79	41,000
28	Liberian	Wafra	1956	50,560	27.2.71	40,000
29	Cyprus	Haven	1973	232,164	11.4.91	39,000
30	Uruguayan	Juan Antonio Lavalleja	1975	131,663	28.12.80	38,000
31	Liberian	Napier	1957	39,180	9.6.73	37,000
32	American	Exxon Valdez	1986	214,861	24.3.89	36,448
33	Liberian	Corinthos	1963	54,988	31.1.75	36,000
34	Greek	Trader	1957	34,545	11.6.72	34,000
35	Cyprus	Athenian Venture	1975	31,016	22.4.88	30,000

Since the publication of this list, the Liberian motor tanker *Sea Empress* spilled approximately 71,000 tonnes of crude oil off Milford Haven in February 1996. The *Sea Empress* would therefore appear at No. 14 on the list. Her entry would read as follows:

14 Liberian Sea Empress 1993 147,273 15.2.96 71,000— —Off Milford Haven Grounded.

Approximate Amount Lost/ Spilled: Barrels	*Approximate Amount Lost/ Spilled: Gallons*	*Location*	*Type of Casualty*
1,890,000	65,792,000	Off Trinidad	Collision/fire/sank
1,760,000	61,184,000	Off Cape Town	Fire and explosions/sank
1,628,000	56,576,000	Off Brittany	Grounded
925,000	33,792,000	Mid-Atlantic	Broke in two/fire/sank
886,930	30,976,000	Gulf of Oman	Collision/fire/sank
864,940	27,848,000	Off Isles of Scilly	Grounded
742,000	25,856,000	300 miles W of Hawaii	Hull cracked/explosion/fire/sank
696,000	24,320,000	Bosporus, Turkey	Collision/fire/explosion
670,000	23,296,000	Corunna	Grounded/fire/sank
636,427	22,227,200	Off Sumburgh Head, Shetland Is	Grounded
600,000	20,992,000	Off Pylos	Explosion/fire/sank
560,000	19,456,000	Off Morocco	Heavy weather damage/fire/explosion
531,425	18,560,000	Corunna	Grounded/fire/explosions
513,100	17,920,000	Off Kharg Island	Collision
427,000	14,848,000	NW of Puerto Rico	Fire in engine-room/hull cracked
410,480	14,336,000	300 miles off Brazil	Explosion/fire/sank
375,000	13,312,000	Tokyo Bay	Collision/fire/explosion/sank
370,000	12,800,000	Off Muscat	Fire/explosions/sank
357,000	12,480,000	900 miles W of Luanda	Explosion/fire/sank
350,000	12,288,000	Off NW Spain	Hull cracked/explosion/fire
337,000	11,776,000	90 miles off Durban	Broke in two/fire/sank
330,000	11,520,000	West of Iwojima	Leaking/sank
330,000	11,520,000	Strait of Magellan	Grounded
324,000	11,264,000	200 miles off Doha	Fire/explosions/sank
300,000	10,496,000	340 miles off Oregon	Collision/fire
300,000	10,496,000	Off Leixoes	Grounded/fire/explosions
300,000	10,496,000	Galveston Bay	Collision/fire/explosions
293,200	10,240,000	Off Cape Agulhas	Grounded
289,770	9,984,000	Genoa Roads	Explosions/fire/sank
280,000	9,728,000	Arzew	Grounded
270,000	9,472,000	Chile	Grounded/broke in two
268,000	9,330,688	Prince William Sound, Alaska	Grounded
264,000	9,216,000	Philadelphia	Collision/explosions/fire
249,220	8,704,000	Off SW Greece	Leakage/sank
219,900	7,680,000	800 miles SE of Nova Scotia	Explosion/fire/broke in two/sank

Source: LMIS

APPENDIX V

CONTRACTING STATES TO THE 1969 CLC AND 1971 FC

Contracting State	CLC	Deposit of Instrument	Entry into Force
Albania	A	6 April 1994	5 July 1994
Algeria	A	14 June 1974	19 June 1975
Antigua and Barbuda	A	23 June 1997	20 September 1997
Australia	R	7 November 1983	5 February 1984
Bahamas	A	22 July 1976	20 October 1976
Bahrain	A	3 May 1996	1 August 1996
Barbados	A	6 May 1994	4 August 1994
Belgium	R	12 January 1977	12 April 1977
Belize	A	2 April 1991	1 July 1991
Benin	A	1 November 1985	30 January 1986
Brazil	R	17 December 1976	17 March 1977
Brunei Darussalam	A	29 September 1992	28 December 1992
Cambodia	A	28 November 1994	26 February 1995
Cameroon	R	14 May 1984	12 August 1984
Canada	A	24 January 1989	24 April 1989
Chile	A	2 August 1977	31 October 1977
China	A	30 January 1980	29 April 1980
China (Hong Kong Special Admin Region)			1 April 1976
Colombia	A	26 March 1990	24 June 1990
Côte d'Ivoire	R	21 June 1973	19 June 1975
Croatia	S		8 October 1991
Cyprus	A	19 June 1989	17 September 1989
Denmark	A	2 April 1975	19 June 1975
Djibouti	A	1 March 1990	30 May 1990
Dominican Republic	R	2 April 1975	19 June 1975
Ecuador	A	23 December 1976	23 March 1977
Egypt	A	3 February 1989	4 May 1989
Equatorial Guinea	A	24 April 1996	23 July 1996
Estonia	A	1 December 1992	1 March 1993
Fiji	A	15 August 1972	19 June 1975
Finland	R	10 October 1980	8 January 1981
France	R	17 March 1975	19 June 1975
Gabon	A	21 January 1982	21 April 1982
Gambia	A	1 November 1991	30 January 1992
Georgia	A	19 April 1994	18 July 1994
Germany	R	20 May 1975	18 August 1975
Ghana	R	20 April 1978	19 July 1978
Greece	A	29 June 1976	27 September 1976
Guatemala	AA	20 October 1982	18 January 1983
Iceland	R	17 July 1980	15 October 1980
India	A	1 May 1987	30 July 1987
Indonesia	R	1 September 1978	30 November 1978
Ireland	R	19 November 1992	17 February 1993
Italy	R	27 February 1979	28 May 1979
Japan	A	3 June 1976	1 September 1976
Kazakhstan	A	7 March 1994	5 June 1994
Kenya	A	15 December 1992	15 March 1993
Kiribati	P	1 April 1976	1 April 1976
Kuwait	A	2 April 1981	1 July 1981
Latvia	A	10 July 1992	8 October 1992
Lebanon	A	9 April 1974	19 June 1975
Liberia	A	25 September 1972	19 June 1975
Luxembourg	A	14 February 1991	15 May 1991
Malaysia	A	6 January 1995	6 April 1995

Fund	*Deposit of Instrument*	*Entry into Force*
A	6 April 1994	5 July 1994
R	2 June 1975	16 October 1978
A	23 June 1997	20 September 1997
A	10 October 1994	5 January 1995
A	22 July 1976	16 October 1978
A	3 May 1996	1 August 1996
A	6 May 1994	4 August 1994
R	1 December 1994	1 March 1995
A	1 November 1985	30 January 1986
A	29 September 1992	28 December 1992
A	14 May 1984	12 August 1984
A	24 January 1989	24 April 1989
A	13 March 1997	11 June 1997
		16 October 1978
A	5 October 1987	3 January 1988
S		8 October 1991
A	26 July 1989	24 October 1989
A	2 April 1975	16 October 1978
A	1 March 1990	30 May 1990
A	1 December 1992	1 March 1993
A	4 March 1983	2 June 1983
R	10 October 1980	8 January 1981
A	11 May 1978	16 October 1978
A	21 January 1982	21 April 1982
A	1 November 1991	30 January 1992
R	30 December 1976	16 October 1978
R	20 April 1978	16 October 1978
A	16 December 1986	16 March 1987
A	17 July 1980	15 October 1980
A	10 July 1990	8 October 1990
A	1 September 1978	30 November 1978
R	19 November 1992	17 February 1993
A	27 February 1979	28 May 1979
R	7 July 1976	16 October 1978
A	15 December 1992	15 March 1993
A	2 April 1981	1 July 1981
A	25 September 1972	16 October 1978
A	6 January 1995	6 April 1995

Contracting State	*CLC*	*Deposit of Instrument*	*Entry into Force*
Maldives	A	16 March 1981	14 June 1981
Malta	A	27 September 1991	26 December 1991
Marshall Islands	A	24 January 1994	24 April 1994
Mauritania	A	19 November 1995	17 February 1996
Mauritius	A	6 April 1995	5 July 1995
Mexico	A	13 May 1994	11 August 1994
Monaco	R	21 August 1975	19 November 1975
Morocco	A	11 April 1974	19 June 1975
Mozambique	A	23 December 1996	23 March 1997
Netherlands	R	9 September 1975	8 December 1975
New Zealand	A	27 April 1976	26 July 1976
Nicaragua	A	4 June 1996	2 September 1996
Nigeria	A	7 May 1981	5 August 1981
Norway	A	21 March 1975	19 June 1975
Oman	A	24 January 1985	24 April 1985
Panama	R	7 January 1976	6 April 1976
Papua New Guinea	A	12 March 1980	10 June 1980
Peru	A	24 February 1987	25 May 1987
Poland	R	18 March 1976	16 June 1976
Portugal	R	26 November 1976	24 February 1977
Qatar	A	2 June 1988	31 August 1988
Republic of Korea	A	18 December 1978	18 March 1979
Russian Federation	S		26 December 1991
Saint Kitts and Nevis	A	14 September 1994	13 December 1994
Saint Vincent and Grenadines	A	19 April 1989	18 July 1989
Saudi Arabia	A	15 April 1993	14 July 1993
Senegal	A	27 March 1972	19 June 1975
Seychelles	A	12 April 1988	11 July 1988
Sierra Leone	A	13 August 1993	11 November 1993
Singapore	A	16 September 1981	15 December 1981
Slovenia	S		25 June 1991
Solomon Islands	P	1 April 1976	1 April 1976
South Africa	A	17 March 1976	15 June 1976
Spain	R	8 December 1975	7 March 1976
Sri Lanka	A	12 April 1983	11 July 1983
Sweden	R	17 March 1975	19 June 1975
Switzerland	R	15 December 1987	14 March 1988
Syrian Arab Republic	A	6 February 1975	19 June 1975
Tonga	A	1 February 1996	1 May 1996
Tunisia	A	4 May 1976	2 August 1976
Tuvalu	S		1 October 1978
United Arab Emirates	A	15 December 1983	14 March 1984
United Kingdom	R	17 March 1975	19 June 1975
Vanuatu	A	2 February 1983	3 May 1983
Venezuela	A	21 January 1992	20 April 1992
Yemen	A	6 March 1979	4 June 1979
Yugoslavia	R	18 June 1976	16 September 1976
TOTALS (excluding Provisional)	98		

A = Accession AA = Acceptance AAA = Approval P = Provisional R = Ratification
S = Succession.

Fund	*Deposit of Instrument*	*Entry into Force*
A	16 March 1981	14 June 1981
A	27 September 1991	26 December 1991
A	30 November 1994	28 February 1995
A	19 November 1995	17 February 1996
A	6 April 1995	5 July 1995
A	13 May 1994	11 August 1994
A	23 August 1979	21 November 1979
A	31 December 1992	31 March 1993
A	23 December 1996	23 March 1997
AAA	3 August 1982	1 November 1982
A	22 November 1996	20 February 1997
A	11 September 1987	10 December 1987
R	21 March 1975	16 October 1978
A	10 May 1985	8 August 1985
A	12 March 1980	10 June 1980
R	16 September 1985	15 December 1985
R	11 September 1985	10 December 1985
A	2 June 1988	31 August 1988
A	8 December 1992	8 March 1993
S		26 December 1991
A	14 September 1994	13 December 1994
A	12 April 1988	11 July 1988
A	13 August 1993	11 November 1993
S		25 June 1991
A	8 October 1981	6 January 1982
A	12 April 1983	11 July 1983
R	17 March 1975	16 October 1978
R	4 July 1996	2 October 1996
A	6 February 1975	16 October 1978
A	1 February 1996	1 May 1996
A	4 May 1976	16 October 1978
S		16 October 1978
A	15 December 1983	14 March 1984
R	2 April 1976	16 October 1978
A	13 January 1989	13 April 1989
A	21 January 1992	20 April 1992
R	16 March 1978	16 October 1978
74		

In the case of the following states, the Conventions are extended to certain dependent territories.

a. For the following dependent territories the relevant dates are the same as for the parent state:

Australia:
Ashmore and Cartier Islands
Australian Antarctic Territory
Christmas Islands
Cocos (Keeling) Islands
Coral Sea Islands
Heard and McDonald Islands
Norfolk Island

Denmark:
Faroe Islands
Greenland

France:
French Guiana
Guadeloupe
Martinique
Réunion
French Polynesia
New Caledonia
Various Indian Ocean islands including Tromelin, Juan de Nova, Bassas de India, Glorieuse and Europa
Southern and Antarctic Territories
Wallis and Futuna Islands
Mayotte
St. Pierre and Miquelon
Clipperton

Norway:
Jan Mayen Island
Svalbard archipelago (including Spitsbergen and other islands) Norwegian Antarctic Territories

Portugal:
The Azores
Madeira

b. For the United Kingdom the relevant dates are as follows:

Dependent territory	*CLC Entry into Force*	*Fund Entry into Force*
Anguilla	1 September 1984	1 September 1984
Bailiwick of Guernsey	1 February 1976	16 October 1978
Bailiwick of Jersey	1 February 1976	16 October 1978
Isle of Man	1 February 1976	16 October 1978
Bermuda	3 February 1976	16 October 1978
British Indian Ocean Territory	1 April 1976	16 October 1978
British Virgin Islands	1 April 1976	16 October 1978
Cayman Islands	1 April 1976	16 October 1978
Falkland Islands and Dependencies	1 April 1976	16 October 1978
Gibraltar	1 April 1976	16 October 1978
Montserrat	1 April 1976	16 October 1978
Pitcairn	1 April 1976	16 October 1978
St. Helena and Dependencies	1 April 1976	16 October 1978
Turks and Caicos Islands	1 April 1976	16 October 1978
Akrotiri and Dhekelia	1 April 1976	16 October 1978

Further information can be obtained from:
International Maritime Organization (IMO)
4 Albert Embankment
London SE1 7SR
Tel: 0171-735 7611 Fax: 0171-587 3210

International Oil Pollution Compensation Fund (IOPC Fund)
4 Albert Embankment
London SE1 7SR
Tel: 0171-582 2606 Fax: 0171-735 0326

Status of Hong Kong[1]
On 30 June 1997 Hong Kong ceased to be a dependent territory of the United Kingdom. From 1 July 1997 Hong Kong became a Special Administrative Region of the People's Republic of China. The change in Hong Kong's status brought two questions into consideration. First, whether the 1971 Fund Convention would apply to pollution damage caused after 30 June 1997 in the Hong Kong Special Administrative Region. Secondly, whether contributions are payable to the 1971 Fund in respect of oil received in the Hong Kong Special Administrative Region after 30 June 1997. The Director submitted these questions to the Assembly for consideration.

The 1971 Fund Convention entered into force for the United Kingdom on 16 October 1978. The ratification of the Convention was effective in respect of certain dependent territories, including Hong Kong. The People's Republic of China has not acceded to the 1971 Fund Convention. Although the People's Republic of China is not party to the Convention, it is recognised that the governments of the United Kingdom and the People's Republic of China have agreed that the Convention should continue to apply to the Hong Kong Special Administrative Region. According to the Joint Declaration signed between these countries, the 1971 Fund Convention and the Protocol to this Convention adopted

1. I.O.P.C.F. 71 FUND/A.20/27, 7 October 1997, pp.1–4.

on 19 December 1976, which applies to Hong Kong at present, will continue to apply to the Hong Kong Special Administrative Region with effect from 1 July 1997.

In its 20th Session the Assembly took the view that it would be beneficial to the compensation system, as well as to the 1971 Fund, if the 1971 Fund were to continue to apply to oil pollution damage in Hong Kong and if contributions were to be payable in respect of contributing oil received in Hong Kong after 30 June 1997.

The questions that arose in respect of the 1971 Fund Convention do not arise in respect of the 1992 Fund Convention as the United Kingdom government did not declare its ratification of the 1992 Fund Protocol to be effective for Hong Kong.

APPENDIX VI

STATES PARTY TO THE 1992 PROTOCOLS TO THE CLC AND FC

Contracting State	*CLC*	*Deposit of Instrument*	*Entry into Force*
Australia	A	9 October 1995	9 October 1996
Bahamas*	A	1 April 1997	1 April 1998
Bahrain	A	3 May 1996	3 May 1997
Cyprus*	A	12 May 1997	12 May 1998
Denmark	R	30 May 1995	30 May 1996
Egypt	A	21 April 1995	30 May 1996
Finland	AA	24 November 1995	24 November 1996
France	AAA	29 September 1994	30 May 1996
Germany	R	29 September 1994	30 May 1996
Greece	R	9 October 1995	9 October 1996
Ireland*	A	15 May 1997	16 May 1998
Jamaica*	A	6 June 1997	6 June 1998
Japan	A	24 August 1994	30 May 1996
Liberia	A	5 October 1995	5 October 1996
Marshall Islands	A	16 October 1995	16 October 1996
Mexico	A	13 May 1994	30 May 1996
Monaco*	R	8 November 1996	8 November 1997
Netherlands*	A	15 November 1996	15 November 1997
Norway	R	3 April 1995	30 May 1996
Oman	A	8 July 1994	30 May 1996
Philippines*	A	7 June 1997	7 June 1998
Singapore*	A	18 September 1997	18 September 1998
South Korea*	A	7 March 1997	16 May 1998
Spain*	A	6 July 1995	6 July 1996
Sweden	R	25 May 1995	30 May 1996
Switzerland	A	4 July 1996	4 July 1997
Tunisia*	AAA	29 January 1997	29 January 1998
United Kingdom	A	29 September 1994	30 May 1996
Uruguay*	A	9 July 1997	9 July 1998
TOTALS	29		

Codes: A = Accession AA = Acceptance AAA = Approval R = Ratification

* States for which an instrument has been deposited for either of the Conventions but which is not yet in force for that State.

Contracting State	*Fund*	*Deposit of Instrument*	*Entry into Force*
Australia	A	9 October 1995	9 October 1996
Bahamas*	A	1 April 1997	1 April 1998
Bahrain	A	3 May 1996	3 May 1997
Cyprus*	A	12 May 1997	12 May 1998
Denmark	R	30 May 1995	30 May 1996
Finland	AA	24 November 1995	24 November 1996
France	AAA	29 September 1994	30 May 1996
Germany	R	29 September 1994	30 May 1996
Greece	R	9 October 1995	9 October 1996
Ireland*	A	15 May 1997	16 May 1998
Jamaica*	A	6 June 1997	6 June 1998
Japan	A	24 August 1994	30 May 1996
Liberia	A	5 October 1995	5 October 1996
Marshall Islands	A	16 October 1995	16 October 1996
Mexico	A	13 May 1994	30 May 1996
Monaco*	R	8 November 1996	8 November 1997
Netherlands*	A	15 November 1996	15 November 1997
Norway	R	3 April 1995	30 May 1996
Oman	A	8 July 1994	30 May 1996
Philippines*	A	7 June 1997	7 June 1998
South Korea*	A	7 March 1997	16 May 1998
Spain*	A	6 July 1995	16 May 1998
Sweden	R	25 May 1995	30 May 1996
Tunisia*	AAA	29 January 1997	29 January 1998
United Kingdom	A	29 September 1996	30 May 1996
Uruguay*	A	9 July 1997	9 July 1998
	26		

*States for which an instrument has been deposited for either of the Conventions but which is not yet in force for that State.

In the case of the following States, the Conventions are extended to certain dependent territories.

Australia:
Ashmore and Cartier Islands
Australian Antarctic Territory
Christmas Islands
Cocos (Keeling) Islands
Coral Sea Islands
Heard and McDonald Islands
Norfolk Island

Denmark:
Faroe Islands
Greenland

France:
French Guiana
Guadeloupe

Martinique
Réunion
French Polynesia
New Caledonia
Southern and Antarctic Territories
Wallis and Futuna Islands
Mayotte
St. Pierre and Miquelon
Clipperton
Various Indian Ocean islands including Tromelin, Juan de Nova, Bassas de India, Glorieuse and Europa

Norway:
Jan Mayen Island
Svalbard archipelago (including Spitsbergen and other islands)
Norwegian Antarctic Territories

United Kingdom:
Bailiwick of Jersey
Isle of Man
Falkland Islands and Dependencies
Montserrat
South Georgia and the South Sandwich Islands

Further information can be obtained from:
International Maritime Organization (IMO)
4 Albert Embankment
London SE1 7SR
Tel: 0171-735 7611 Fax: 0171-587 3210

International Oil Pollution Compensation Fund (IOPC Fund)
4 Albert Embankment
London SE1 7SR
Tel: 0171-582 2606 Fax: 0171-735 0326

APPENDIX VII

1992 CIVIL LIABILITY CONVENTION*

(INTERNATIONAL CONVENTION ON CIVIL LIABILITY FOR OIL POLLUTION DAMAGE 1992—CONSOLIDATED TEXT)

* Text of Conventions on Liability and Compensation for Oil Pollution Damage can be obtained from the International Oil Pollution Compensation Fund.

INTERNATIONAL CONVENTION ON CIVIL LIABILITY FOR OIL POLLUTION DAMAGE, 1992 (CLC 1992)

The States Parties to the present Convention,

CONSCIOUS of the dangers of pollution posed by the worldwide maritime carriage of oil in bulk,

CONVINCED of the need to ensure that adequate compensation is available to persons who suffer damage caused by pollution resulting from the escape or discharge of oil from ships,

DESIRING to adopt uniform international rules and procedures for determining questions of liability and providing adequate compensation in such cases,

HAVE AGREED as follows:

Article I

For the purposes of this Convention:

1. "Ship" means any sea-going vessel and sea-borne craft of any type whatsoever constructed or adapted for the carriage of oil in bulk as cargo, provided that a ship capable of carrying oil and other cargoes shall be regarded as a ship only when it is actually carrying oil in bulk as cargo and during any voyage following such carriage unless it is proved that it has no residues of such carriage of oil in bulk aboard.

2. "Person" means any individual or partnership or any public or private body, whether corporate or not, including a State or any of its constituent subdivisions.

3. "Owner" means the person or persons registered as the owner of the ship or, in the absence of registration, the person or persons owning the ship. However in the case of a ship owned by a State and operated by a company which in that State is registered as the ship's operator, "owner" shall mean such company.

4. "State of the ship's registry" means in relation to registered ships the State of registration of the ship, and in relation to unregistered ships the State whose flag the ship is flying.

5. "Oil" means any persistent hydrocarbon mineral oil such as crude oil, fuel oil, heavy diesel oil and lubricating oil, whether carried on board a ship as cargo or in the bunkers of such a ship.

6. "Pollution damage" means:

(a) loss or damage caused outside the ship by contamination resulting from the escape or discharge of oil from the ship, wherever such escape or discharge may occur, provided that compensation for impairment of the environment other than loss of profit from such impairment shall be limited to costs of reasonable measures of reinstatement actually undertaken or to be undertaken;

(b) the costs of preventive measures and further loss or damage caused by preventive measures.

7. "Preventive measures" means any reasonable measures taken by any person after an incident has occurred to prevent or minimise pollution damage.

8. "Incident" means any occurrence, or series of occurrences having the same origin, which causes pollution damage or creates a grave and imminent threat of causing such damage.

9. "Organization" means the International Maritime Organization.

10. "1969 Liability Convention" means the International Convention on Civil Liability for Oil Pollution Damage, 1969. For States Parties to the Protocol of 1976 to that Convention, the term shall be deemed to include the 1969 Liability Convention as amended by that Protocol.

Article II

This Convention shall apply exclusively:

(a) to pollution damage caused:

(i) in the territory, including the territorial sea, of a Contracting State; and

(ii) in the exclusive economic zone of a Contracting State, established in accordance with international law, or, if a Contracting State has not established such a zone, in an area beyond and adjacent to the territorial sea of that State determined by that State in accordance with international law and extending not more than 200 nautical miles from the baselines from which the breadth of its territorial sea is measured;

(b) to preventive measures, wherever taken, to prevent or minimise such damage.

Article III

1. Except as provided in paragraphs 2 and 3 of this Article, the owner of a ship at the time of an incident, or, where the incident consists of a series of occurrences, at the time of the first such occurrence, shall be liable for any pollution damage caused by the ship as a result of the incident.

2. No liability for pollution damage shall attach to the owner if he proves that the damage:

(a) resulted from an act of war, hostilities, civil war, insurrection or a natural phenomenon of an exceptional, inevitable and irresistible character, or

(b) was wholly caused by an act or omission done with intent to cause damage by a third party, or

(c) was wholly caused by the negligence or other wrongful act of any Government or other authority responsible for the maintenance of lights or other navigational aids in the exercise of that function.

3. If the owner proves that the pollution damage resulted wholly or partially either from an act or omission done with intent to cause damage by the person who suffered the damage or from the negligence of that person, the owner may be exonerated wholly or partially from his liability to such person.

4. No claim for compensation for pollution damage may be made against the owner otherwise than in accordance with this Convention. Subject to paragraph 5 of this Article, no claim for compensation for pollution damage under this Convention or otherwise may be made against:

(a) the servants or agents of the owner or the members of the crew;

(b) the pilot or any other person who, without being a member of the crew, performs services for the ship;

(c) any charterer (howsoever described, including a bareboat charterer), manager or operator of the ship;

(d) any person performing salvage operations with the consent of the owner or on the instructions of a competent public authority;

(e) any person taking preventive measures;

(f) all servants or agents of persons mentioned in sub-paragraphs (c), (d) and (e);

unless the damage resulted from their personal act or omission, committed with the intent to cause such damage, or recklessly and with knowledge that such damage would probably result.

5. Nothing in this Convention shall prejudice any right of recourse of the owner against third parties.

Article IV

When an incident involving two or more ships occurs and pollution damage results therefrom, the owners of all the ships concerned, unless exonerated under Article III, shall be jointly and severally liable for all such damage which is not reasonably separable.

Article V

1. The owner of a ship shall be entitled to limit his liability under this Convention in respect of any one incident to an aggregate amount calculated as follows:

(a) 3 million units of account for a ship not exceeding 5,000 units of tonnage;

(b) for a ship with a tonnage in excess thereof, for each additional unit of tonnage, 420 units of account in addition to the amount mentioned in sub-paragraph (a);

provided, however, that this aggregate amount shall not in any event exceed 59.7 million units of account.

2. The owner shall not be entitled to limit his liability under this Convention if it is proved that the pollution damage resulted from his personal act or omission, committed with the intent to cause such damage, or recklessly and with knowledge that such damage would probably result.

3. For the purpose of availing himself of the benefit of limitation provided for in paragraph 1 of this Article, the owner shall constitute a fund for the total sum representing the limit of his liability with the Court or other competent authority of any one of the Contracting States in which action is brought under Article IX or, if no action is brought, with any Court or other competent authority in any one of the Contracting States in which an action can be brought under Article IX. The fund can be constituted either by depositing the sum or by producing a bank guarantee or other guarantee, acceptable under the legislation of the Contracting State where the fund is constituted, and considered to be adequate by the Court or other competent authority.

4. The fund shall be distributed among the claimants in proportion to the amounts of their established claims.

5. If before the fund is distributed the owner or any of his servants or agents or any person providing him insurance or other financial security has as a result of the incident in question paid compensation for pollution damage, such person shall, up to the amount he has paid, acquire by subrogation the rights which the person so compensated would have enjoyed under this Convention.

6. The right of subrogation provided for in paragraph 5 of this Article may also be exercised by a person other than those mentioned therein in respect of any amount of compensation for pollution damage which he may have paid but only to the extent that such subrogation is permitted under the applicable national law.

7. Where the owner or any other person establishes that he may be compelled to pay at a

later date in whole or in part any such amount of compensation, with regard to which such person would have enjoyed a right of subrogation under paragraphs 5 or 6 of this Article, had the compensation been paid before the fund was distributed, the Court or other competent authority of the State where the fund has been constituted may order that a sufficient sum shall be provisionally set aside to enable such person at such later date to enforce his claim against the fund.

8. Claims in respect of expenses reasonably incurred or sacrifices reasonably made by the owner voluntarily to prevent or minimise pollution damage shall rank equally with other claims against the fund.

9(a). The "unit of account" referred to in paragraph 1 of this Article is the Special Drawing Right as defined by the International Monetary Fund. The amounts mentioned in paragraph 1 shall be converted into national currency on the basis of the value of that currency by reference to the Special Drawing Right on the date of the constitution of the Fund referred to in paragraph 3. The value of the national currency, in terms of the Special Drawing Right, of a Contracting State which is a member of the International Monetary Fund shall be calculated in accordance with the method of valuation applied to the International Monetary Fund in effect on the date in question for its operations and transactions. The value of the national currency, in terms of the Special Drawing Right, of a Contracting State which is not a member of the International Monetary Fund shall be calculated in a manner determined by that State.

9(b). Nevertheless, a Contracting State which is not a member of the International Monetary Fund and whose law does not permit the application of the provisions of paragraph 9(a) may, at the time of ratification, acceptance, approval of or accession to this Convention or at any time thereafter, declare that the unit of account referred to in paragraph 9(a) shall be equal to 15 gold francs. The gold franc referred to in this paragraph corresponds to sixty-five and a half milligrammes of gold of millesimal fineness nine hundred. The conversion of the gold franc into the national currency should be made according to the law of the State concerned.

9(c). The calculation mentioned in the last sentence of paragraph 9(a) and the conversion mentioned in paragraph 9(b) shall be made in such manner as to express in the national currency of the Contracting State as far as possible the same real value for the amounts in paragraph 1 as would result from the application of the first three sentences of paragraph 9(a). Contracting States shall communicate to the depositary the manner of calculation pursuant to paragraph 9(a), or the result of the conversion in paragraph 9(b) as the case may be, when depositing an instrument of ratification, acceptance, approval of or accession to this Convention and whenever there is a change in either.

10. For the purpose of this Article the ship's tonnage shall be the gross tonnage calculated in accordance with the tonnage measurement regulations contained in Annex I of the International Convention on Tonnage Measurement of Ships, 1969.

11. The insurer or other person providing financial security shall be entitled to constitute a fund in accordance with this Article on the same conditions and having the same effect as if it were constituted by the owner. Such a fund may be constituted even if, under the provisions of paragraph 2, the owner is not entitled to limit his liability, but its constitution shall in that case not prejudice the rights of any claimant against the owner.

Article VI

1. Where the owner, after an incident, has constituted a fund in accordance with Article V, and is entitled to limit his liability,

(a) no person having a claim for pollution damage arising out of that incident shall be entitled to exercise any rights against any other assets of the owner in respect of such claims;

(b) the Court or other competent authority of any Contracting State shall order the release of any ship or other property belonging to the owner which has been arrested in respect of a claim for pollution damage arising out of that incident, and shall similarly release any bail or other security furnished to avoid such arrest.

2. The foregoing shall, however, only apply if the claimant has access to the Court administering the fund and the fund is actually available in respect of his claim.

Article VII

1. The owner of a ship registered in a Contracting State and carrying more than 2,000 tons of oil in bulk as cargo shall be required to maintain insurance or other financial security, such as the guarantee of a bank or a certificate delivered by an international compensation fund, in the sums fixed by applying the limits of liability prescribed in Article V, paragraph 1 to cover his liability for pollution damage under this Convention.

2. A certificate attesting that insurance or other financial security is in force in accordance with the provisions of this Convention shall be issued to each ship after the appropriate authority of a Contracting State has determined that the requirements of paragraph 1 have been complied with. With respect to a ship registered in a Contracting State such certificate shall be issued or certified by the appropriate authority of the State of the ship's registry; with respect to a ship not registered in a Contracting State it may be issued or certified by the appropriate authority of any Contracting State. This certificate shall be in the form of the annexed model and shall contain the following particulars:

(a) name of ship and port of registration;

(b) name and principal place of business of owner;

(c) type of security;

(d) name and principal place of business of insurer or other person giving security and, where appropriate, place of business where the insurance or security is established;

(e) period of validity of certificate which shall not be longer than the period of validity of the insurance or other security.

3. The certificate shall be in the official language or languages of the issuing State. If the language used is neither English nor French, the text shall include a translation into one of these languages.

4. The certificate shall be carried on board the ship and a copy shall be deposited with the authorities who keep the record of the ship's registry or, if the ship is not registered in a Contracting State, with the authorities of the State issuing or certifying the certificate.

5. An insurance or other financial security shall not satisfy the requirements of this Article if it can cease, for reasons other than the expiry of the period of validity of the insurance or security specified in the certificate under paragraph 2 of this Article, before three months have elapsed from the date on which notice of its termination is given to the authorities referred to in paragraph 4 of this Article, unless the certificate has been surrendered to these authorities or a new certificate has been issued within the said period. The foregoing provisions shall similarly apply to any modification which results in the insurance or security no longer satisfying the requirements of this Article.

6. The State of registry shall, subject to the provisions of this Article, determine the conditions of issue and validity of the certificate.

7. Certificates issued or certified under the authority of a Contracting State in accordance with paragraph 2 shall be accepted by other Contracting States for the purposes of this Convention and shall be regarded by other Contracting States as having the same force as certificates issued or certified by them even if issued or certified in respect of a ship not registered in a Contracting State. A Contracting State may at any time request consultation with the issuing or certifying State should it believe that the insurer or guarantor named in the certificate is not financially capable of meeting the obligations imposed by this Convention.

8. Any claim for compensation for pollution damage may be brought directly against the insurer or other person providing financial security for the owner's liability for pollution damage. In such case the defendant may, even if the owner is not entitled to limit his liability according to Article V, paragraph 2, avail himself of the limits of liability prescribed in Article V, paragraph 1. He may further avail himself of the defences (other than the bankruptcy or winding up of the owner) which the owner himself would have been entitled to invoke. Furthermore, the defendant may avail himself of the defence that the pollution damage resulted from the wilful misconduct of the owner himself, but the defendant shall not avail himself of any other defence which he might have been entitled to invoke in proceedings brought by the owner against him. The defendant shall in any event have the right to require the owner to be joined in the proceedings.

9. Any sums provided by insurance or by other financial security maintained in accordance with paragraph 1 of this Article shall be available exclusively for the satisfaction of claims under this Convention.

10. A Contracting State shall not permit a ship under its flag to which this Article applies to trade unless a certificate has been issued under paragraph 2 or 12 of this Article.

11. Subject to the provisions of this Article, each Contracting State shall ensure, under its national legislation, that insurance or other security to the extent specified in paragraph 1 of this Article is in force in respect of any ship, wherever registered, entering or leaving a port in its territory, or arriving at or leaving an offshore terminal in its territorial sea, if the ship actually carries more than 2,000 tons of oil in bulk as cargo.

12. If insurance or other financial security is not maintained in respect of a ship owned by a Contracting State, the provisions of this Article relating thereto shall not be applicable to such ship, but the ship shall carry a certificate issued by the appropriate authorities of the State of the ship's registry stating that the ship is owned by that State and that the ship's liability is covered within the limits prescribed by Article V, paragraph 1. Such a certificate shall follow as closely as practicable the model prescribed by paragraph 2 of this Article.

Article VIII

Rights of compensation under this Convention shall be extinguished unless an action is brought thereunder within three years from the date when the damage occurred. However, in no case shall an action be brought after six years from the date of the incident which caused the damage. Where this incident consists of a series of occurrences, the six years' period shall run from the date of the first such occurrence.

Article IX

1. Where an incident has caused pollution damage in the territory, including the territorial sea or an area referred to in Article II, of one or more Contracting States or

preventive measures have been taken to prevent or minimise pollution damage in such territory including the territorial sea or area, actions for compensation may only be brought in the Courts of any such Contracting State or States. Reasonable notice of any such action shall be given to the defendant.

2. Each Contracting State shall ensure that its Courts possess the necessary jurisdiction to entertain such actions for compensation.

3. After the fund has been constituted in accordance with Article V the Courts of the State in which the fund is constituted shall be exclusively competent to determine all matters relating to the apportionment and distribution of the fund.

Article X

1. Any judgment given by a Court with jurisdiction in accordance with Article IX which is enforceable in the State of origin where it is no longer subject to ordinary forms of review, shall be recognised in any Contracting State, except:

(a) where the judgment was obtained by fraud; or

(b) where the defendant was not given reasonable notice and a fair opportunity to present his case.

2. A judgment recognised under paragraph 1 of this Article shall be enforceable in each Contracting State as soon as the formalities required in that State have been complied with. The formalities shall not permit the merits of the case to be re-opened.

Article XI

1. The provisions of this Convention shall not apply to warships or other ships owned or operated by a State and used, for the time being, only on government non-commercial service.

2. With respect to ships owned by a Contracting State and used for commercial purposes, each State shall be subject to suit in the jurisdictions set forth in Article IX and shall waive all defences based on its status as a sovereign State.

Article XII

This Convention shall supersede any International Conventions in force or open for signature, ratification or accession at the date on which the Convention is opened for signature, but only to the extent that such Conventions would be in conflict with it; however, nothing in this Article shall affect the obligations of Contracting States to non-Contracting States arising under such International Conventions.

Article XII*bis*

Transitional provisions

The following transitional provisions shall apply in the case of a State which at the time of an incident is a Party both to this Convention and to the 1969 Liability Convention:

(a) where an incident has caused pollution damage within the scope of this Convention, liability under this Convention shall be deemed to be discharged if, and to the extent that, it also arises under the 1969 Liability Convention;

(b) where an incident has caused pollution damage within the scope of this Convention, and the State is a Party both to this Convention and to the International Convention on the Establishment of an International Fund for Compensation for Oil Pollution Damage, 1971, liability remaining to be discharged after the application of subparagraph (a) of this Article shall arise under this Convention only to the extent that pollution damage remains uncompensated after application of the said 1971 Convention;

(c) in the application of Article III, paragraph 4, of this Convention the expression "this Convention" shall be interpreted as referring to this Convention or the 1969 Liability Convention, as appropriate;

(d) in the application of Article V, paragraph 3, of this Convention the total sum of the fund to be constituted shall be reduced by the amount by which liability has been deemed to be discharged in accordance with subparagraph (a) of this Article.

Article XIIter

Final Clauses

The final clauses of this Convention shall be Articles 12 to 18 of the Protocol of 1992 to amend the 1969 Liability Convention. References in this Convention to Contracting States shall be taken to mean references to the Contracting States of that Protocol.

Final Clauses of the Protocol of 1992 to amend the 1969 Civil Liability Convention

Article 12

Signature, Ratification, Acceptance, Approval and Accession

1. This Protocol shall be open for signature at London from 15 January 1992 to 14 January 1994 by all States.

2. Subject to paragraph 4, any State may become a Party to this Protocol by:

(a) signature subject to ratification, acceptance or approval followed by ratification, acceptance or approval; or

(b) accession.

3. Ratification, acceptance, approval or accession shall be effected by the deposit of a formal instrument to that effect with the Secretary-General of the Organization.

4. Any Contracting State to the International Convention on the Establishment of an International Fund for Compensation for Oil Pollution Damage, 1971, hereinafter referred to as the 1971 Fund Convention, may ratify, accept, approve or accede to this Protocol only if it ratifies, accepts, approves or accedes to the Protocol of 1992 to amend that Convention at the same time, unless it denounces the 1971 Fund Convention to take effect on the date when this Protocol enters into force for that State.

5. A State which is a Party to this Protocol but not a party to the 1969 Liability Convention shall be bound by the provisions of the 1969 Liability Convention as amended by this Protocol in relation to other States Parties hereto, but shall not be bound by the provisions of the 1969 Liability Convention in relation to States Parties thereto.

6. Any instrument of ratification, acceptance, approval or accession deposited after the entry into force of an amendment to the 1969 Liability Convention as amended by this Protocol shall be deemed to apply to the Convention so amended, as modified by such amendment.

Article 13

Entry into Force

1. This Protocol shall enter into force twelve months following the date on which ten States including four States each with not less than one million units of gross tanker tonnage have deposited instruments of ratification, acceptance, approval or accession with the Secretary-General of the Organization.

2. However, any Contracting State to the 1971 Fund Convention may, at the time of the deposit of its instrument of ratification, acceptance, approval or accession in respect of this Protocol, declare that such instrument shall be deemed not to be effective for the purposes of this Article until the end of the six-month period in Article 31 of the Protocol of 1992 to amend the 1971 Fund Convention. A State which is not a Contracting State to the 1971 Fund Convention but which deposits an instrument of ratification, acceptance, approval or accession in respect of the Protocol of 1992 to amend the 1971 Fund Convention may also make a declaration in accordance with this paragraph at the same time.

3. Any State which has made a declaration in accordance with the preceding paragraph may withdraw it at any time by means of notification addressed to the Secretary-General of the Organization. Any such withdrawal shall take effect on the date the notification is received, provided that such State shall be deemed to have deposited its instrument of ratification, acceptance, approval or accession in respect of this Protocol on that date.

4. For any State which ratifies, accepts, approves or accedes to it after the conditions in paragraph 1 for entry into force have been met, this Protocol shall enter into force twelve months following the date of deposit by such State of the appropriate instrument.

Article 14

Revision and Amendment

1. A Conference for the purpose of revising or amending the 1992 Liability Convention may be convened by the Organization.

2. The Organization shall convene a Conference of Contracting States for the purpose of revising or amending the 1992 Liability Convention at the request of not less than one-third of the Contracting States.

Article 15

Amendments of Limitation Amounts

1. Upon the request of at least one-quarter of the Contracting States any proposal to amend the limits of liability laid down in Article V, paragraph 1, of the 1969 Liability Convention as amended by this Protocol shall be circulated by the Secretary-General to all Members of the Organization and to all Contracting States.

2. Any amendment proposed and circulated as above shall be submitted to the Legal Committee of the Organization for consideration at a date at least six months after the date of its circulation.

3. All Contracting States to the 1969 Liability Convention as amended by this Protocol, or not Members of the Organization, shall be entitled to participate in the proceedings of the Legal Committee for the consideration and adoption of amendments.

4. Amendments shall be adopted by a two-thirds majority of the Contracting States present and voting in the Legal Committee, expanded as provided for in paragraph 3, on condition that at least one-half of the Contracting States shall be present at the time of voting.

5. When acting on a proposal to amend the limits, the Committee shall take into account the experience of incidents and in particular the amount of damage resulting therefrom, changes in the monetary values and the effect of the proposed amendment on the cost of insurance. It shall also take into account the relationship between the limits in Article V, paragraph 1, of the 1969 Liability Convention as amended by this Protocol and those in paragraph 4 of Article 4 of the International Convention on the Establishment of an International Fund for Compensation for Oil Pollution Damage, 1992.

6. (a) No amendment of the limits of liability under this Article may be considered before 15 January 1998 nor less than five years from the date of entry into force of a previous amendment under this Article. No amendment under this Article shall be considered before this Protocol has entered into force.

(b) No limit may be increased so as to exceed an amount which corresponds to the limit laid down in the 1969 Liability Convention as amended by this Protocol increased by 6 per cent per year calculated on a compound basis from 15 January 1993.

(c) No limit may be increased so as to exceed an amount which corresponds to the limit laid down in the 1969 Liability Convention as amended by this Protocol multiplied by 3.

7. Any amendment adopted in accordance with paragraph 4 shall be notified by the Organization to all Contracting States. The amendment shall be deemed to have been accepted at the end of a period of eighteen months after the date of notification, unless within that period not less than one-quarter of the States that were Contracting States at the time of the adoption of the amendment by the Legal Committee have communicated to the Organization that they do not accept the amendment in which case the amendment is rejected and shall have no effect.

8. An amendment deemed to have been accepted in accordance with paragraph 7 shall enter into force eighteen months after its acceptance.

9. All Contracting States shall be bound by the amendment, unless they denounce this Protocol in accordance with Article 16, paragraphs 1 and 2, at least six months before the amendment enters into force. Such denunciation shall take effect when the amendment enters into force.

10. When an amendment has been adopted by the Legal Committee but the eighteen-month period for its acceptance has not yet expired, a State which becomes a Contracting State during that period shall be bound by the amendment if it enters into force. A state which becomes a Contracting State after that period shall be bound by an amendment which has been accepted in accordance with paragraph 7. In the cases referred to in this paragraph, a State becomes bound by an amendment when that amendment enters into force, or when this Protocol enters into force for that State, if later.

Article 16

Denunciation

1. This Protocol may be denounced by any Party at any time after the date on which it enters into force for that Party.

2. Denunciation shall be effected by the deposit of an instrument with the Secretary-General of the Organization.

3. A denunciation shall take effect twelve months, or such longer period as may be specified in the instrument of denunciation, after its deposit with the Secretary-General of the Organization.

4. As between the Parties to this Protocol, denunciation by any of them of the 1969 Liability Convention in accordance with Article XVI thereof shall not be construed in any way as a denunciation of the 1969 Liability Convention as amended by this Protocol.

5. Denunciation of the Protocol of 1992 to amend the 1971 Fund Convention by a State which remains a Party to the 1971 Fund Convention shall be deemed to be a denunciation of this Protocol. Such denunciation shall take effect on the date on which denunciation of the Protocol of 1992 to amend the 1971 Fund Convention takes effect according to Article 34 of that Protocol.

Article 17

Depositary

1. This Protocol and any amendments accepted under Article 15 shall be deposited with the Secretary-General of the Organization.

2. The Secretary-General of the Organization shall:

(a) inform all States which have signed or acceded to this Protocol of:

(i) each new signature or deposit of an instrument together with the date thereof;

(ii) each declaration and notification under Article 13 and each declaration and communication under Article V, paragraph 9, of the 1992 Liability Convention;

(iii) the date of entry into force of this Protocol;

(iv) any proposal to amend limits of liability which has been made in accordance with Article 15, paragraph 1;

(v) any amendment which has been adopted in accordance with Article 15, paragraph 4;

(vi) any amendment deemed to have been accepted under Article 15, paragraph 7, together with the date on which that amendment shall enter into force in accordance with paragraphs 8 and 9 of that Article;

(vii) the deposit of any instrument of denunciation of this Protocol together with the date of deposit and the date on which it takes effect;

(viii) any denunciation deemed to have been made under Article 16, paragraph 5;

(ix) any communication called for by any Article of this Protocol.

(b) transmit certified true copies of this Protocol to all Signatory States and to all States which accede to this Protocol.

3. As soon as this Protocol enters into force, the text shall be transmitted by the Secretary-General of the Organization to the Secretariat of the United Nations for registration and publication in accordance with Article 102 of the Charter of the United Nations.

Article 18

Languages

This Protocol is established in a single original in the Arabic, Chinese, English, French, Russian and Spanish languages, each text being equally authentic.]

ANNEX

CERTIFICATE OF INSURANCE OR OTHER FINANCIAL SECURITY IN RESPECT OF CIVIL LIABILITY FOR OIL POLLUTION DAMAGE

Issued in accordance with the provisions of Article VII of the International Convention on Civil Liability for Oil Pollution Damage, 1992.

Name of ship	Distinctive number or letters	Port of registry	Name and address of owner

This is to certify that there is in force in respect of the above-named ship a policy of insurance or other financial security satisfying the requirements of Article VII of the International Convention on Civil Liability for Oil Pollution Damage, 1992.

Type of Security ....

....

Duration of Security ....

....

Name and Address of the Insurer(s) and/or Guarantor(s)

Name ....

Address ....

This certificate is valid until ....

Issued or certified by the Government of ....

....

(Full designation of the State)

At .... On ....

(Place) (Date)

....

Signature and Title of issuing or certifying official

Explanatory Notes:

1. If desired, the designation of the State may include a reference to the competent public authority of the country where the certificate is issued.
2. If the total amount of security has been furnished by more than one source, the amount of each of them should be indicated.
3. If security is furnished in several forms, these should be enumerated.
4. The entry "Duration of Security" must stipulate the date on which such security takes effect.

APPENDIX VIII

1992 FUND CONVENTION*

(INTERNATIONAL CONVENTION ON THE ESTABLISHMENT OF AN INTERNATIONAL FUND FOR COMPENSATION FOR OIL POLLUTION DAMAGE, 1992—CONSOLIDATED TEXT)

* Text of Conventions on Liability and Compensation for Oil Pollution Damage can be obtained from the International Oil Pollution Compensation Fund.

INTERNATIONAL CONVENTION ON THE ESTABLISHMENT OF AN INTERNATIONAL FUND FOR COMPENSATION FOR OIL POLLUTION DAMAGE, 1992 (1992 FC)[1]

The States Parties to the present Convention,

BEING PARTIES to the International Convention on Civil Liability for Oil Pollution Damage, adopted at Brussels on 29 November 1969,

CONSCIOUS of the dangers of pollution posed by the world-wide maritime carriage of oil in bulk,

CONVINCED of the need to ensure that adequate compensation is available to persons who suffer damage caused by pollution resulting from the escape or discharge of oil from ships,

CONSIDERING that the International Convention of 29 November 1969, on Civil Liability for Oil Pollution Damage, by providing a régime for compensation for pollution damage in Contracting States and for the costs of measures, wherever taken, to prevent or minimize such damage, represents a considerable progress towards the achievement of this aim,

CONSIDERING HOWEVER that this régime does not afford full compensation for victims of oil pollution damage in all cases while it imposes an additional financial burden on shipowners,

CONSIDERING FURTHER that the economic consequences of oil pollution damage resulting from the escape or discharge of oil carried in bulk at sea by ships should not exclusively be borne by the shipping industry but should in part be borne by the oil cargo interests,

CONVINCED of the need to elaborate a compensation and indemnification system supplementary to the International Convention on Civil Liability for Oil Pollution Damage with a view to ensuring that full compensation will be available to victims of oil pollution incidents and that the shipowners are at the same time given relief in respect of the additional financial burdens imposed on them by the said Convention,

TAKING NOTE of the Resolution on the Establishment of an International Compensation Fund for Oil Pollution Damage which was adopted on 29 November 1969 by the International Legal Conference on Marine Pollution Damage,

HAVE AGREED as follows:

General Provisions

Article 1

For the purposes of this Convention:

1. "1992 Liability Convention" means the International Convention on Civil Liability for Oil Pollution Damage, 1992.

1 bis. "1971 Fund Convention" means the International Convention on the Establishment of an International Fund for Compensation for Oil Pollution Damage, 1971. For States Parties to the Protocol of 1976 to that Convention, the term shall be deemed to include the 1971 Fund Convention as amended by that Protocol.

2. "Ship", "Person", "Owner", "Oil", "Pollution Damage", "Preventive Measures",

1. 1992 Fund Convention.

"Incident" and "Organization" have the same meaning as in Article I of the 1992 Liability Convention.

3. "Contributing Oil" means crude oil and fuel oil as defined in sub-paragraphs (a) and (b) below:

(a) "Crude Oil" means any liquid hydrocarbon mixture occurring naturally in the earth whether or not treated to render it suitable for transportation. It also includes crude oils from which certain distillate fractions have been removed (sometimes referred to as "topped crudes") or to which certain distillate fractions have been added (sometimes referred to as "spiked" or "reconstituted" crudes).

(b) "Fuel Oil" means heavy distillates or residues from crude oil or blends of such materials intended for use as a fuel for the production of heat or power of a quality equivalent to the "American Society for Testing and Materials' Specification for Number Four Fuel Oil (Designation D 396-69)", or heavier.

4. "Unit of account" has the same meaning as in Article V, paragraph 9, of the 1992 Liability Convention.

5. "Ship's tonnage" has the same meaning as in Article V, paragraph 10, of the 1992 Liability Convention.

6. "Ton", in relation to oil, means a metric ton.

7. "Guarantor" means any person providing insurance or other financial security to cover an owner's liability in pursuance of Article VII, paragraph 1, of the 1992 Liability Convention.

8. "Terminal Installation" means any site for the storage of oil in bulk which is capable of receiving oil from waterborne transportation, including any facility situated off-shore and linked to such site.

9. Where an incident consists of a series of occurrences, it shall be treated as having occurred on the date of the first such occurrence.

Article 2

1. An International Fund for compensation for pollution damage, to be named "The International Oil Pollution Compensation Fund 1992" and hereinafter referred to as "The Fund", is hereby established with the following aims:

(a) to provide compensation for pollution damage to the extent that the protection afforded by the 1992 Liability Convention is inadequate:

(b) to give effect to the related purposes set out in this Convention.

2. The Fund shall in each Contracting State be recognised as a legal person capable under the laws of that State of assuming rights and obligations and of being a party in legal proceedings before the courts of that State. Each Contracting State shall recognise the Director of the Fund (hereinafter referred to as "The Director") as the legal representative of the Fund.

Article 3

This Convention shall apply exclusively:

(a) to pollution damage caused:

(i) in the territory, including the territorial sea, of a Contracting State, and

(ii) in the exclusive economic zone of a Contracting State, established in accordance with international law, or, if a Contracting State has not established such a zone, in an area beyond and adjacent to the territorial sea of that State determined by that State in accordance with international law and extending not more than 200 nautical miles from the baselines from which the breadth of its territorial sea is measured;

(b) to preventive measures, wherever taken, to prevent or minimise such damage.

Compensation

Article 4

1. For the purpose of fulfilling its function under Article 2, paragraph 1(a), the Fund shall pay compensation to any person suffering pollution damage if such person has been unable to obtain full and adequate compensation for the damage under the terms of the 1992 Liability Convention,

(a) because no liability for the damage arises under the 1992 Liability Convention;

(b) because the owner liable for the damage under the 1992 Liability Convention is financially incapable of meeting his obligations in full and any financial security that may be provided under Article VII of that Convention does not cover or is insufficient to satisfy the claims for compensation for the damage; an owner being treated as financially incapable of meeting his obligations and a financial security being treated as insufficient if the person suffering the damage has been unable to obtain full satisfaction of the amount of compensation due under the 1992 Liability Convention after having taken all reasonable steps to pursue the legal remedies available to him;

(c) because the damage exceeds the owner's liability under the 1992 Liability Convention as limited pursuant to Article V, paragraph 1, of that Convention or under the terms of any other international Convention in force or open for signature, ratification or accession at the date of this Convention.

Expenses reasonably incurred or sacrifices reasonably made by the owner voluntarily to prevent or minimise pollution damage shall be treated as pollution damage for the purposes of this Article.

2. The Fund shall incur no obligation under the preceding paragraph if:

(a) it proves that the pollution damage resulted from an act of war, hostilities, civil war or insurrection or was caused by oil which has escaped or been discharged from a warship or other ship owned or operated by a State and used, at the time of the incident, only on Government non-commercial service; or

(b) the claimant cannot prove that the damage resulted from an incident involving one or more ships.

3. If the Fund proves that the pollution damage resulted wholly or partially either from an act or omission done with the intent to cause damage by the person who suffered the damage or from the negligence of that person, the Fund may be exonerated wholly or partially from its obligation to pay compensation to such person. The Fund shall in any event be exonerated to the extent that the shipowner may have been exonerated under Article III, paragraph 3, of the 1992 Liability Convention. However, there shall be no such exoneration of the Fund with regard to preventive measures.

4. (a) Except as otherwise provided in sub-paragraphs (b) and (c) of this paragraph, the aggregate amount of compensation payable by the Fund under this Article shall in

respect of any one incident be limited, so that the total sum of that amount and the amount of compensation actually paid under the 1992 Liability Convention for pollution damage within the scope of application of this Convention as defined in Article 3 shall not exceed 135 million units of account.

(b) Except as otherwise provided in sub-paragraph (c), the aggregate amount of compensation payable by the Fund under this Article for pollution damage resulting from a natural phenomenon of an exceptional, inevitable and irresistible character shall not exceed 135 million units of account.

(c) The maximum amount of compensation referred to in sub-paragraphs (a) and (b) shall be 200 million units of account with respect to any incident occurring during any period when there are three Parties to this Convention in respect of which the combined relevant quantity of contributing oil received by persons in the territories of such Parties, during the preceding calendar year, equalled or exceeded 600 million tons.

(d) Interest accrued on a fund constituted in accordance with Article V, paragraph 3, of the 1992 Liability Convention, if any, shall not be taken into account for the computation of the maximum compensation payable by the Fund under this Article.

(e) The amounts mentioned in this Article shall be converted into national currency on the basis of the value of that currency by reference to the Special Drawing Right on the date of the decision of the Assembly of the Fund as to the first date of payment of compensation.

5. Where the amount of established claims against the Fund exceeds the aggregate amount of compensation payable under paragraph 4, the amount available shall be distributed in such a manner that the proportion between any established claim and the amount of compensation actually recovered by the claimant under this Convention shall be the same for all claimants.

6. The Assembly of the Fund may decide that, in exceptional cases, compensation in accordance with this Convention can be paid even if the owner of the ship has not constituted a fund in accordance with Article V, paragraph 3, of the 1992 Liability Convention. In such case paragraph 4 (e) of this Article applies accordingly.

7. The Fund shall, at the request of a Contracting State, use its good offices as necessary to assist that State to secure promptly such personnel, material and services as are necessary to enable the State to take measures to prevent or mitigate pollution damage arising from an incident in respect of which the Fund may be called upon to pay compensation under this Convention.

8. The Fund may on conditions to be laid down in the Internal Regulations provide credit facilities with a view to the taking of preventive measures against pollution damage arising from a particular incident in respect of which the Fund may be called upon to pay compensation under this Convention.

Article 5

(*deleted*)

Article 6

Rights to compensation under Article 4 shall be extinguished unless an action is brought thereunder or a notification has been made pursuant to Article 7, paragraph 6, within three years from the date when the damage occurred. However, in no case shall an action be brought after six years from the date of the incident which caused the damage.

Article 7

1. Subject to the subsequent provisions of this Article, any action against the Fund for compensation under Article 4 of this Convention shall be brought only before a court competent under Article IX of the 1992 Liability Convention in respect of actions against the owner who is or who would, but for the provisions of Article III, paragraph 2, of that Convention, have been liable for pollution damage used by the relevant incident.

2. Each Contracting State shall ensure that its courts possess the necessary jurisdiction to entertain such actions against the Fund as are referred to in paragraph 1.

3. Where an action for compensation for pollution damage has been brought before a court competent under Article IX of the 1992 Liability Convention against the owner of a ship or his guarantor, such court shall have exclusive jurisdictional competence over any action against the Fund for compensation under the provisions of Article 4 of this Convention in respect of the same damage. However, where an action for compensation for pollution damage under the 1992 Liability Convention has been brought before a court in a State Party to the 1992 Liability Convention but not to this Convention, any action against the Fund under Article 4 of this Convention shall at the option of the claimant be brought either before a court of the State where the Fund has its headquarters or before any court of a State Party to this Convention competent under Article IX of the 1992 Liability Convention.

4. Each Contracting State shall ensure that the Fund shall have the right to intervene as a party to any legal proceedings instituted in accordance with Article IX of the 1992 Liability Convention before a competent court of that State against the owner of a ship or his guarantor.

5. Except as otherwise provided in paragraph 6, the Fund shall not be bound by any judgment or decision in proceedings to which it has not been a party or by any settlement to which it is not a party.

6. Without prejudice to the provisions of paragraph 4, where an action under the 1992 Liability Convention for compensation for pollution damage has been brought against an owner or his guarantor before a competent court in a Contracting State, each party to the proceedings shall be entitled under the national law of that State to notify the Fund of the proceedings. Where such notification has been made in accordance with the formalities required by the law of the court seized and in such time and in such a manner that the Fund has in fact been in a position effectively to intervene as a party to the proceedings, any judgment rendered by the court in such proceedings shall, after it has become final and enforceable in the State where the judgment was given, become binding upon the Fund in the sense that the facts and findings in that judgment may not be disputed by the Fund even if the Fund has not actually intervened in the proceedings.

Article 8

Subject to any decision concerning the distribution referred to in Article 4, paragraph 5, any judgment given against the Fund by a court having jurisdiction in accordance with

Article 7, paragraphs 1 and 3, shall, when it has become enforceable in the State of origin and is in that State no longer subject to ordinary forms of review, be recognised and enforceable in each Contracting State on the same conditions as are prescribed in Article X of the 1992 Liability Convention.

Article 9

1. The Fund shall, in respect of any amount of compensation for pollution damage paid by the Fund in accordance with Article 4, paragraph 1, of this Convention, acquire by subrogation the rights that the person so compensated may enjoy under the 1992 Liability Convention against the owner or his guarantor.

2. Nothing in this Convention shall prejudice any right of recourse or subrogation of the Fund against persons other than those referred to in the preceding paragraph. In any event the right of the Fund to subrogation against such person shall not be less favourable than that of an insurer of the person to whom compensation has been paid.

3. Without prejudice to any other rights of subrogation or recourse against the Fund which may exist, a Contracting State or agency thereof which has paid compensation for pollution damage in accordance with provisions of national law shall acquire by subrogation the rights which the person so compensated would have enjoyed under this Convention.

Contributions

Article 10

1. Annual contributions to the Fund shall be made in respect of each Contracting State by any person who, in the calendar year referred to in Article 12, paragraphs 2(a) or (b), has received in total quantities exceeding 150,000 tons:
   - (a) in the ports or terminal installations in the territory of that State contributing oil carried by sea to such ports or terminal installations; and
   - (b) in any installations situated in the territory of that Contracting State contributing oil which has been carried by sea and discharged in a port or terminal installation of a non-Contracting State, provided that contributing oil shall only be taken into account by virtue of this sub-paragraph on first receipt in a Contracting State after its discharge in that non-Contracting State.

2. (a) For the purposes of paragraph 1, where the quantity of contributing oil received in the territory of a Contracting State by any person in a calendar year when aggregated with the quantity of contributing oil received in the same Contracting State in that year by any associated person or persons exceeds 150,000 tons, such person shall pay contributions in respect of the actual quantity received by him notwithstanding that that quantity did not exceed 150,000 tons.
   - (b) "Associated person" means any subsidiary or commonly controlled entity. The question whether a person comes within this definition shall be determined by the national law of the State concerned.

Article 11

(deleted)

Article 12

1. With a view to assessing the amount of annual contributions due, if any, and taking account of the necessity to maintain sufficient liquid funds, the Assembly shall for each calendar year make an estimate in the form of a budget of:

(i) Expenditure

(a) costs and expenses of the administration of the Fund in the relevant year and any deficit from operations in preceding years;
(b) payments to be made by the Fund in the relevant year for the satisfaction of claims against the Fund due under Article 4, including repayment on loans previously taken by the Fund for the satisfaction of such claims, to the extent that the aggregate amount of such claims in respect of any one incident does not exceed four million units of account;
(c) payments to be made by the Fund in the relevant year for the satisfaction of claims against the Fund due under Article 4, including repayments on loans previously taken by the Fund for the satisfaction of such claims, to the extent that the aggregate amount of such claims in respect of any one incident is in excess of four million units of account.

(ii) Income

(a) surplus funds from operations in preceding years, including any interest;
(b) annual contributions, if required to balance the budget;
(c) any other income.

2. The Assembly shall decide the total amount of contributions to be levied. On the basis of that decision, the Director shall, in respect of each Contracting State, calculate for each person referred to in Article 10 the amount of his annual contribution:

(a) in so far as the contribution is for the satisfaction of payments referred to in paragraph 1(i)(a) and (b) on the basis of a fixed sum for each ton of contributing oil received in the relevant State by such persons during the preceding calendar year; and
(b) in so far as the contribution is for the satisfaction of payments referred to in paragraph 1(i)(c) of this Article on the basis of a fixed sum for each ton of contributing oil received by such person during the calendar year preceding that in which the incident in question occurred, provided that State was a party to this Convention at the date of the incident.

3. The sums referred to in paragraph 2 above shall be arrived at by dividing the relevant total amount of contributions required by the total amount of contributing oil received in all Contracting States in the relevant year.

4. The annual contribution shall be due on the date to be laid down in the Internal Regulations of the Fund. The Assembly may decide on a different date of payment.

5. The Assembly may decide, under conditions to be laid down in the Financial Regulations of the Fund, to make transfers between funds received in accordance with Article 12.2(a) and funds received in accordance with Article 12.2(b).

Article 13

1. The amount of any contribution due under Article 12 and which is in arrears shall bear interest at a rate which shall be determined in accordance with the Internal Regulations of the Fund, provided that different rates may be fixed for different circumstances.

2. Each Contracting State shall ensure that any obligation to contribute to the Fund arising under this Convention in respect of oil received within the territory of that State is fulfilled and shall take any appropriate measures under its law, including the imposing of such sanctions as it may deem necessary, with a view to the effective execution of any such obligation; provided however, that such measures shall only be directed against those persons who are under an obligation to contribute to the Fund.

3. Where a person who is liable in accordance with the provisions of Articles 10 and 12 to make contributions to the Fund does not fulfil his obligations in respect of any such contribution or any part thereof and is in arrear, the Director shall take all appropriate action against such person on behalf of the Fund with a view to the recovery of the amount due. However, where the defaulting contributor is manifestly insolvent or the circumstances otherwise so warrant, the Assembly may, upon recommendation of the Director, decide that no action shall be taken or continued against the contributor.

Article 14

1. Each Contracting State may at the time when it deposits its instrument of ratification or accession or at any time thereafter declare that it assumes to itself obligations that are incumbent under this Convention on any person who is liable to contribute to the Fund in accordance with Article 10, paragraph 1, in respect of oil received within the territory of that State. Such declaration shall be made in writing and shall specify which obligations are assumed.

2. Where a declaration under paragraph 1 is made prior to the entry into force of this Convention in accordance with Article 40, it shall be deposited with the Secretary-General of the Organization who shall after the entry into force of the Convention communicate the declaration to the Director.

3. A declaration under paragraph 1 which is made after the entry into force of this Convention shall be deposited with the Director.

4. A declaration made in accordance with this Article may be withdrawn by the relevant State giving notice thereof in writing to the Director. Such notification shall take effect three months after the Director's receipt thereof.

5. Any State which is bound by a declaration made under this Article shall, in any proceedings brought against it before a competent court in respect of any obligation specified in the declaration, waive any immunity that it would otherwise be entitled to invoke.

Article 15

1. Each Contracting State shall ensure that any person who receives contributing oil within its territory in such quantities that he is liable to contribute to the Fund appears on a list to be established and kept up to date by the Director in accordance with the subsequent provisions of this Article.

2. For the purposes set out in paragraph 1, each Contracting State shall communicate, at a time and in the manner to be prescribed in the Internal Regulations, to the Director the

name and address of any person who in respect of the State is liable to contribute to the Fund pursuant to Article 10, as well as data on the relevant quantities of contributing oil received by any such person during the preceding calendar year.

3. For the purposes of ascertaining who are, at any given time, the persons liable to contribute to the Fund in accordance with Article 10, paragraph 1, and of establishing, where applicable, the quantities of oil to be taken into account for any such person when determining the amount of his contribution, the list shall be *prima facie* evidence of the facts stated therein.

4. Where a Contracting State does not fulfil its obligations to submit to the Director the communication referred to in paragraph 2 and this results in a financial loss for the Fund, that Contracting State shall be liable to compensate the Fund for such loss. The Assembly shall, on the recommendation of the Director, decide whether such compensation shall be payable by that Contracting State.

Organisation and Administration

Article 16

The Fund shall have an Assembly and a Secretariat headed by a Director.

Assembly

Article 17

The Assembly shall consist of all Contracting States to this Convention.

Article 18

The functions of the Assembly shall be:

1. to elect at each regular session its Chairman and two Vice-Chairmen who shall hold office until the next regular session;
2. to determine its own rules of procedure, subject to the provisions of this Convention;
3. to adopt Internal Regulations necessary for the proper functioning of the Fund;
4. to appoint the Director and make provisions for the appointment of such other personnel as may be necessary and determine the terms and conditions of service of the Director and other personnel;
5. to adopt the annual budget and fix the annual contributions;
6. to appoint auditors and approve the accounts of the Fund;
7. to approve settlements of claims against the Fund, to take decisions in respect of the distribution among claimants of the available amount of compensation in accordance with Article 4, paragraph 5, and to determine the terms and conditions according to which provisional payments in respect of claims shall be made with a view to ensuring that victims of pollution damage are compensated as promptly as possible;

8. (*deleted*);
9. to establish any temporary or permanent subsidiary body it may consider to be necessary, to define its terms of reference and to give it the authority needed to perform the functions entrusted to it; when appointing the members of such body, the Assembly shall endeavour to secure an equitable geographical distribution of members and to ensure that the Contracting States, in respect of which the largest quantities of contributing oil are being received, are appropriately represented; the Rules of Procedure of the Assembly may be applied, *mutatis mutandis*, for the work of such subsidiary body;
10. to determine which non-Contracting States and which inter-governmental and international non-governmental organisations shall be admitted to take part, without voting rights, in meetings of the Assembly and subsidiary bodies;
11. to give instructions concerning the administration of the Fund to the Director and subsidiary bodies;
12. (*deleted*);
13. to supervise the proper execution of the Convention and of its own decisions;
14. to perform such other functions as are allocated to it under the Convention or are otherwise necessary for the proper operation of the Fund.

Article 19

1. Regular sessions of the Assembly shall take place once every calendar year upon convocation by the Director.

2. Extraordinary sessions of the Assembly shall be convened by the Director at the request of at least one-third of the members of the Assembly and may be convened on the Director's own initiative after consultation with the Chairman of the Assembly. The Director shall give members at least thirty days' notice of such sessions.

Article 20

A majority of the members of the Assembly shall constitute a quorum for its meetings.

(*heading deleted*)

Articles 21–27

(*deleted*)

Secretariat

Article 28

1. The Secretariat shall comprise the Director and such staff as the administration of the Fund may require.
2. The Director shall be the legal representative of the Fund.

Article 29

1. The Director shall be the chief administrative officer of the Fund. Subject to the instructions given to him by the Assembly, he shall perform those functions which are assigned to him by this Convention, the Internal Regulations of the Fund and the Assembly.

2. The Director shall in particular:

(a) appoint the personnel required for the administration of the Fund;

(b) take all appropriate measures with a view to the proper administration of the Fund's assets;

(c) collect the contributions due under this Convention while observing in particular the provisions of Article 13, paragraph 3;

(d) to the extent necessary to deal with claims against the Fund and carry out the other functions of the Fund, employ the services of legal, financial and other experts;

(e) take all appropriate measures for dealing with claims against the Fund within the limits and on conditions to be laid down in the Internal Regulations, including the final settlement of claims without the prior approval of the Assembly where these Regulations so provide;

(f) prepare and submit to the Assembly the financial statements and budget estimates for each calendar year;

(g) prepare, in consultation with the Chairman of the Assembly, and publish a report of the activities of the Fund during the previous calendar year;

(h) prepare, collect and circulate the papers, documents, agenda, minutes and information that may be required for the work of the Assembly and subsidiary bodies.

Article 30

In the performance of their duties the Director and the staff and experts appointed by him shall not seek or receive instructions from any Government or from any authority external to the Fund. They shall refrain from any action which might reflect on their position as international officials. Each Contracting State on its part undertakes to respect the exclusively international character of the responsibilities of the Director and the staff and experts appointed by him, and not to seek to influence them in the discharge of their duties.

Finances

Article 31

1. Each Contracting State shall bear the salary, travel and other expenses of its own delegation to the Assembly and of its representatives on subsidiary bodies.

2. Any other expenses incurred in the operation of the Fund shall be borne by the Fund.

Voting

Article 32

The following provisions shall apply to voting in the Assembly:

(a) each member shall have one vote;

(b) except as otherwise provided in Article 33, decisions of the Assembly shall be by a majority vote of the members present and voting;

(c) decisions where a three-fourths or a two-thirds majority is required shall be by a three-fourths or two-thirds majority vote, as the case may be, of those present;

(d) for the purpose of this Article the phrase "members present" means "members present at the meeting at the time of the vote", and the phrase "members present and voting" means "members present and casting an affirmative or negative vote". Members who abstain from voting shall be considered as not voting.

Article 33

The following decisions of the Assembly shall require a two-thirds majority:

(a) a decision under Article 13, paragraph 3, not to take or continue action against a contributor;

(b) the appointment of the Director under Article 18, paragraph 4;

(c) the establishment of subsidiary bodies, under Article 18, paragraph 9, and matters relating to such establishment.

Article 34

1. The Fund, its assets, income, including contributions, and other property shall enjoy in all Contracting States exemption from all direct taxation.

2. When the Fund makes substantial purchases of movable or immovable property, or has important work carried out which is necessary for the exercise of its official activities and the cost of which includes indirect taxes or sales taxes, the Government of Member States shall take, whenever possible, appropriate measures for the remission or refund of the amount of such duties and taxes.

3. No exemption shall be accorded in the case of duties, taxes or dues which merely constitute payment for public utility services.

4. The Fund shall enjoy exemption from all customs duties, taxes and other related taxes on articles imported or exported by it or on its behalf for its official use. Articles thus imported shall not be transferred either for consideration or gratis on the territory of the country into which they have been imported except on conditions agreed by the Government of that country.

5. Persons contributing to the Fund and victims and owners of ships receiving compensation from the Fund shall be subject to the fiscal legislation of the State where they are taxable, no special exemption or other benefit being conferred on them in this respect.

6. Information relating to individual contributors supplied for the purpose of this Convention shall not be divulged outside the Fund except in so far as it may be strictly necessary to enable the Fund to carry out its functions including the bringing and defending of legal proceedings.

7. Independently of existing or future regulations concerning currency or transfers,

Contracting States shall authorise the transfer and payment of any contribution to the Fund and of any compensation paid by the Fund without any restriction.

Transitional Provisions

Article 35

Claims for compensation under Article 4 arising from incidents occurring after the date of entry into force of this Convention may not be brought against the Fund earlier than the one hundred and twentieth day after that date.

Article 36

The Secretary-General of the Organization shall convene the first session of the Assembly. This session shall take place as soon as possible after entry into force of this Convention and, in any case, not more than thirty days after such entry into force.

Article 36 *bis*

The following transitional provisions shall apply in the period, hereinafter referred to as the transitional period, commencing with the date of entry into force of this Convention and ending with the date on which the denunciations provided for in Article 31 of the 1992 Protocol to amend the 1971 Fund Convention take effect:

(a) In the application of paragraph 1(a) of Article 2 of this Convention, the reference to the 1992 Liability Convention shall include reference to the International Convention on Civil Liability for Oil Pollution Damage, 1969, either in its original version or as amended by the Protocol thereto of 1976 (referred to in this Article as "the 1969 Liability Convention"), and also the 1971 Fund Convention.

(b) Where an incident has caused pollution damage within the scope of this Convention, the Fund shall pay compensation to any person suffering pollution damage only if, and to the extent that, such person has been unable to obtain full and adequate compensation for the damage under the terms of the 1969 Liability Convention, the 1971 Fund Convention and the 1992 Liability Convention provided that, in respect of pollution damage within the scope of this Convention in respect of a Party to this Convention but not a Party to the 1971 Fund Convention, the Fund shall pay compensation to any person suffering pollution damage only if, and to the extent that, such person would have been unable to obtain full and adequate compensation had that State been party to each of the above-mentioned Conventions.

(c) In the application of Article 4 of this Convention, the amount to be taken into account in determining the aggregate amount of compensation payable by the Fund shall also include the amount of compensation actually paid under the 1969 Liability Convention, if any, and the amount of compensation actually paid or deemed to have been paid under the 1971 Fund Convention.

(d) Paragraph 1 of Article 9 of this Convention shall also apply to the rights enjoyed under the 1969 Liability Convention.

Article 36 *ter*

1. Subject to paragraph 4 of this Article, the aggregate amount of the annual contributions payable in respect of contributing oil received in a single Contracting State during a calendar year shall not exceed 27.5% of the total amount of annual contributions pursuant to the 1992 Protocol to amend the 1971 Fund Convention, in respect of that calendar year.

2. If the application of the provisions in paragraphs 2 and 3 of Article 12 would result in the aggregate amount of the contributions payable by contributors in a single Contracting State in respect of a given calendar year exceeding 27.5% of the total annual contributions, the contributions payable by all contributors in that State shall be reduced *pro rata* so that their aggregate contributions equal 27.5% of the total annual contributions to the Fund in respect of that year.

3. If the contributions payable by persons in a given Contracting State shall be reduced pursuant to paragraph 2 of this Article, the contributions payable by persons in all other Contracting States shall be increased *pro rata* so as to ensure that the total amount of contributions payable by all persons liable to contribute to the Fund in respect of the calendar year in question will reach the total amount of contributions decided by the Assembly.

4. The provisions in paragraphs 1 to 3 of this Article shall operate until the total quantity of contributing oil received in all Contracting States in a calendar year has reached 750 million tons or until a period of 5 years after the date of entry into force of the said 1992 Protocol has elapsed, whichever occurs earlier.

Article 36 *quater*

Notwithstanding the provisions of this Convention, the following provisions shall apply to the administration of the Fund during the period in which both the 1971 Fund Convention and this Convention are in force:

(a) The Secretariat of the Fund, established by the 1971 Fund Convention (hereinafter referred to as "the 1971 Fund"), headed by the Director, may also function as the Secretariat and the Director of the Fund.

(b) If, in accordance with sub-paragraph (a), the Secretariat and the Director of the 1971 Fund also perform the function of Secretariat and Director of the Fund, the Fund shall be represented, in cases of conflict of interests between the 1971 Fund and the Fund, by the Chairman of the Assembly of the Fund.

(c) The Director and the staff and experts appointed by him, performing their duties under this Convention and the 1971 Fund Convention, shall not be regarded as contravening the provisions of Article 30 of this Convention in so far as they discharge their duties in accordance with this Article.

(d) The Assembly of the Fund shall endeavour not to take decisions which are incompatible with decisions taken by the Assembly of the 1971 Fund. If differences of opinion with respect to common administrative issues arise, the Assembly of the Fund shall try to reach a consensus with the Assembly of the 1971 Fund, in a spirit of mutual co-operation and with the common aims of both organizations in mind.

(e) The Fund may succeed to the rights, obligations and assets of the 1971 Fund if the Assembly of the 1971 Fund so decides, in accordance with Article 44, paragraph 2, of the 1971 Fund Convention.

(f) The Fund shall reimburse to the 1971 Fund all costs and expenses arising from administrative services performed by the 1971 Fund on behalf of the Fund.

Article 36 *quinquies*

Final Clauses

The final clauses of this Convention shall be Articles 28 to 39 of the 1992 Protocol to amend the 1971 Fund Convention. References in this Convention to Contracting States shall be taken to mean references to the Contracting States of that Protocol.

Final Clauses of the Protocol of 1992 to amend the 1971 Fund Convention

Article 28

Signature, Ratification, Acceptance,
Approval and Accession

1. This Protocol shall be open for signature at London from 15 January 1993 to 14 January 1994 by any State which has signed the 1992 Liability Convention.
2. Subject to paragraph 4, this Protocol shall be ratified, accepted or approved by States which have signed it.
3. Subject to paragraph 4, this Protocol is open for accession by States which did not sign it.
4. This Protocol may be ratified, accepted, approved or acceded to, only by States which have ratified, accepted, approved or acceded to the 1992 Liability Convention.
5. Ratification, acceptance, approval or accession shall be effected by the deposit of a formal instrument to that effect with the Secretary-General of the Organization.
6. A State which is a Party to this Protocol but is not a Party to the 1971 Fund Convention shall be bound by the provisions of the 1971 Fund Convention as amended by this Protocol in relation to other parties hereto, but shall not be bound by the provisions of the 1971 Fund Convention in relation to Parties thereto.
7. Any instrument of ratification, acceptance, approval or accession deposited after the entry into force of an amendment to the 1971 Fund Convention as amended by this Protocol shall be deemed to apply to the Convention so amended, as modified by such amendment.

Article 29

Information on Contributing Oil

1. Before this Protocol comes into force for a State, that State shall, when depositing an instrument referred to in Article 28, paragraph 5, and annually thereafter at a date to be determined by the Secretary-General of the Organization, communicate to him the name and address of any person who in respect of that State would be liable to contribute to the Fund pursuant to Article 10 of the 1971 Fund Convention as amended by this Protocol as well as data on the relevant quantities of contributing oil received by any such person in the territory of that State during the preceding calendar year.

2. During the transitional period, the Director shall, for Parties, communicate annually to the Secretary-General of the Organization data on quantities of contributing oil received by persons liable to contribute to the Fund pursuant to Article 10 of the 1971 Fund Convention as amended by this Protocol.

Article 30

Entry into Force

1. This Protocol shall enter into force twelve months following the date on which the following requirements are fulfilled:
   (a) at least eight States have deposited instruments of ratification, acceptance, approval or accession with the Secretary-General of the Organization; and
   (b) the Secretary-General of the Organization has received information in accordance with Article 29 that those persons who would be liable to contribute pursuant to Article 10 of the 1971 Fund Convention as amended by this Protocol have received during the preceding calendar year a total quantity of at least 450 million tons of contributing oil.

2. However, this Protocol shall not enter into force before the 1992 Liability Convention has entered into force.

3. For each State which ratifies, accepts, approves or accedes to this Protocol after the conditions in paragraph 1 for entry into force have been met, the Protocol shall enter into force twelve months following the date of the deposit by such State of the appropriate instrument.

4. Any State may, at the time of the deposit of its instrument of ratification, acceptance, approval or accession in respect of this Protocol declare that such instrument shall not take effect for the purpose of this Article until the end of the six-month period in Article 31.

5. Any State which has made a declaration in accordance with the preceding paragraph may withdraw it at any time by means of a notification addressed to the Secretary-General of the Organization. Any such withdrawal shall take effect on the date the notification is received, and any State making such a withdrawal shall be deemed to have deposited its instrument of ratification, acceptance, approval or accession in respect of this Protocol on that date.

6. Any State which has made a declaration under Article 13, paragraph 2, of the Protocol of 1992 to amend the 1969 Liability Convention shall be deemed to have also made a declaration under paragraph 4 of this Article. Withdrawal of a declaration under the said Article 13, paragraph 2, shall be deemed to constitute a withdrawal also under paragraph 5 of this Article.

Article 31

Denunciation of 1969 and 1971 Conventions

Subject to Article 30, within six months following the date on which the following requirements are fulfilled:
   (a) at least eight States have become Parties to this Protocol or have deposited instruments of ratification, acceptance, approval or accession with the Secretary-General of the Organization, whether or not subject to Article 30, paragraph 4, and

(b) the Secretary-General of the Organization has received information in accordance with Article 29 that those persons who are or would be liable to contribute pursuant to Article 10 of the 1971 Fund Convention as amended by this Protocol have received during the preceding calendar year a total quantity of at least 750 million ton of contributing oil;

each Party to this Protocol and each State which has deposited an instrument of ratification, acceptance, approval or accession, whether or not subject to Article 30, paragraph 4 shall, if party thereto, denounce the 1971 Fund Convention and the 1969 Liability Convention with effect twelve months after the expiry of the above-mentioned six month period.

Article 32

Revision and Amendment

1. A conference for the purpose of revising or amending the 1992 Fund Convention may be convened by the Organization.

2. The Organization shall convene a Conference of Contracting States for the purpose of revising or amending the 1992 Fund Convention at the request of not less than one-third of all Contracting States.

Article 33

Amendment of Compensation Limits

1. Upon the request of at least one-quarter of the Contracting States, any proposal to amend the limits of amounts of compensation laid down in Article 4, paragraph 4, of the Convention as amended by this Protocol shall be circulated by the Secretary-General to all Members of the Organization and to all Contracting States.

2. Any amendment proposed and circulated as above shall be submitted to the Legal Committee of the Organization for consideration at a date at least six months after the date of its circulation.

3. All Contracting States to the 1971 Fund Convention as amended by this Protocol, whether or not Members of the Organization, shall be entitled to participate in the proceedings of the Legal Committee for the consideration and adoption of amendments.

4. Amendments shall be adopted by a two-thirds majority of the Contracting States present and voting in the Legal Committee, expanded as provided for in paragraph 3, on condition that at least one-half of the Contracting States shall be present at the time of voting.

5. When acting on a proposal to amend the limits, the Committee shall take into account the experience of incidents and in particular the amount of damage resulting therefrom and changes in the monetary values. It shall also take into account the relationship between the limits in Article 4, paragraph 4, of the Convention as amended by this Protocol and those in Article V, paragraph 1 of the International Convention on Civil Liability for Oil Pollution Damage 1992.

6. (a) No amendment of the limits under this Article may be considered before 15 January 1998 nor less than five years from the date of entry into force of a previous amendment under this Article. No amendment under this Article shall be considered before this Protocol has entered into force.

(b) No limit may be increased so as to exceed an amount which corresponds to the limit laid down in the 1971 Fund Convention as amended by this Protocol increased by six per cent per year calculated on a compound basis from 15 January 1993.

(c) No limit may be increased so as to exceed an amount which corresponds to the limit laid down in the Convention as amended by this Protocol multiplied by three.

7. Any amendment adopted in accordance with paragraph 4 shall be notified by the Organization to all Contracting States. The amendment shall be deemed to have been accepted at the end of a period of eighteen months after the date of notification unless within that period not less than one-quarter of the States that were Contracting States at the time of the adoption of the amendment by the Committee have communicated to the Organization that they do not accept the amendment in which case the amendment is rejected and shall have no effect.

8. An amendment deemed to have been accepted in accordance with paragraph 7 shall enter into force eighteen months after its acceptance.

9. All Contracting States shall be bound by the amendment, unless they denounce this Protocol in accordance with Article 34, paragraphs 1 and 2, at least six months before the amendment enters into force. Such denunciation shall take effect when the amendment enters into force.

10. When an amendment has been adopted by the Committee but the eighteen-month period for its acceptance has not yet expired, a State which becomes a Contracting State during that period shall be bound by the amendment if it enters into force. A State which becomes a Contracting State after that period shall be bound by an amendment if it enters into force. A State which becomes a Contracting State after that period shall be bound by an amendment which has been accepted in accordance with paragraph 7. In the cases referred to in this paragraph, a State becomes bound by an amendment when that amendment enters into force, or when this Protocol enters into force for that State, if later.

Article 34

Denunciation

1. This Protocol may be denounced by any Party at any time after the date on which it enters into force for that Party.

2. Denunciation shall be effected by the deposit of an instrument with the Secretary-General of the Organization.

3. A denunciation shall take effect twelve months, or such longer period as may be specified in the instrument of denunciation, after its deposit with the Secretary-General of the Organization.

4. Denunciation of the 1992 Liability Convention shall be deemed to be a denunciation of this Protocol. Such denunciation shall take effect on the date on which denunciation of the Protocol of 1992 to amend the 1969 Liability Convention takes effect according to Article 16 of that Protocol.

5. Any Contracting State to this Protocol which has not denounced the 1971 Fund Convention and the 1969 Liability Convention as required by Article 31 shall be deemed to have denounced this Protocol with effect twelve months after the expiry of the six month

period in that Article. As from the date on which the denunciations provided for in Article 31 take effect, any Party to this Protocol which deposits an instrument of ratification, acceptance, approval or accession to the 1969 Liability Convention shall be deemed to have denounced this Protocol with effect from the date on which such instrument takes effect.

6. As between the Parties to this Protocol, denunciation by any of them of the 1971 Fund Convention in accordance with Article 41 thereof shall not be construed in any way as a denunciation of the 1971 Fund Convention as amended by this Protocol.

7. Notwithstanding a denunciation of this Protocol by a Party pursuant to this Article, any provisions of this Protocol relating to the obligations to make contributions under Article 10 of the 1971 Fund Convention as amended by this Protocol with respect to an incident referred to in Article 12, paragraph 2(b), of that amended Convention and occurring before the denunciation takes effect shall continue to apply.

Article 35

Extraordinary Sessions of the Assembly

1. Any Contracting State may, within ninety days after the deposit of an instrument of denunciation the result of which it considers will significantly increase the level of contributions for the remaining Contracting States, request the Director to convene an extraordinary session of the Assembly. The Director shall convene the Assembly to meet not later than sixty days after receipt of the request.

2. The Director may convene, on his own initiative, an extraordinary session of the Assembly to meet within sixty days after the deposit of any instrument of denunciation, if he considers that such denunciation will result in a significant increase in the level of contributions of the remaining Contracting States.

3. If the Assembly at an extraordinary session convened in accordance with paragraph 1 or 2 decides that the denunciation will result in a significant increase in the level of contributions for the remaining Contracting States, any such State may, not later than one hundred and twenty days before the date on which the denunciation takes effect, denounce this Protocol with effect from the same date.

Article 36

Termination

1. This Protocol shall cease to be in force on the date when the number of Contracting States falls below three.

2. States which are bound by this Protocol on the day before the date it ceases to be in force shall enable the Fund to exercise its functions as described under Article 37 of this Protocol and shall, for that purpose only, remain bound by this Protocol.

Article 37

Winding up of the Fund

1. If this Protocol ceases to be in force, the Fund shall nevertheless:

(a) meet its obligations in respect of any incident occurring before the Protocol ceased to be in force;

(b) be entitled to exercise its rights to contributions to the extent that these contributions are necessary to meet the obligations under sub-paragraph (a), including expenses for the administration of the Fund necessary for this purpose.

2. The Assembly shall take all appropriate measures to complete the winding up of the Fund including the distribution in an equitable manner of any remaining assets among those persons who have contributed to the Fund.

3. For the purposes of this Article the Fund shall remain a legal person.

Article 38

Depositary

1. This Protocol and any amendments accepted under Article 33 shall be deposited with the Secretary-General of the Organization.

2. The Secretary-General of the Organization shall:

(a) inform all States which have signed or acceded to this Protocol of:

(i) each new signature or deposit of an instrument together with the date hereof;

(ii) each declaration and notification under Article 30 including declarations and withdrawals deemed to have been made in accordance with that Article;

(iii) the date of entry into force of this Protocol;

(iv) the date by which denunciations provided for in Article 31 are required to be made;

(v) any proposal to amend limits of amounts of compensation which has been made in accordance with Article 33, paragraph 1;

(vi) any amendment which has been adopted in accordance with Article 33, paragraph 4;

(vii) any amendment deemed to have been accepted under Article 33, paragraph 7, together with the date on which that amendment shall enter into force in accordance with paragraphs 8 and 9 of that Article;

(viii) the deposit of an instrument of denunciation of this Protocol together with the date of the deposit and the date on which it takes effect;

(ix) any denunciation deemed to have been made under Article 34, paragraph 5;

(x) any communication called for by any Article in this Protocol;

(b) transmit certified true copies of this Protocol to all Signatory States and to all States which accede to the Protocol.

3. As soon as this Protocol enters into force, the test shall be transmitted by the Secretary-General of the Organization to the Secretariat of the United Nations for registration and publication in accordance with Article 102 of the Charter of the United Nations.

Article 39

Languages

This Protocol is established in a single original in the Arabic, Chinese, English, French, Russian and Spanish Languages, each text being equally authentic.

APPENDIX IX

THE BRITANNIA P&I CLUB: FINANCIAL RESPONSIBILITY CLAUSE IN RESPECT OF OIL POLLUTION

FINANCIAL RESPONSIBILITY IN RESPECT OF POLLUTION

The following clause is applicable to all self-propelled tank vessels and to non-self-propelled tank vessels carrying more than 2,000 tons of persistent oil in bulk as cargo:

(1) Owners warrant that throughout the currency of this charter they will provide the vessel with the following certificates:

(a) Certificates issued pursuant to the Civil Liability Convention 1969 ("CLC"), and pursuant to the 1992 protocols to the CLC, as and when in force.

(b) Certificates issued pursuant to section 1016(a) of the Oil Pollution Act 1990, and section 108(a), of the Comprehensive Environmental Response, Compensation and Liability Act 1980, as amended in accordance with Part 138 of Coast Guard Regulations 33 CFR, so long as these can be obtained by the owners from or by **(identify the applicable scheme or schemes)**.

(2) Notwithstanding anything whether printed or typed herein to the contrary:

(a) Save as required for compliance with paragraph (1) hereof, owners shall not be required to establish or maintain financial security or responsibility in respect of oil or other pollution damage to enable the vessel lawfully to enter, remain in or leave any port, place, territorial or contiguous waters of any country, state or territory in performance of this charter.

(b) Charterers shall indemnify owners and hold them harmless in respect of any loss, damage, liability or expense (including but not limited to the costs of any delay incurred by the vessel as a result of any failure by the charterers promptly to give alternative voyage orders) whatsoever and howsoever arising which owners may sustain by reason of any requirement to establish or maintain financial security or responsibility in order to enter, remain in or leave any port, place or waters, other than to the extent provided in paragraph (1) hereof.

(c) Owners shall not be liable for any loss, damage, liability or expense whatsoever and howsoever arising which charterers and/or the holders of any bill of lading issued pursuant to this charter may sustain by reason of any requirement to establish or maintain financial security or responsibility in order to enter, remain in or leave any port, place or waters, other than to the extent provided in paragraph (1) hereof.

(3) Charterers warrant that the terms of this clause will be incorporated effectively into any bill of lading issued pursuant to this charter.

The following clause is applicable to all ships other than self-propelled tank vessels and non-self-propelled tank vessels carrying more than 2,000 tons of persistent oil in bulk as cargo:

(1) Owners warrant that throughout the currency of this charter they will provide the vessel with the following certificates:

(a) Certificates issued pursuant to section 311(p) of the US Federal Water Pollution Control Act, as amended (Title 33 US Code, Section 1321(p)) up to **(insert the date upon which such certificate(s) is/are due to expire)**.

(b) Certificates issued pursuant to section 1016(a) of the Oil Pollution Act 1990, and section 108(a) of the Comprehensive Environmental Response, Compensation and Liability Act 1980, as amended in accordance with Part 138 of Coast Guard Regulations 33 CFR, from **(indicate the earliest date upon**

which the owners may be required to deliver the vessel into the charter or, if later, the date inserted in sub-paragraph (a) above), so long as these can be obtained by the owners from or by **(identify the applicable scheme or schemes)**.

(2) Notwithstanding anything whether printed or typed herein to the contrary:

(a) Save as required for compliance with paragraph (1) hereof, owners shall not be required to establish or maintain financial security or responsibility in respect of oil or other pollution damage to enable the vessel lawfully to enter, remain in or leave any port, place, territorial or contiguous waters of any country, state or territory in performance of this charter.

(b) Charterers shall indemnify owners and hold them harmless in respect of any loss, damage, liability or expense (including but not limited to the costs of any delay incurred by the vessel as a result of any failure by the charterers promptly to give alternative voyage orders) whatsoever and howsoever arising which owners may sustain by reason of any requirement to establish or maintain financial security or responsibility in order to enter, remain in or leave any port, place or waters, other than to the extent provided in paragraph (1) hereof.

(c) Owners shall not be liable for any loss, damage, liability or expense whatsoever and howsoever arising which charterers and/or the holders of any bill of lading issued pursuant to this charter may sustain by reason of any requirement to establish or maintain financial security or responsibility in order to enter, remain in or leave any port, place or waters, other than to the extent provided in paragraph (1) hereof.

(3) Charterers warrant that the terms of this clause will be incorporated effectively into any bill of lading issued pursuant to this charter.

APPENDIX X

STATES PARTIES TO REGIONAL AND BILATERAL AGREEMENTS

States Parties to Regional and Bilateral Agreements[1]

Albania (dli[2]: April 1994); Barcelona Convention (with states bordering the Mediterranean).

Algeria (dli: June 1996); Barcelona Convention (with states bordering the Mediterranean).

Anguilla (dli: July 1995); Anguilla is not a party to any regional or bilateral agreements.

Antigua and Barbuda (dli: May 1994); Cartagena Convention (with states of the Wider Caribbean Region).

Argentina (dli: January 1997); a bilateral agreement exists with Uruguay for response in the River Plate. Operative Network for Regional Co-operation among Maritime Authorities of South America, Mexico, Panama and Cuba (ROCRAM). Argentina is involved in the development of the South West Atlantic agreement with Brazil, French Guiana and Uruguay.

Aruba (dli: December 1995); Cartagena Convention (with states of the Wider Caribbean Region).

Australia (dli: November 1996); Noumea Convention (with states of the South Pacific Region). A Memorandum of Understanding exists with New Zealand and others are being formulated with Papua New Guinea and Indonesia.

Bahamas (dli: May 1994); Cartagena Convention (with states of the Wider Caribbean Region).

Bahrain (dli: May 1996); Kuwait Convention (with countries bordering the Gulf), administered by the Regional Organisation for the Protection of the Marine Environment (ROPME), based in Kuwait.

Bangladesh (dli: July 1992); South Asia Co-operative Environment Programme with Sri Lanka, Pakistan, India and the Maldives (established by UNEP).

Barbados (dli: November 1996); Cartagena Convention (with states of the Wider Caribbean Region).

Belgium (dli: May 1996); Bonn Agreement (with countries bordering the North Sea). Member of the European Community Task Force.

Benin (dli: January 1989); Abidjan Convention (with states of the West and Central African Region).

Bermuda (dli: November 1996); an agreement exists with the USA concerning assistance rendered by the US Coast Guard.

Brazil (dli: May 1996); Operative Network for Regional Co-operation among Maritime Authorities of South America, Mexico and Panama (ROCRAM). Brazil is involved in the development of the South West Atlantic agreement with Argentina, French Guiana and Uruguay.

Brunei (dli: July 1995); ASEAN–OSRAP (Association of South East Asian Nations Oil Spill Response Action Plan) with other ASEAN countries and with contributions from Japan. Regional Programme for the Prevention and Management of Marine Pollution in the East Asian Seas with the ASEAN countries, China, Cambodia and Vietnam. A Standard Operating Procedure exists for joint spill combat in the South China Sea including Brunei Bay with Malaysia.

Bulgaria (dli: July 1996); Bucharest Convention (with countries bordering the Black Sea).

1. Source: ITOPF, *Country Profiles, A Summary of Oil Spill Response Arrangements and Resources Worldwide*, March 1997.

2. Date of latest information.

Cameroon (dli: January 1989); Abidjan Convention (with states of the West and Central African Region).

Canada (dli: July 1995); Canada–United States Joint Marine Pollution Contingency Plan. With Denmark for Baffin Bay, Davis Strait and other joint sea areas.

Cape Verde Islands (dli: January 1989/February 1995); Cape Verde is not a party to any agreements.

Cayman Islands (dli: May 1994); Cartagena Convention (with states of the Wider Caribbean Region).

Chile (dli: May 1994/January 1995); Quito Convention (with states bordering the South East Pacific). Operative Network for Regional Co-operation among Maritime Authorities of South America, Mexico, Panama and Cuba (ROCRAM).

China (dli: February 1997); a common communication channel for verbal notification and response exists between the Shekou Harbour Superintendency and Hong Kong authorities. Regional Programme for the Prevention and Management of Marine Pollution in the East Asian Seas with ASEAN countries, Cambodia, PDR of Korea, Republic of Korea and Vietnam.

Colombia (dli: March 1997); Quito Convention (with states bordering the South East Pacific). Cartagena Convention (with states of the Wider Caribbean Region). Operative Network for Regional Co-operation among Maritime Authorities of South America, Mexico, Panama and Cuba (ROCRAM).

Congo Brazzaville (dli: February 1979; Abidjan Convention (with states of the West and Central African Region).

Cook Islands (dli: September 1993); Cook Islands would have access to the resources available within New Zealand. Noumea Convention (with states of the South Pacific Region).

Costa Rica (dli: January 1989/July 1995); Cartagena Convention (with states of the Wider Caribbean Region).

Côte d'Ivoire (dli: April 1993); Abidjan Convention (with states of the West and Central African Region).

Croatia (dli: July 1995); Barcelona Convention (with states bordering the Mediterranean).

Cuba (dli: July 1995); Cartagena Convention (with states of the Wider Caribbean Region). Operative Network for Regional Co-operation among Maritime Authorities of South America, Mexico, Panama and Cuba (ROCRAM).

Cyprus (dli: February 1995); Barcelona Convention (with states bordering the Mediterranean). A trilateral agreement and subregional contingency plan with Egypt and Israel.

Denmark (dli: July 1993/May 1995); Bonn Agreement (with countries bordering the North Sea). Helsinki Convention (with countries bordering the Baltic Sea). Copenhagen Agreement (with Finland, Iceland, Norway and Sweden). DenGer Agreement (Denmark/Germany) covering the Wadden Sea and the Southern Baltic Sea. Bilateral agreement with Canada (covering Greenland and Canada). Member of the European Community Task Force.

Djibouti (dli: March 1994); l'Accord de Djibouti (with Somalia and Yemen).

Dominican Republic (dli: May 1996); Cartagena Convention (with states of the Wider Caribbean Region).

Ecuador (dli: May 1994); Operative Network for Regional Co-operation among Maritime Authorities of South America, Mexico, Panama and Cuba (ROCRAM). Quito Convention (with states bordering the South East Pacific).

Egypt (dli: October 1996); Jeddah Convention (with states of the Red Sea and Gulf of

Aden). Upper Gulf of Aqaba sub-regional contingency plan with Israel and Jordan. Barcelona Convention (with states bordering the Mediterranean). Trilateral agreement and sub-regional contingency plan with Israel and Cyprus. Member of the European Community Task Force.

El Salvador (dli: November 1995); El Salvador is not a party to any agreements.

Estonia (dli: February 1996); Helsinki Convention (with countries bordering the Baltic Sea). Trilateral agreement with Russian Federation and Finland.

Fiji (dli: September 1993); Noumea Convention (with states of the South Pacific Region).

Finland (dli: March 1997); Helsinki Convention (with countries bordering the Baltic Sea). Copenhagen Agreement (with Denmark, Iceland, Sweden and Norway). Trilateral agreement with the Russian Federation and Estonia.

France (dli: September 1996); Barcelona Convention (with states bordering the Mediterranean). Bonn Agreement (with countries bordering the North Sea). Lisbon Agreement (with Portugal, Spain and Morocco). Manche Plan (a bilateral agreement with United Kingdom). Mediplan (a bilateral agreement with Italy) covering the Tyrrhenian Sea and Ligurian Sea. RAMOGE (a trilateral agreement with Italy and Monaco). Member of the European Community Task Force.

French Antilles (dli: May 1996); Cartagena Convention (with states of the Wider Caribbean Region).

French Guiana (dli: November 1996); French Guiana is involved in the development of the South West Atlantic agreement with Argentina, Brazil and Uruguay.

French Polynesia (dli: May 1996); Noumea Convention (with states of the South Pacific Region).

Gabon (dli: January 1989); Abidjan Convention (with states of the West and Central African Region).

Gambia, the (dli: January 1989); Abidjan Convention (with states of the West and Central African Region).

Georgia (dli: November 1995); Bucharest Convention on the Protection of the Black Sea against Pollution (with countries bordering the Black Sea).

Germany (dli: March 1995); Bonn Agreement (countries bordering the North Sea). Helsinki Convention (countries bordering the Baltic Sea). DenGer Agreement (Denmark/Germany) covering the Wadden Sea, parts of the North Sea and the Western Baltic Sea. NethGer Agreement (Netherlands/Germany) covering the Wadden Sea and parts of the North Sea. German–Swedish pollution response plan, covering parts of the South-Western Baltic. Member of the European Community Task Force.

Ghana (dli: January 1989); Abidjan Convention (with states of the West and Central African Region).

Greece (dli: August 1996); Barcelona Convention (with states bordering the Mediterranean). Bilateral agreement with Italy (covering the Ionian Sea). Member of the European Community Task Force.

Greenland (dli: May 1996); With Canada for Baffin Bay, Davis Strait and other joint sea areas.

Grenada (dli: March 1994); Cartagena Convention (with states of the Wider Caribbean Convention).

Guam (dli: September 1993); Noumea Convention (with states of the South Pacific Region).

Guatemala (dli: January 1989); Cartagena Convention (with states of the Wider Caribbean Region).

Guinea (dli: January 1989); Abidjan Convention (with states of the West and Central African Region).

Guinea-Bissau (dli: January 1989); Guinea-Bissau is not a party to any agreements.

Guyana (dli: January 1989); Operative Network for Regional Co-operation among Maritime Authorities of South America, Mexico, Panama and Cuba (ROCRAM).

Honduras (dli: January 1989); Cartagena Convention (with states of the Wider Caribbean Region).

Hong Kong (dli: February 1995); a common communication channel for verbal notification and response exists with Shekou Harbour Superintendency, China.

Iceland (dli: September 1995); Copenhagen Agreement (with Denmark, Finland, Norway and Sweden).

India (dli: July 1995); South Asia Co-operative Environment Programme with Pakistan, Sri Lanka, the Maldives and Bagladesh (established by UNEP).

Indonesia (dli: August 1996); bilateral plan with Australia covering bordering seas. Tripartite Plan with Malaysia and Singapore covering the Straits of Malacca and Singapore. Sulawesi Sea Network co-operative agreement covering the Sulawesi Sea with Malaysia and the Philippines. ASEAN–OSRAP (Association of South East Asian Nations Oil Spill Response Action Plan) with the other ASEAN countries and with contributions from Japan. Regional Programme for the Prevention and Management of Marine Pollution in the East Asian Seas with the ASEAN countries, China, PDR of Korea, Republic of Korea, Cambodia and Vietnam.

Ireland (dli: May 1996); a bilateral agreement exists with the United Kingdom covering the Irish Sea. Member of the European Community Task Force.

Israel (dli: December 1996); Barcelona Convention (with states bordering the Mediterranean). A trilateral agreement and sub-regional contingency plan with Egypt and Cyprus. Upper Gulf of Aqaba sub-regional contingency plan with Egypt and Jordan.

Italy (dli: April 1994); Barcelona Convention (with states bordering the Mediterranean). Mediplan (a bilateral agreement with France) covering the Tyrrhenian Sea and Ligurian Sea. RAMOGE (a trilateral agreement with France and Monaco). Bilateral agreement with Greece (covering the Ionian Sea). Member of the European Community Task Force.

Jamaica (dli: May 1994); Cartagena Convention (with states of the Wider Caribbean Region).

Japan (dli: November 1996); The ASEAN–OSRAP (Association of South East Asian Nations Oil Spill Response Action Plan) with ASEAN countries. A Memorandum of Understanding exists with South Korea for the Sea of Japan. A Memorandum of Understanding exists with the USA on pollution preparedness and response.

Kenya (dli: May 1993; Nairobi Convention (with states of the East African Region).

Kiribati (dli: September 1993); Kiribati is not a party to any agreements.

Korea, Republic of (South Korea) (dli: May 1996); a Memorandum of Understanding exists with Japan for the Sea of Japan. Regional Programme for the Prevention and Management of Marine Pollution in the East Asian Seas with the ASEAN countries, China, Cambodia, PDR Korea and Vietnam.

Kuwait (dli: April 1994); Kuwait Convention (with countries bordering the Gulf) administered by the Regional Organisation for the Protection of the Marine Environment (ROPME) based in Kuwait.

Latvia (dli: February 1996); Helsinki Convention (with countries bordering the Baltic Sea).

Lebanon (dli: June 1996); Barcelona Convention (with states bordering the Mediterranean).

Liberia (dli: November 1995); Abidjan Convention (with states of the West and Central African Region).

Libya (dli: September 1995); Barcelona Convention (with states bordering the Mediterranean).

Lithuania (dli: February 1996); Helsinki Convention (with countries bordering the Baltic Sea).

Madagascar (dli: May 1993); Commission de l'Océan Indien (Indian Ocean Commission) together with Mauritius, Seychelles, the Comores and La Réunion. Nairobi Convention (with states of the East African Region).

Malaysia (dli: March 1997); the Singapore Revolving Fund with Singapore and Indonesia for the Straits of Malacca, Lombok–Macassar Contingency Plan with Indonesia. Proposed Bay of Brunei Contingency Plan with Brunei Darussalam. The ASEAN–OSRAP (Association of South East Asian Nations Oil Spill Response Action Plan) with the other ASEAN countries and with contribution from Japan. Regional Programme for the Prevention and Management of Marine Pollution in the East Asian Seas with the ASEAN countries, Cambodia, China, the People's Democratic Republic of Korea, Republic of Korea and Vietnam.

Malta (dli: April 1994); Barcelona Convention (with states bordering the Mediterranean)

Marshall Islands (dli: November 1995); Noumea Convention (with states of the South Pacific Region).

Mauritania (dli: November 1995); Abidjan Convention (with states of West and Central Africa).

Mauritius (dli: July 1995); Commission de l'Océan Indien (Indian Ocean Commission) together with Madagascar, Seychelles, the Comores and La Réunion.

Mexico (dli: November 1996); the Agreement on Co-operation on Pollution of the Marine Environment with the USA for the Mexican Gulf and associated coastlines. Cartagena Convention (with states of the Wider Caribbean Region). Operative Network for Regional Co-operation among Maritime Authorities of South America, Mexico, Panama and Cuba (ROCRAM).

Micronesia (Caroline Islands) (dli: September 1993); Noumea Convention (with states of the South Pacific Region).

Monaco (dli: December 1996); Barcelona Convention (with states bordering the Mediterranean). RAMOGE (a trilateral agreement with France and Italy).

Montserrat (dli: July 1995); Montserrat is not a party to any regional or bilateral agreements.

Morocco (dli: June 1996); Lisbon Agreement (with Portugal, Spain and France). Barcelona Convention (with states bordering the Mediterranean).

Mozambique (dli: December 1996); Nairobi Convention (with states of the East African Region).

Netherlands (dli: December 1996); Bonn Agreement (with countries bordering the North Sea). NethGer Agreement (Netherlands/Germany) for Wadden and North Sea. Member of the European Community Task Force.

Netherlands Antilles (dli: October 1995); Cartagena Convention (with states of the Wider Caribbean Region). A bilateral agreement exists with Venezuela.

New Zealand (dli: December 1996); Noumea Convention (with states of the South Pacific Region). A Memorandum of Understanding exists with the Australian MSA.

Nicaragua (dli: May 1996); Cartagena Convention (with states of the Wider Caribbean Region).

Nigeria (dli: November 1994); Abidjan Convention (with states of the West and Central African Region).

Northern Mariana Islands (dli: May 1996); Noumea Convention (with states of the South Pacific Region).

Norway (dli: April 1995); Bonn Agreement (countries bordering the North Sea). Norbrit Plan (a bilateral contingency plan with the UK). Copenhagen Agreement (with Denmark, Finland, Iceland and Sweden). Bilateral agreement with the Russian Federation for the Barents Sea.

Oman (dli: February 1995); Kuwait Convention (with the countries bordering the Gulf) administered by the Regional Organisation for the Protection of the Marine Environment (ROPME) based in Kuwait.

Pakistan (dli: January 1995); South Asia Co-operative Environment Programme with Sri Lanka, India, Bangladesh and the Maldives (established by UNEP).

Palau (dli: September 1993); as a UN Trust Territory administered by the USA, formal arrangements are in place with the USCG, Guam, for spill response. Noumea Convention (with states of the South Pacific Region).

Panama (dli: May 1996); Quito Convention (with states bordering the South East Pacific). Cartagena Convention (with states of the Wider Caribbean Region). Operative Network for Regional Co-operation among Maritime Authorities of South America, Mexico, Panama and Cuba (ROCRAM).

Papua New Guinea (dli: July 1995); Noumea Convention (with states of the South Pacific Region). A Memorandum of Understanding is being formulated with Australia.

Peru (dli: January 1994); Quito Convention (with states bordering the South East Pacific). Operative Network for Regional Co-operation among Maritime Authorities of South America, Mexico, Panama and Cuba (ROCRAM).

Philippines (dli: May 1996); the Sulawesi Sea Oil Spill Response Network, a co-operative agreement with Indonesia and Malaysia. The ASEAN–OSRAP (Association of South East Asian Nations Oil Spill Response Action Plan) with other ASEAN countries and with contribution from Japan. Regional Programme for the Prevention and Management of Marine Pollution in the East Asian Seas with the ASEAN countries, Cambodia, China, People's Democratic Republic of Korea, Republic of Korea and Vietnam.

Poland (dli: February 1996); Helsinki Convention (with countries bordering the Baltic Sea). Bilateral agreement with Russia. Bilateral agreement with Germany.

Portugal (dli: February 1995); Lisbon Agreement (with France, Spain and Morocco). Member of the European Community Task Force.

Puerto Rico (dli: May 1994); Cartagena Convention (with states of the Wider Caribbean Region).

Qatar (dli: February 1995); Kuwait Convention (with countries bordering the Gulf) administered by the Regional Organisation for the Protection of the Marine Environment (ROPME), based in Kuwait.

Romania (dli: July 1996); Bucharest Convention (with countries bordering the Black Sea).

Russian Federation (dli: July 1995); Helsinki Agreement (with countries bordering the

Baltic Sea). Bucharest Convention (with countries bordering the Black Sea). Bilateral agreement with the USA (covering the Bering Sea and Chukchi Sea). Bilateral agreement with Norway for the Barents Sea. Trilateral agreement with Finland and Estonia.

St Kitts and Nevis (dli: May 1994); Cartagena Convention (with states of the Wider Caribbean Region).

St Lucia (dli: May 1996); Cartagena Convention (with states of the Wider Caribbean Region).

St Vincent (dli: March 1994); Cartagena Convention (with states of the Wider Caribbean Region).

Samoa (American) (dli: September 1993); Noumea Convention (with states of the South Pacific Region).

Samoa (Western) (dli: September 1993); Noumea Convention (with states of the South Pacific Region).

Sao Tomé and Principe (dli: January 1989); Sao Tomé and Principe is not a party to any agreements.

Saudi Arabia (dli: May 1996); Jeddah Convention (with states bordering the Red Sea and Gulf of Aden). Kuwait Convention (with countries bordering the Gulf).

Senegal (dli: March 1997); Abidjan Convention (with states of the West and Central African Region).

Seychelles (dli: February 1995); Commission de l'Océan Indien (Indian Ocean Commission) together with Madagascar, Mauritius, the Comores and La Réunion. Nairobi Convention (with states of the East African Region).

Sierra Leone (dli: November 1994); Sierra Leone is not a party to any agreements.

Singapore (dli: May 1996); a Standard Operating Procedure for the Malacca Straits with Malaysia and Indonesia. The ASEAN–OSRAP (Association of South East Asian Nations Oil Spill Response Action Plan) with the other ASEAN countries and with contribution from Japan. Regional Programme for the Prevention and Management of Marine Pollution in the East Asian Seas with the ASEAN countries, Cambodia, China, People's Democratic Republic of Korea, Republic of Korea and Vietnam.

Slovenia (dli: January 1997); Barcelona Convention (with states bordering the Mediterranean).

South Africa (dli: March 1997); South Africa is not a party to any regional agreements.

Spain (dli: July 1995); Barcelona Convention (with states bordering the Mediterranean). The Lisbon Agreement (with Portugal, Morocco and France). Member of the European Community Task Force. Spain has observer status under the Bonn Agreement.

Sri Lanka (dli: May 1996); South Asia Co-operative Environment Programme with Pakistan, India, Bangladesh and the Maldives (established by UNEP).

Sudan (dli:); Jeddah Convention (with states bordering the Red Sea and Gulf of Aden).

Sweden (dli: May 1995); Bonn Agreement (with countries bordering the North Sea). Helsinki Convention (with countries bordering the Baltic Sea). Copenhagen Agreement (with Denmark, Finland, Iceland and Norway). German–Swedish pollution response plan, covering parts of the South-Western Baltic.

Syria (dli: June 1996); Barcelona Convention (with states bordering the Mediterranean).

Taiwan (dli: September 1994); Taiwan is not a party to any agreements.

Tanzania (dli: May 1993); Nairobi Convention (with states of the East African Region).

Thailand (dli: November 1994); ASEAN–OSRAP (Association of South East Asian Nations Oil Spill Response Action Plan) with the other ASEAN countries and with contribution from Japan. Regional Programme for the Prevention and Management of

Marine Pollution in the East Asian Seas with the ASEAN countries, Cambodia, China, People's Democratic Republic of Korea, Republic of Korea and Vietnam.

Togo (dli: January 1989); Abidjan Convention (with states of the West and Central African Region).

Tonga (dli: February 1996); Noumea Convention (with states of the South Pacific Region).

Trinidad and Tobago (dli: October 1995); Cartagena Convention (with states of the Wider Caribbean Region). A bilateral agreement exists with Venezuela.

Tunisia (dli: January 1997); Barcelona Convention (with states bordering the Mediterranean).

Turkey (dli: October 1996); Barcelona Convention (with states bordering the Mediterranean). Bucharest Convention (with countries bordering the Black Sea).

Ukraine (dli: July 1994); Bucharest Convention (with countries bordering the Black Sea).

United Arab Emirates (dli: May 1996); Kuwait Convention (with countries bordering the Gulf) administered by the Regional Organisation for the Protection of the Marine Environment (ROPME) based in Kuwait.

United Kingdom (dli: August 1996); Bonn Agreement (with countries bordering the North Sea). Manche Plan (a bilateral agreement with France). Norbrit Plan (a bilateral contingency plan with Norway). A bilateral agreement exists with Ireland covering the Irish Sea. Arrangements exist to aid the Channel Islands and Isle of Man if requested. Member of the European Community Task Force.

United States of America (dli: February 1997); agreements for co-operation in spill response exist between the USA and Bermuda, Canada, Japan, Mexico and the Russian Federation.

Uruguay (dli: January 1989); Operative Network for Regional Co-operation among Maritime Authorities of South America, Mexico, Panama and Cuba (ROCRAM). A bilateral agreement exists with Argentina.

Vanuatu (dli: July 1995); Vanuatu is not a party to any agreements.

Venezuela (dli: November 1996); Cartagena Convention (with states of the Wider Caribbean Region). Operative Network for Regional Co-operation among Maritime Authorities of South America, Mexico, Panama and Cuba (ROCRAM). Bilateral agreements exist with Colombia, the Netherlands Antilles and Trinidad and Tobago.

Vietnam (dli: June 1996); Regional Programme for the Prevention and Management of Marine Pollution in the East Asian Seas with the ASEAN countries, Cambodia, China, People's Democratic Republic of Korea and Republic of Korea.

Virgin Islands (UK) (dli: July 1995); Cartagena Convention (with states of the Wider Caribbean Region).

Virgin Islands (US) (dli: May 1994); Cartagena Convention (with states of the Wider Caribbean Region).

Republic of Yemen (dli: March 1994); Jeddah Convention (with states bordering the Red Sea and Gulf of Aden). l'Accord de Djibouti (with Djibouti and Somalia).

Zaïre (now Democratic Republic of Congo) (dli: January 1989); Zaïre is not a party to any agreements.

APPENDIX XI

STATUS OF CONVENTIONS

	IMO Convention 48	IMO amendments 91	IMO amendments 93	SOLAS Convention 74	SOLAS Protocol 78	SOLAS Protocol 88	Stockholm Agreement 96	LOAD LINES Convention 66	LOAD LINES Protocol 88	TONNAGE Convention 69	COLREG Convention 72	CSC Convention 72	CSC amendments 93	SFV Protocol 93	STCW Convention 78	STCW-F Convention 95	SAR Convention 79	STP Agreement 71	STP Protocol 73	INMARSAT Convention 76	INMARSAT amendments 94	INMARSAT OA 76	INMARSAT OA amendments 94	FACILITATION Convention 65	MARPOL 73/78	MARPOL Annex III	MARPOL Annex IV	MARPOL Annex V	LC Convention 72	LC Protocol 96	INTERVENTION Convention 69	INTERVENTION Protocol 73	CLC Convention 69	CLC Protocol 76	CLC Protocol 92	FUND Convention 71	FUND Protocol 76	FUND Protocol 92	NUCLEAR Convention 71	PAL Convention 74	PAL Protocol 76	PAL Protocol 90	LLMC Convention 76	LLMC Protocol 96	SUA Convention 88	SUA Protocol 88	SALVAGE Convention 89	OPRC Convention 90	HNS Convention 96
Afghanistan												X																	X																				
Albania	X																																X	X		X	X												
Algeria	X		X	X	X			X		X	X				X		X			X		X		X	X	X	X	X					X			X													
Andorra																																																	
Angola	X			X	X			X			X				X																																		
Antigua & Barbuda	X			X	X			X		X	X				X										X	X	X	X	X				X	X		X													
Argentina	X		X	X	X	X		X	X	X	X	X			X		X			X		X		X	X	X	X	X	X		X								X	X	X				X			X	
Armenia																																																	
Australia	X	X	X	X	X	X		X	X	X	X	X	X		X		X			X	X	X	X	X	X	X		X	X		X	X	X	X	X	X	X	X					X		X	X	X	X	
Austria	X			X	X			X		X	X	X			X									X	X	X	X	X																	X	X			
Azerbaijan	X			X				X		X	X				X														X																				
Bahamas	X			X	X			X		X	X	X			X					X		X		X	X	X		X			X	X	X	X	X	X	X	X		X	X		X						
Bahrain	X			X				X		X	X				X					X	X	X	X										X	X	X	X	X	X											
Bangladesh	X			X				X		X	X				X			X	X	X		X									X																		
Barbados	X			X	X			X		X	X	X			X		X							X	X	X		X	X		X	X	X	X		X	X			X	X		X		X	X			
Belarus				X				X			X	X								X		X			X	X	X	X	X																				
Belgium	X	X		X	X			X		X	X	X			X		X			X	X	X	X	X	X	X	X	X	X		X	X	X	X		X	X		X	X	X		X						
Belize	X		X	X	X			X		X	X				X										X	X	X	X					X	X															
Benin	X			X				X		X	X	X			X									X							X		X			X							X						
Bhutan																																																	
Bolivia	X														X																																		
Bosnia & Herzegovina	X																																																
Botswana																																																	
Brazil	X	X	X	X	X			X		X	X	X			X		X			X		X		X	X				X				X																
Brunei Darussalam	X			X	X			X		X	X				X					X		X			X								X	X		X													
Bulgaria	X	X	X	X	X			X		X	X	X			X					X		X			X	X	X	X			X																		
Burkina Faso																																																	
Burundi																																																	
Cambodia	X			X	X			X		X	X														X	X	X	X					X																
Cameroon	X	X		X				X			X				X		X			X		X		X							X		X	X		X													
Canada	X	X	X	X				X		X	X	X			X		X			X		X		X	X				X				X	X		X	X								X	X	X	X	
Cape Verde	X			X				X			X				X									X					X																				
Central African Republic																																																	
Chad																																																	
Chile	X	X		X	X	X		X	X	X	X	X			X		X			X	X	X	X	X	X	X	X		X		X	X	X												X	X		X	
China	X	X	X	X	X	X		X	X	X	X	X			X		X			X		X		X	X	X		X	X		X	X	X	X						X	X				X	X	X		
Colombia	X			X	X			X		X	X				X					X		X		X	X	X	X	X					X	X		X	X												
Comoros																																																	
Congo	X			X				X			X																																						
Costa Rica	X																			X		X							X																				
Cote d'Ivoire	X			X	X			X		X	X				X		X							X	X	X	X	X	X		X		X			X													
Croatia	X			X	X			X		X	X	X			X					X		X		X	X	X	X	X	X		X	X	X			X							X						
Cuba	X	X	X	X	X			X		X	X	X		X	X					X		X		X	X				X		X																		
Cyprus	X	X	X	X	X			X		X	X	X			X		X	X	X	X	X	X	X		X			X	X				X	X	X	X	X	X											
Czech Republic	X			X	X	*		X	*	X	X	X			X					X		X		X	X	X	X	X																					
Dem. People's Rep. Korea	X		X	X	X			X		X	X	X			X									X	X	X	X	X																					
Dem. Rep. of the Congo	X							X			X				X														X																				
Denmark	X	X	X	X	X	X	X	X	X	X	X	X			X		X			X	X	X	X	X	X	X	X	X	X	X	X	X	X	X	X	X	X	X	X				X		X	X	X	X	
Djibouti	X			X				X			X														X						X		X			X													
Dominica	X		X																																														
Dominican Republic	X			X				X			X													X					X		X		X																
Ecuador	X			X				X		X	X				X		X							X	X	X	X	X			X		X																
Egypt	X	X	X	X	X	X		X	X		X				X			X	X	X		X		X	X	X	X	X	X		X	X	X	X	X					X		X	X		X	X	X	X	
El Salvador	X									X	X																																					X	
Equatorial Guinea	X			X	X	X		X	X	X	X				X										X	X	X	X			X		X							X			X						
Eritrea	X			X				X		X	X				X																																		
Estonia	X	X	X	X	X			X		X	X	X	X		X										X	X	X	X					X			X													
Ethiopia	X			X	X			X		X	X				X																																		
Fiji	X			X				X		X	X				X									X							X		X			X													
Finland	X	X	X	X	X		X	X		X	X				X		X			X	X	X	X	X	X	X	X	X	X		X	X	X	X	X	X	X	X	X				X					X	
France	X	X		X	X	X		X	X	X	X	X			X		X	X	X	X	X	X		X	X	X	X	X	X		X	X	X	X	X	X	X	X	X				X		X	X		X	
Gabon	X			X				X			X				X					X		X			X	X	X	X	X		X		X			X			X										
Gambia	X			X				X		X	X				X		X							X	X	X	X	X					X			X									X				
Georgia	X		X	X		X		X		X	X	X			X		X							X	X	X	X	X			X	X	X	X						X	X		X				X	X	
Germany	X		X	X	X	X	X	X	X	X	X	X			X		X			X	X	X	X	X	X	X	X	X	X		X	X	X	X	X	X	X	X	X				X		X	X		X	
Ghana	X		X	X	X			X		X	X				X					X		X		X	X						X		X			X													

*/ Clarification has been sought in respect of the succession by the Czech Republic to these treaties

	IMO Convention 48	IMO amendments 91	IMO amendments 93	SOLAS Convention 74	SOLAS Protocol 78	SOLAS Protocol 88	Stockholm Agreement 96	LOAD LINES Convention 66	LOAD LINES Protocol 88	TONNAGE Convention 69	COLREG Convention 72	CSC Convention 72	CSC amendments 93	SFV Protocol 93	STCW Convention 78	STCW-F Convention 95	SAR Convention 79	STP Agreement 71	STP Protocol 73	INMARSAT Convention 76	INMARSAT amendments 94	INMARSAT OA 76	INMARSAT OA amendments 94	FACILITATION Convention 65	MARPOL 73/78	MARPOL Annex III	MARPOL Annex IV	MARPOL Annex V	LC Convention 72	LC Protocol 96	INTERVENTION Convention 69	INTERVENTION Protocol 73	CLC Convention 69	CLC Protocol 76	CLC Protocol 92	FUND Convention 71	FUND Protocol 76	FUND Protocol 92	NUCLEAR Convention 71	PAL Convention 74	PAL Protocol 76	PAL Protocol 90	LLMC Convention 76	LLMC Protocol 96	SUA Convention 88	SUA Protocol 88	SALVAGE Convention 89	OPRC Convention 90	HNS Convention 96
Greece	X	X	X	X	X	X		X	X	X	X	X			X		X	X	X	X		X		X	X	X	X	X	X				X	X	X	X	X	X		X	X		X		X	X	X	X	
Grenada																																																	
Guatemala	X			X				X			X																		X				X																
Guinea	X			X				X		X	X	X			X									X																									
Guinea-Bissau	X																																																
Guyana	X																																																
Haiti	X			X				X		X					X														X																				
Holy See																																																	
Honduras	X			X	X			X			X	X			X														X																				
Hungary	X			X	X			X		X	X	X			X					X		X		X	X	X	X	X	X																X	X			
Iceland	X			X	X			X		X	X	X			X		X			X		X		X	X	X		X	X		X		X	X		X	X											X	
India	X	X	X	X	X			X		X	X	X			X			X	X	X		X		X	X								X	X		X	X										X		
Indonesia	X	X	X	X	X			X		X	X	X			X			X	X	X		X			X								X			X													
Iran (Islamic Republic of)	X		X	X				X		X	X				X		X			X		X		X					X		X	X															X		
Iraq	X			X						X										X		X		X																									
Ireland	X			X	X		X	X		X	X				X		X							X	X			X	X		X	X	X	X	X	X	X	X									X		
Israel	X			X	X			X		X	X	X			X					X	X	X		X	X	X																							
Italy	X			X	X	X		X	X	X	X	X			X		X			X		X		X	X	X	X	X	X		X	X	X	X		X	X		X						X	X	X		
Jamaica	X			X				X			X				X		X								X	X	X	X	X		X	X			X			X											
Japan	X			X	X	X		X	X	X	X	X			X		X			X		X			X	X	X	X	X		X		X	X	X	X	X	X					X					X	
Jordan	X			X						X														X					X											X							X		
Kazakhstan	X			X	X			X		X	X	X			X										X	X	X	X					X																
Kenya	X							X		X	X				X		X								X	X	X	X	X				X			X													
Kiribati															X														X																				
Kuwait	X		X	X	X			X		X	X									X	X	X	X								X		X	X		X													
Kyrgyzstan																																																	
Lao People's Dem. Rep.																																																	
Latvia	X			X				X			X				X					X		X			X	X	X	X					X																
Lebanon	X		X	X	X			X		X					X					X		X			X	X	X	X			X		X												X	X			
Lesotho																																																	
Liberia	X		X	X	X	X		X	X	X	X	X			X					X		X		X	X	X		X			X	X	X	X	X	X	X	X	X	X	X		X		X	X		X	
Libyan Arab Jamahiriya	X			X	X			X							X														X																				
Liechtenstein																																																	
Lithuania	X			X	X			X		X	X	X			X										X	X	X	X																					
Luxembourg	X			X	X	X		X	X	X	X	X			X		X							X	X	X	X	X	X				X	X						X	X								
Madagascar	X		X	X				X							X									X																									
Malawi	X			X											X																									X									
Malaysia	X			X	X			X		X	X				X					X		X			X			X					X			X												X	
Maldives	X			X				X		X	X				X																		X	X		X													
Mali																																																	
Malta	X		X	X	X			X		X	X				X					X		X			X				X				X	X		X	X												
Marshall Islands				X	X	X		X	X	X	X	X	X		X					X		X		X	X	X	X	X			X	X	X	X	X	X	X	X		X	X		X		X	X	X	X	
Mauritania	X							X			X				X																		X	X		X													
Mauritius	X		X	X				X		X	X				X					X		X		X	X	X	X	X					X	X		X	X												
Mexico	X		X	X	X	X		X	X	X	X	X			X		X			X	X	X	X	X	X				X		X	X	X	X	X	X	X	X					X		X	X	X	X	
Micronesia (Fed. States of)																																																	
Monaco	X		X	X				X		X	X						X			X	X	X	X	X	X	X	X	X	X		X		X		X	X		X											
Mongolia	X																																																
Morocco	X	X	X	X				X		X	X	X			X										X	X	X	X	X		X		X			X	X												
Mozambique	X			X				X		X	X				X		X			X		X											X			X													
Myanmar	X			X	X			X		X	X				X										X																								
Namibia	X																																																
Nauru																													X																				
Nepal	X																																																
Netherlands	X	X	X	X	X	X	X	X	X	X	X	X	X		X		X			X	X	X	X	X	X	X		X	X		X	X	X	X	X	X	X	X	X				X		X	X		X	
New Zealand	X			X	X			X		X	X	X			X		X			X		X		X					X		X		X			X							X						
Nicaragua	X							X	X	X																					X	X	X	X															
Niger																																																	
Nigeria	X		X	X	X			X		X	X				X					X		X		X					X				X			X											X	X	
Norway	X	X		X	X	X	X	X	X	X	X	X		X	X		X	X	X	X	X	X	X	X	X	X		X	X		X	X	X	X	X	X	X	X	X				X		X	X	X	X	
Oman	X			X	X	X		X	X	X	X				X		X			X	X	X	X		X	X	X	X	X		X	X	X	X	X	X		X							X	X	X		
Pakistan	X			X	X			X		X	X	X			X		X			X		X			X	X	X	X	X		X	X																X	
Palau																																																	
Panama	X		X	X	X			X		X	X				X					X		X			X	X	X	X	X		X		X																
Papua New Guinea	X			X				X		X	X				X		X								X	X	X	X	X		X		X			X													
Paraguay	X																																																
Peru	X	X	X	X	X			X		X	X	X			X		X			X		X		X	X	X	X	X					X	X															
Philippines	X			X				X		X					X			X		X		X							X						X			X											
Poland	X		X	X	X			X		X	X	X			X		X			X		X		X	X	X	X	X	X		X	X	X	X		X	X			X	X		X		X	X			

	IMO Convention 48	IMO amendments 91	IMO amendments 93	SOLAS Convention 74	SOLAS Protocol 78	SOLAS Protocol 88	Stockholm Agreement 96	LOAD LINES Convention 66	LOAD LINES Protocol 88	TONNAGE Convention 69	COLREG Convention 72	CSC Convention 72	CSC amendments 93	SFV Protocol 93	STCW Convention 78	STCW-F Convention 95	SAR Convention 79	STP Agreement 71	STP Protocol 73	INMARSAT Convention 76	INMARSAT amendments 94	INMARSAT OA 76	INMARSAT OA amendments 94	FACILITATION Convention 65	MARPOL 73/78	MARPOL Annex III	MARPOL Annex IV	MARPOL Annex V	LC Convention 72	LC Protocol 96	INTERVENTION Convention 69	INTERVENTION Protocol 73	CLC Convention 69	CLC Protocol 76	CLC Protocol 92	FUND Convention 71	FUND Protocol 76	FUND Protocol 92	NUCLEAR Convention 71	PAL Convention 74	PAL Protocol 76	PAL Protocol 90	LLMC Convention 76	LLMC Protocol 96	SUA Convention 88	SUA Protocol 88	SALVAGE Convention 89	OPRC Convention 90	HNS Convention 96
Portugal	X			X	X			X		X	X	X			X		X			X		X		X	X	X	X	X	X		X	X	X	X		X	X								X	X			
Qatar	X			X				X		X	X									X	X	X	X								X		X	X		X													
Republic of Korea	X	X	X	X	X	X		X	X	X	X	X			X		X			X	X	X	X		X	X		X	X				X	X	X	X		X											
Republic of Moldova																																																	
Romania	X			X				X		X	X	X			X					X		X			X			X																	X	X			
Russian Federation	X	X	X	X	X			X		X	X	X			X	X	X			X		X		X	X	X	X	X	X		X	X	X	X		X	X			X	X								
Rwanda																																																	
Saint Kitts and Nevis																																	X			X													
Saint Lucia	X																												X																				
St. Vincent & Grenadines	X			X	X			X		X	X				X			X	X						X	X	X	X					X																
Samoa	X			X	X			X			X				X																																		
San Marino																																																	
Sao Tome & Principe	X																																																
Saudi Arabia	X		X	X	X			X		X	X	X			X			X	X	X	X	X	X										X	X													X		
Senegal	X			X	X			X		X	X				X		X			X		X		X	X	X	X	X			X		X															X	
Seychelles	X	X		X	X	X		X	X		X				X									X	X				X				X			X									X	X		X	
Sierra Leone	X			X				X							X																		X			X													
Singapore	X	X	X	X	X			X		X	X				X		X			X	X	X	X	X	X	X							X	X	X														
Slovakia	X	X	X	X	X	X		X	X	X	X	X			X					X	X	X	X	X	X	X	X	X																					
Slovenia	X			X	X			X		X	X	X			X									X	X	X	X	X	X		X	X	X			X													
Solomon Islands	X										X				X														X																				
Somalia	X							X																																									
South Africa	X			X	X			X		X	X	X			X		X			X	X	X	X		X	X		X	X		X	X	X																
Spain	X	X	X	X	X	X		X	X	X	X	X			X		X			X		X		X	X	X	X	X	X		X	X	X	X	X	X	X	X	X	X	X	X	X		X	X		X	
Sri Lanka	X			X				X		X	X				X			X	X	X		X			X	X	X	X			X		X			X													
Sudan	X			X				X							X																																		
Suriname	X			X				X																X	X	X	X	X	X		X																		
Swaziland																																																	
Sweden	X	X	X	X	X	X	X	X	X	X	X	X			X		X	X	X	X		X		X	X	X	X	X	X		X	X	X	X	X	X	X	X	X				X		X	X	X	X	
Switzerland	X		X	X	X			X		X	X				X					X		X		X	X	X		X	X		X	X	X	X	X	X				X	X		X		X	X	X	X	
Syrian Arab Republic	X							X		X	X													X	X						X		X			X													
Tajikistan																																																	
Thailand	X	X	X	X				X		X	X				X					X		X		X																									
the former Yugoslav Republic of Macedonia	X																																																
Togo	X			X	X			X		X	X				X										X	X	X	X																					
Tonga				X				X		X	X				X										X	X	X	X	X		X	X	X			X				X								X	
Trinidad & Tobago	X	X	X	X				X		X	X				X		X							X																					X	X			
Tunisia	X		X	X	X			X			X				X					X		X		X	X	X	X	X	X		X	X	X		X	X		X										X	
Turkey	X			X				X		X	X				X		X			X		X			X			X																					
Turkmenistan	X																																																
Tuvalu				X				X		X	X				X										X	X	X	X					X			X													
Uganda																																																	
Ukraine	X			X	X			X		X	X	X			X		X			X	X	X	X	X	X	X	X	X	X		X									X	X				X	X			
United Arab Emirates	X		X	X	X			X		X	X				X		X			X		X							X		X		X	X		X											X		
United Kingdom	X	X	X	X	X		X	X		X	X	X			X		X	X	X	X	X	X	X	X	X	X	X	X	X		X	X	X	X	X	X	X	X		X	X		X		X	X	X	X	
United Rep. of Tanzania	X							X							X																																		
United States	X			X	X	X		X	X	X	X	X			X		X			X	X	X	X	X	X	X		X	X		X	X													X	X	X	X	
Uruguay	X			X	X			X		X	X				X		X							X	X	X	X	X							X			X										X	
Uzbekistan																																																	
Vanuatu	X			X	X	X		X	X	X	X	X			X		X							X	X	X		X	X		X	X	X	X		X	X			X	X		X						
Venezuela	X			X				X		X	X				X		X								X	X	X	X					X	X		X	X											X	
Viet Nam	X			X	X			X		X	X				X										X																								
Yemen	X			X				X		X	X	X						X	X					X							X	X	X	X					X	X	X		X						
Yugoslavia	X			X	X			X		X	X	X			X					X		X		X	X	X	X	X	X		X	X	X			X													
Zambia								X																X																									
Zimbabwe																																																	
Hong Kong, China (Associate Member)	X			X	X			X		X	X	X			X		X	X	X	X				X	X	X		X	X		X	X	X	X		X	X			X	X		X				X		

SUMMARY OF STATUS OF CONVENTIONS

Convention	Entry into force date	No. of Contracting States	% world tonnage*
IMO Convention	17 Mar 58	155	95.91
1991 amendments	—	32	—
1993 amendments	—	50	—
SOLAS 1974	25 May 80	135	98.26
SOLAS Protocol 1978	01 May 81	88	91.95
SOLAS Protocol 1988	—	28	41.73
Stockholm Agreement 1996	01 Apr 97	8	9.34
LL 1966	21 Jul 68	140	98.19
LL Protocol 1988	—	28	41.69
TONNAGE 1969	18 Jul 82	117	97.50
COLREG 1972	15 Jul 77	130	96.20
CSC 1972	06 sep 77	63	64.47
1993 amendments	—	4	—
SFV Protocol 1993	—	2	4.35
STCW 1978	28 Apr 84	129	97.55
STCW-F 1995	—	1	2.71
SAR 1979	22 Jun 85	56	49.11
STP 1971	02 Jan 74	15	23.95
SPACE STP 1973	02 Jun 77	14	22.17
INMARSAT C 1976	16 Jul 79	81	93.07
1994 amendments	—	25	—
INMARSAT OA 1976	16 Jul 79	81	—
1994 amendments	—	23	—
FAL 1965	05 Mar 67	78	56.73
MARPOL Annex I/II	02 Oct 83	100	93.47
MARPOL Annex III	01 Jul 92	81	78.21
MARPOL Annex IV	—	66	41.46
MARPOL Annex V	31 Dec 88	83	82.02
LC 1972	30 Aug 75	77	67.64
1978 amendments	—	20	—
LC Protocol 1996	—	1	1.18

Convention	Entry into force date	No. of Contracting States	% world tonnage*
INTERVENTION 1969	06 May 75	70	66.57
INTERVENTION Protocol 1973	30 Mar 83	40	44.20
CLC 1969	19 Jun 75	98	88.54
CLC Protocol 1976	08 Apr 81	55	67.26
CLC Protocol 1992	30 May 96	29	50.29
FUND 1971	16 Oct 78	74	62.02
FUND Protocol 1976	22 Nov 94	35	57.00
FUND Protocol 1992	30 May 96	26	46.73
NUCLEAR 1971	15 Jul 75	14	23.01
PAL 1974	28 Apr 87	23	35.59
PAL Protocol 1976	30 Apr 89	18	35.33
PAL Protocol 1990	—	2	0.57
LLMC 1976	01 Dec 86	28	42.60
LLMC Protocol 1996	—	0	—
SUA 1988	01 Mar 92	33	39.23
SUA Protocol 1988	01 Mar 92	31	39.11
SALVAGE 1989	14 Jul 96	23	26.65
OPRC 1990	13 May 95	33	37.91
HNS Convention 1996	—	0	—

* *Source: Lloyd's Register of Shipping/World Fleet Statistics as at 31 December 1996.*

BIBLIOGRAPHY

Abecassis, David W. (ed.), *Oil Pollution From Ships, International, United Kingdom and United States Law and Practice* (2nd ed., London, Stevens, 1985).

Akehurst, Michael, *A Modern Introduction to International Law* (6th ed., London, Unwin Hayman, 1987).

Alcantara, Leonard F., and Cox, Mary A., "OPA 90 Certificates of Financial Responsibility" (1992) 23 *Journal of Maritime Law and Commerce*, pp.369–386.

Asariotis, Regina, "Contracts for the Carriage of Goods by Sea and Conflict of Laws: Some Questions Regarding the Contracts (Applicable Law) Act 1990" (1995) 26 *Journal of Maritime Law and Commerce*, pp.293–314.

Astle, W.E., *Shipping and the Law* (London, Fairplay Publications, 1980).

—— *Limitation of Liability; Merchant Shipping Act, 1894; section 503 and subsequent amendments* (London, Fairplay Publications, 1985).

Ayorinde, Akintayo A., "Inconsistencies Between OPA 90 and MARPOL 73/78: What is the Effect on Legal Rights and Obligations of the United States and Other Parties to MARPOL 73/78?" (1994) 25 *Journal of Maritime Law and Commerce*, pp.55–94.

Bates, John H., and Benson, Charles, *Marine Environment* (London, Lloyd's of London Press, 1993).

Beatson, J., and Cooper, J.J., "Rights of Suit in Respect of Carriage of Goods by Sea" [1991] *Lloyd's Maritime and Commercial Law Quarterly*, pp.196–208.

Beck, Andrew, "Floating Choice of Law Clauses" [1987] *Lloyd's Maritime and Commercial Law Quarterly*, pp.523–533.

Bennett, T.W., and Dean, W.H.B., *et al.*, *The Law of the Sea* (Cape Town, Juta & Co. Ltd. 1986).

Berlingieri, F., "Jurisdiction and an Overview of the Concept of Fault and its Apportionment" (1977) 51 *Tulane Law Review*, pp.866–879.

Boczek, B.A., *Flags of Convenience, An International Legal Study* (Cambridge, Mass., Harvard University Press, 1962).

Bovet, David, "Certificates of Financial Responsibility: Can the Train Wreck be Averted?" 1995 International Oil Spill Conference, 27 February–2 March 1995, California.

Brandon, H.V., "Apportionment of Liability in British Courts under Maritime Conventions Act of 1911" (1977) 51 *Tulane Law Review*, pp.1025–1046.

Braekhus, S., and Rein, A., *Handbook on P&I Insurance* (2nd ed., Arendal, Norway, Assuranceforeningen Gard, 1979).

Brice, Geoffrey, "Maritime Claims: The European Judgments Convention" [1987] *Lloyd's Maritime and Commercial Law Quarterly*, pp.281–304.

Briggs, Adrian, "The Validity of 'Floating' Choice of Law and Jurisdiction Clauses" [1986] *Lloyd's Maritime and Commercial Law Quarterly*, pp.508–518.

Britannia Steamship Insurance Association Limited, *TOVALOP and CRISTAL Explanatory Notes* (Revised Edition, February 1994).

Brodecki, Z., "New Definition of Pollution Damage" [1985] *Lloyd's Maritime and Commercial Law Quarterly*, pp.382–391.

Brown, E.D., *The International Law of the Sea*, Volume I, Introductory Manual; Volume II, Documents, Cases and Tables (Aldershot, Dartmouth, 1994).

Brown, Robert H., *Training Notes for Brokers, The Insurance of Merchant Ships & Shipowners' Interests* (London, Witherby, 1990).

—— *The Institute Time Clauses Hulls 1995, Analysis of Marine Insurance Clauses* (London, Witherby and Co., 1996).

Brown, Scott R., and Savage, Ian, "The Economics of Double Hulled Tankers" (1996) 23 *Maritime Policy Management*, pp.167–175.

Brubaker, Douglas, *Marine Pollution and International Law, Principles and Practice* (London, Belhoven Press, 1993).

Buglass, Leslie, J., "Limitation of Liability from a Marine Insurance Viewpoint" (1979) 53 *Tulane Law Review*, pp.1364–1387.

Carver, Thomas Gilbert, *Carver's Carriage by Sea* (ed. Raoul Colinvaux., 13th ed., London, Stevens, 1982).

Chalos, Michael G., "A Perspective on the *Exxon Valdez* and the Consequences for Vessel Oil Pollution Liability in the United States", Liverpool & London P&I 2nd International Conference, Malta, 24–26 April 1990.

Cheka, C.N., "Conduct Barring Limitation" (1987) 18 *Journal of Maritime Law and Commerce*, pp.487–498.

Chorley, Robert Samuel-Theodore, *Chorley & Giles' Shipping Law* (ed. N.J.J. Gaskell, C. Debattista, Richard J. Swatton, 8th ed., Pitman, London, 1988).

Churchill, R.R., and Lowe, A.V., *The Law of the Sea* (2nd ed., Manchester, Manchester University Press, 1988).

Clark, Bessemer A.F., "The US Oil Pollution Act of 1990" [1991] *Lloyd's Maritime and Commercial Law Quarterly*, pp.247–256.

CMI, *Guidelines on Oil Pollution Damage*, Part II: Economic Loss, 8 October 1994.

Cockcroft, A.N., and Lameijer, J.N.F., *A Guide to the Collision Avoidance Rules, International Regulations for Preventing Collisions at Sea* (4th ed., Heinemann Newness, 1990).

Corbet, A.G., "Who is the Shipowner?", *Seaways* (The Journal of Nautical Institute, December 1993), p.16.

Cottrill, Ken, "Campaigning Against Flagging Out", *The Baltic*, Vol. 5, Issue 3, May 1996, pp.62–64.

CRISTAL, *Memorandum of Explanation of the Contract Regarding a Supplement to Tanker Liability for Oil Pollution*, Revised, 20 February, 1994.

De La Rue, Colin M. (ed.), *Liability for Damage to the Marine Environment* (London, Lloyd's of London, Press 1993).

—— "Oil Pollution from Ships: Problems of Economic Loss and Environmental Damage", Conference: Managing Oil Pollution Risks in Shipping Operations, Managing Spill Liabilities: Claims and Compensation, 15 December 1995, London.

Dicey, A.V. and Carlile, J.H., *Dicey and Morris on the Conflict of Laws* (ed. Lawrance Collins, 12th ed., London, Sweet and Maxwell, 1993).

Dinstein, Yoram, "Oil Pollution by Ships and Freedom of the High Seas" (1972) 3 *Journal of Maritime Law and Commerce*, pp.363–374.

Dixon, Martin, *Textbook on International Law* (3rd ed., London, Blackstone, 1996).

Dockray, Martin, *Cases and Materials on the Carriage of Goods By Sea* (Abingdon, Professional Books, 1987).

Doganis, S.R., and Metaxas, B.N., *The Impact of Flags of Convenience*, Research Reports/Polytechnic of Central London, Transport Studies Group; No.3, 1976.

Donovan, J. James, Admiralty Law Institute: Symposium on Limitation of Liability. "The Origins and Development of Limitation of Shipowners' Liability" (1979) 53 *Tulane Law Review*, pp.999–1045.

Dover, V., *A Handbook to the Marine Insurance, incorporating revisions by Robert H. Brown* (8th ed., London, Witherby, 1975).

Elliot, Sharon, "The Relevance of the United States' 'Superfund Law' to Vessel Owners and Operators", *Gard News*, Issue 129, April 1993, pp.12–14.

Emanuelli, Claude C., "The Right of Intervention of Coastal States on the High Seas in Cases of Pollution Casualties" (1976) XXV *University of Brunswick Law Journal*, pp.79–96.

Ewing, K., "Union Action against Flags of Convenience—The Legal Position in Great Britain" (1980) 11 *Journal of Maritime Law and Commerce*, pp.503–508.

Franciani, Francesco and Scourazzi, Tullio, *International Responsibility for Envrionmental Damage* (London, Graham and Trotman, 1991).

Gard News, "The Collision Aspect of P&I", Issue 141, April 1996, pp.11–20.

Gold, Edgar, *Handbook on Marine Pollution* (Arendal, Norway, Assuranceforeningen Gard, 1985).

—— "International Convention on Oil Pollution Preparedness, Response and Co-operation", *Gard News*, Issue 121, June 1991, pp.3–4.

—— "Marine Pollution Liability After 'Exxon Valdez': The US. 'All-Or-Nothing' Lottery!" (1991) *Journal of Maritime Law and Commerce*, 22, pp.423–444.

Gray, William O., "OPA 90—The Independent Tanker Owners' Perspective", *Bimco Bulletin*, 6/92, November/December, pp.5–9.

Griggs, Patrick, and Williams, Richard, *Limitation of Liability for Maritime Claims* (2nd ed., London, Lloyd's of London Press, 1991).

Grime, R., *Shipping Law* (2nd ed, London, Sweet and Maxwell, 1991).

Grumbles, Benjamin. "Federal Oil Spill Legislation in the Wake of Exxon Valdez", *MTS Journal*, Vol. 24, No. 4, pp.5–11.

Grunsey, K., *Flags of Convenience in 1978*, Transport Studies Group, Discussion Paper No. 8, 1978.

Harley, S.T., "Collision Cases 1974" [1975] *Lloyd's Maritime and Commercial Law Quarterly*, pp.156–165.

Hazelwood, Steven, "Gaps in the action *in rem*" [1982] *Lloyd's Maritime and Commercial Law Quarterly*, pp.210–217.

—— "Gaps in the action *in rem* plugged?" [1982] *Lloyd's Maritime and Commercial Law Quarterly*, pp.422–426.

—— *P&I Clubs. Law and Practice* (London, Lloyd's of London Press, 1994).

Healy, N.J., and Sweeney, J.C., "Basic Principles of the Law of Collision" (1991) 22 *Journal of Maritime Law and Commerce*, pp.359–404.

—— Paulsen, Gordon W., and Marion, Matthew A., "The United States Oil Pollution Act of 1990", *Bimco Bulletin*, 5/90, September/October, pp.27–33.

Hill, Christopher, *Maritime Law* (4th ed., London, Lloyd's of London Press, 1995).

—— "Oil Pollution", *P&I International*, October 1995, pp.194–196.

Hodges, Susan, *Law of Marine Insurance* (London, Cavendish, 1996).

Howard, M.N., "Floating Choice of Law Clauses, Case and Comment" [1995] *Lloyd's Maritime and Commercial Law Quarterly*, pp.1–8.

—— "The Carriage of Goods by Sea Act 1992" (1993) 24 *Journal of Maritime Law and Commerce*, pp.181–190.

Hudges, A.D., *Casebook on Carriage of Goods By Sea* (London, Blackstone Press, 1994).

Hudson, Cynthia A., "Vessel Response Plans Under OPA 1990", *Bimco Bulletin*, 6/92, November/December, pp.10–13.

Hudson, N.G. and J.C. Allen, *The Institute Clauses Handbook* (London, Lloyd's of London Press, 1986).

Imaizumi, Takatada, *Transition of Shipowners' Liability Laws and Changes of P&I Clubs in the UK* (Tokyo, Seinbundo Publishing Co. Ltd., 1992).

Institute of Maritime Law, University of Southampton (comp), *The Limitation of Shipowners' Liability. The New Law* (London, Sweet and Maxwell, 1986).

Insurance Institute of London, Report of the Advanced Study Group No. 109, *The History and Development of Protecting and Indemnity Clubs*.

International Maritime Organisation, *Oil Pollution Liability and Compensation, Course-+Compendium, Model Course, 3.16* (1991).

International Oil Spill Conference, 27 February–2 March 1995. Long Beach, California, American Petroleum Institute Publication No. 4620, Port City Press Inc., Maryland.

The International Two Day Forum on Oil Pollution 1997, 17–18 February 1997, London.

ISM, "Guide to International Ship Registers and Management Services" (1996).

ITF (9597 release), "European week of action." http://www.itf.org.uk/english7/97ewal.html(30 May 1997).

ITF (12 May 1997), "Successful start to action week", http://www.itf.org.uk/english/woa1.html (30 May 1997).

ITF (13 May 1997), "European Actions Proliferate as Week Heats Up", http://www.itf.org.uk/english/woa2.html (30 May 1997).

ITF (14 May 1997), "ITF Calls for Action over Crew Abandonment. Rights to Boycott in Germany Legally Upheld." http://www.itf.org.uk/english/woa4.html (30 May 1997).

ITOPF, *Country Profiles, A Summary of Oil Spill Response Arrangements and Resources Worldwide* (May 1996).

International Oil Pollution Compensation Fund, *Annual Report 1995.*

International Oil Pollution Compensation Fund, *Annual Report 1996.*

International Oil Pollution Compensation Fund, *Claims Manual* (4th ed., June 1995).

International Oil Pollution Compensation Fund, *General Information on Liability and Compensation for Oil Pollution Damage* (July 1995).

Jackson, D.C., "Admiralty Jurisdiction—The Supreme Court Act 1981" [1982] *Lloyd's Maritime and Commercial Law Quarterly*, pp.236–248.

—— *Enforcement of Maritime Claims* (London, Lloyd's of London Press, 2nd ed., 1996).

Joyner, Christopher C., "The 1991 Madrid Environmental Protection Protocol", Contributions to Marine Pollution Law (1996) 20 *Marine Policy*, pp.183–197.

Kaye, Peter, *The New International Law of Contract of the European Community; Implementation of the EEC's Contractual Obligations Convention in England and Wales under the Contracts (Applicable Law) Act 1990* (Brookfield, Vt., Dartmouth Pub. Co., 1993).

Lambeth, R.J., *Templeman on Marine Insurance, Its Principles and Practice* (6th ed., London, Pitman, 1986).

Law Commission and the Scottish Law Commission, Rights of Suit in Respect of Carriage of Goods at Sea, 19 March 1991.

Lee, Luke T., "The Law of the Sea Convention and Third States" (1983) 77 *American Journal of International Law*, pp.541–568.

Mankabady, S., *Collision at Sea, A Guide to the Legal Consequences* (Amsterdam; New York; Oxford, North-Holland Publishing Company, 1978).

—— *The Law of Collision at Sea* (Amsterdam, Elsevier Science Publishers, 1987).

Marine Cargo and Hull Liability Conference, 1–2 June 1994, Sydney.

Marshall, Enid A., "Incorporation of Arbitration Clauses into Charterparty Bills of Lading", (1982) *Journal of Busines Law*, pp.478–484.

Mayss, Abla J. *Conflict of Laws* (London, Cavendish Publishing, 1994).

Meeson, Nigel, *Admiralty Jurisdiction and Practice* (London, Lloyd's of London Press, 1993).

—— "The Effect of Brussels Convention 1968 on the English Action 'in rem'", *P&I International*, April 1992, pp.5–7.

Metaxas, B.N., *Flags of Convenience, A Study of Internationalization* (Aldershot, Gower, 1985).

Meyers, H., *The Nationality of Ships* (The Hague, Martinus Nijhoff, 1967).

Mitchell, Ronald B., *International Oil Pollution at Sea* (Cambridge, Mass., MIT Press, 1994).

Mocatta, Sir Alan Abraham, Mustill, Sir Michael J., *et seq.*, *Scrutton on Charterparties and Bills of Lading* (19th ed., London, Sweet and Maxwell, 1984).

National Oceanic and Atmospheric Administration, Natural Resource Damage Assessments, *Final Rule*, 15 CFR Part 1990, 5 January 1996.

Noble, John M., "Some issues of the HNS Convention addressed", *P&I International*, March 1994, pp.4–5.

North, Peter, *Private International Law Problems in Common Law Jurisdictions* (Dordrecht, London, Martinus Nijhoff, 1993).

Northrup, Herbert R., and Scrase, Peter B., "The International Transport Workers' Federation Flag of Convenience Shipping Campaign: 1983–1995" (1996) 23 *Transportation Law Journal*, pp.369–423.

Noyes, John E., "The US Oil Pollution Act of 1990" (1992) 7 *International Journal of Estuarine and Coastal Law*, pp.43–56.

O'May, Donald, *On Marine Insurance Law and Policy* (ed. and co-author. Julian Hill, Sweet and Maxwell, 1993).

Owen. D.R., "The Origins and Development of Marine Collision Law" (1977) 51 *Tulane Law Review*, pp.759–809.

Oxman, Bernard H., "United States Interest in the Law of the Sea Convention" (1994) 88 *American Journal of International Law*, pp.167–178.

Paulsen, Gordon W., "US Oil Pollution Legislation Status as at mid July 1990", *Bimco Bulletin*, 4/90, July/August 1990, pp.4–7.

Peet, Gerard, "The MARPOL Convention: Inplementation and Effectiveness" (1992) *International Journal of Estuarine and Coastal Law*, pp.277–295.

Petroleum Industry Research Foundation, Inc., *Transporting U.S. Oil Imports: The Impact of Oil Spill Legislation on the Tanker Market*, Final Report, June 1992.

Phillips, Nevil, *Merchant Shipping Act 1995. An Annotated Guide* (Colchester, Lloyd's of London Press, 1996).

Ready, N.P., *Ship Registration* (2nd ed., London, Lloyd's of London Press, 1994).

Rein, Alex, "International Variations on Concepts of Limitation of Liability" (1979) 53 *Tulane Law Review*, pp.1259–1276.

Rodriguez, Antonio J., and Jaffe, Paul A.C., "The Oil Pollution Act of 1990" (1990) 15 *Tulane Maritime Law*, pp.1–28.

Schmidt, Dagmar Etkin, *The Financial Costs of Oil Spills, Oil Spill Intelligence Report* (Arlington MA, Cutter Information Corp., 1994).

Selvig, Erling, "The 1976 Limitation Convention and Oil Pollution Damage" [1979] *Lloyd's Maritime and Commercial Law Quarterly*, pp.21–25.

Shaw, M.N., *International Law* (2nd ed., Cambridge, Grotius, 1986).

Shearer, I.A., *Starke's International Law* (11th ed., London, Butterworths, 1994).

Shell Briefing Service, *The Tanker Industry in the 1990s*, Number 5, 1991.

Sheeman, Daniel F., *Economic Impact of Vessel Certificates of Financial Responsibility* (Department of Transportation, US Coast Guard).

Singh, N., and Colinvaux, R., *Shipowners, British Shipping Laws* (Vol. 13, London, Stevens, 1967).

Singh, Nagendra, *Maritime Flag and International Law* (Leyden, Sijthoff, 1978).

Smith, Raymond, *Conflict of Laws* (London, Cavendish Publishing, 1993).

Sohn, Louis B., "Law of the Sea Forum: The 1994 Agreement on Implementation of the Seabed Provisions of the Convention on the Law of the Sea, The 1994 Agreement and the Convention" (1994) 88 *American Journal of International Law*, pp.687–705.

Starke, J.G., *Starke's International Law* (ed. I.A. Shearer, and Ivan Anthony, 11th ed., London, Butterworths, 1994).

Stephenson, M.A., "Vessel-source Pollution Under the Law of the Sea Convention—an Analysis of the Prescriptive Standards" (1992) 11 *University of Queensland Law Journal*, pp.117–134.

Stolls, Amy M., "Vessel Liability and Financial Responsibility in the US Coastal States", the International Oil Spill Conference, pp.255–260.

Sturmey, S.G., "The United Nations Convention on Conditions for Registration of Ships" [1987] *Lloyd's Maritime and Commercial Law Quarterly*, pp.97–117.

Sturt, R.H.B., *The Collision Regulations; the Application and Enforcement of the Merchant Shipping (Distress Signals and Prevention of Collisions) Regulations 1989/illustrations by Ann L. Sturt* (3rd ed., London, Lloyd's of London Press, 1991).

Swanson, Steven R., "Federalism, the Admiralty and Oil Spills" (1996) 27 *Journal of Maritime Law and Commerce*, pp.379–416.

Tabbush, S.J., "Arrest of Ships Owned by Charterers" [1982] *Lloyd's Maritime and Commercial Law Quarterly*, pp.585–591.

Tetley, William, "Attachment, the Mareva injunction and *saisie conservatoire*" [1985] *Lloyd's Maritime and Commercial Law Quarterly*, pp.58–80.

—— "Limitation, Non-Responsibility and Disclaimer Clauses", *The Maritime Lawyer*.

—— "Shipowners' Limitation of Liability and Conflicts of Law: The Properly Applicable Law" [1992] 23 *Journal of Maritime Law and Commerce*, p.585.

—— "The Law of the Flag, 'Flag Shopping', and Choice of Law" (1993) 17 *Tulane Maritime Law and Journal*, pp.139–179.

Tettenborn, A.M., "The Time Charterer, the One Ship Company and the Sister Ship Action *in rem*" [1981] *Lloyd's Maritime and Commercial Law Quarterly*, pp.507–509.

Thomas, D.R., "The Sister Ship Action *in rem*" [1979] *Lloyd's Maritime and Commercial Law Quarterly*, pp.158–168.

—— *Maritime Liens, British Shipping Laws* (Vol. 14, London, Stevens and Sons, 1980).

Thomas, Michael, "British Concepts of Limitation of Liability" (1979) 53 *Tulane Law Review*, pp.1205–1258.

Todd, Paul, *Contracts for the Carriage of Goods by Sea* (Oxford, BSP Professional Books, 1988).

—— *Modern Bills of Lading* (2nd ed., Oxford, BSP Professional, 1990).

Tolofari, S.R., *Open Registry Shipping, A Comprehensive Study of Costs and Freight Rates* (New York, London, Gordon and Breach, 1989).

TOVALOP and CRISTAL, *A Guide to Oil Spill Compensation*.

Turner, H., *The Principles of Marine Insurance* (7th ed., London, Stone and Cox, 1986).

UNCTAD, "Legal and Documentary Aspects of Marine Insurance Contract", 1982. TD/BC.4/ISL/27 Rev.1.

United Nations, *Guidelines for Maritime Legislation, Economic and Social Commission for Asia and the Pacific*, Vol. I, 1991.

—— (ed.), *Report on the Fourth Regional Seminar on Maritime Legislation, Rights and Responsibilities of Shippers and Carriers in the Transport of Goods by Sea*, 16–19 December 1991.

US Coast Guard, National Pollution Funds Center, *Claimant's Information Guide*, September 1993.

—— *FY94 Annual Report*, July 1995.

—— *FY95 Annual Report*, March 1996.

The US Oil Pollution Act of 1990: An Assessment One Year Later, Proceedings, OPA Conference, Oslo, Norway, 17–18 September 1991.

Wagner, Thomas J., "The Oil Pollution Act of 1990: An Analysis" (1990) 21 *Journal of Maritime Law and Commerce*, pp.569–587.

Warot, J.A., "A Comment on the Lisbon Rules on Compensation for Damages in Collision Cases", (1978) 18 *Journal of Maritime Law and Commerce*, pp.583–587.

Wetterstein, Peter, "Trends in Maritime Environmental Impairment Liability" [1994] *Lloyd's Maritime and Commercial Law Quarterly*, pp.230–247.

Williams, Richard, "The New English Carriage of Goods by Sea Act 1992 (COGSA 1992)", *Gard News*, Issue 127, September 1992, pp.7–12.

Wood, Philip J., "OPA 90" (1995) 22 *Maritime Policy Management*, pp.201–208.

[illegible]

INDEX

£115.00